CATACLYSMIC VARIABLES AND RELATED OBJECTS

ASTROPHYSICS AND SPACE SCIENCE LIBRARY

VOLUME 208

CATACLYSMIC VARIABLES AND RELATED OBJECTS

PROCEEDINGS OF THE 158TH COLLOQUIUM OF THE INTERNATIONAL ASTRONOMICAL UNION, HELD AT KEELE, UNITED KINGDOM, JUNE 26–30, 1995

Edited by

A. EVANS

and

JANET H. WOOD

Keele University, United Kingdom

KLUWER ACADEMIC PUBLISHERS

DORDRECHT / BOSTON / LONDON

A C.I.P. Catalogue record for this book is available from the Library of Congress.

ISBN-13: 978-94-010-6632-7 e-ISBN-13: 978-94-009-0325-8
DOI: 10.1007/978-94-009-0325-8

Published by Kluwer Academic Publishers,
P.O. Box 17, 3300 AA Dordrecht, The Netherlands.

Kluwer Academic Publishers incorporates
the publishing programmes of
D. Reidel, Martinus Nijhoff, Dr W. Junk and MTP Press.

Sold and distributed in the U.S.A. and Canada
by Kluwer Academic Publishers,
101 Philip Drive, Norwell, MA 02061, U.S.A.

In all other countries, sold and distributed
by Kluwer Academic Publishers Group,
P.O. Box 322, 3300 AH Dordrecht, The Netherlands.

Printed on acid-free paper

CONTENTS

Section headings and reviews are listed in boldface type

Preface

In 1993 we began to consider the possibility of holding a conference on Catacysmic Variables (CVs) at Keele University. There have been several meetings in the area of CVs recently (e.g. Eilat, Abano-Padova, Capetown). However as preparations for the Keele meeting progressed we realized that, while there had been a number of IAU meetings devoted to related and to peripheral topics (such as IAU Colloquium 122 on Classical Novae in 1989, IAU Colloquium 129 on Accretion Disks in 1990), there had been no IAU-sponsored conferences in the area of cataclysmic variables (CVs) for a number of years.

We felt therefore that the time was ripe to have an IAU meeting devoted to an overview of CVs and related objects and the SOC organized the conference such that there was an emphasis on invited reviews of the most recent advances in the field. The conference covered both CVs and LMXBs and the inter-relations between them. The meeting was held at a time when powerful satellite observatories, and rapid improvements in ground based instrumentation, had led to many advances in both CV and LMXB research. The conference provided a forum to review observations from ASCA, EUVE, ROSAT, Ginga and the recently-refurbished HST. Photometric, spectroscopic and polarimetric observations of CVs and LMXBs have thrown new light on the distribution of matter and the nature of the stellar components in these systems. There have also been a number of important recent developments in our theoretical understanding of the CVs and LMXBs themselves, and in the inter-relationship between the various classes of interacting binary stars.

This volume contains the proceedings of IAU Colloquium 158 *Cataclysmic Variables and Related Objects*, which was held at Keele University in 1995 June 26 – 30. We greatly appreciate the prompt submission of papers by the contributors, and the willingness with which they agreed to modifications and revisions. We also appreciate the co-operation of contributors in the restrictions imposed by the editors, including an excellent paper submitted by la Dous & Howell on the delights of Pecks Restaurant, where the conference dinner was held (this paper can be accessed on the World Wide Web at http://www.astro.keele.ac.uk/~ftp/pub/cvconf/pecks.html).

We are particularly indebted to the International Astronomical Union, The European Commission for Science, Research & Development, The Royal Astronomical Society, Kluwer Academic Publishers, Keele University, The Institute of Physics, and Barclays and Midland Banks for their generous sponsorship of the meeting.

The members of the Local Organizing Committee did an excellent job in the preparations for the meeting. However other members of the Keele Astrophysics Group also did a great deal to help in the day-to-day running of the meeting; our thanks go to Jon Lockley, Mark Somers, Derryck Case, Greg Beekman and Andrew Scott. We also wish to record our appreciation for the magnificent support provided by the staff of Keele University Physics Department, in particular Dave Bowyer, Andy Brammer, Mike Daniels, Michelle Evans, Jeanette Lucas, Helen Moors, Bill Robinson, Dave Sammons, Neil Rodgers. Natasha Papousek-Welsh, Gurnos Jones and Lyndon Mitchell did a superb job in organizing the guest excursions.

A. Evans
J. H. Wood

List of participants

Ivan Andronov	*Odessa*	**Jean Gunther**	*AFOEV*
Philip Armitage	*IoA*	**Emilios Harlaftis**	*St. Andrews*
Raymundo Baptista	*STScI*	**Amos Harpaz**	*Haifa*
S. Bard	*Tromsoe*	**Barbara Hassall**	*Central Lancs.*
Orly Barziv	*Amsterdam*	**Carole Haswell**	*Columbia*
Greg Beekman	*Keele*	**Coel Hellier**	*Keele*
Gaetano Belvedere	*Catania*	**Paul Heptinstall**	*Newcastle*
Paul Bennie	*St. Andrews*	**Margarita Hernanz**	*CEAB*
Klaus Beuermann	*Goettingen*	**Frederic Hessman**	*Goettingen*
Ralf Bocksch	*Tuebingen*	**Robert Hjellming**	*NRAO*
Mike Bode	*LJMU*	**Keith Horne**	*St. Andrews*
Howard Bond	*STScI*	**Steve Howell**	*PSI*
Stella Bradbury	*MPI Physik*	**Min Huang**	*Villanova*
Mike Brough	*Keele*	**Wolfgang Hummel**	*Brussels*
Albert Bruch	*Muenster*	**Guy Hurst**	*BAA*
David Buckley	*SAAO*	**T. Iijima**	*Asiago*
Vadim Burwitz	*Goettingen*	**Manabu Ishida**	*ISAS*
John Cannizzo	*GSFC*	**Anatoli Iyudin**	*MPE*
Jorge Casares	*Oxford*	**Joni J. Johnson**	*Wisconsin*
Derryck Case	*Keele*	**Derek Jones**	*RGO*
Sandi Catalan	*St. Andrews*	**Peter Kahabka**	*Amsterdam*
Fuhua Cheng	*Maryland*	**Taichi Kato**	*Kyoto*
Lidia Chinarova	*Odessa*	**Soon-Wook Kim**	*Texas*
Drahomir Chochol	*Slovak Acad. Sci.*	**Andrew King**	*Leicester*
Kathryn Clayton	*Leicester*	**Christian Knigge**	*Oxford*
James DeYoung	*USNO*	**Ulrich Kolb**	*Lick/UCSC*
Vik Dhillon	*La Palma*	**Attay Kovetz**	*Tel Aviv*
Marcos Diaz	*Penn State*	**Erik Kuulkers**	*ESA/ESTEC*
Rachel Dickinson	*UCL*	**Constanze la Dous**	*VILSPA*
Elena Dmitrienko	*Crimea*	**Stefan Larsson**	*Stockholm*
Christine Done	*Durham*	**Jean-Pierre Lasota**	*Meudon*
Sean Dougherty	*LJMU*	**Laurits Leedjarv**	*Tartu*
Simon Duck	*Oxford*	**Elia Leibowitz**	*Tel Aviv*
Mike Eracleous	*STScI*	**Mario Livio**	*STScI*
Ene Ergma	*Tartu*	**Huw Lloyd**	*LJMU*
Nye Evans	*Keele*	**Jon Lockley**	*Keele*
Stewart Eyres	*NRAL*	**Knox Long**	*STScI*
Michael Friedjung	*Paris*	**James MacDonald**	*Delaware*
Boris Gaensicke	*Goettingen*	**Rosemary Mardling**	*Monash*
Michael Garcia	*SAO*	**Peter Marks**	*Sussex*
Mark Garlick	*Sussex*	**Tom Marsh**	*Southampton*
Ami Glasner	*Hebrew University*	**Paul Mason**	*Case Western*
Patrick Godon	*JPL*	**Christopher Mauche**	*LLNL*
Paul Groot	*Amsterdam*	**Ronald Mennickent**	*Concepcion*

Freidrich Meyer	*MPIA*	**Klaus Schenker**	*MPIA*
E. Meyer-Hofmeister	*MPIA*	**Axel Schwope**	*Potsdam*
Joanna Mikolajewska	*Copernicus*	**Andrew Scott**	*Keele*
Maciej Mikolajewski	*Copernicus*	**Allen Shafter**	*SDSU*
Shin Mineshige	*Kyoto*	**N. Shakhovsky**	*Crimea*
Sophia Mitronova	*SAO RAS*	**Sergei Shugarov**	*Sternberg*
Claudio Moreno	*La Palma*	**Andy Silber**	*Washington*
Koji Mukai	*GSFC*	**Edward Sion**	*Villanova*
James Murray	*CITA*	**Jozef Smak**	*Copernicus*
Lorne Nelson	*Quebec*	**Alan Smale**	*GSFC*
Stewart Newell	*NRAL*	**Barry Smalley**	*Keele*
Daisaku Nogami	*Kyoto*	**Robert C. Smith**	*Sussex*
Tim O'Brien	*LJMU*	**Keith Sohl**	*Leicester*
Marina Orio	*Wisconsin*	**Jan-Erik Solheim**	*Tromsoe*
Yoji Osaki	*Tokyo*	**Loretta Solmi**	*Bologna*
Julian Osborne	*Leicester*	**Mark Somers**	*Keele*
Manfred Pakull	*Strasbourg*	**Karen Southwell**	*Oxford*
Joe Patterson	*Columbia*	**Ralph Spencer**	*NRAL*
Elena Pavlenko	*Crimea*	**Lee Nigel Sproats**	*MSSL*
Kevin Pearson	*Leicester*	**Danny Steeghs**	*Utrecht*
Daniel Pequignot	*Meudon*	**Martin Still**	*St. Andrews*
Nicolas Pereyra	*GSFC*	**Paula Szkody**	*Washington*
Karol Petrik	*Comenius*	**John Thorstensen**	*Dartmouth*
Shlomi Pistinner	*Ben-Gurion*	**Chris Tout**	*Cambridge*
Marina Popova	*SAO RAS*	**Gaghik Tovmassian**	*UNAM*
Stephen Potter	*MSSL*	**Alexander Tutukov**	*Russian Acad. Sci.*
Gary Poyner	*BAA*	**Ana Ulla**	*NBI*
Dina Prialnik	*Tel Aviv*	**Zdenek Urban**	*Slovak Acad. Sci.*
Hannah Quaintrell	*Sussex*	**Horst Vaeth**	*Kiel*
Jonathan Rainger	*Leicester*	**Jan van Paradijs**	*Amsterdam*
Gavin Ramsay	*Utrecht*	**A. van Teeseling**	*Goettingen*
Tod Ramseyer	*Gettysburg*	**Irina Voloshina**	*Sternberg*
Saul Rappaport	*MIT*	**Sonja Vrielmann**	*Goettingen*
Klaus Reinsch	*Goettingen*	**Saeqa Vrtilek**	*Maryland*
Alon Retter	*Tel Aviv*	**Richard Wade**	*Penn State*
Hayley Richman	*Columbia*	**Brian Warner**	*Capetown*
Fred Ringwald	*Keele*	**Bill Welsh**	*Keele*
Hans Ritter	*MPIA*	**Peter Wheatley**	*Utrecht*
Edward L. Robinson	*Texas*	**Glen Williams**	*CMU*
St. John Robinson	*Keele*	**Janet Wood**	*Keele*
Simon Rosen	*Leicester*	**Bill Worraker**	*BAA*
Diane Roussel-Dupre	*LANL*	**Graham Wynn**	*Leicester*
Rene Rutten	*La Palma*	**Boris Yudin**	*Sternberg*
Marek Sarna	*Copernicus*	**Lev Yungelson**	*Russian Acad. Sci.*
Karen Schaefer	*STScI*	**Luca Zangrilli**	*Padua*
Susanne Schandl	*MPIA*	**Tomaz Zwitter**	*Ljubljana*

NON-MAGNETIC CATACLYSMIC VARIABLES

THE NOVA-LIKE VARIABLES

VIK DHILLON
Royal Greenwich Observatory, La Palma

Abstract. We review optical observations and theoretical models of the non-magnetic nova-like variables (UX UMa, VY Scl and SW Sex stars). A brief discussion of the classification scheme is followed by a statistical overview of the observed properties. The most important features of each of the sub-classes are then reviewed, concluding with a summary of the theoretical models invoked to understand these systems.

1. Classification

Nova-like variables are defined as cataclysmic variable stars (CVs) which have never been observed to undergo nova or dwarf nova-type outbursts. Such a crude definition encompasses a wide variety of objects which can be divided into two distinct groups; those which are believed to accrete via magnetic field lines – AM Her stars (or polars), DQ Her stars and intermediate polars – and those which are believed to accrete via an accretion disc – UX UMa stars, VY Scl stars (or anti-dwarf novae), SW Sex stars and AM CVn stars (or double-degenerates). The magnetic nova-likes have been reviewed elsewhere in this volume. The subject of this review will be the optical characteristics of non-magnetic nova-likes (NMNLs), with the exception of the AM CVn stars which are outside the scope of this paper.

There is little agreement on the classification scheme for NMNLs in the literature. Perhaps the most generally accepted scheme adheres to the following rules: if a nova-like shows no evidence for magnetic accretion and is not recognized to be a double-degenerate, it is classed as a UX UMa star. A UX UMa star which is observed to show states of low brightness becomes a VY Scl star. A number of UX UMa and VY Scl stars with high inclinations and periods of 3...4 h, show single-peaked emission lines which remain (largely) unobscured during primary eclipse and which exhibit transient absorption features and distorted radial velocity curves. These systems

A. Evans and J. H. Wood (eds.), Cataclysmic Variables and Related Objects, 3–12.

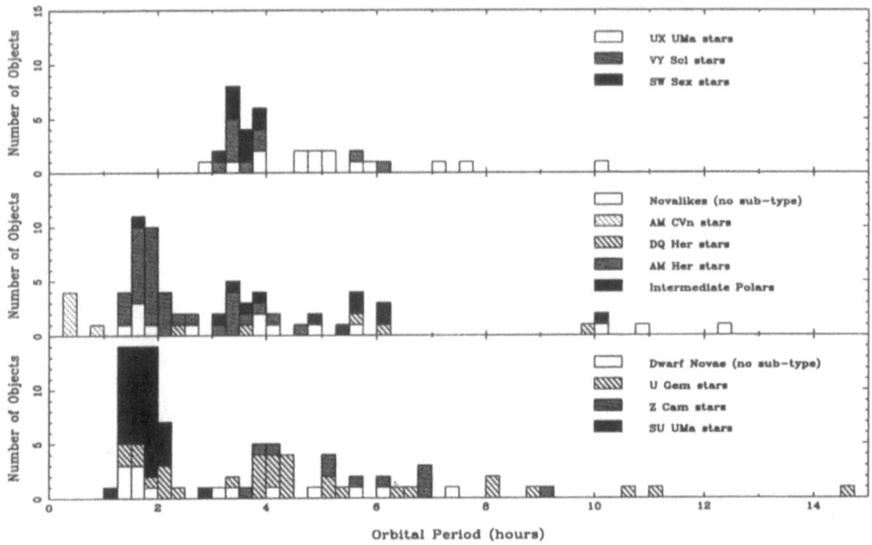

Figure 1. **The orbital period distribution of CVs. Data from [31].**

were classed as SW Sex stars by Thorstensen et al. [44]. This classification is controversial, however, since SW Sex stars can only be recognized by their spectroscopic properties, whereas other classes of CV are traditionally classified by their photometric variations. Amongst other things, this can lead to confusion, since a star can then be both a VY Scl star and an SW Sex star. In addition, it is possible that the SW Sex stars are simply the high inclination counterparts of other NMNLs, in which case a separate classification is not justified. Bearing these caveats in mind, it is nevertheless sometimes convenient to be able to refer to these objects as SW Sex stars and this term will be used in the present review.

2. Statistical overview

Of the 703 CVs known, 104 are classed as nova-like variables [9]. Of these, 21 are UX UMa stars and 13 VY Scl stars. Arguably, up to 6 of these UX UMa stars and 3 of these VY Scl stars exhibit the SW Sex phenomenon [31]. When plotted in Galactic coordinates [23], it is immediately apparent that nova-likes are found in equal numbers in the Galactic plane and at high galactic latitudes, with no tendency for their numbers to increase towards the Galactic centre. The implication is that, like the dwarf novae, nova-likes are nearby and, given their low apparent brightness, they must also be intrinsically faint. These conclusions are confirmed by determinations of distances (of order hundreds of parsecs) and absolute magnitudes ($M_V \sim 4$) of nova-likes [47]. The actual Galactic distributions and space densities are, however, much more contentious subjects: see la Dous [23].

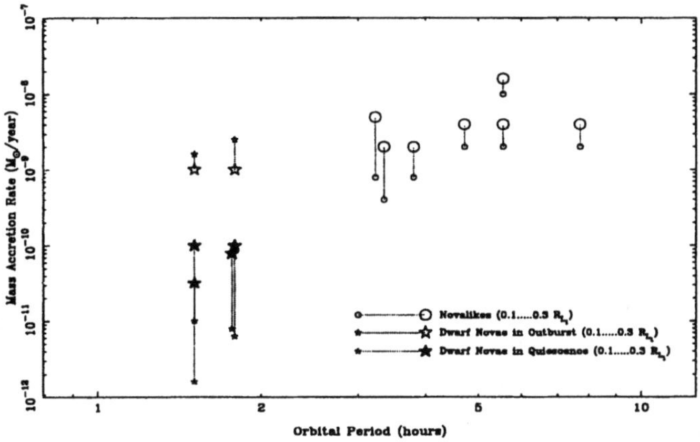

Figure 2. The dependence of mass accretion rate at disc radii $0.1R_{L_1}$ and $0.3R_{L_1}$ on orbital period for dwarf novae (in outburst and quiescence) and NMNLs.

Fig. 1 shows the orbital period distribution of CVs. It can be seen that the NMNLs are tightly grouped in the 3...4 h period range and there are no NMNLs below the 2...3 h period gap. The dwarf novae, on the other hand, are found both above and below the period gap and there appears to be a dearth of dwarf novae in the 3...4 h period range favoured by the NMNLs. The implications of this result will be discussed in Sect. 4.

One of the most important distinctions between NMNLs and other CVs can be found by comparing their mass accretion rates. Following Horne [22], Fig. 2 plots the mass accretion rate derived from eclipse mapping experiments as a function of binary period for a number of NMNLs and dwarf novae in outburst and quiescence. It is evident that both NMNLs and dwarf novae in outburst accrete mass at similar rates and at a much higher rate than dwarf novae in quiescence. The accretion rate also appears to increase slightly with orbital period.

A discrepancy between steady-state accretion (in which the mass accretion rate is constant throughout the disc) and the observed accretion rates can also be inferred from Fig. 2. The mass accretion rate appears to decrease slightly with radius in the dwarf novae discs during outburst. This is expected since material should be draining from the disc as it declines from outburst. The NMNLs, however, all show mass accretion rates that increase with radius in the disc. The increase is low for the longer period NMNLs, but becomes significant for systems in the 3...4 h period range. The latter result has been inferred from the eclipse maps of Rutten, van Paradijs & Tinbergen [36], which show that the radial temperature dependences in the inner disc regions of SW Sex stars are much flatter than the $T \propto R^{-3/4}$ relation predicted by steady-state accretion theory. The implication is that the central regions of these discs may be releasing some of the accretion energy in a non-radiative form, and it seems likely that this result is linked to the SW Sex phenomenon which requires an extra component of light that remains visible during primary eclipse (see Sect. 4).

3. Observed properties

The long-term light curves of NMNLs are as varied and complex as those exhibited by novae and dwarf novae. For example, some VY Scl stars (e.g. V794 Aql [19]) show deep low states of $\sim$ 3 mag, preceded by a number of shallower low states with saw-toothed light curves. Other VY Scl stars (e.g. DW UMa – also an SW Sex star [18]), show even deeper low states of much longer duration (typically years). Inevitably, there are also VY Scl stars which show both types of behaviour (e.g. MV Lyr [34]). As well as low states, however, some NMNLs also show outbursts; Honeycutt et al. [20] observed a 1 mag outburst in the VY Scl star KR Aur, while Still, Dhillon & Jones [42], observed the UX UMa star RW Tri during a 3.5 mag outburst. It is not known how often such eruptions occur in NMNLs, and whether or not they are related to nova or dwarf nova-type outbursts is unclear.

Other types of long-term photometric variations have also been observed in NMNLs. For example, RW Tri was observed by Honeycutt et al. [20] to show sinusoidal-like brightness variations of 0.5 mag with a period of 25 d. RW Tri and UX UMa also show aperiodic variations in their orbital period [35], [42]. There have been suggestions that these phenomena might be the result of magnetic cycles within the secondary star: see [1] and references therein.

Moving to variations on orbital time-scales, the light curves of high inclination NMNLs exhibit smooth, round-bottomed eclipses, often with a pronounced egress shoulder and orbital hump due to the bright spot (see Fig. 5). Similar orbital light curves are exhibited by dwarf novae in outburst. The eclipse light curves of NMNLs can be variable from cycle to cycle, both in depth and in shape (e.g. PX And [44]). Eclipse maps show that the discs are generally symmetric, much brighter in their centres than their edges, and have bright spots whose strengths vary considerably from object to object [36].

On time-scales of minutes, all NMNLs show stochastic flickering with amplitudes of up to several tenths of a magnitude. By analysing the behaviour of the flickering in RW Tri during eclipse, Horne & Stiening [21] showed that the entire disc participates in the flickering; it is not known whether this flickering is the result of irradiation from near the white dwarf, or energy released at localized sites in the disc. On even shorter time-scales, quasi-periodic oscillations (QPOs) with periods of tens of seconds are sometimes seen in the light curves of NMNLs (and dwarf novae during outburst),

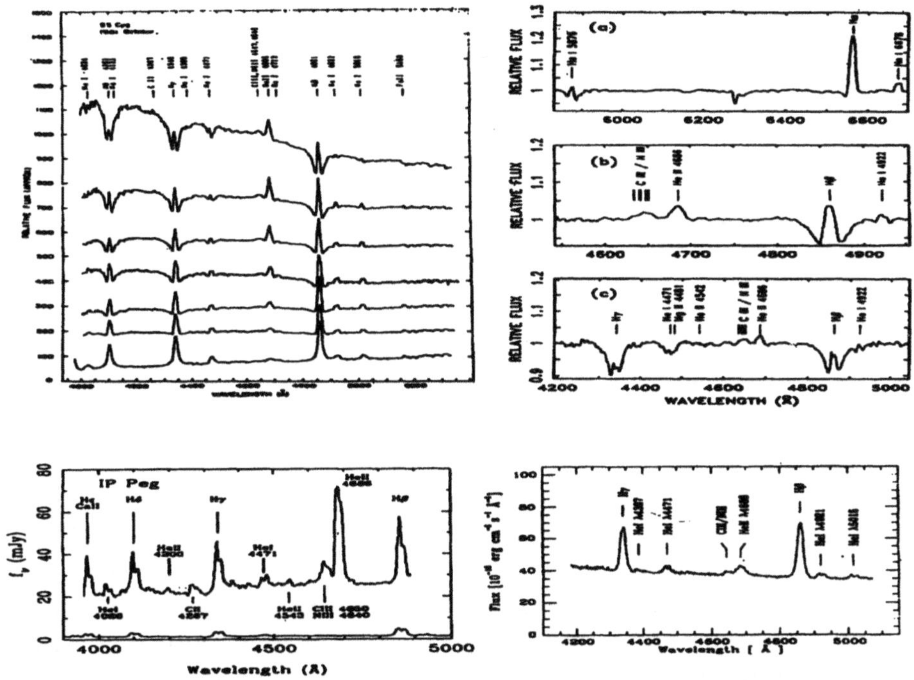

Figure 3. A comparison of the spectra of dwarf novae and NMNLs. Clockwise from top left: the low inclination dwarf nova SS Cyg in outburst and quiescence [15]; the low inclination UX UMa star IX Vel [2]; the high inclination SW Sex star WX Ari [3]; the high inclination dwarf nova IP Peg in outburst and quiescence [27].

e.g. V3885 Sgr [46] and UX UMa [28]. Petterson [30] successfully modelled the systematic changes observed in the QPO phases during eclipse with an accretion disc reflecting radiation from a rotating source on or near the surface of the white dwarf. Although various models for the source of radiation have been invoked (e.g. bright spots on the white dwarf; non-radial pulsations; bright spots at the inner edge of the accretion disc), their true nature remains uncertain.

NMNLs exhibit the same variety in their optical spectra as do dwarf novae in outburst and quiescence, ranging from pure emission line to almost pure absorption line spectra. In Fig. 3, representative spectra of high and low inclination dwarf novae in outburst and quiescence are displayed alongside spectra of high and low inclination NMNLs. The similarities are striking. Low inclination dwarf novae in outburst show strong absorption lines, as do low inclination NMNLs. High inclination dwarf novae in outburst show strong emission lines, as do high inclination NMNLs. This strong correlation between orbital inclination and the strength of the emission lines has been demonstrated for classical novae by Warner [47] and has been noted for dwarf novae in outburst by Robinson, Marsh & Smak [33].

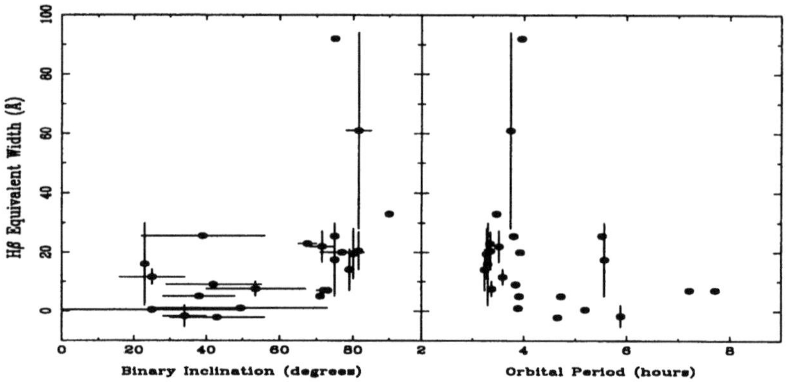

Figure 4. Binary inclination and orbital period versus the equivalent width of Hβ in a sample of 24 UX UMa, VY Scl and SW Sex-type nova-like variables.

Fig. 4, which plots the equivalent width of Hβ versus the orbital inclination, shows that the same correlation exists in NMNLs; low inclination NMNLs generally show absorption lines (often with strong emission cores, which is why the equivalent widths are rarely negative), whereas high inclination NMNLs show emission lines. The reason that this correlation exists is due to dilution: at high inclinations the flux in the optically-thick continuum of the accretion disc is reduced by projection and limb darkening, so there is less dilution of the emission line flux. Fig. 4 also shows that the equivalent width of Hβ increases with decreasing orbital period in NMNLs. Hessman [14] and Echevarria [11] found a similar correlation exists for all CVs. This effect can be understood in terms of the correlation between line strength and absolute magnitude found by Patterson [29]: as the accretion disc becomes brighter the emission lines are diluted by the optically thick continuum flux from the disc and hence their equivalent widths decrease. Since CVs with longer orbital periods tend to have higher rates of mass transfer (see Fig. 2), their emission line strengths decrease.

The remainder of this section will review the spectroscopic properties specific to the UX UMa, SW Sex and VY Scl sub-classes. Perhaps the best understood class of NMNLs are the UX UMa stars. That this is the case has been verified in some detail by the spectral eclipse-mapping results of Rutten et al. [37]. They reconstructed the spectra of different parts of the accretion disc in UX UMa and found an inner disc and bright spot with a blue continuum and Balmer absorption lines and an outer disc with a red continuum and Balmer emission lines. This is qualitatively what is expected from accretion discs which are hot and optically thick in their centres, and cool and optically thin in their outer regions [48]. In addition, theoretical spectra now appear to be giving encouraging fits to the observed continua of UX UMa stars [40].

The same agreement with theory cannot be said to exist for SW Sex

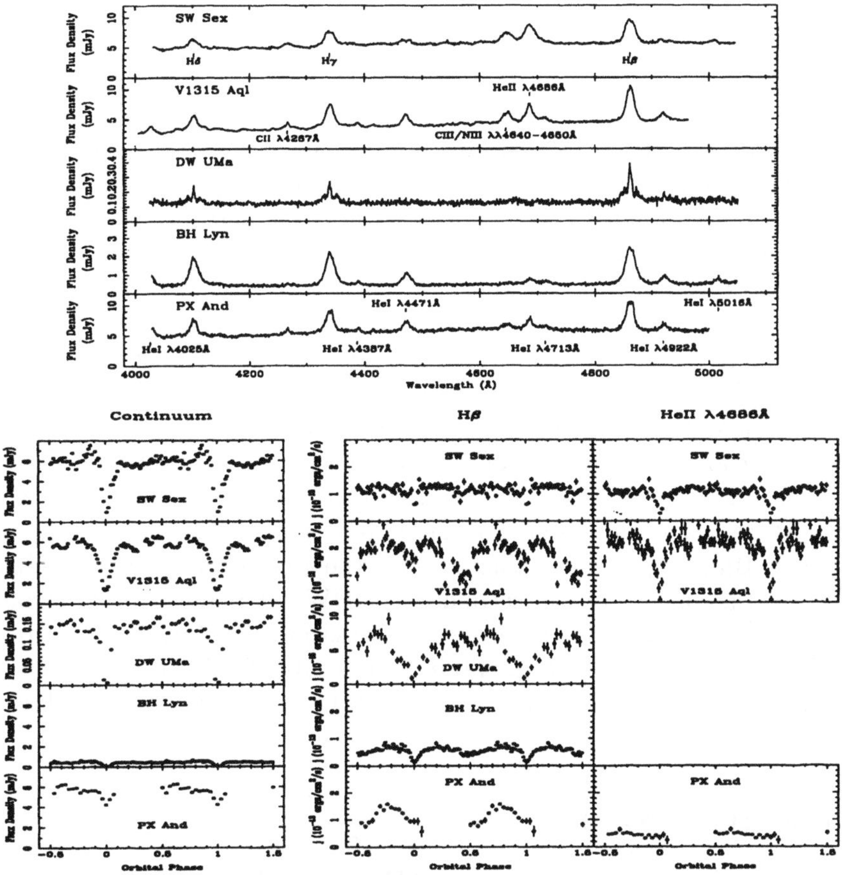

Figure 5. Spectra and light curves of 5 SW Sex stars. From top to bottom: SW Sex [4], V1315 Aql [5], DW UMa (low state [7]), BH Lyn (low state [6]) and PX And [43].

stars. This group of stars are all high inclination NMNLs[1] with periods in the range 3...4 h. The spectra of 5 SW Sex stars and their continuum and emission line light curves are displayed in Fig. 5. Note that the spectra and light curves of DW UMa and BH Lyn were obtained during low states (see below). All SW Sex stars exhibit strong, single-peaked Balmer, He I and He II emission lines which, with the exception of He II, remain largely unobscured during primary eclipse. This is in stark contrast to standard accretion disc theory which predicts that emission lines from high inclination discs should appear double peaked and be eclipsed once every orbital period. In addition, the emission lines in SW Sex stars show strong absorption features around phase 0.5 and their radial velocity curves exhibit significant phase shifts relative to photometric conjunction. As shall be discussed in

[1] Even WX Ari, previously thought to be low inclination, shows grazing eclipses [13]

Sect. 4, these observational constraints have yet to be fully satisfied by a single theoretical model.

The VY Scl stars in their normal state are indistinguishable from UX UMa or SW Sex stars. In their low states, however, which are due to a decrease in the mass accretion rate, VY Scl stars completely change in appearance. An example of typical VY Scl-star behaviour has been presented by Dhillon, Jones & Marsh [7] , who observed DW UMa in a $\sim$ 3 mag low state; the resulting spectrum and light curves are shown in Fig. 5. It can be seen that, in contrast to the normal state, there are no high-excitation emission lines present. The Balmer lines are dominated by strong, narrow emission spikes superposed upon faint wings. Dhillon et al. [7] showed that the emission spikes originate on the secondary star and the line wings show evidence for an accretion disc origin. By measuring the continuum eclipse width, they also found that the accretion disc was smaller than during the normal state. Using this same data in conjunction with Roche tomography, Rutten & Dhillon [38] were able to map the Balmer line intensity distribution on the surface of the secondary in DW UMa and showed strongly enhanced emissivity on the inner face of the secondary star. A similar result was also obtained by Still et al. [42] in their study of RW Tri during a $\sim$ 3.5 mag *high* state, suggesting that irradiation by the accretion disc plays an important role in NMNLs.

4. Theoretical models

The observed similarities between NMNLs and dwarf novae in outburst suggest that NMNLs might be successfully modelled by canonical CVs with steady-state accretion discs in which the mass transfer rate is sufficiently high to prevent disc instability-type outbursts. But can this model explain all of the observed NMNL phenomena, and in particular, the low states, the orbital period distribution, the departures from steady-state accretion in the inner disc regions and the SW Sex phenomenon? In this final section we will show that a number of modifications to the canonical model are required in order to approach an understanding of the NMNLs, a conclusion which has also been reached by la Dous [24].

Turning first to the SW Sex phenomenon, it is clear that a successful model will have to include some component of line emission from above the orbital plane in order to explain the lack of eclipse in the emission lines. For this reason, the Stark broadening model of Lin, Williams & Stover [25] and the bright-spot overflow model of Hellier & Robinson [12] are unlikely to be correct. However, the latter model is very successful in explaining many of the other properties of SW Sex stars, such as the phase 0.5 absorption. This suggests that a model with a bright-spot overflow operating in conjunction with some form of accretion disc wind ([8]; [16]; [17]), magnetic accretion column [49], or magnetically driven outflow ([45]; [50]) might be able to explain all of the observed phenomena. The requirement that some mechanism drives material out of the orbital plane is also consistent with the observed departures from steady-state accretion in the inner discs of these systems and provides an explanation for the redder UV spectra of NMNLs compared to dwarf novae in outburst [45].

The first attempt to explain the observation of low states in VY Scl stars was made by Robinson et al. [32]. They noted the period grouping of VY Scl stars and speculated that the low states are a consequence of their imminent entry into the period gap. In this model, the magnetic braking which drives mass transfer ceases as a result of the secondary star becoming fully convective at periods immediately above the gap [41]. However, Livio & Pringle [26] showed that this model cannot account for the low states in VY Scl stars due to the disparate time-scales – it takes VY Scl stars of order 10...100 d to enter a low state whereas it would take 10 000 yr or more to respond to a sudden cessation in the mechanism driving the mass transfer. They proposed an alternative model where star spots covering the L_1 point block the mass transfer and hence cause low states. It may be possible to test this model using Roche tomography [38].

Using the eruptive characteristics of CVs to infer relative mass transfer rates, Shafter [39] concluded that magnetic braking models have severe difficulties accounting for the orbital period distribution of CVs, in particular the observed dearth of dwarf novae with orbital periods immediately above the period gap and the dominance of NMNLs in this same period range. Intriguingly, the star spot model of Livio & Pringle [26] can also be used to explain this orbital period distribution. The dwarf novae, with lower mass transfer rates, are more easily interrupted by the star spot mechanism because the density and pressure are correspondingly lower at the L_1 point. As the periods decrease, the magnetic fields and their covering factors might be expected to increase, so that if mass transfer occurs it can only do so at a relatively high rate. In their picture, therefore, the low mean mass transfer rate systems (dwarf novae) become detectable as high mass transfer rate systems (NMNLs) as they approach the period gap.

But why do NMNLs have higher mean mass transfer rates than dwarf novae? This is ultimately the most important question of all, the answer to which remains uncertain. According to the hibernation model for the cyclical evolution of CVs (see [10] and references therein), it is due to the fact that NMNLs are post-novae whose outbursts we have missed. Only time will tell if this theory is correct.

References

1. Applegate, J.H., 1992, Ap. J., **385**, 621

2. Beuermann, K., Thomas, H.-C., 1990, A&A, **230**, 326
3. Beuermann, K., Thorstensen, J.R., Schwope, A.D., et al., 1992, A&A, **256**, 442
4. Dhillon, V.S., 1990, D.Phil. thesis, University of Sussex
5. Dhillon, V.S., Marsh, T.R., Jones, D.H.P., 1991, MNRAS, **252**, 342
6. Dhillon, V.S., Jones, D.H.P., Marsh, T.R., Smith, R.C., 1992, MNRAS, **258**, 225
7. Dhillon, V.S., Jones, D.H.P., Marsh, T.R., 1994, MNRAS, **266**, 859
8. Dhillon, V.S., Rutten, R.G.M., 1995, MNRAS, **277**, 777
9. Downes, R.A., Shara, M.M., 1993, PASP, **105**, 127
10. Duerbeck, H.W., 1993, in "Cataclysmic Variables and Related Physics", eds O. Regev, G. Shaviv, Ann. Israel Phys. Soc., **10**, 77
11. Echevarria, J., 1988, MNRAS, **233**, 513
12. Hellier, C., Robinson, E.L., 1994, Ap. J., **431**, L107
13. Hellier, C., Ringwald, F.A., Robinson, E.L., 1995, A&A, **289**, 148
14. Hessman, F.V., 1985, Ph.D. thesis, University of Texas at Austin
15. Hessman, F.V., Robinson, E.L., Nather, R.E., Zhang, E., 1984, Ap. J., **286**, 747
16. Hoare, M.G., 1994, MNRAS, **267**, 153
17. Honeycutt, R.K., Schlegel, E.M., Kaitchuck, R.H., 1986, Ap. J., **302**, 388
18. Honeycutt, R.K., Livio, M., Robertson, J.W., 1993, PASP, **105**, 922
19. Honeycutt, R.K., Cannizzo, J.K., Robertson, J.W., 1994, Ap. J., **425**, 835
20. Honeycutt, R.K., Robertson, J.W., Turner, G.W., et al., 1994, in "Interacting Binary Stars", ed. A. Shafter, ASP Conf. Ser. **56**, 277
21. Horne, K., Stiening, R.F., 1985, MNRAS, **216**, 933
22. Horne, K., 1993, in "Acc. Disks in Comp. Stell. Sys.", ed. J.C. Wheeler, p117
23. la Dous, C., 1993a, in 'CVs and Related Objects', NASA Monograph, p14
24. la Dous, C., 1993b, in "Cataclysmic Variables and Related Physics", eds O. Regev, G. Shaviv, Ann. Israel Phys. Soc., **10**, 39
25. Lin, D.N.C, Williams, R.E., Stover, R.J., 1988, Ap. J., **327**, 234
26. Livio, M., Pringle, J.E., 1994, Ap. J., **427**, 956
27. Marsh, T.R., Horne, K., 1990, Ap. J., **349**, 593
28. Nather, R.E., Robinson, E.L., 1974, Ap. J., **190**, 637
29. Patterson, J., 1984, Ap. J. Supp., **54**, 443
30. Petterson, J.A., 1980, Ap. J., **241**, 247
31. Ritter, H., Kolb, U., 1995, in "X-ray Binaries", eds. Lewin, W.H.G., et al., CUP, Cambridge, p578
32. Robinson, E.L., Barker, E.S., Cochran, A.L., et al., 1981, Ap. J., **251**, 611
33. Robinson, E.L., Marsh, T.R., Smak. J.I., 1993, in "Acc. Disks in Comp. Stell. Sys.", ed. J.C. Wheeler, p75
34. Rosino, L., Romano, G., Marziani, P., 1993, PASP, **105**, 51
35. Rubenstein, E.P., Patterson, J., Africano, J.L., 1991, PASP, **103**, 1258
36. Rutten, R.G.M., van Paradijs, J., Tinbergen, J., 1992, A&A, **260**, 213
37. Rutten, R.G.M., Dhillon, V.S., Horne, K., Kuulkers, E., 1994, A&A, **283**, 441
38. Rutten, R.G.M., Dhillon, V.S., 1994, A&A, **288**, 773
39. Shafter, A.W., 1992, Ap. J., **394**, 268
40. Shaviv, G., Wehrse, R., 1993, in "Acc. Disks in Comp. Stell. Sys.", ed. J.C. Wheeler, p148
41. Spruit, H.C., Ritter, H., 1983, A&A, **124**, 267
42. Still, M.D., Dhillon, V.S., Jones, D.H.P., 1995, MNRAS, **273**, 849
43. Still, M.D., Dhillon, V.S., Jones, D.H.P., 1995, MNRAS, **273**, 863
44. Thorstensen, J.R., Ringwald, F.A., Wade, R.A., et al., 1991, AJ, **102**, 272
45. Tout, C.A., Pringle, J.E., la Dous, C., 1993, MNRAS, **265**, 5p
46. Warner, B., 1973, MNRAS, **162**, 189
47. Warner, B., 1987, MNRAS, **227**, 23
48. Williams, R.E., 1980, Ap. J., **235**, 939
49. Williams, R.E., 1989, AJ, **97**, 1752
50. Wynn, G.A., King, A.R., 1996, these proceedings, p161

THE NOVA-LIKES AC CNC, RW TRI, UX UMA AND THE NOVA DQ HER: COMMON AND DIFFERENT PROPERTIES

E. S. DMITRIENKO
Crimean Astrophysical Observatory, 334413, Crimea, Ukraine
e-mail: dmitrien@crao.crimea.ua

Abstract. An analysis of the main photometric properties of the nova-like variables AC Cnc, RW Tri and UX UMa, and Nova Her 1934 (DQ Her) is presented. The analysis is based on simultaneous high-speed *UBVRI* photometry, covering several dozens of orbital cycles evenly distributed over 14 years (1982 ... 1995) for DQ Her, and over 11 years (1985 ... 1995) for the other three stars. All data have been obtained with the 1.25 m telescope at the Crimean Astrophysical Observatory.

1. Discussion

According to our analysis of the photometric data for AC Cnc, RW Tri and UX UMa [1] an essential distinction between the accretion discs of AC Cnc and RW Tri on the one hand, and of UX UMa on the other, may exist. The possibility of a difference between the secondary components of these three systems should also not be excluded.

These three nova-like stars have not undergone a nova explosion, unlike DQ Her. However the latter shares many common properties with them. The main common trait is the existence of *UBVRI* brightness variability on time-scales of days, with amplitudes of less than 2 mag, when an increase in brightness is accompanied by a decrease of the colour indices $(U-B)$, $(B-V)$, $(V-R)$ and $(V-I)$ and vice versa.

In Fig. 1 (left) we present the time dependence of the uneclipsed light, $\mathcal{L}$, of DQ Her in the B−band (in $UVRI$ the light curves are similar). For every night, the value of $\mathcal{L}$ was calculated by averaging several dozen 'uneclipsed' data points. The errors in $\mathcal{L}$ are negligible compared to the physical brightness changes (due to flickering etc.). The light variation on time-scales of years, as well as the the change in $(O-C)$ on this time-scale, may be caused by activity of the red dwarf, secondary star.

A. Evans and J. H. Wood (eds.), Cataclysmic Variables and Related Objects, 13–14.

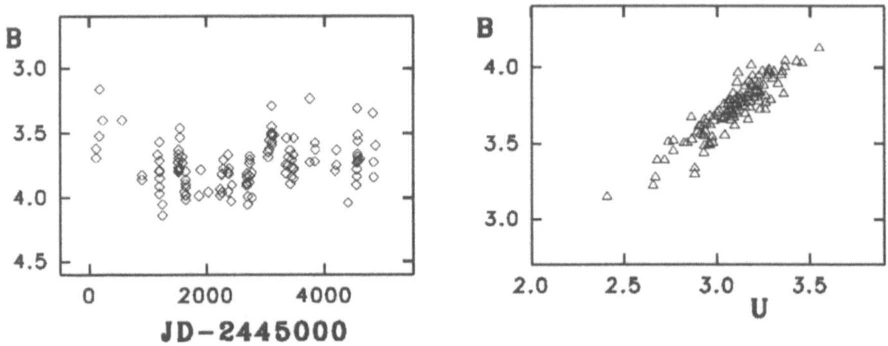

Figure 1. Left: uneclipsed B–magnitude of DQ Her. Right: uneclipsed U– versus B–magnitude for DQ Her. Magnitudes are differential with respect to a comparison star having $U = 11.08$ mag, $B = 11.07$ mag. Observations were obtained during 1982 ... 1995.

The *UBVRI* brightness of AC Cnc, RW Tri and UX UMa was found to change with a period of about 11, 5 and 2 d, respectively [1]. The analysis of the existence of any periodic *UBVRI* variability in DQ Her we intend to make in the future.

For AC Cnc, RW Tri and UX UMa the values of *UBVRI* brightness and colour indices $(U - B)$, $(B - V)$, $(V - R)$, $(V - I)$ almost uniformly fill the interval between the maximum and minimum limits corresponding to the most active and quiescent states [1]. The same situation holds in the case of DQ Her (Fig. 1 right) with the exception of 1982 July 19, when the value of $\mathcal{L}$ is separated from the remainder by about 0.3 mag (averaged *UBVRI*).

In DQ Her, AC Cnc and RW Tri the *UBVRI* light variability is caused by a change in the light of the eclipsed source – the accretion disc around the white dwarf and the part of the secondary facing them. The variations in the emission (averaged *UBVRI*) are no more than a factor of two, whereas in dwarf novae in outbursts it is of a factor of 10...100. So this variation may be due to the effect of the precession of the accretion disc, for example. On the other hand in UX UMa there is a variation of the uneclipsed source – the outer areas of the accretion disc and part of the secondary star facing away from it.

References

1. Dmitrienko, E. S., 1995, Astronomy Letters, **21**, 171

TEMPERATURE STRUCTURE OF THE DISK IN V1315 AQL

C. MORENO, R. G. M. RUTTEN, V. S. DHILLON
Royal Greenwich Observatory, Isaac Newton Group, La Palma, Spain

Abstract. We present a progress report on the analysis of time-resolved spectrophotometric observations of the cataclysmic variable V1315 Aql. The spectral range of the data covers the entire Balmer series up to and including the Balmer jump. We aim to use these data to test the steady-state accretion disk model.

1. Introduction

First studied in detail spectroscopically by Downes et al. [1] and widely studied since, V1315 Aql is a high-inclination nova-like system that undergoes deep (~ 1.5 mag) eclipses. Previous studies of the intensity structure of the disk in V1315 Aql have shown the disk to be highly symmetric without evidence for a hot spot.

From broad band photometric eclipse studies it was deduced that the disks radial temperature profile is much flatter than expected from standard accretion disk theory [2]. In order to study the disk structure in more detail we obtained time-resolved spectra through several eclipses.

2. Results

All studies carried out on V1315 Aql to date show symmetrical eclipses [2, 3]. Our data, however, exhibit large variations from one eclipse light curve to the next, and the pre-eclipse and post-eclipse light levels vary by substantial amounts. From our time resolved spectra we obtained narrow band light curves from $\lambda\lambda$ 6770 Å to 5220 Å. Although the emission line flux remains largely uneclipsed [1], the continuum light curves may be used to map the disk brightness distribution using the eclipse mapping technique. The brightness temperature as a function of radial distance from the centre

A. Evans and J. H. Wood (eds.), Cataclysmic Variables and Related Objects, 15–16.

is deduced from these maps and shown in Fig. 1 for two wavelengths. From Fig. 1 we find that the mass accretion rate is $\sim 0.3\,10^{-8}\,M_{\odot}\,yr^{-1}$.

The brightness temperature profiles in the outer disk appear to follow that of the canonical disk model with $T(R)$ closely following the predicted relation for a constant $\dot{M}$. In the inner disk however, $T(R)$ is much flatter than predicted, in accordance with previous studies [2].

A full analysis of the disk temperature, as well as the behaviour of the emission lines and the light curve shape will be published elsewhere.

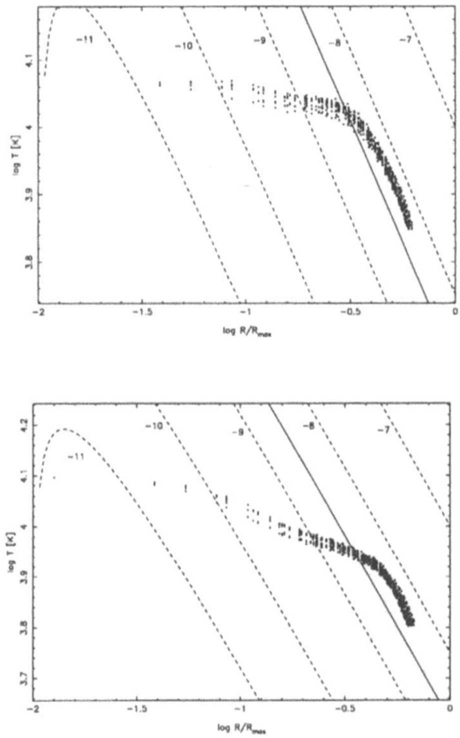

Figure 1. Fits of radial temperature profile for $\lambda\lambda$ 6770 Å (top panel) and 5220 Å (bottom) as a function of distance from the centre of the disc, the best fit to the given temperature distribution is shown by the solid line.

References

1. Downes, R.A., Mateo, M., Szkody, P., et al., 1986, Ap. J., **301**, 240
2. Rutten, R.G.M., van Paradijs, J., Tingbergen, J., 1992, A&A, **260**, 213
3. Dhillon, V.S., Marsh, T.R., Jones, D.H.P., 1991, MNRAS, **252**, 342

SPECTRALLY RESOLVED MAPS OF OPTICALLY THICK ACCRETION DISKS

R. BAPTISTA[1,2], K. HORNE[2], I. HUBENY[3], K. LONG[1],
C. W. MAUCHE[4], R. G. M. RUTTEN[5], R. A. WADE [6]
1. Space Telescope Science Institute, USA
2. University of St. Andrews, Scotland, UK
3. Goddard Space Flight Center, USA
4. Lawrence Livermore Nat. Lab., USA
5. Royal Greenwich Observatory, Tenerife, Canary Islands
6. Pennsylvania State University, USA

Abstract. We report first results of the HST/FOS time-resolved, low-resolution spectroscopy of the eclipsing cataclysmic variable UX UMa, including multi-wavelength mapping of its accretion disk in the ultraviolet region.

1. Introduction

The *Faint Object Spectrograph* on HST was used to monitor the nova-like UX UMa during eclipse. The basic aim of the project is to use eclipse mapping techniques to derive the structure of an optically thick accretion disk and to test our models of accretion disk atmospheres and chromospheres by comparison with spatially-resolved disk spectra.

2. Observations and data analysis

Time-resolved spectroscopy centered on eclipses of UX UMa was obtained with HST/FOS in 1994 August (G160L, 1100...2500 Å) and 1994 November (PRISM, 1600...8500 Å) at a time resolution of 5.3 s, in a total of 4 data sets and 3122 spectra.

The G160L out-of-eclipse spectrum (Fig. 1) shows prominent emission lines (C III λ1176, Lyα, N V λ1240, Si IV λ1400, C IV λ1550) as well as many absorption features and possibly broad absorption bands, particularly near 1900 Å and 2400 Å. He II λ1640 appears as a weak emission line. At mid-eclipse the continuum flux is reduced by a factor of ~ 3, while the emission

A. Evans and J. H. Wood (eds.), Cataclysmic Variables and Related Objects, 17–20.

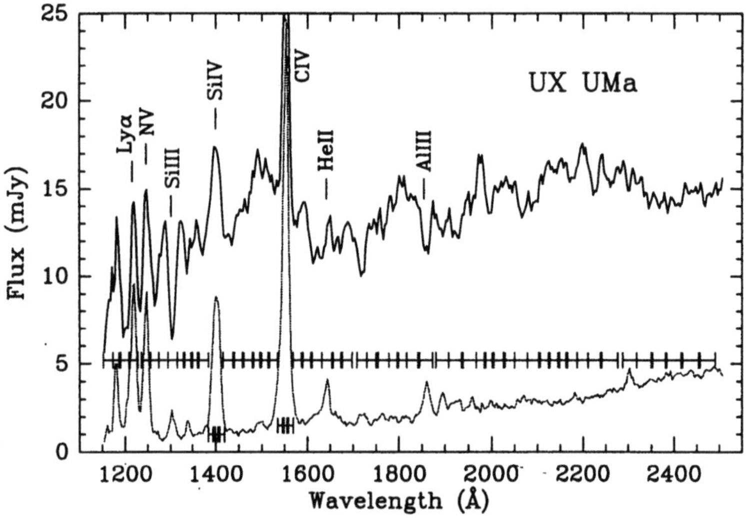

Figure 1. Average out-of-eclipse (dark) and mid-eclipse (light) spectra of UX UMa. Horizontal bars indicate the passbands used to extract eclipse light curves.

lines are much less deeply eclipsed and some of the lines that are seen in absorption in the out-of-eclipse spectra appear in emission. UV continuum light curves show significant flickering activity outside of eclipse, relatively much stronger than observed in the optical range. The eclipses are deep and steep in the continuum with an asymmetry in their egress portion, however no clear evidence of a compact bright spot is seen.

The calibrated spectra were divided into 59 (G160L) and 36 (PRISM) passbands (15...30 Å wide in the continuum and $\sim$ 3000 km s^{-1} in the emission lines) and light curves were constructed for each one. Fig. 2 shows average light curves at selected passbands. Maximum entropy eclipse mapping techniques (Horne 1985; Baptista & Steiner 1993) were used to solve for a map of the disk brightness distribution and for the flux of an additional uneclipsed component in each band (Fig. 2).

3. Results

Spatially resolved disk spectra of the 1994 August data (G160L) show that the inner disk in the UV is characterized by a blue continuum filled with absorption lines and bands, which change to emission with increasing disk radius (Fig. 3). The spectrum of the infalling gas stream, compared with the disk spectrum at the same distance from disk center, shows pronounced emission lines of C II λ1336, N IV+Fe II λ1488, Fe II λ2367, and possibly a blend of Si I lines at 1977...2015 Å. The spectrum of the uneclipsed

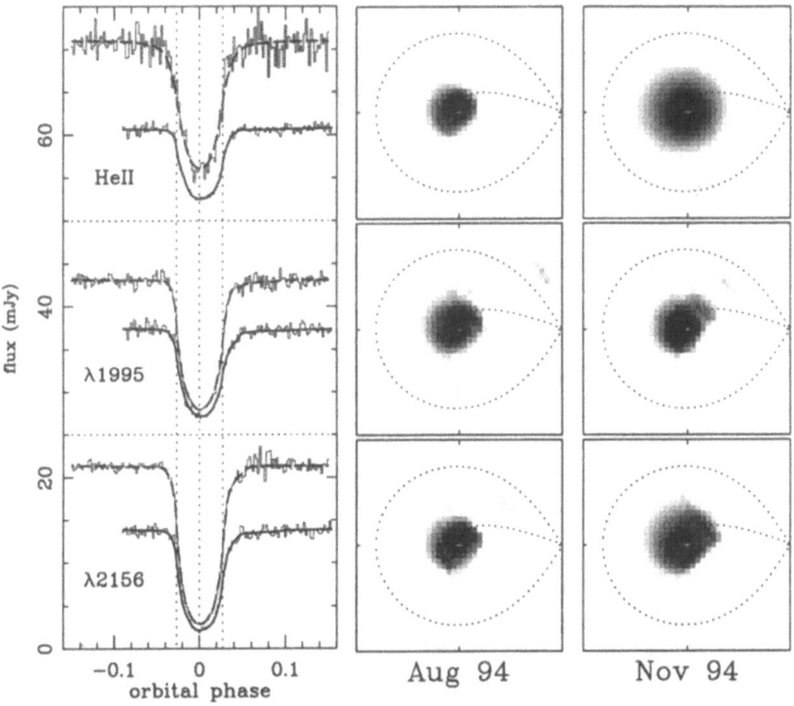

Figure 2. Left: Eclipse light curves (light gray) and fitted models [solid (Aug) and dashed (Nov) lines] for selected passbands. Right: eclipse maps in logarithmic grayscale. Dotted curves show the projection of the primary Roche lobe onto the orbital plane; the secondary star is to the right of each panel and the stars rotate counter-clockwise.

component shows strong emission lines of Lyα, N V $\lambda1240$, Si IV $\lambda1400$, and C IV $\lambda1550$, a weak He II $\lambda1640$ component, and a continuum rising towards longer wavelengths.

UX UMa was brighter (by up to 50% at 2000 ... 3000 Å) in 1994 November than in 1994 August. Quasi-periodic oscillations (QPOs) of period 20...30 s are easily seen in the 1994 November light curves. The comparison of the 1994 August and November eclipse maps shows a significant ($\gtrsim$ 50%) increase in brightness in the inner disk regions at the later epoch.

The radial temperature profiles of the August continuum maps are well described by a steady-state disk model ($\dot{M} = 10^{-8.4\pm0.3}\ M_\odot\,yr^{-1}$), while the November maps give an excellent fit to a steady disk model of $\dot{M} = 10^{-8.2\pm0.3}\ M_\odot\,yr^{-1}$, indicating an increase in the mass accretion rate compatible with the observed increase in brightness. Since the UX UMa disk seems to be in a high-viscosity state in both cases, this result suggests variations in the mass transfer rate from the companion star of considerable magnitude ($\gtrsim$ 50%).

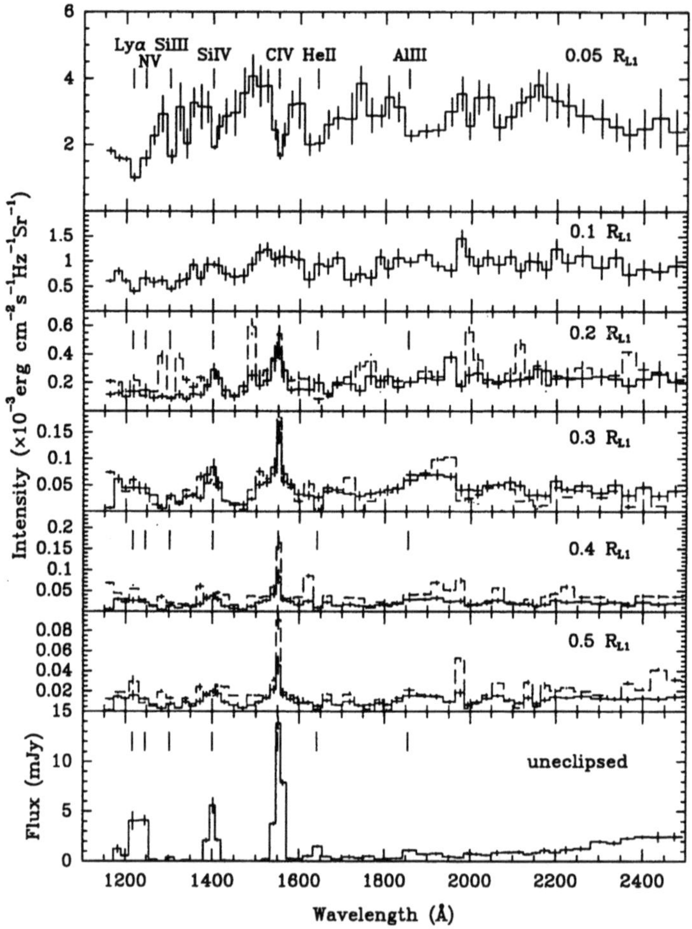

Figure 3. Spatially resolved disk spectra of UX UMa. The spectra were computed for a set of concentric annular sections (labeled values in units of the distance from disk center to the inner Lagrangian point). The spectra of the disk section containing the gas stream are shown as dashed lines.

Further analysis will extend the eclipse mapping study up to 8500 Å, study and map the flickering and QPOs, and model the spatially-resolved disk spectra with state-of-the-art disk atmosphere models.

Acknowledgment. This work was supported by NASA grant GO-5488 from the STScI (which is operated by AURA under NASA contract NAS5-26555).

References

Baptista, R., Steiner, J.E., 1993, A&A, **277**, 331
Horne, K., 1985, MNRAS, **213**, 129

ROCHE TOMOGRAPHY OF THE COOL STAR IN IP PEG

R. G. M. RUTTEN[1,2], V. S. DHILLON[2]
1. Netherlands Foundation for Research in Astronomy
2. Royal Greenwich Observatory, La Palma, Spain

Abstract. We analyse orbitally-resolved spectra of the Na I doublet in the dwarf nova IP Peg, using the newly developed technique of Roche tomography. We show that Roche tomography can be successfully employed to image the secondary stars in CVs, and that it also serves as an independent aid for determining binary system parameters. Our preliminary results suggest that the mass of the white-dwarf primary is $1.2\,M_\odot$, significantly higher than previous estimates.

1. Introduction

A complete understanding of CVs is impossible without understanding the cool secondary stars (typically K or M dwarfs). From studies of the secondary star we may obtain insight into, for example, the binary star parameters, magnetic activity on its surface (and thus on the evolution of CVs), on the mass-transfer process, and the disk's radiation field through the study of irradiation. This makes detailed study of the secondary stars worthwhile, although their faintness in comparison to the usually much brighter accretion disk renders them difficult objects to study.

IP Peg is a dwarf nova, and arguably one of the best studied cataclysmic variables (see Dhillon & Rutten 1995 and reference therein). The secondary star in IP Peg has been detected through the presence of ellipsoidal variations (Martin, Jones & Smith 1987). Also, spectral features in IP Peg, such as the Na I doublet at $\lambda\lambda$ 8183, 8195 Å, are formed in the photosphere of the secondary star, as is shown by radial velocity studies (Martin et al. 1989). These lines can be used to study secondary star characteristics, and, when observed with sufficiently high resolution and as a function of orbital phase, can be used to map the surface of the secondary star, as we will show in this paper. The results presented here are of a preliminary nature; a full analysis will be published elsewhere.

A. Evans and J. H. Wood (eds.), Cataclysmic Variables and Related Objects, 21–24.

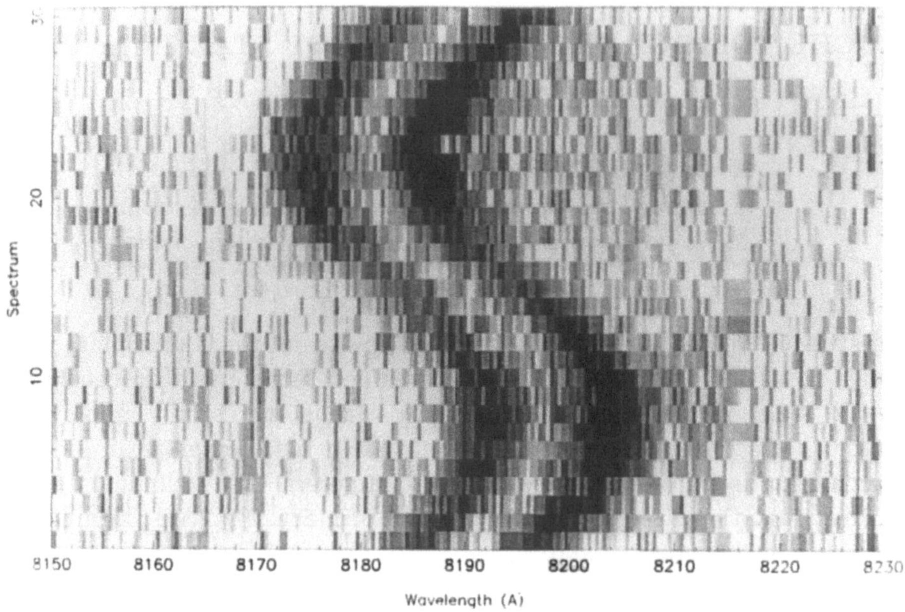

Figure 1. Trailed spectrum of the Na I doublet in IP Peg.

2. Observations

The phase-resolved spectra of the Na I doublet of IP Peg in quiescence were obtained on two nights in 1994 August with the ISIS dual-beam spectrograph at the Cassegrain focus of the 4.2 m William Herschel Telescope on the island of La Palma, Spain. The phase-folded spectra of the Na I doublet, averaged into 30 phase bins, are presented in Fig. 1 as a trailed spectrum. Clearly visible are the orbital radial velocity variations, and variations in line strength as well as line width.

3. Roche tomography

The observed line profile at a given orbital phase is the sum of line contributions from all visible parts of the star. The orbital motion of the star introduces Doppler shifts of the line features as a whole, and changes the line strength and width. The observed line profile will also depend on how the line strength varies over the stellar surface, and its detailed signature depends on the geometry of the binary. By modeling the observed line shape as a function of orbital phase, information may be deduced as to how the line strength varies over the stellar surface; this is Roche tomography (Rutten & Dhillon 1994).

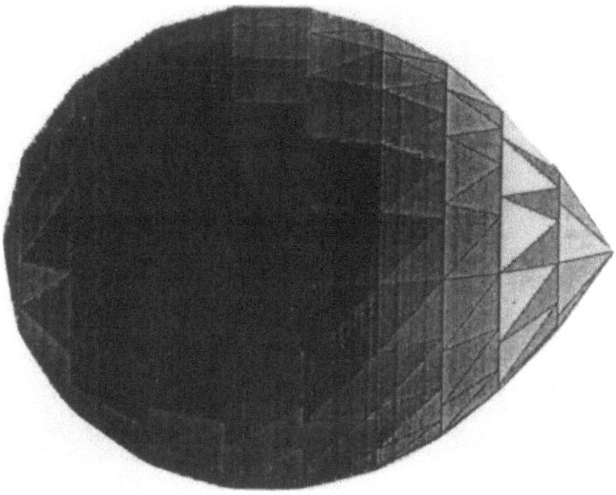

Figure 2. Representation of the optimum Na I brightness map on the surface of the secondary star in IP Peg. Dark areas indicate stronger Na I absorption.

In our implementation of Roche tomography the stellar surface is modelled by a large number of triangular tiles. The photosphere coincides with the critical Roche surface, and the star is locked into orbital motion. Furthermore, it is assumed that the line shape does not change over the stellar surface, but its strength may. It is this strength of the line for each surface element which is adjusted when fitting the data. Optimization of the fit is obtained using the maximum-entropy criterion, which arrives at the smoothest possible solution which fits the data to within the errors.

4. Analysis and results

The brightness map of the Na I doublet on the surface of the secondary star of IP Peg, computed from the data in Fig. 1, is shown in Fig. 2. This map shows that the Na I line is less strong on the hemisphere facing the white dwarf. Presumably heating by irradiation causes a reduced line strength.

So far, we have assumed that the binary star geometry is known. However, published binary star parameters are not in agreement over what the masses of the components are, which renders the results from Roche tomography less certain. To overcome this problem, rather than picking just one set of binary star masses, we reconstruct the Na I brightness map of the secondary star using a large number of combinations of masses, and from the set of solutions choose the one which, in line with the maximum entropy criterion, has the least structure. This method basically extends the entropy optimization by including the masses as additional free param-

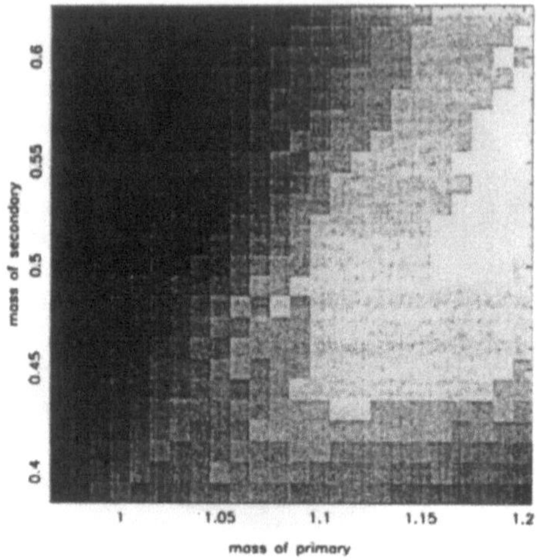

Figure 3. **Entropy landscape for IP Peg where the smoothness parameter (entropy) of the fit is plotted against the masses of the binary star components. The brightest point indicates the smoothest map.**

eters. Fig. 3 shows the so-called entropy landscape, which plots entropy as a function of masses of the primary and secondary star (see examples in Rutten & Dhillon 1994). Our preliminary results suggest the optimum fit is reached for $M_1 = 1.2\,M_\odot$ and $M_2 = 0.55\,M_\odot$. The radial velocities corresponding to this solution agree well with those measured from our data, as well as with published radial velocities. Note that Roche tomography automatically takes irradiation effects into account, and hence no *a posteriori* corrections are needed as in the case of masses determined from radial velocities only. Hence a mass determination from Roche tomography is potentially more reliable.

5. Conclusions

It is shown that Roche tomography is capable of producing surface maps of the secondary stars in cataclysmic variables. The same technique also results in a new and independent method for estimating binary star parameters, the strength of which is that it does not just attempt to fit one aspect of the line profile, e.g. the radial velocity, but rather fits the whole line profile and its variations as a function of orbital phase.

References

Dhillon, V.S., Rutten, R.G.M., 1995, MNRAS, **274**, 27
Martin, J.S., Friend, M.T., Smith, R.C., Jones, D.H.P., 1989, MNRAS, **240**, 519
Martin, J.S., Jones, D.H.P., Smith, R.C., 1987, MNRAS, **224**, 1031
Rutten, R.G.M., Dhillon, V.S., 1994, A&A, **288**, 773

APPLICATIONS OF INDIRECT ACCRETION DISC IMAGING TO ORBITAL SOLUTIONS FOR PRIMARY STARS

MARTIN STILL[1,2]
1. University of St. Andrews, Physics & Astronomy, North Haugh, St. Andrews, Fife KY16 9SS, UK
2. Astronomy Centre, University of Sussex, Falmer, Brighton BN1 9QH, UK

The general inability of optical observations of low mass interacting binaries to provide accurate radial velocity measures for the stellar components is a problem for which there is no universally applicable solution. Velocity measurements are handicapped because the accretion disc about the primary often dominates the optical spectrum. Therefore spectroscopists resort to measuring line profile symmetry in individual spectra of the accretion disc in order to provide information on the primary's motion. However, discs are not axisymmetric structures and the standard approach generally fails because the asymmetry-free inner disc emission is usually dominated by statistical noise.

We adopt an algorithm which tests for axisymmetry in tomograms. This approach is an improvement on the standard method because (i) the form of symmetry search has changed to include the whole data set, and (ii) asymmetries are removed, providing they are not too dominant, by median averaging annuli about test velocities. Primary velocities are given by the minima in velocity maps of a summed residual distribution which is the result of removing the computed symmetric profile from the data.

In Fig. 1 two test cases for the algorithm are plotted. The first is a model of optically thin line emission from a disc and bright spot, about a primary with a projected radial velocity semi-amplitude of $K = 100\,\mathrm{km\,s^{-1}}$. The standard method provides a 'best' solution of $K = 122 \pm 3\,\mathrm{km\,s^{-1}}$, where the phase lag between systemic velocity crossing and primary superior conjunction is $\phi_0 = 18° \pm 3°$. Clearly this is unacceptable. However, the minimum in the residual distribution presented in Fig. 2 provides an orbital solution of improved accuracy: $K = 97 \pm 6\,\mathrm{km\,s^{-1}}$ and $\phi_0 = 0.3° \pm 0.1°$. The second case concerns Balmer emission from the non-eclipsing quiescent dwarf nova LY Hya. The standard approach provides $K = 133 \pm 4\,\mathrm{km\,s^{-1}}$. However the phasing of this solution indicates that the bright spot location is not

A. Evans and J. H. Wood (eds.), Cataclysmic Variables and Related Objects, 25–26.

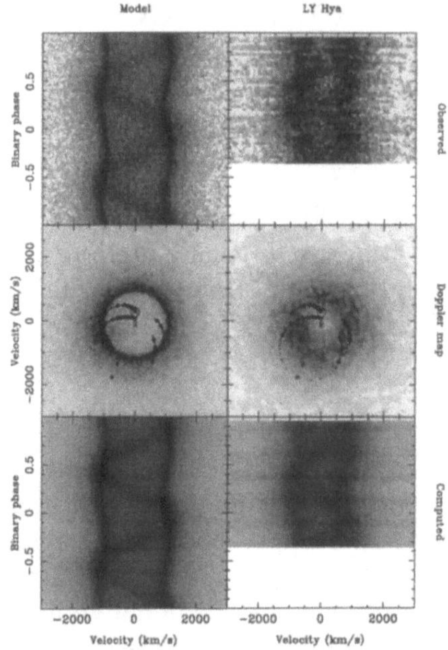

Figure 1. Data and tomograms of our two test cases – optically thin line emission from a disc and bright spot and Balmer emission from the quiescent dwarf nova LY Hya.

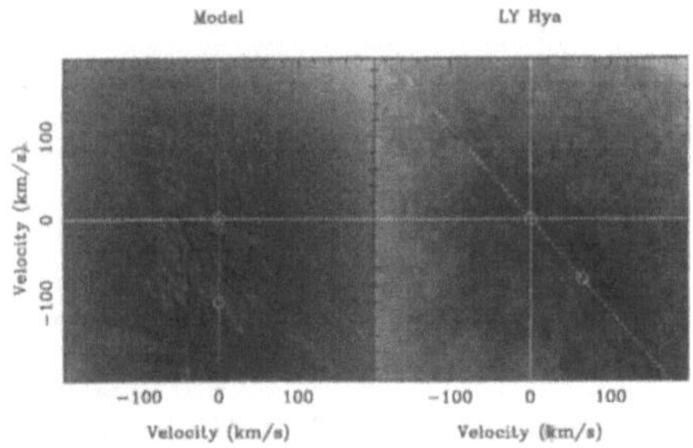

Figure 2. Velocity maps of the summed residual after removing axisymmetric emission from the line profile. The minima in these distributions predict the primary velocity, circles represent centres of mass and predicted primary velocities.

compatible with flow off the canonical gas stream trajectory (Fig. 1). Either the solution is inaccurate or a physical model is required to explain the peculiar bright spot distribution. Our algorithm provides $K = 97 \pm 31\,\mathrm{km\,s^{-1}}$, suggesting that the previous determination was biased. More significantly, the phasing has been redetermined (dashed line, Fig. 2), providing a more canonical location for the bright spot emission.

DISTANCES AND ABSOLUTE MAGNITUDES OF A SAMPLE OF FAINT CATACLYSMIC VARIABLES

L. N. SPROATS[1], S. B. HOWELL[2], K. O. MASON[1]
1. Mullard Space Science Laboratory, Holmbury St. Mary, Dorking, Surrey, RH5 6NT, England
2. Astrophysics Group, Planetary Science Institute, 620 N. 6th Ave, Tucson, AZ., USA

Our current knowledge of the basic characteristics of CVs, such as their distance and luminosity, are largely based on observations of optically bright ($V < 16$ mag) systems. This has tended to favour the more luminous members ($M_V \sim 7 \ldots 10$ mag) lying nearby in the Galactic plane ($z < 400$ pc). In order to alleviate this bias, Howell & Szkody (1990) compiled a list of ~90 CVs that were faint enough and were at high galactic latitude to potentially place them well outside the Galactic plane, in the halo. We have obtained infrared J and K band images of a number of the CVs listed by Howell & Szkody (1990) in order to establish whether these systems are truely at large distances, or are instead a population of intrinsically faint objects lying relatively nearby. The distances were determined using Bailey's method (1982) and incorporated the recent S_K calibrations by Ramseyer (1994).

We find that the majority of our sample are *not* at great distances but have $z < 400$ pc with a distribution which is indistinguishable from their brighter well studied counterparts (from the compilation of Warner 1987; see Fig. 1). Only two stars in our sample (DO Leo and RU LMi) were found to have sufficiently large z distances to constitute genuine halo candidates. The implication of this finding is that the CV space density in the Solar neighbourhood may be higher than previously anticipated.

While we find that the long period systems in our sample have M_V (min) consistent with the range derived by Warner (1987; see Fig. 2), the short period ($P < 2$ h) CVs are up to 4 mag *fainter* in absolute magnitude when compared to those from Warner. The likely reason for their low luminosity is low mass transfer rates during minimum. However, the mean luminosity, averaged over their outburst cycle, is probably comparable to other systems of similar orbital period (M_V (mean) $\sim 7 \ldots 9$ mag).

A. Evans and J. H. Wood (eds.), Cataclysmic Variables and Related Objects, 27–28.

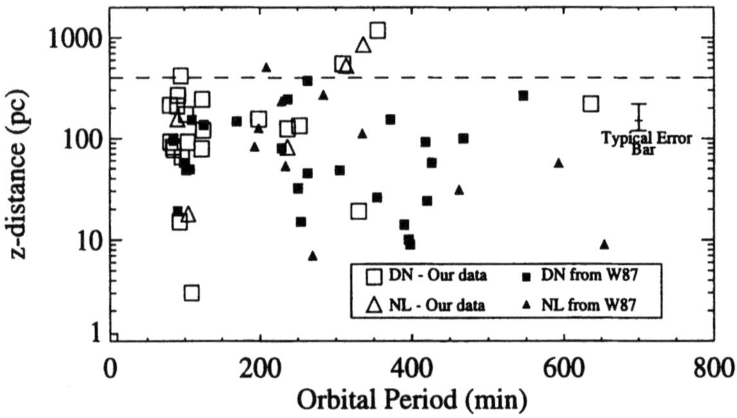

Figure 1. Distance above the Galactic plane for our sample of CVs plotted against orbital period. The mean z distance of the old disk-halo population of stars as well as the bright CVs from Warner (1987) are also shown.

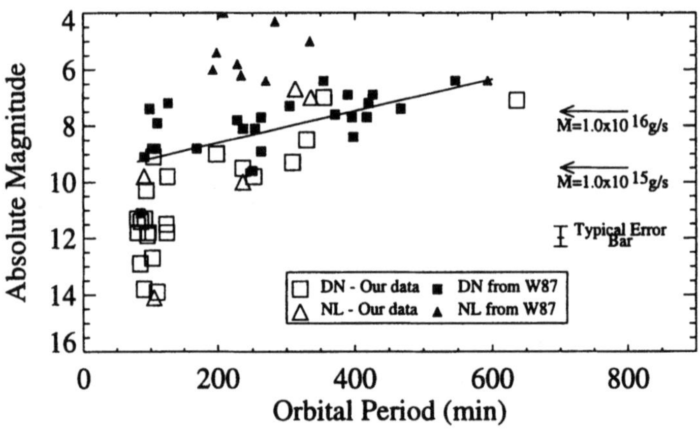

Figure 2. Absolute magnitude at minimum plotted against orbital period. Also plotted are approximate rates of mass transfer through the disk and the linear fit to dwarf novae from the Warner (1987) compilation.

References

Bailey, J., 1982, MNRAS, **197**, 31
Howell, S.B., Szkody, P., 1990, Ap. J., **356**, 623
Ramseyer, T.F., 1994, Ap. J., **425**, 243
Warner, B., 1987, MNRAS, **245**, 208

STUDIES IN FLICKERING

W. F. WELSH[1], JANET H. WOOD[1], K. HORNE[2]
[1] *Keele University* [2] *University of St.Andrews*

Abstract. Flickering is a poorly understood phenomenon associated with accretion processes, but this does not suffice to make it interesting. Why then should we bother studying this 'noise'? Three reasons come to mind: (i) flickering is a fundamental signature of accretion, to the point of being a necessary characteristic (if it doesn't flicker, it's not a CV); (ii) energetically, flickering is not a small effect and can contribute up to a few tens of percent of the total luminosity of the system (hence it is often more luminous than the entire secondary star); (iii) flickering is inherently a time–dependent phenomenon and it is hoped that it can provide clues to the nature of the disc viscosity, something that time–independent theory cannot provide.

Study 1: Eclipse mapping the flickering

To determine where the flickering originates, it is best to examine an eclipsing system. Because the amplitude of the flickering decreases during eclipse, it has been known for a long time that the flickering is associated with the white dwarf, disc and/or hot spot. By use of the eclipse mapping method [3] the mean eclipse light curve can be used to spatially resolve structure in the disc. The concept is trivially extended to mapping the flickering. The difficulty lies in determining the eclipse light curve of the flickering.

Two possible methods are available for determining the flickering light curves, which we will call the 'single' and the 'ensemble' methods. In both cases, many light curves are needed to remove sensitivity to individual flickers and to get a statistically significant result. In the absence of any changes in the state of the CV (i.e. the steady plus random variations are statistically stationary), these methods are identical. Obviously things like dwarf nova outbursts violate this criterion, but changes in observing conditions (e.g. different detector sensitivity or bandpasses) can also cause problems. To minimize these contaminating effects, it is best to observe all the light curves as close together in time as possible. The 'single' method (championed by Bruch [2]) computes the deviation of the observed light curve from a running mean. This is done for many light curves and the results

A. Evans and J. H. Wood (eds.), Cataclysmic Variables and Related Objects, 29–32.

averaged. Its main disadvantage is the difficulty of separating the rapid flickering fluctuations from the rapid variations due to the eclipse itself. Just as flickering contaminates the mean eclipse light curve when studying the steady disc, so too will the disc eclipse contaminate the study of the flickering; in the end, one must exercise extreme care. Horne & Stiening [4] laid the foundation of the 'ensemble' method: the flickering light curve is defined as the rms of the residuals of the individual light curves relative to the mean light curve at each orbital phase. The advantages of this method are (i) because many observations are used the mean is well defined and so there is no need to approximate it (via a running mean or spline or polynomial fit); (ii) each orbital phase is treated independently (thus there are no correlations between adjacent bins). The major disadvantage is the sensitivity to violations of stationarity. This latter problem is very serious, as it requires that there are no substantial changes in the disc structure (or detector/observing conditions) over the course of obtaining the data. If this is violated, the flickering eclipse light curves can be misleading (one is seeing the effects of long–term structural changes in the disc, or instrumental effects, and not flickering). Luckily it is easy to test if stationarity has been violated and so one knows when compensations for non-stationarity are needed. Unfortunately, these corrections are not easy to determine and the final results are sensitive to any assumptions that go into them. However, with care they can be derived (e.g. [1]) and the ensemble of light curves used. The alternative is to use only those light curves which lie within some tolerance of the mean. We prefer the ensemble method because it has no free parameters (no need to fit to the light curves). However the single method has the advantage of being able to cope more readily with observational non-stationarity. Both these methods will play important roles in determining the light curve of the flickering.

Analysis of the flickering in HT Cas, including eclipse mapping, is in progress [5]. Preliminary results reveal that in the U–band, the flickering is highly concentrated toward the center of the disc, but not confined to the immediate area around the white dwarf. The flickering amplitude increases towards the center faster than the mean (steady) light. The hot spot contributes a negligible amount of flickering.

Study 2: The spectrum of the flickering

Predictions of a simple model

Assuming that flickering arises from dynamical time-scale processes, we have attempted to model the flickering by extending the standard steady–state accretion disc model in the simplest possible way, giving it a time dependency. This was accomplished by allowing small sinusoidal temperature variations about the local steady–state temperature:

$T(r,t) = T_0(r)[1 + A\sin(2\pi t/\tau_{\rm Kep})]$,

where $T_0(r) \propto r^{-3/4}$ and $\tau_{\rm Kep}$ is the local Keplerian time-scale ($\propto r^{3/2}$). Because each annulus in the disc radiates like a blackbody, and because a blackbody has an intensity maximum which varies inversely with temperature and has an exponential cutoff at wavelengths shorter than this (Wien's law), we have the following predictions about the power-spectrum: a high-frequency wavelength–independent cutoff, and a low-frequency wavelength–dependent roll over (roll over frequency $\propto \lambda^{-2}$). The main virtue of this model is its simplicity — the predictions are clear and unambiguous. By examining the power-spectrum at each wavelength (a 'power-spectrum spectrum'), we can test these predictions.

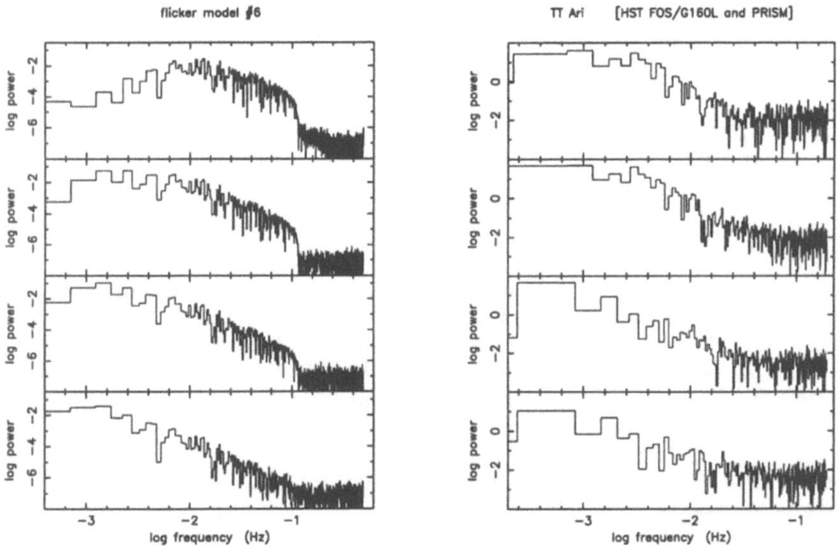

Figure 1. **Left:** Power spectra from flickering model at λ1000 Å (top), 2000, 3000 and 6000 Å (bottom). The high-frequency break occurs at the Keplerian time-scale of the innermost orbit (~10 s) and is independent of wavelength. The wavelength–dependent low–frequency rollover occurs at time-scales longer than ~4 m. **Right:** TT Ari power spectra at λ1250–1400 Å (top), 2000, 3000 and ~6000 Å (bottom). No low–frequency rollover or high–frequency break is seen.

A first look at some HST data on flickering

High–speed HST FOS spectrophotometry of TT Ari and V603 Aql were examined and compared with these predictions. No low-frequency rollover is seen, and if any high-frequency cutoff exists, we have yet to dig it out of the noise. These preliminary results indicate that our simple model fails, or at least confines the parameters to an uncomfortably small region of parameter space. As the predictions of the model are inescapable, what

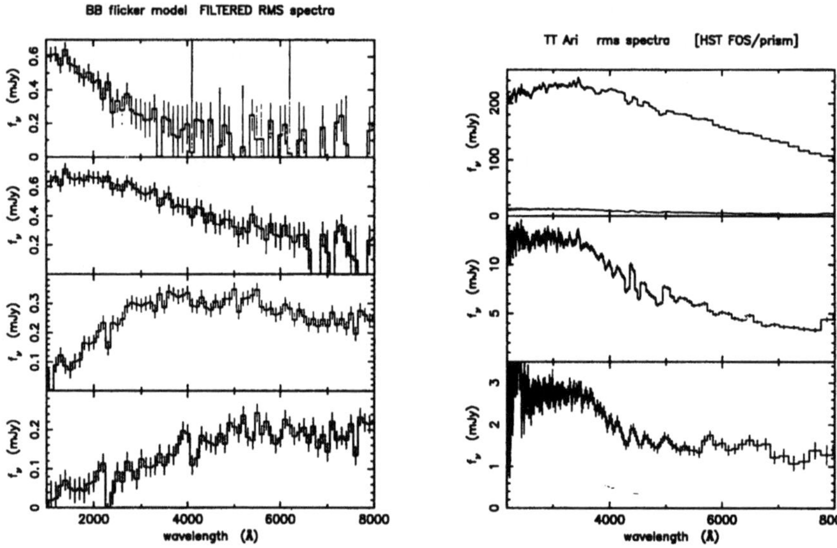

Figure 2. **Left:** Model flickering spectra (top to bottom): fastest flickering (τ <100 s), faster than 300 s, slower than 300 s, and the slowest flickering attainable from the 1000 s of simulated data. Note the strong time-scale dependence. **Right** (top to bottom): TT Ari mean, rms, and high–pass filtered rms spectra. The bottom panel shows the spectrum of the fastest flickering, revealing a large Balmer jump in emission.

has gone wrong? In essence, this failure indicates that our intuition about flickering is incorrect.

Delving deeper into the data, we find where our intuition breaks down by examining the actual spectrum of the flickering (determined by computing a 'debiased rms spectrum'). It appears that the spectrum of the fastest flickering contains a large Balmer jump in emission, indicating optically thin gas. Thus we have learned that the flickering cannot be treated as an extension of standard (optically thick) accretion disc theory.

References

1. Bennie, P.J., Hilditch, R.W., Horne, K., 1996, these proceedings, p33
2. Bruch, A., 1996, these proceedings, p35
3. Horne, K., 1985, MNRAS, **213**, 129
4. Horne, K., Stiening, R.F., 1985, MNRAS, **216**, 933
5. Welsh, W.F., Wood, J.H., 1995, in "Flares & Flashes" – Proc. of IAU Colloquium 151, eds. J. Greiner, H.W. Duerbeck, R.E. Gershberg, Springer–Verlag, p300

DECOMPOSING INTERACTING BINARY LIGHT CURVES: THE ECLIPSES OF THE MEAN LIGHT, SECULAR VARIABILITY AND FLICKERING IN RW TRI

PAUL J. BENNIE, R. W. HILDITCH, KEITH HORNE
School of Physics and Astronomy
University of St Andrews, KY16 9SS, UK

Abstract. We describe a new method of orbital light curve decomposition which is applicable to long-term photometry of interacting binaries. This method determines the orbital dependence (including eclipses) of the mean light, secular variability and RMS flickering. We identify the true line of centres of RW Tri and find that the accretion disc is a location of secular variability and a source of flickering.

1. Introduction

The 13th magnitude nova-like cataclysmic variable RW Tri is an eclipsing binary which displays photometric variability on differing time-scales. The secular brightness variations, which can be sinusoidal (Honeycutt et al. 1994), have an amplitude of 0.5 mag. Horne & Stiening (1985) determined the flickering to be compactly eclipsed through phases ± 0.04.

A consistent set of light curves was obtained spanning 1993 November to 1995 March comprising differential photometry from the CCD equipped 0.9 m James Gregory telescope (St. Andrews). Thirty binary orbits were observed which yielded 19 $R-$band eclipses covering 2100 cycles. The eclipse shapes were broad, rounded and variable to the extent typical of nova-likes.

2. Decomposition method and discussion.

The eclipses of the mean light, secular variability and RMS flickering may be determined by decomposing ensembles of phase binned eclipse light curves. Real variations in the disc flux between light curves are used to quantify the amounts by which that flux is lessened through eclipse. These real variations of the disc flux are determined by calculating a mean flux value across each light curve, excluding phases where the flux is attenuated (eclipse) or

A. Evans and J. H. Wood (eds.), Cataclysmic Variables and Related Objects, 33–34.

enhanced (bright spot). These mean flux values are the reference fluxes. We then plot, for an individual phase bin, the observed flux versus the reference flux for each light curve. Fig. 1a shows plots for three such phase bins through eclipse egress, where the 19 data points from the 19 light curves are clearly linearly related.

Each linear fit to a flux-flux plot yields a mean light value, and a gradient which represents the visibility of the secular variability at that phase. Fits for all phases yield the eclipse light curves of the mean light, and the secular variability (Fig. 1b). The minimum of the secular variability occurs earlier and may represent the true line of centres if we equate the centres of light variability and mass. The eclipse shape of the secular variability is clearly dissimilar to that of the mean light, indicating differing disc distributions.

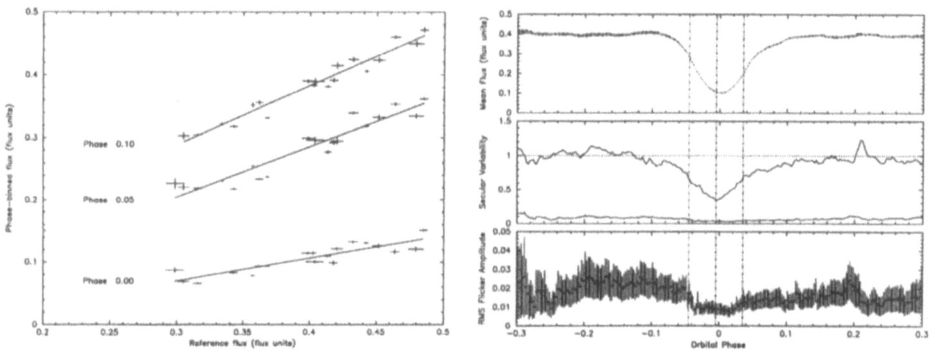

Figure 1. (a) Three flux-flux plots at phases through eclipse egress. (b) The eclipses of the mean light (top), secular variability (middle) and flickering (bottom) in RW Tri.

The residuals about the fit for each phase in Fig. 1a are larger than the plotted measurement uncertainties because they include contributions from orbital flickering and observational noise. The RMS flickering amplitude is simply isolated by manipulating these distributions in quadrature and is presented in Fig. 1b. The eclipse of the flickering source is coincident with the inner disc but the ingress appears steeper than egress and the eclipse may not be of the dominant system flickering. This result may be sensitive to our preliminary technique, but is insensitive to biasing by the secular variability. Our eclipse results may be eclipse mapped (Horne 1985).

Acknowledgements. PJB is funded by a PPARC research studentship.

References

Honeycutt, R.K., Robertson, R.W., Turner, G.W., Vesper, D.N., 1994, in "Interacting Binary Stars", ed. A. Shafter, ASP Conf. Ser. **56**, 277

Horne, K., 1985, MNRAS, **213**, 129

Horne, K., Stiening, R.F., 1985, MNRAS, **216**, 933

THE PLACE OF ORIGIN OF THE FLICKERING IN Z CHA

ALBERT BRUCH
Astronomisches Institut
Wilhelm-Klemm-Str. 10
D-48149 Münster, FRG

The variability in the observable light curve of a CV can be attributed to (i) orbital and long-term variations; (ii) flickering, (iii) photon noise, (iv) scintillation (negligible) and (v) any residual noise. Calculating the rms-scatter of the count rates in a light curve as a function of orbital phase, ϕ, through accretion disk eclipse of high inclination CVs – after subtraction of a smoothed light curve to remove orbital and long-term variations, and applying a correction for photon noise – permits one to confine the place of origin of the flickering if its visibility is phase dependent or if eclipses take place (assuming any residual source of noise to be constant in time). The rms-scatter curve is then defined as the function $\sigma(\phi) = \sqrt{\sigma_f^2 + \sigma_r^2}$, where σ_f^2 is the variance due to flickering and σ_r^2 the variance due to residual noise.

This method was applied to a large number of light curves of Z Cha in quiescence, normal maximum and super-maximum. The mean of the normalized rms-scatter curves of the individual light curves is shown in Fig. 1. The dashed and dotted vertical lines indicate the phases of the inner contact points of the white dwarf eclipse and the hot spot eclipse, respectively. Due to details of the subtraction of orbital and long-term variations the applied method is sensitive only to low-amplitude, small time-scale (< 1 m) flickering. The phase dependence of the rms-scatter clearly indicates eclipses of the flickering light source. There are, however, significant differences depending on the photometric state of Z Cha.

Quiescence: The rms-scatter curve has a significant minimum centred on the total eclipse of the white dwarf. Furthermore, an enhanced scatter is evident during the phases of the orbital hump at $-0.26 \leq \phi \leq -0.02$, while it is somewhat lower between $0.02 \leq \phi \leq 0.3$, i.e. at phases when the hot spot is not visible. This suggest the presence of two different sources of flickering, one residing in the hot spot, the other close to the white dwarf.

A. Evans and J. H. Wood (eds.), Cataclysmic Variables and Related Objects, 35–36.

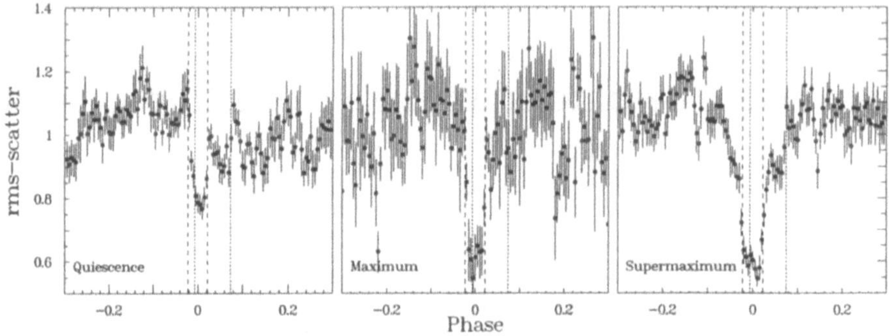

Figure 1. **Rms-scatter curves of Z Cha in different photometric states**

During the orbital hump phase both sources are visible, leading to enhanced scatter compared to the phases when the hot spot is invisible. When both white dwarf and hot spot are eclipsed the scatter assumes a minimum. A gradual decline of the rms-curve when the white dwarf – but not the hot spot – is totally eclipsed also argues for the contribution of hot spot flickering. A rapid increase of the scatter when the white dwarf emerges from the eclipse indicates that the white dwarf flickering is well localized in its immediate vicinity (the boundary layer?).

Maximum: In this case the rms-scatter curve attains a sharp, flat-bottomed minimum during the phases of invisibility of the white dwarf, while at all other phases it remains on a level which – considering the strong noise – can be considered as constant. In particular, there is no significant difference before and after the eclipse. Thus, it can be concluded that the flickering is dominated by a contribution from the immediate vicinity of the white dwarf. Flickering of the hot spot, seen in quiescence, is completely outshone if it exists at all during outburst.

Super-maximum: Again the rms-scatter has a minimum centred on the eclipse phase of the white dwarf. However, in particular the egress of this minimum is more gradual, indicating the flickering light source to be more extended than during quiescence and normal outburst. Superposed is a broad asymmetrical feature with a gradual decrease of the rms-scatter in the interval $-0.07 \leq \phi \leq -0.03$ and a sudden increase (after a constant phase) close to $\phi = 0.07$. Its broadness suggests that it is associated with an extended structure at the rim of the accretion disk. A more detailed analysis shows that the broad feature is present only when the superhump light source is not shining (superhump phase $0.25 < \Phi < 0.75$), while during superhump phases $-0.25 < \Phi < 0.25$, i.e. if the superhump maximum is roughly centred on orbital eclipse, the broad feature disappears and the ingress and egress of the eclipse of the flickering light source become significantly more gradual, indicating a large extension.

TT ARI–94: A STUDY OF 1.6...60 MINUTE VARIABILITY

I. L. ANDRONOV[1], K. ARAI[2], L. L. CHINAROVA[1], N. I. DOROKHOV[1], T. A. DOROKHOVA[1], A. DUMITRESCU[3], D. NOGAMI[4], A. LEPARDO[5], P. A. MASON[6], K. MATUMOTO[7], G. OPRESCU[3], G. PAJDOSZ[8], R. PASSUELO[5], L. PATKOS[9], G. SOSTERO[5], J. TREMKO[10], S. ZOLA[8,11]

[1] *Department of Astronomy, Odessa State University, T.G.Shevchenko Park, 270014 Odessa, Ukraine*
[2] *Kuki–Hokuyo Senior High School, Kuki Saitama, 346 Japan*
[3] *Astronomical Institute of the Romanian Academy 75212 Bucharest 28, Romania*
[4] *Department of Astronomy, Faculty of Science, Kyoto University, Oiwake–cho, Kitashirakawa Sakyo–ku, Kyoto 606–01, Japan*
[5] *Stazione Astronomica, 33047 Remanzacco (ud), via S.Stefano 31, Italy*
[6] *Case Western Reserve University, Department of Astronomy, Cleveland, OH 44106, USA*
[7] *Osaka Kyoiku University, B4–212, 698–1, asahigoaka chome kashiwara–shi, Osaka 582, Japan*
[8] *Mt.Suhora Observatory, Pedagogical University, ul.Podchorążych 2, 30–084 Kraków, Poland*
[9] *Konkoly Observatory, P.O. Box 67, 1525 Budapest, Hungary*
[10] *Astronomical Institute of the Slovak Academy of Sciences, 059 60 Tatranska Lomnica, Slovakia*
[11] *Astronomical Observatory, Jagiellonian University ul.Orla 171, 30–244 Kraków, Poland*

We report photometric results from 44 runs at 11 observatories during the international campaign 'TT Ari–94'. No coherent oscillations in the frequency range 10...900 cycle d^{-1} are found. The highest peaks in the power spectrum cover the wide range of 28...139 cycle d^{-1}. Variations occur at a few preferred time-scales rather than at one cycle length, with

A. Evans and J. H. Wood (eds.), Cataclysmic Variables and Related Objects, 37–40.

a possible secular decrease. In the frequency range 90...900 cycle d^{-1} the power spectrum obeys a power law with slope γ ranging from 0.8 to 2.6 for different runs.

TT Ari is one of the brightest cataclysmic variables and remains one of the most interesting objects of this class. It exhibits a variety of phenomena observed at time-scales from seconds to months. A recent detailed photometric study of this object and a bibliographical overview may be found in Tremko et al. (1996). Tremko et al. (1992, 1993, 1994) discuss aspects of the TT Ari–88 campaign. Our campaign TT Ari–94 was unprecedented, as the observations were longitude–dispersed, from Japan through Turkmenia, to Europe and the American continent. Moreover, our optical data on October 7 partially overlap with the HST observations obtained by Horne & Welsh (1995, private communication). The numerical results of observations obtained during these nights are shown in Table 1.

During TT Ari–94 we tried to obtain as many nights as possible at different observatories. Most successful were the observations obtained at the Dushak mountain station of the Odessa State University located in Turkmenia. A total of 16 nights containing 13 879 observations with mainly 20 s integrations were gathered. At the Mt. Suhora station near Krakow 8 nights with 19 355 observations with mainly 5 s integrations were obtained.

To study variability on time-scales shorter than the primary '3–h period' and a possible longer secondary wave (see Tremko et al. 1992, 1995 for discussion), we have computed a low–frequency trend by using the method of 'running parabolae' (Andronov 1990) with a filter half-width $\Delta t = 0.05$ d. The $(O - C)$ deviations from this fit were used for further analysis.

Periodograms were computed by the program FOUR-1 (Andronov 1994) using a one–frequency harmonic fit with unknown mean. In Table 1 the following parameters of the individual runs are listed: observatory O (numbered as per affiliation on title page, 8 means 8 and 11), filter F (W = white light, i.e. unfiltered); Julian date (JD – 2 449 600), number of points N; r.m.s. deviation in mmag σ_O of the signal from the mean, and σ_{O-C} from the running parabolae fit; the Nyquist frequency F_N in cycle d^{-1} ; the 'best- fit' amplitude r in mmag and frequency f in the range 10...200 cycle d^{-1} ; the value $LP = -\ln(Pr)$, where Pr is the 'false alarm probability'; and the best-fit slope γ.

Tremko et al. (1993, 1995) and Efimov et al. (1995) have found that the power spectrum obeys the power law $S(f) \propto f^{-\gamma}$ in the frequency range 90...900 cycle d^{-1} . It may be noted that the values of γ for the runs with $f_N < 900$ cycle d^{-1} are to be neglected because the interval being fit extends beyond the Nyquist frequency. For these runs the apparent values are closer to zero than those data-sets with a time resolution of 1...20 s. For some runs we have binned the counts to obtain 20–s resolution. The

TABLE 1. Results of the periodogram analysis

O	F	JD	N	σ_O	σ_{O-C}	$r \pm \sigma_r$		$f \pm \sigma_f$		f_N	LP	$\gamma \pm \sigma_\gamma$	
8	B	23	1116	77	70	57	2	50.8	.3	7377	94	1.81	.17
8	W	24	715	52	32	21	2	122.7	.9	8171	35	2.56	.48
8	B	24	1394	55	51	31	2	44.5	.4	8387	57	2.01	.17
8	B	26	3204	62	53	25	1	32.4	.1	8111	75	2.01	.12
8	B	31	498	50	39	25	2	50.2	1.7	8387	24	1.71	.32
1	B	31	501	71	44	29	2	59.1	.4	2009	24	2.17	.22
4	B	31	294	72	64	40	5	40.3	.5	942	11	0.94	.12
4	B	31	639	75	58	24	3	66.0	.3	924	10	1.52	.10
1	B	32	1429	67	43	22	1	76.2	.2	3680	42	1.86	.16
1	B	33	353	62	49	43	3	91.1	.8	3849	35	1.80	.17
1	B	33	6218	72	63	36	1	95.1	.1	25205	231	1.32	.13
5	B	34	1203	72	61	32	2	85.2	.2	3272	39	1.81	.15
1	B	35	847	66	52	32	2	58.1	.3	3697	34	1.56	.15
1	B	36	765	61	50	29	2	64.9	.2	1862	27	1.70	.12
8	B	37	3560	80	59	31	1	68.4	.1	8331	110	1.79	.10
1	B	37	1529	95	58	38	2	68.7	.1	3885	77	1.41	.13
1	B	38	950	62	50	29	2	52.4	.2	1834	34	1.78	.11
1	B	39	106	55	48	45	5	74.3	2.3	1812	10	1.66	.33
1	B	39	738	59	47	23	2	58.8	.3	1815	17	1.88	.15
1	B	40	917	67	54	27	2	59.4	.2	1775	23	1.83	.17
6	R	46	58	62	34	25	5	33.9	.9	160	1	-0.32	.17
6	R	47	49	48	39	20	7	121.1	1.6	160	0	-0.06	.17
1	B	51	1096	66	53	24	2	36.2	.2	1872	22	1.80	.11
1	B	52	1116	84	57	27	2	26.2	.2	1830	27	1.76	.13
1	B	53	776	63	50	35	2	58.1	.2	1838	44	1.60	.12
1	B	54	112	74	58	57	6	58.6	2.0	1815	13	1.90	.28
1	B	54	1052	77	55	27	2	30.5	.2	1849	26	1.83	.11
1	B	56	534	60	48	30	3	62.7	.4	1860	23	1.87	.15
8	B	59	2501	64	47	21	1	70.5	.2	8343	52	2.03	.16
1	B	59	1476	70	47	23	2	27.8	.1	1877	37	2.06	.11
8	B	60	3199	65	53	24	1	57.3	.2	8452	74	1.87	.13
1	B	60	1409	73	48	21	2	36.3	.1	1832	28	1.97	.08
8	B	61	3883	62	46	20	1	49.4	.1	8383	76	1.93	.11
3	B	61	134	74	62	38	7	104.2	.7	485	4	1.46	.18
10	B	62	338	62	53	26	4	68.8	.4	832	7	1.35	.14
9	B	63	265	63	48	22	4	28.8	.4	509	4	1.11	.11
9	V	63	266	58	50	28	4	28.3	.3	511	7	1.14	.10
3	B	63	118	67	54	30	7	85.9	.9	435	2	0.51	.16
10	B	63	451	59	50	19	3	79.0	.4	890	5	1.23	.11
7	V	64	248	65	47	26	4	72.5	.5	641	6	1.45	.16
7	V	64	358	61	42	20	3	128.8	.3	618	7	1.11	.09
2	B	64	440	75	50	34	3	37.4	.2	846	22	0.81	.16
2	V	90	95	106	63	45	7	139.2	.4	178	2	0.16	.13
2	B	90	320	100	53	30	4	85.9	.2	606	10	0.96	.20

corresponding values of γ changed by only 0.02, i.e. much below the error estimate. The values for B, V obtained quasi–simultaneously at JD 63 are very close, similar to the results of the *UBVRI* photometry by Efimov et al. (1995).

The periodograms often show a few peaks of comparable height, arguing for a set of preferred time-scales, rather than a secular decrease like the change from 27 m in 1962 to 17 m in 1985 suggested by Semeniuk et al. (1987). Characteristics of the highest peaks for 45 runs obtained in 1988 are presented by Tremko et al. (1994). Our present results support their suggestion that, in TT Ari, we observe contributions from several instability mechanisms with similar time-scales. It will be very fruitful to obtain spectra with time resolution sufficient to study 20–m variability and to locate the source of its origin.

On 1994 October 7, observations with 1–s resolution were carried out at the 80 cm telescope of the Odessa State University. The best fit cycle length was equal to 15.80 ± 0.14 m with the extremely low 'false alarm probability' 10^{-241}. Such low false alarm probabilities are an indication that that the noise might be dominated by red power-law shot-noise rather than white noise. For a discussion of red noise modeling, see van der Klis (1989). The apparently low value $\gamma = 1.32$ listed in Table 1 is due to lower signal-to-noise ratio for shorter exposures. Reduced to a standard 20–s resolution we find a corresponding $\gamma = 1.85 \pm 0.16$, which is very consistent within error estimates with the values obtained on the days before and after.

References

Andronov, I.L., 1990, Kinematics and Phys. Celest. Bodies (Kiev) **6**, 87

Andronov, I.L., 1994, Odessa Astron. Publ. **7**, 49

Efimov, Yu.S., Shakhovskoy, N.M., Andronov, I.L., Kolesnikov, S.V., 1995, Izvestija Krim. Astrofiz. Obs., in press

Semeniuk, I., Schwarzenberg–Czerny, A., Duerbeck, H., et al., 1987, A&SS, **130**, 167

Tremko, J., Andronov, I.L., Luthardt, R., et al., 1992, IBVS, 3763

Tremko, J., Andronov, I.L., Chinarova, L.L., et al., 1993, in "Cataclysmic Variables and Related Physics", eds O. Regev, G. Shaviv, Ann. Israel Phys. Soc., **10**, 324

Tremko, J., Andronov, I.L., Chinarova, L.L., et al., 1994, Odessa Astron. Publ., **7**, 132

Tremko, J., Andronov, I.L., Chinarova, L.L., et al., 1996, A&A, submitted

van der Klis, M., 1989, ARAA, **27**, 517

TIME RESOLVED OPTICAL SPECTROSCOPY OF V795 HER

R. J. DICKINSON[1], R. K. PRINJA[1], S. R. ROSEN[2], K. HORNE[3]
1. Department of Physics and Astronomy, University College London, Gower Street, London, WC1E 6BT, UK
2. Department of Physics, University of Leicester, University Road, Leicester, LE1 7RH, UK
3. Department of Physics and Astronomy, University of St Andrews, North Haugh, St Andrews, KY16 9SS, Scotland

Abstract. We present high resolution optical spectroscopy of the cataclysmic variable (CV) V795 Her. The time- and phase- dependent behaviour of the emission source is examined using two-dimensional image analysis and Doppler reconstruction techniques on the orbital profiles. A clear S-wave modulation of a weak emission feature in the Hβ and Hγ line wings (extending to $\sim$1800 km s^{-1}) is present, with complex asymmetric properties.

1. Introduction

Preliminary results are presented from an optical spectroscopic study of the CV V795 Her, based on $\sim$1 Å resolution data secured using the 2.5 m *Isaac Newton Telescope* during 6 nights in 1994 June. V795 Her is an important system since (i) it has a mooted orbital period of $\sim$2.6 h which places it inside the CV 'period gap', (ii) its UV spectrum exhibits broad blue-shifted absorption line troughs, indicative of a fast outflow (Prinja, Drew & Rosen 1992) and (iii) its precise classification remains debated.

Our analysis is based on the inspection of 241 individual exposures ($\lambda\lambda$ 4080–4970) and appropriate phase binning. The data provide constraints on localised emission regions residing on the accretion disk of V795 Her.

2. Results

Fourier analysis confirms that the total observed emission strengths of the Balmer lines, and the He I λ4471 absorption equivalent width, modulate

A. Evans and J. H. Wood (eds.), Cataclysmic Variables and Related Objects, 41–42.

on the 2.6 h orbital period of V795 Her. The double-peaked emission core of Hβ is variable in strength and width, but does not exhibit any obvious radial velocity modulation. Significant variability is also seen in the shallow extended wings of Hβ.

Phase-folded, trailed spectral images of the binned Hβ profiles were constructed to examine systematic profile changes as a function of the 2.6 h orbital phase. A prominent S-wave modulation is seen extending to $\sim$1800 km s^{-1}, as previously observed by Haswell et al. (1994). Their lower velocity S-wave is not obvious in our data, though an enhancement of the blueward and redward peaks is noted at orbital phase $\sim$0.4 and $\sim$0.9, respectively.

Our higher resolution data also reveal some complex and previously unrecognised properties of the outer S-wave. We identify two possible asymmetries: (i) the blue Hβ wing shows greater enhancement in emission compared to the red wing, and (ii) the motion of the feature on the red side does not represent an obvious continuation of its blue counterpart. A similar analysis carried out on the Hγ profiles reveals the same emission enhancements at phases $\sim$0.4 and $\sim$0.9 and a corresponding, asymmetric outer S-wave.

The Balmer line profiles show a redward absorption feature, extending to $\sim$1000 km s^{-1}, which acts to subdue the redward portion of the S-wave. It is stronger in the first half of the orbit i.e., from phases 0.0...0.5.

Doppler mapping of the Hβ profiles suggests an emission ring from gas circulating near the outer rim of the accretion disk. It has a $\sim$600 km s^{-1} diameter which corresponds to the splitting of the double peaks in the emission line profile, and implies a projected Keplerian velocity of $\sim$300 km s^{-1} at the disk's outer rim. The image has a greater extent in azimuth than the one reported by Haswell et al. (1994). Enhanced Hβ emission is seen at, and downstream from, the point of impact of the gas stream, corresponding to the inner S-wave feature that boosts the blue and red emission peaks at around phases 0.4 and 0.9, respectively. Doppler maps based on data from phases 0.5...1.0 avoid the redward absorption feature in the first half of the orbit, and reveal a pattern corresponding to the outer S-wave.

Our ongoing analysis aims to constrain the accretion stream geometry and the potential roles of inner disk and rim hotspots. We also plan a comparison to *Hubble Space Telescope* (FOS) UV spectra of V795 Her taken during the same week in 1994 June (see Rosen et al. 1996).

References

Haswell, C.A., Horne, K., Thomas, G., et al. 1994, in "Interacting Binary Stars", ed. A. Shafter, ASP Conf. Ser. **56**, 268

Prinja, R.K., Drew, J.E., Rosen, S.R., 1992, MNRAS, **256**, 219

Rosen, S., Prinja, R.K., Drew, J.E., 1996, these proceedings, p43

HST UV OBSERVATIONS OF THE DISK AND WIND OF V795 HER

S. R. ROSEN[1], R. K. PRINJA[2], J. E. DREW[3], K. O. MASON[4]
S. B. HOWELL[5]
1. Leicester University, Leicester, LE1 7RH, UK
2. University College London, London, WC1E 6BT, UK
3. Imperial College London, London, SW7 2BZ, UK
4. MSSL, Holmbury St. Mary, RH5 6NT, UK
5. Planetary Sciences Institute, Tuscon, AZ, USA

Abstract. HST UV observations of V795 Her reveal a strong 2.6-h orbital variation in the prominent UV lines, in contrast to earlier (IUE) evidence of a 4.8-h period. Only the C IV line contains a strong blue-shifted, wind formed absorption component. Several lines exhibit a 'narrow' absorption feature near rest velocity which may originate in the disk, and a blue-shifted emission feature which accounts for most of the line profile variability.

1. Introduction

Previous UV (IUE) observations of V795 Her found that the strengths and profiles of the C IV and Si IV UV resonance lines were modulated on a probable period of 4.8 h, rather than the 2.6-h binary period [1,2,3]. To examine this phenomenon further, we obtained high time-resolution HST UV spectra of V795 Her which have yielded important new results.

2. Observations and results

V795 Her was observed by HST in 1994 June using the Faint Object Spectrograph (FOS + G130H grating) in rapid mode (Δt=20 s). We obtained 1450 spectra covering the 1150...1600 Å range at ≈1 Å resolution.

A key result is that in the HST data, all the strong UV line fluxes vary on the 2.6-h binary period. There is no sign of the previously witnessed 4.8-h period [2,3]. The continuum is not varying on either period. The line profile variability is concentrated over about 40% of the cycle and is manifested by broadening and deepening of each profile around phase 0.75 on the

A. Evans and J. H. Wood (eds.), Cataclysmic Variables and Related Objects, 43–44.

ephemeris of [4]. The C IV line profile can be decomposed into (i) a broad, asymmetric, blue-shifted absorption trough, (ii) a *blue-shifted* emission peak and (iii) a narrower absorption core near rest velocity. The Si IV doublet, and indeed other measurable lines, appears to contain the latter two components but not the broad trough. We estimate the FWHM of the 'narrow' absorption component in the Si IV doublet to be $\sim$2000 km s^{-1}. The phase-resolved analysis supports this conclusion and suggests that changes in the emission peak alone can account for most of the variability between phase 0.6 and 1.0.

3. Discussion and conclusions

Whilst the apparently different behaviour of V795 Her during these HST observations (cf. our earlier IUE data) is perplexing, the orbitally related variability seen here does at least place the star back amongst the more conventional wind-driving nova-like systems. However, two key new results emerge from these high quality data. Firstly, several lines show a 'narrow' absorption feature. The breadth of this feature, its relative constancy, and the fact that it is located near rest velocity are consistent with an origin in the inner accretion disk. A simple comparison with stellar atmosphere models suggests the Si IV absorption doublet may arise in gas at about 21 000 K, the C IV feature in slightly hotter material. Secondly, the line profile variability around phase 0.75 arises principally from the disappearance and reappearance of the emission peak. This component appears to exhibit low amplitude ($\sim$100 km s^{-1}) *orbital* radial velocity excursions about a centroid that is blue-shifted by about 500 km s^{-1}. This apparent net blue-shift excludes association with the receding volume of any wind outflow. Similarly, it can not readily be ascribed to emission on the disk, for example a disk bright spot. A possible origin for this emission feature remains to be established. Finally, these data confirm that, apart perhaps from C III λ1176, only C IV contains the blue-shifted, extended (presumably wind-formed) absorption trough.

This new view of the line profiles in V795 Her may have important consequences for our understanding of the line behaviour in similar systems.

References

1. Prinja, R.K., Rosen, S.R., Suppelli, K., 1991, MNRAS, **248**, 40
2. Prinja, R.K., Drew, J.E., Rosen, S.R., 1992, MNRAS, **256**, 219
3. Prinja, R.K., Rosen, S.R., 1993, MNRAS, **262**, 37p
4. Shafter, A.W., 1990, Ap. J., **354**, 708

DWARF NOVA OUTBURSTS AND SUPEROUTBURSTS

J. SMAK
N. Copernicus Astronomical Center,
Polish Academy of Sciences,
Bartycka 18, 00-716 Warsaw, Poland
e-mail: jis@camk.edu.pl

1. The viscosity parameter α

Dwarf nova outbursts provide an almost unique opportunity of getting an insight into the nature of viscosity in accretion disks or, within the α-disk approach, of putting some constraints on α. In particular, the strong dependence of the viscous time-scale on viscosity itself permits us to estimate α almost directly from the observed time-scales. In the case of the hot branch of the $\Sigma - T_e$ relation, the most reliable estimates (α_{hot}) are based on the rate of decline following the dwarf nova outburst. From a comparison with model light curves calculated with different αs one gets: $\alpha_{hot} \approx 0.2$ (e.g. Smak 1984b). An independent, but much cruder, estimate can be obtained from the widths of *normal* outbursts, by assuming that the duration of an outburst represents the travel time of an instability wave across the disk. The result is similar: $\alpha_{hot} \approx 0.2$ (Gicger 1987).

In the case of the cool branch of the $\Sigma - T_e$ relation, estimates of α (α_{cool}) can be obtained from theoretical relation(s), based on models of dwarf nova outbursts, which predict the observed length of the outburst cycle to be a function of α_{cool}, the mass transfer rate and other system parameters (see Cannizzo 1993). This was done by Livio & Spruit (1991) who applied such a relation from Cannizzo, Shafter & Wheeler (1988) to a large number of dwarf novae with reliable system parameters (Warner 1987). For the purpose of this presentation, these estimates have been repeated with a somewhat updated set of data (but with no attempt to update the theoretical part with more recent model calculations). The results are shown in Fig. 1 as squares. Another set of estimates of α_{cool} can be obtained from much cruder considerations, applicable to type B outbursts (Smak 1993,

A. Evans and J. H. Wood (eds.), Cataclysmic Variables and Related Objects, 45–54.

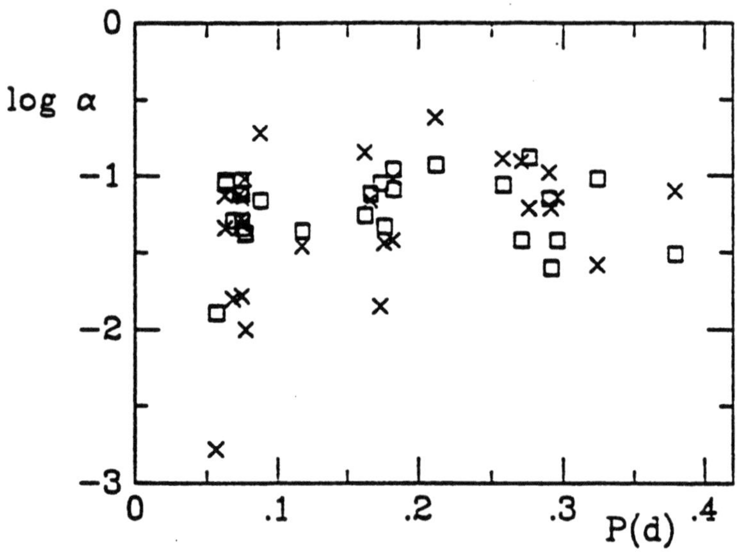

Figure 1. Viscosity parameter $\alpha_{\rm cool}$, estimated from a sample of dwarf novae, plotted against the orbital period. See text for details.

equation 23); note that, in the case of type A outbursts, such estimates give only an upper limit to α. The results are shown in Fig. 1 as crosses.

The following points can be made after inspecting Fig. 1: (i) the two types of estimates agree reasonably well; (ii) there is no correlation with the orbital period (but there might be a weak correlation with the mass ratio; Livio & Spruit 1991); (iii) the 'mean' value of $\alpha_{\rm cool}$ is, roughly, about 0.05; (iv) the scatter does not exceed a factor of 2 or 3 and must be, at least partly, due to uncertainties involved, so that no firm conclusion about its intrinsic value can be made.

With estimates of $\alpha_{\rm hot}$ and $\alpha_{\rm cool}$, as presented above, we get $\alpha_{\rm hot}/\alpha_{\rm cool} \approx 4$, a value which is independently suggested (in fact: strongly required) by model calculations for dwarf nova outbursts (see Cannizzo 1993).

The picture presented so far, however, is not complete. It does not include dwarf novae belonging to the SU UMa subtype, which show only the superoutbursts (group C of Vogt 1993) with very long superoutburst cycles and very large superoutburst amplitudes (the TOADs, Howell, Szkody & Cannizzo 1995). The very long outburst cycles, lasting years or decades, clearly imply that either the viscosity ($\alpha_{\rm cool}$), or the mass transfer rate (or both) must be much lower than in the case of other dwarf novae. Concerning the mass transfer rate we must note that there is a natural lower limit set by the effects of the gravitational radiation. These qualitative considerations are confirmed by results available for WZ Sge, the most extreme and the best documented member of the group: the mass transfer rate is indeed very low ($\dot{M} \approx 10^{15}\,\mathrm{g\,s^{-1}}$) and consistent with theoretical predictions (Paczyński & Sienkiewicz 1983), while the viscosity in its disk during quiescence is 2 or 3 orders of magnitude lower than in other dwarf novae (Osaki 1995; Smak 1993).

It is not clear at present whether other objects of this group have also systematically lower viscosity or, more likely, form a continous transition from 'normal' dwarf novae to the extreme case of WZ Sge. In the latter case, we could probably expect α_{cool} to be correlated with the length of the superoutburst cycle (note, for example, that the lowest point in Fig. 1 is SW UMa with a superoutburst period of 570 d). But, if so, the main question would be: what are the basic system parameters related to, or responsible for, those lower values of α_{cool}? An answer to that question might be crucial for our understanding of the nature of viscosity. Unfortunately, except for WZ Sge, reliable determinations of basic system parameters are not available for members of this group and this appears to be one of the most important and urgent goals for observers.

2. Two types of dwarf nova outbursts

Models of dwarf nova outbursts predicted two types of outbursts: type A, when the thermal instability occurs first in the outer parts of the disk and propagates inward and type B, when it begins in the inner parts and propagates outward. Generally, type B outbursts occur at lower mass transfer rates. There are now 15 dwarf novae with positively identified outburst types (Smak 1987; Warner 1995, Table 3.6). Using this sample we can conclude that:

(i) both types of outbursts are represented among dwarf novae of the U Gem subtype;

(ii) dwarf novae of the SU UMa subtype show only type A outbursts. This implies that the lower limit to the mass transfer rate set by the effects of the gravitational radiation is sufficiently high to inhibit the occurence of type B outbursts;

(iii) dwarf novae of the Z Cam subtype show only type B outbursts. This implies that the mass transfer rate during the dwarf nova phase is *considerably* lower than the *mean* mass transfer rate.

3. Effects of dwarf nova outbursts

3.1. HEATING OF THE WHITE DWARF

The first dwarf nova for which clear photometric evidence for substantial heating of the white dwarf during outburst and its subsequent cooling during quiescence was available some 30 years ago, namely U Gem (Paczyński

1965). Thanks to the effort of many observers and particularly, the availability of the Hubble Space Telescope, we now have very extensive photometric and spectroscopic coverage for several dwarf novae (OY Car: Marsh, Horne & Cheng 1993; Z Cha: Wood et al. 1993; U Gem: Huang, Sion & Szkody 1995; WZ Sge: Smak 1993, Sparks et al. 1993), including their outbursts and superoutbursts.

Theoretical interpretation of these data (e.g. Sparks et al. 1993; Sparks 1996) gives us deeper insight into the structure of the boundary layer and of the outer layers of the accreting white dwarf. It may be worth noting here that, in the case of non-magnetic CVs, the heating of the white dwarf and the accretion onto its surface both occur in the equatorial regions (while in the case of magnetic CVs this happens in the polar caps). Better understanding of these phenomena is needed for studies involving meridional circulation and mixing, i.e. processes of crucial importance for the theory of nova outbursts.

3.2. HEATING OF THE SECONDARY

There is ample observational evidence showing this effect. The best documented examples are: SS Cyg (Hessman et al. 1984; Echevarria et al. 1989) and IP Peg (Martin et al. 1989; Marsh & Horne 1990), while the most dramatic case was probably the 1978 superoutburst of WZ Sge, during which the temperature of its secondary exceeded 10 000 K (Smak 1993).

With this evidence being now available and likely to be extended by future observations, the problem is ripe for theoretical studies (e.g. Sarna 1990; Brett & Smith 1993; Hameury, King & Lasota 1993). One of the most important questions here concerns the mechanism(s) responsible for the resulting enhanced mass outflow from the surface of the secondary (see next Section). Heating by radiation affects *directly* only those parts of the star's hemisphere facing the primary which are not shielded by the disk. Its equatorial parts, including the vicinity of L_1, are not irradiated. The question is then: does the heating affect *indirectly* local conditions around L_1 in a sufficient degree to increase significantly the rate of outflow, or should we rather look for another mechanism?

3.3. ENHANCED MASS TRANSFER RATE

Modulation of the mass tranfer rate, due to the time dependent heating of the secondary, was an essential ingredient of models proposed to explain the Z Cam behavior (Meyer & Meyer-Hofmeister 1983) and the superoutbursts of SU UMa stars (Osaki 1985). At that time the only observational evidence for a substantial increase of $\dot{M}$ during outbursts was due to Vogt (1983), who found that the hot spot in VW Hyi becomes considerably brighter,

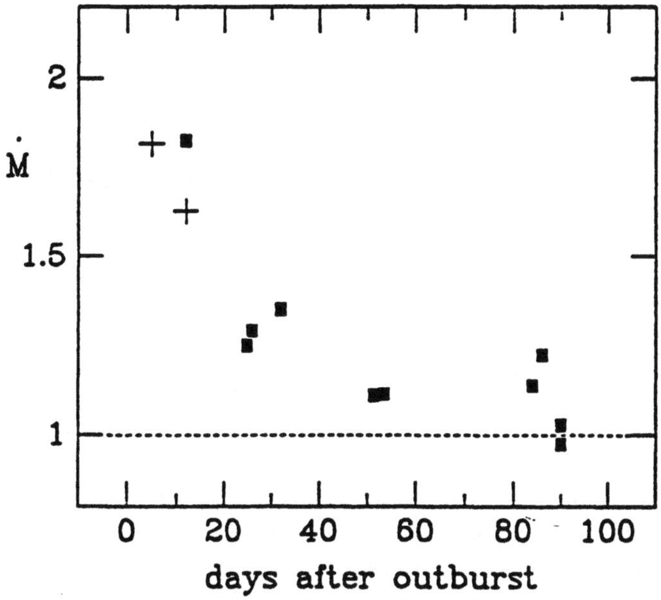

Figure 2. Variations in the mass transfer rate in U Gem, normalized to $\dot{M} = 1$ just before outburst (from Smak 1995). Filled squares are based on luminosities of the hot spot measured during its eclipse. The crosses – on luminosities corresponding to the amplitude of the hump.

indicating an increase in $\dot{M}$, during those outbursts which occur less than 40 d before the next superoutburst.

We note in passing that, while the luminosity of the hot spot is an obvious measure of the mass transfer rate, it must be remembered that it also depends on the impact parameter, i.e. on the relative velocity of collision between the stream and the outer parts of the disk. When the radius of the disk varies within the outburst cycle (Sect. 3.4), the effect of such variations should be taken into account in the determination of $\dot{M}$.

Possible evidence for an enhanced $\dot{M}$ during a *superoutburst* came from a single eclipse observed early in a superoutburst of Z Cha (Warner & O'Donoghue 1988), the shape of which could be interpreted as indicative of a very bright hot spot. An important point here is that, generally, it is difficult to separate the hot spot contribution from superhumps and photometric effects connected with the heating of the secondary. Furthermore, if superoutbursts are due to an enhanced mass transfer, then it is only during the initial phase of presumably rapid increase of $\dot{M}$ that the hot spot can be relatively bright. Consequently, it is very difficult to identify possible effects of enhanced $\dot{M}$ during *superoutbursts* and the absence of more convincing observational evidence cannot be treated as proof that such an enhancement occurs only sporadically, or does not occur at all.

Good evidence is now available (Smak 1995) for modulated mass transfer rate during normal outburst cycles in two dwarf novae: U Gem (Fig. 2) and Z Cha. In both of them the mass transfer rate increases during outbursts by a factor of 2, with the subsequent decline in U Gem taking, roughly, one-half of its outburst cycle. We can conclude that the modulation of the mass transfer rate within the outburst cycle is probably a common phenomenon among dwarf novae.

3.4. DISK RADIUS VARIATIONS

Variations of the disk radius during the dwarf nova cycle are best documented in the case of two eclipsing dwarf novae: U Gem (Smak 1984a) and IP Peg (Wood et al. 1989; Wolf et al. 1993). In both cases the radius increases during outburst by about 30...40%, and then decreases slowly during quiescence; in both cases variations repeat quite well from cycle to cycle. Detailed comparison between model-predicted and observed disk radius variations can serve not only as a test for models of dwarf nova outbursts but may also permit an independent estimate of viscosity. It is clear that, in order to make such a comparison more meaningful, the models should include the effect of variable $\dot{M}$ within the dwarf nova cycle.

Data are also available for two eclipsing dwarf novae of the SU UMa subtype. In the case of Z Cha (O'Donoghue 1986; Zoła 1989) disk radius variations, when plotted against time since the last outburst, show considerable scatter due, in part, to the poor repetition from cycle to cycle. But when the *phase* of the outburst cycle is used instead of time, the characteristic pattern, similar to that in U Gem, becomes visible. In the case of OY Car (Fig. 3) no regular variations of the type described above can be seen in either of the two plots.

The evidence available for Z Cha and, particularly, for OY Car should *not* be considered as disappointing. On the contrary, it may help us look for effects which, so far, have not been included in our considerations. The most obvious difference between Z Cha and OY Car on the one side, and U Gem and IP Peg on the other, is that the former two, as belonging to the SU UMa subtype, show superoutbursts. To that end, simple models of superoutbursts, involving either the enhanced mass transfer, or the tidal instability of an eccentric disk, predict very regular behavior of R_d, of the type observed in U Gem (Osaki 1989, Smak 1991). Clearly then, the mechanism of superoutbursts is likely to include an additional factor responsible for observed irregularities. The case of Z Cha may be particularly instructive in this respect: it suggests that the length of the outburst cycles within the supercycle and R_d are likely to be controlled by the same unknown factor. At the same time the difference between Z Cha and OY Car may

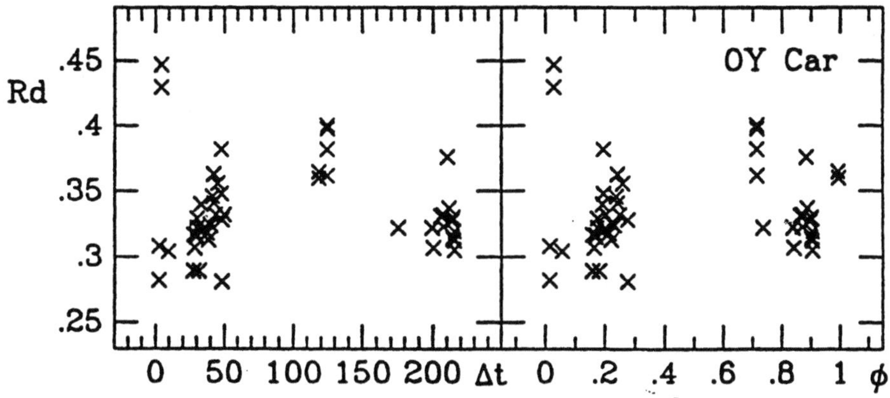

Figure 3. Disk radius variations in OY Car plotted against time Δt since the last outburst (left) and the phase ϕ of the outburst cycle (right). All data available in the literature are used in these diagrams.

somehow reflect their different outburst/superoutburst patterns: in OY Car there are fewer normal outbursts within each superoutburst cycle.

4. The superoutbursts

4.1. THE ENHANCED MASS TRANSFER (EMT) MODEL

This is the model by Osaki (1985) involving a feedback between the accretion rate, the irradiation of the secondary, and the mass transfer rate, which – under certain conditions – may result in an instability leading to a limit cycle behavior. The evidence now available (Sect. 3.3) clearly shows that, indeed, the mass transfer rate undergoes enhancement during *normal* outbursts and, very likely, during *superoutbursts.*

Osaki (1985) proposed also a model for superhumps, combining the EMT model with that of a slowly precessing eccentric disk (Vogt 1981). In this model the irradiation of the secondary was modulated with the synodic period due to varying extent of the eccentric disk's shadow, causing the mass transfer rate and, consequently, the luminosity of the hot spot, to vary with that same period. This model was substantially modified by Mineshige (1988) who included the effects of heating of the outer parts of the disk thereby offering a potential explanation for the variable azimuthal distribution and extension of the superhump source (e.g. Warner & O'Donoghue 1988).

4.2. THE THERMAL AND TIDAL INSTABILITY (TTI) MODEL

This model (Osaki 1989) combines the effects of the thermal instability responsible for dwarf nova outbursts, with the tidal instability which occurs – due to the 3:1 resonance – for large disks in systems with $q < q_{\rm crit} \approx 0.25$, and leads to a development of an eccentric, precessing disk (Whitehurst 1988; Hirose & Osaki 1990; Whitehurst & King 1991; Osaki, Hirose & Ichikawa 1993).

This model successfully reproduces light curves of ordinary SU UMa stars with normal outbursts and superoutbursts and also, of those of the WZ Sge subgroup which exhibit only superoutbursts (Osaki 1995). Moreover the eccentric, precessing disk naturally explains the superhump periods (e.g. Hirose & Osaki 1993; Molnar & Kobulnicky 1992; Whitehurst & King 1991).

4.3. PROBLEMS

The two models predict very different variations of the disk radius and of the lengths of normal cycles within a supercycle (Ichikawa, Hirose & Osaki 1993; Smak 1991). Unfortunately, the observational evidence, which seems to favor marginally the EMT model, is not adequate for testing these predictions in a decisive way.

Both models have their weak points. The EMT model requires an eccentric, precessing disk to account for superhumps but does not explain *why* it should become eccentric to begin with. The TTI model, while offering a convincing mechanism for an eccentricity to appear, encounters problems related to its growth. Osaki (1989) and Hirose & Osaki (1990) maintain that the tidal instability is *intrinsic* to the disk itself, and that the eccentricity can grow to a finite value within a day or two after the critical disk radius has been exceeded. On the other hand, Whitehurst (1988) and Whitehurst & King (1991) argue that a strong disturbance by the enhanced mass transfer is needed to accelerate the, otherwise too slow, growth of the eccentricity. The role of the enhanced mass transfer is, however, more complex. Lubow (1994) showed that (i) with variable $\dot{M}$ the stream can indeed induce an eccentricity but it is likely to be small, but (ii) when $\dot{M} \approx$ constant, the stream causes the eccentricity to *decrease*. Furthermore, during the phase of enhanced mass transfer the radius of the disk contracts and becomes much smaller than the critical radius at which the tidal instability can develop (Ichikawa et al. 1993; Smak 1991).

Perhaps the weakest point of the TTI model is connected with the sequence of events responsible for superoutbursts and superhumps. In the TTI model it begins with a tidal instability, leading to the formation of an eccentric disk and causing a major enhancement of the accretion rate.

Accordingly, the superhumps should appear at the very early phase of a superoutburst. In most cases, however, they appear one or two days *after* the superoutburst maximum.

4.4. A HYBRID EMT + TTI MODEL?

As the two models appear to complement each other in many respects it is tempting to consider a hybrid model, with a possible scenario which could look like this:

(i) the basic mechanism of a superoutburst and, in particular, the *clock* controlling the length of a supercycle is that of the EMT model;

(ii) the superoutburst begins with a major enhancement of the mass transfer rate caused by irradiation of the secondary during the preceding *normal* outburst (which acts as a trigger). At this phase, due to a high $\dot{M}$, the disk contracts and its outer parts become geometrically much thicker;

(iii) during the subsequent phase, due to decreasing irradiation of the secondary caused by a more effective screening by the disk, the mass transfer rate declines;

(iv) at this phase the disk expands considerably (Ichikawa et al. 1993; Smak 1991) and becomes vulnerable to the tidal instability leading to the development of an eccentric, precessing disk. It is then at this phase (but not earlier!) that the superhumps appear;

(v) as both models appear to offer plausible mechanisms for the origin of superhumps, it is not unlikely that a combination of the two mechanisms might provide the best explanation.

Note added after the Colloquium. Jean-Pierre Lasota, in his review talk, brought to the attention of those present at the *Colloquium* a very important fact concerning the occurence of superoutbursts (which, strangely enough, remained largely unknown!): In 1985 October/November U Gem underwent a *superoutburst* lasting for 45 d (Mason et al. 1988). For the models of superoutbursts, which were tailored *specifically* to explain their occurence *exclusively* in the very short period, very low mass ratio dwarf novae of the SU UMa subtype, the superoutburst of U Gem (unless it can be explained as a completely different kind of event) implies that *all of them may actually be wrong!*

References

Brett, J.M., Smith, R.C., 1993, MNRAS, **264**, 641

Cannizzo, J.K., 1993, in "Accretion Disks in Compact Stellar Systems", ed. J.C. Wheeler, Singapore: World Scientific Publishing, p6

Cannizzo, J.K., Shafter, A.W., Wheeler, J.C., 1988, Ap. J., **333**, 227

Echevarria, J., Diego, F., Tapia, M., et al., 1989, MNRAS, **240**, 975

Gicger, A., 1987, Acta Astr., **37**, 29
Hameury, J.M., King, A.R., Lasota, J.-P. 1993, in "Accretion Disks in Compact Stellar Systems", ed. J.C. Wheeler, Singapore: World Scientific Publishing, p360
Hessman, F.V., Robinson, E.L., Nather, R.E., Zhang, E.-H., 1984, Ap. J., **286**, 747
Hirose, M., Osaki, Y., 1990, PASJ, **42**, 135
Hirose, M., Osaki, Y., 1993, PASJ, **45**, 595
Howell, S.B., Szkody, P., Cannizzo, J.K., 1995, Ap. J., **439**, 337
Huang, M., Sion, E.M., Szkody, P., 1995, AJ, **109**, 1746
Ichikawa, S., Hirose, M., Osaki, Y., 1993, PASJ, **45**, 243
Livio, M., Spruit, H.C., 1991, A&A, **252**, 189
Lubow, S.H., 1994, Ap. J., **432**, 224
Marsh, T.R., Horne, K., 1990, Ap. J., **349**, 593
Marsh, T.R., Horne, K., Cheng, F., 1993, in "Cataclysmic Variables and Related Physics", eds O. Regev, G. Shaviv, Ann. Israel Phys. Soc., **10**, 7
Martin, J.S., Friend, M.T., Smith, R.C., Jones, D.H.P., 1989, MNRAS, **240**, 519
Mason, K.O., Cordova, F.A., Watson, M.G., King, A.R., 1988, MNRAS, **232**, 779
Meyer, F., Meyer-Hofmeister, E., 1983, A&A, **121**, 29
Mineshige, S., 1988, Ap. J., **335**, 881
Molnar, L., Kobulnicky, H.A., 1992, Ap. J., **392**, 678
O'Donoghue, D., 1986, MNRAS, **220**, 23P
Osaki, Y., 1985, A&A, **144**, 369
Osaki, Y., 1989, PASJ, **41**, 1005
Osaki, Y., 1995, PASJ, **47**, 47
Osaki, Y., Hirose, M., Ichikawa, S., 1993, in "Accretion Disks in Compact Stellar Systems", ed. J.C. Wheeler, Singapore: World Scientific Publishing, p272
Paczyński, B., 1965, Acta Astr., **15**, 305
Paczyński, B., Sienkiewicz, R., 1983, Ap. J., **268**, 825
Sarna, M., 1990, A&A, **239**, 163
Smak, J., 1984a, Acta Astr., **34**, 93
Smak, J., 1984b, *Acta Astr.*, **34**, 161
Smak, J., 1987, A&SS, **131**, 497
Smak, J., 1991, Acta Astr., **41**, 269
Smak, J., 1993, Acta Astr., **43**, 101
Smak, J., 1995, Acta Astr., **45**, 355
Sparks, W.M. 1996, private communication
Sparks, W.M., Sion, E.M., Starrfield, S.G., Austin, S., 1993, in "Cataclysmic Variables and Related Physics", eds O. Regev, G. Shaviv, Ann. Israel Phys. Soc., **10**, 96
Vogt, N., 1981, Ap. J., **252**, 653
Vogt, N., 1983, A&A, **118**, 95
Vogt, N., 1993, in "Cataclysmic Variables and Related Physics", eds O. Regev, G. Shaviv, Ann. Israel Phys. Soc., **10**, 63
Warner, B., 1987, MNRAS, **227**, 23
Warner, B., 1995, Cataclysmic Variable Stars, Cambridge: Cambridge Univ.Press.
Warner, B., O'Donoghue, D., 1988, MNRAS, **233**, 705
Whitehurst, R., 1988, MNRAS, **232**, 35
Whitehurst, R., King, A.R., 1991, MNRAS, **249**, 25
Wolf, S., Mantel, K.H., Horne, K., et al., 1993, A&A, **273**, 160
Wood, J.H., Marsh, T.R., Robinson, E.L., et al., 1989, MNRAS, **239**, 809
Wood, J.H., Robinson, E.L., Bless, R.C., et al., 1993, in "Cataclysmic Variables and Related Physics", eds O. Regev, G. Shaviv, Ann. Israel Phys. Soc., **10**, 19
Zoła, S., 1989, Acta Astr., **39**, 45

IUE AND OPTICAL SPECTRA OF RZ LMI AND ER UMA THROUGHOUT THEIR 19 AND 43 DAY CYCLES

PAULA SZKODY[1], ANDREW SILBER[1], R. K. HONEYCUTT[2], J. W. ROBERTSON[2], D. W. HOARD[1], LORA PASTWICK[1]
1. Dept. of Astronomy, University of Washington, Seattle, WA 98195, USA
2. Astronomy Dept., Indiana U, Bloomington, IN 47405, USA

Abstract. IUE and optical data were obtained throughout the supercycles of RZ LMi and ER UMa during 1993/94. While most of the spectral and photometric characteristics are consistent with disk changes in systems with a relatively high accretion rate and low inclination, the cause of the high accretion rate in this subset of SU UMa stars is not known.

1. Introduction

In the past year, it has become apparent that the 3 CVs RZ LMi, ER UMa and V1159 Ori exhibit a unique pattern of behavior similar to the normal SU UMa stars, but at extreme minima of time-scale and amplitude (Robertson, Honeycutt & Turner 1995). These 3 objects show very short (19...45 day) supercycles, have very low $\dot{P}$ (10^{-3}) and very low amplitude (3 mag) superoutbursts. Between superoutbursts, they have regular 2 mag outbursts at about 4 day intervals. In order to determine if these objects can be interpreted within the scenario of disk instability and tidal changes invoked to explain the systems below the period gap (Osaki 1989), we obtained IUE and optical spectroscopic and photometric observations throughout the supercycles of RZ LMi and ER UMa.

2. Observations

The photometric cycles of RZ LMi and ER UMa were monitored with the Roboscope (Honeycutt & Turner 1992) as well as with CCD frames obtained coincident with the optical spectra. The University of Washington (UW) 30 inch telescope at Manastash Ridge Observatory (MRO) was used on 4 nights to obtain time-resolved photometry of RZ LMi during decline

A. Evans and J. H. Wood (eds.), Cataclysmic Variables and Related Objects, 55–58.

from superoutburst and at minimum following a normal outburst. The IUE satellite was used to obtain spectra at 4 times during the cycle of RZ LMi and 7 times during the cycle of ER UMa. Optical spectra were obtained with the UW 3.5 m telescope at Apache Point Observatory (APO), using the Double Imaging Spectrograph (DIS). The flexible scheduling available on this telescope allowed for 1 h runs on different nights at low resolution (10 Å) covering 3900...9000 Å to obtain data throughout the cycles. A few nights were scheduled in high resolution mode (2 Å) to obtain radial velocity curves for orbital period determination. Overall, 10 phase points in the superoutburst period were sampled for RZ LMi and 7 for ER UMa.

3. Results

The IUE data on ER UMa show large flux changes throughout the cycle with corresponding large line changes (Fig. 1). The continuum flux at 1500 Å increased by a factor of 13 from quiescence to supermaximum, while the optical flux at 5500 Å increased by a factor of 8. The slope of the IUE flux distribution (in log wavelength units) was −2.1 during the 5th minimum (April 17) and steepened to −2.6 during supermaximum (May 5). P Cygni profiles for C IV are evident in most of the spectra near outburst and superoutburst (March 26, April 3, 12, 25). The lines change from absorption at outburst to emission at quiescence (April 17). While these changes are 'normal' for SU UMa stars, the amplitude of the UV flux increase is much less than in systems such as SW UMa (where the UV increases by a factor of 1000 and the optical by a factor of 400; Szkody, Osborne & Hassall 1988). In RZ LMi, the IUE spectral distribution shows no change in slope in data obtained during the faintest magnitudes compared with data near superoutburst and the flux changes are even smaller than for ER UMa (a factor of 4 at 1500 Å and a factor of 6 at 5500 Å).

In the optical, the Hβ and Hγ lines of ER UMa undergo a transition from absorption at superoutburst, to absorption with emission cores during the normal outbursts, to pure emission at the minimum magnitudes (Fig. 2). Except for a brief hint of absorption wings at superoutburst, Hα is always in emission, with increased strength at fainter magnitudes. The lines in RZ LMi are in absorption at the superoutburst and normal outbursts, and only turn into emission during minima. In addition, the emission lines in RZ LMi are very narrow, being only half the full width zero intensity of ER UMa and V1159 Ori (Jablonski & Cieslinski 1992). In general, the equivalent widths of the emission (at mimina times) of Hβ in both ER UMa and RZ LMi are factors of 4...5 smaller than typical for SU UMa stars (Szkody 1987).

The high resolution spectra were used to construct radial velocity curves,

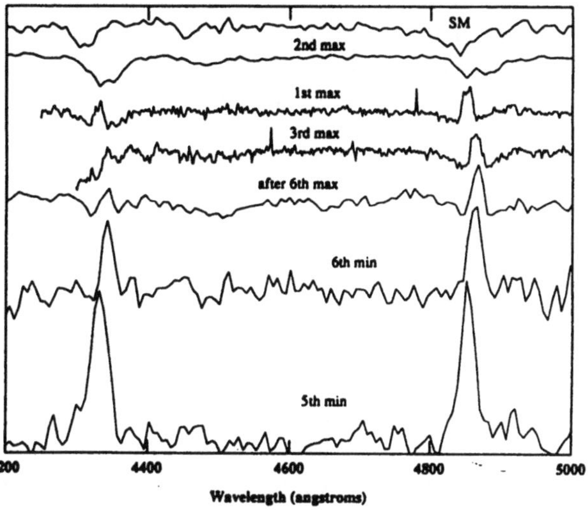

Figure 1. Representative IUE spectra of ER UMa plotted with arbitrary vertical offsets

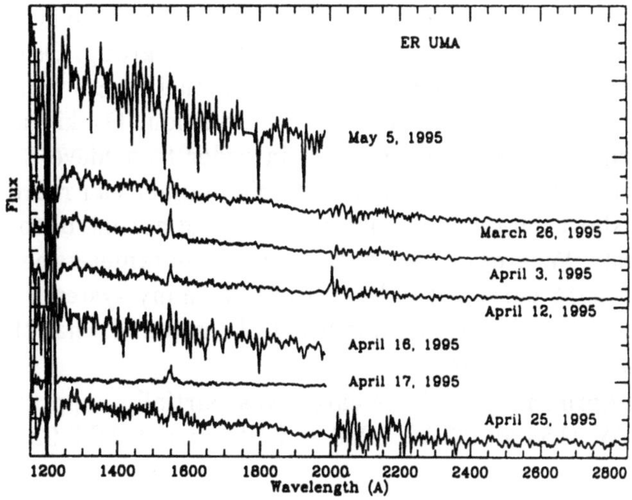

Figure 2. Representative low and high resolution blue spectra of ER UMa

using a double Gaussian fitting procedure on the emission line wings. A sine fit to the ER UMa data determines an orbital period of 90±3 m and a semi-amplitude, K, of 57±8 $\mathrm{km\,s^{-1}}$. The data on RZ LMi could not be fit with any sine curve, to a K limit near 20 $\mathrm{km\,s^{-1}}$. However, the MRO photometry at minimum showed a periodic modulation over 2 nights at a period of 80.9±0.1 m. As these periods are shorter than the superhump periods

(95 m for ER UMa and 85.6 m for RZ LMi; Robertson et al. 1995), and are close to those expected from the superhump-orbital period relationships of SU UMa stars in general (Howell & Hurst 1994), it is reasonable to assume that these are the observed orbital periods.

4. Conclusions

In comparison to other SU UMa stars, our IUE and optical data on ER UMa and RZ LMi have shown similar IUE slopes and spectra, optical spectral changes similar to SU UMa quiescent, outburst and superoutburst states, and orbital periods that are shorter than the superhump periods. In these respects, this group of 3 objects are typical of SU UMa stars. However, distinct differences remain. The identical 3 mag amplitude from minimum to superoutburst for all 3 is very low. The 20...45 d for the supercycles are extremely short. The supercycle times are very stable compared to normal SU UMa stars. The radial velocity amplitudes of all 3 are very low and the line equivalent widths are abnormally small. The IUE and optical flux changes during outburst are very small compared to normal SU UMa-type behavior. Whereas the line widths, small equivalent widths, and low radial velocity amplitudes could be explained by low inclinations, it is not clear why this is such a strong selection effect for this group. Osaki (1995) and Kato & Kunjaya (1995) have suggested that their mass transfer rates are larger than for typical SU UMa stars. In addition, Osaki had to decrease the tidal torque operating in RZ LMi in order to achieve a supercycle as short as 19 d. The models produced in this manner can reproduce the low amplitudes, short periods and correct number of normal outbursts between superoutbursts. Whereas these are the physical parameters needed to match the observed light curves, it is less clear how many systems really exist at this high mass transfer rate regime below the gap and what the mechanism providing this high rate is.

Acknowledgements. This research was partially supported by NASA grants NAGW 3158 and NAG 5236 and NSF grant AST 9217911 to PS.

References

Honeycutt, R.K., Turner, G.W., 1992, in "Robotic Telescopes in the 1990's", ed. A. Filippenko (San Francisco, ASP), p77
Howell, S.B., Hurst, G.M., 1994, IBVS, 4043
Jabłoński, F.J., Cieslinski, D., 1992, A&A, **259**, 198
Kato, T., Kunjaya, C., 1995, PASJ, **47**, 163
Osaki, Y., 1989, PASJ, **41**, 1005
Osaki, Y., 1995, PASJ, **47**, L25
Robertson, J.W., Honeycutt, R.K., Turner, G.W., 1995, PASP, **107**, 443
Szkody, P., 1987, Ap. J. Supp., **63**, 685
Szkody, P., Osborne, J., Hassall, B.J.M., 1988, Ap. J., **328**, 243

NEW OBSERVATIONAL FRONTIERS OF ER UMA (= PG 0943+521) TYPE DWARF NOVAE

D. NOGAMI[1], T. KATO[1], S. MASUDA[1], R. HIRATA[1],
K. MATSUMOTO[2], K. TANABE[2], T. YOKOO[2]
1. Department of Astronomy, Faculty of Science, Kyoto University, Sakyo-ku, Kyoto 606-01, Japan
2. Astronomical Institute, Osaka Kyoiku University, Asahigaoka, Kashiwara, Osaka 582, Japan

The ER UMa-type, including ER UMa, V1159 Ori and RZ LMi, is a subgroup of SU UMa-type dwarf novae. Outbursts of these stars are characterized by: (i) the extremely short recurrence time of the superoutburst (~ 40 d in ER UMa and V1159 Ori, ~ 20 d in RZ LMi), (ii) short outbursts with short recurrence times (~ 4 d) between the bright states, and (iii) extremely long duty cycles (~ 0.5). Assuming that the mass transfer rate from the secondary is ten times higher than that of ordinary SU UMa stars, which was invoked by Kato & Kunjaya (1995), and a weak tidal torque in the case of RZ LMi, Osaki (1996) showed that the light curves of these stars can be reproduced by the disk instability model, which does not require mass transfer bursts. These indicate that ER UMa, V1159 Ori and RZ LMi are not on the natural extension of SU UMa-type dwarf novae.

The discovery of ER UMa stars has made it natural to interpret that nova-likes showing light variations with photometric periods longer than orbital are in superoutburst all the time. These objects are called 'permanent superhumpers' (PSHs). The mass transfer rates of PSHs can be considered to be even higher than those of ER UMa stars. In the following discussion, we regard ER UMa stars and PSHs as a group of high mass transfer objects.

Let us consider the evolutionary status of these stars. The lifetime of dwarf novae with orbital periods below the period gap is roughly proportional to the inverse of the mass transfer rate from the secondary stars. Thus, the lifetime of ER UMa stars and PSHs is 0.1 times smaller, or less, than that of ordinary SU UMa stars. Since the probability of finding the stars in this stage in CV evolution is proportional to the lifetime, the probability of encountering these stars is estimated as 0.1 of that of ordinary SU UMa stars, or less. However, the observed number ratio of these stars

A. Evans and J. H. Wood (eds.), Cataclysmic Variables and Related Objects, 59–60.

(6) to SU UMa stars ($\sim$ 50) has already exceeded that probability and seems to be growing. There are at least two possibilities for the evolutionary status of these stars: (a) these stars are an entirely different population from the ordinary SU UMa stars, and (b) some ordinary SU UMa stars spend some fraction of their life time as ER UMa stars or PSHs.

Suppose that the former possibility, (a), is the case, that is SU UMa stars and a group of ER UMa stars and PSHs have different evolutionary paths and different progenitors. Thus, the number of progenitor ER UMa stars and PSHs may be larger than that of ordinary SU UMa stars. In other words, an important course of the evolution of CVs may have been missed.

In general it is predicted that if the loss of angular momentum is high, that is, if the mass transfer rate is high, then the minimum orbital period should be longer. However, the orbital period of RZ LMi is identical to the minimum value expected from the theory of gravitational-wave radiation. In addition, the orbital periods of ER UMa stars and PSHs seem to be uniformly distributed below the upper edge of the well-known period gap ($\sim$2.8 h). No period gap seems to exist in their orbital period distribution. One possibility to explain this curious distribution is that these stars evolved from pre-CVs to CVs inside the period gap.

In the latter case, (b), even if all of the SU UMa stars experience secular variation of the mass transfer rate as suggested in the hibernation scenario, the fraction of SU UMa stars in the ER UMa star phase or PSH phase to ordinary SU UMa stars is estimated to be 0.2 from the ratio of the observed number of ER UMa stars and PSHs to that of SU UMa stars. To be more realistic, the masses of white dwarfs in SU UMa stars will not always allow SU UMa stars to cause nova explosions. If only some of the SU UMa stars change their mass transfer rates as predicted by the hibernation scenario, the fraction mentioned above becomes still larger, which suggests that the mean mass transfer rate averaged over the lifetime of these systems will far exceed that expected from gravitational-wave radiation. This result is the same as that expected from the former possibility. In either case, these stars are expected to show a different distribution of the orbital period from other SU UMa stars.

The present discussion is based on the data hitherto observed; however it is likely that the data are biased in some way. More precise observational determination of the orbital-period distribution of ER UMa stars and PSHs is highly encouraged, as well as theoretical work on CV evolution taking into account the present discussion.

References

Kato, T., Kunjaya, C., 1995, PASJ, **47**, 163
Osaki, Y., 1996, these proceedings, p127

SUPERHUMPS IN THE SU UMA STAR AK CNC

R. MENNICKENT[1], D. NOGAMI[2], T. KATO[2],
W. WORRAKER[3]
*1. Dpto. de Física, Instituto de Cs. Físicas y Matemáticas,
University de Concepción, Casilla 4009, Concepción, Chile
2. Dept. Astron., Faculty of Sci., Kyoto University,
Oiwake-cho, Kitashirakawa, Saykyo-ku, Kyto 606-01, Japan
3. British Astronomical Association, 65 Wantage Road,
Didcot, Oxon, OX11 0AE, UK*

Abstract. We have observed AK Cnc during the 1995 March superoutburst, by means of differential $V - U$ photometry. The observations were obtained at Las Campanas (Chile), Ouda Station (Kyoto, Japan) and in England. Superhumps with amplitude 0.2 mag appeared at least 4 d after maximum. They evolved from a single peak structure, developing a secondary peak during decline, until they reached a double-wave modulation with 0.05 mag amplitude, after 8 d. After removing linear trends in our data, the following ephemeris was found:

$$UT_{\mathrm{max}}\ (1995\ \mathrm{March}) = 12.152(1) + 0.06749(1)N\ \mathrm{d}. \qquad (1)$$

This superhump period was confirmed, independently, with our homogeneous set of 462 V magnitudes, obtained at LCO, then adding 63 U data points and finally considering all the dataset, consisting of 789 $V + U$ magnitudes. The superhump period is slightly larger than that reported by Kato (1995) during the 1992 January superoutburst (0.06735(5) d). Our results imply an orbital period $P_{\mathrm{orb}} = 0.065(2)$ d, using the relationship given by Howell & Hurst (1994). The eruption duration was $\gtrsim 10$ d, and its amplitude > 6 mag. An abrupt and short luminosity decrease occured $\approx$ 12 d after maximum. Eleven outbursts have been observed since 1992 January, being the minimum reported lag between successive eruptions: 40 d. More details will be given in a future publication.

References

Howell, S., Hurst, G., 1994, IBVS, 4043
Kato, T., 1995, IBVS, 4136

A. Evans and J. H. Wood (eds.), Cataclysmic Variables and Related Objects, 61–62.

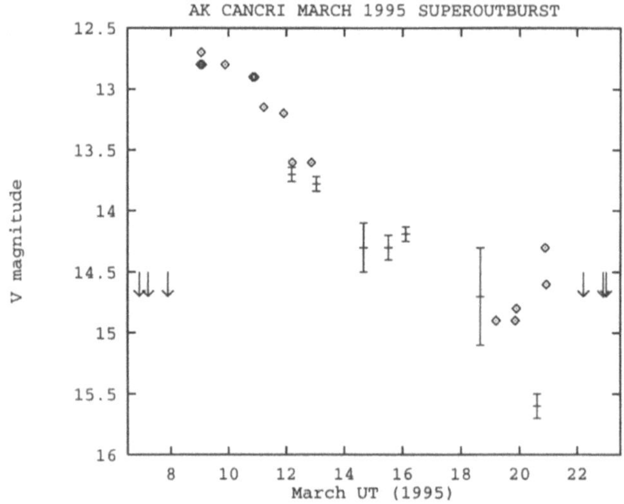

Figure 1. V light curve of AK Cnc during the 1995 March superoutburst. Data are from the British Astronomical Association Variable Star Section database (◇) and from our own observations (+). The reported quiescent magnitude is $V \approx 19$ mag. Note the sudden brightness drop on March 20.

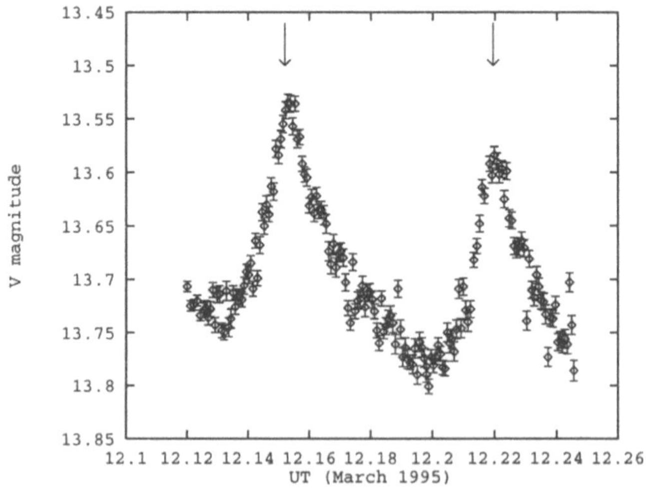

Figure 2. CCD V light curve of AK Cnc. Arrows indicate times of maximum derived from equation (1). The star is fading at rate 0.4 ± 0.1 mag d^{-1}. Note the quasi-periodic signal during decline (time-scale $T \approx 13.3$ m).

OUTBURST ACTIVITY DATA ON SELECTED CATACLYSMIC VARIABLES

T. VANMUNSTER[1], S. B. HOWELL[2]
1. Vereniging voor Sterrenkunde, Werkgroep Veranderlijke Sterren, Walhostraat 1A, B-3401 Landen, Belgium
2. Planetary Science Institute, Astrophysics Group, 620 N. 6th Ave, Tucson, AZ 85705, USA

Abstract. We have created a database with outburst activity for 269 selected cataclysmic variables. It includes nearly 1000 outburst measurements discovered through 1995 June. The database and a complete paper describing the observations and the sources of information that were consulted are accessible over the Internet.

1. Introduction and history

The Outburst Activity Database on Selected Cataclysmic Variables originated in 1992 as an attempt to collect extensive details of recent outburst data for dwarf novae, observed by members of the Belgian Astronomical Association VVS (Vereniging voor Sterrenkunde) working group 'Veranderlijke Sterren'. The objective was to focus this database and to establish a list of dwarf novae for which little or no observational data existed. After a few months of operation, we learned about similar initiatives, set up by amateur astronomers in other countries, the most notable being the 'Recurrent Objects Catalogue' issued by Hurst (1989). Two limitations of this database are: (i) the database contains only a few objects with a negative declination and (ii) the object must be visible (at maximum magnitude) in large amateur telescopes (i.e. become brighter than visual magnitude 15 or so). However, some exceptions to these rules do exist in the database.

2. Structure of the database

The Outburst Activity Database currently contains nearly 1000 entries on 269 selected cataclysmic variables. For some objects, detailed outburst mea-

A. Evans and J. H. Wood (eds.), Cataclysmic Variables and Related Objects, 63–64.

surements are available, whereas for others very little or no information at all could be retrieved. The database includes a large number of known and suspected tremendous outburst amplitude dwarf novae (TOADs; Howell, Szkody & Cannizzo 1995). The tabular format, used to represent the outburst data, is described in detail in the internet version of this paper.

3. Variable star alert programmes

We conclude this short paper by briefly discussing the variable star alert programmes. The major contribution to visual observations of dwarf novae comes from a few very active and enthusiastic observers. Most of them are very experienced variable star observers, with proven skills in 'traditional' variable star work. Lots of practice and perseverance has turned them into well-trained amateurs, who are capable of memorising tens of star fields. On every clear night, they visually check these fields, hunting for 'new' stars. Using this technique, some amateurs are capable of visually inspecting between 50 and 100 star fields in one single night. Special observing programmes have been set up, to direct the attention of these amateurs towards the more exotic cataclysmic variables. Two such examples are discussed in the electronic version of this paper: the Recurrent Objects Programme (The Astronomer Group, UK) and the Cataclysmic Variables Alert Programme (VVS, Belgium). When activity of a suspected star is noticed, other members of the network are informed (by telephone) in order to obtain immediate confirmation. Positive identifications of outbursts result in alert calls (electronic circulars), that are issued to amateurs all over the world and to interested professional astronomers. The Outburst Activity Database and a detailed paper describing it are both available on line at the WWW site of the second author (homepage http://www.psi.edu/sbhgen.html).

Acknowledgements. We would like to thank Paul Van Cauteren, Director of the Belgian working group 'Veranderlijke Sterren' and Gary Poyner, Coordinator of the Recurrent Objects Programme, for their continuous support and cooperation. Special thanks also to the staff members of the Uccle Observatory, Belgium.

References

Howell, S., Szkody, P., Cannizzo, J., 1995, Ap. J., **439**, 337
Hurst, G., 1989, UK Nova/Supernova Patrol, Recurrent Objects Catalogue, 1-7

ECLIPSE OBSERVATIONS OF IP PEG DURING OUTBURST

W. J. WORRAKER[1], T. NAYLOR[2], G. POYNER[3], S. J. ROBINSON[4]

1. British Astronomical Association, 65 Wantage Road, Didcot, Oxon., OX11 0AE, UK
2. Department of Physics, Keele University, Staffs, ST5 5BG, UK
3. British Astronomical Association, 67 Ellerton Road, Kingstanding, Birmingham, B44 0QE, UK
4. 178 Fearns Avenue, Bradwell, Stoke-on-Trent, Staffs, UK

1. Introduction

IP Peg is a U Gem-type dwarf nova with a very high orbital inclination such that the secondary star eclipses the white dwarf primary, accretion disc and hot spot in each orbit. An observing project was set up by 'The Astronomer' (TA) magazine and the British Astronomical Association Variable Star Section to make visual and CCD eclipse observations of IP Peg during outbursts in the 1994/95 season.

2. Observations

Two outbursts of IP Peg were observed during the 1994/95 season. The first was detected at $V = 12.7$ mag by Yamada on August 26.717 UT. The time of minimum light was estimated for 11 eclipses, although the extent and quality of coverage varied considerably.

The second outburst was detected by Worraker at $V = 14.3$ mag on December 05.855 UT. This outburst was less well covered, and times of minimum light were estimated for only four eclipses.

Visual estimates were in general based on the chart and comparison sequence for IP Peg produced by the editor of TA (latest revision GMH 940130). The maximum visual observing rate was about 1 estimate per minute.

Platt's CCD counts were referenced to star D of the TA sequence at magnitude 12.7 (see [2]), which gave better agreement with visual results

A. Evans and J. H. Wood (eds.), Cataclysmic Variables and Related Objects, 65–70.

than the TA value of $V = 12.2\,\mathrm{mag}$; independent checks suggest that the latter is too bright. Platt's Starlight Xpress camera was used unfiltered, which means that his counts include a possibly significant contribution from the near-infrared part of the spectrum. The integration time was 40 s early in the August outburst, 80 s later when IP Peg was fainter.

The Keele CCD counts were referenced to star B at magnitude $V = 11.0$ [2]. The ST-6 camera was in most cases used with a V-filter, which gave good agreement with visual results. A B–filter was used instead on two occasions late in the August outburst. The integration time was normally 10 s with a dead time of 5 s.

Samples of the resulting eclipse profiles are shown in Fig. 1. The times above the figures refer to predicted times of minimum light. The upper panel shows the best-covered eclipse of the year, while the lower panel shows the eclipse at the end of the August outburst, which was covered by both CCD observers.

3. Eclipse modelling

Since the eclipses are quite shallow, using them to derive the temperature distribution across a thin disc will result in a radial temperature distribution which is much flatter than steady state. Since the eclipse profiles are stable on a relatively long time-scale (typically 1...2 weeks) compared with the viscous time-scale of the disc (probably around 1 d), such a non-steady state could not be maintained for the duration of the plateau phase of the outburst. It is more likely that the mass flux through the disc is close to a steady state distribution, so we have assumed this whilst modelling the eclipses using the code described in [3]. The most important feature of this model is that the accretion disc has an 'opening angle', i.e. flares towards its edge (see Fig. 2).

Most of the parameters of the model are either unimportant (e.g. the disc is so bright that the temperature of the secondary is irrelevant), or fixed by previous workers (e.g. the mass ratio and distance). The only parameters we can vary which have a significant effect on the light curve are the mass transfer rate, disc opening angle, and disc radius. The mass transfer rate fixes the out-of-eclipse magnitude, the disc opening angle the eclipse depth, and the disc radius the eclipse width.

Fig. 3 shows a typical model eclipse overlayed on the data. The depth of the eclipse is well matched. The implied semi-opening angle of the disc is 14° (for a system inclination of 80°). The disc radius used for this model is the tidal radius, and as the model eclipse is a little too narrow, this implies that the disc has grown over the tidal radius for some fraction of the outburst.

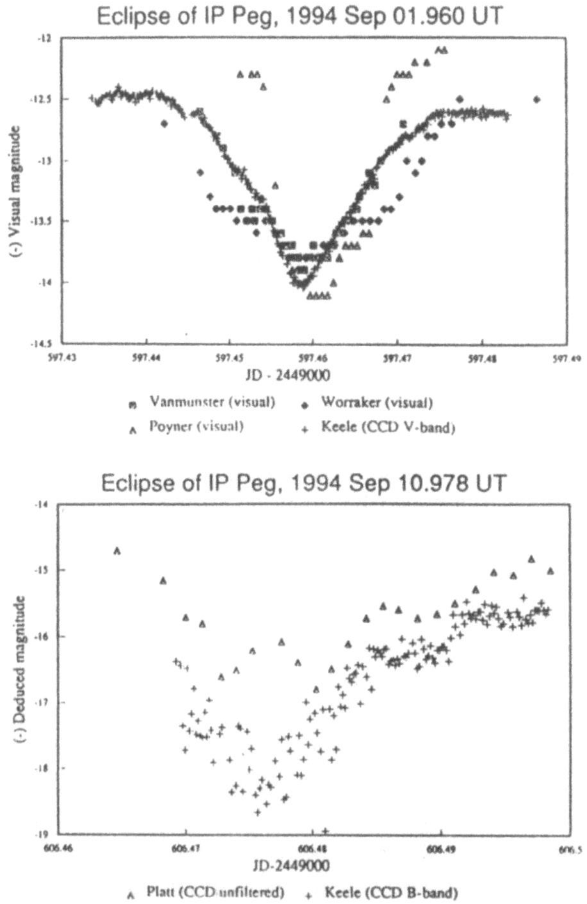

Figure 1. Typical outburst eclipses

Figure 2. The V−band light distribution from the model projected onto the plane of the sky. Note that the intensity scale is logarithmic.

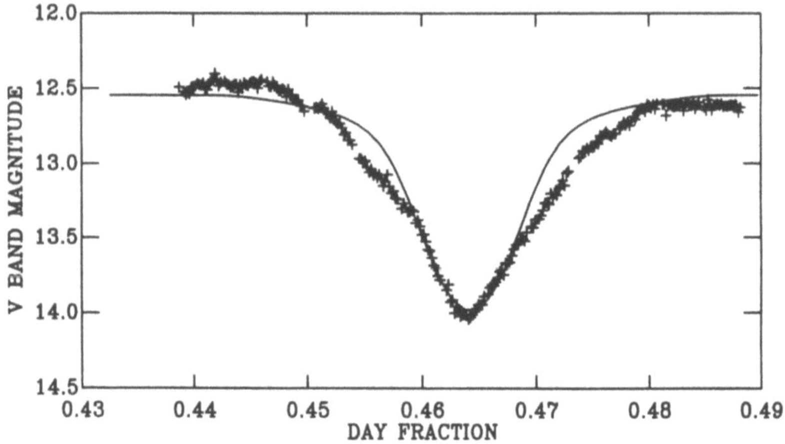

Figure 3. The eclipse resulting from the model shown in Fig. 2 overlayed on a typical eclipse. Note that the real eclipse is wider than the model.

4. Discussion

(i) Comparison of observing techniques: The largest errors in the visual estimates are due to the comparison sequence. Thus Poyner's brightest estimates in the upper panel of Fig. 1 are based on star D being at magnitude 12.2; the Skiff value of $V = 12.7$ mag [2] gives better agreement with other results. The CCD results show a number of subtle features on the ingress and egress curves, e.g. changes of slope, which are often missed in the visual results. In some cases this is simply because visual estimates are being made too slowly to give adequate time resolution. On the positive side, the eclipse depth, the time of minimum light and the beginning and end of an eclipse are generally well represented by the visual data.

The plots shown in the lower panel of Fig. 1 combine the results of CCD measurements in different wavebands; Keele were using a $B-$band filter while Platt was using no filter. It seems that the hump observed by Platt around mid-eclipse must have been due to material emitting mainly red and infrared light which would have been filtered out in the Keele system. Whilst interesting, this comparison suggests that filtering of CCD systems is necessary in this kind of work to standardise results; Johnson $V-$ or $B-$band filters are recommended.

(ii) Interpretation of the light curves: Most of the eclipses are broadly similar in duration and are about 1.5 mag deep. However the Ouda station team observed an eclipse 2.6 mag deep on August 26.789 UT, which was very early in the outburst [4]. Combining these results suggests that the main light source in the August outburst was initially very compact but subsequently expanded and stabilised. Later (day 10 onwards) the eclipses became deeper again as the outburst light began to fade significantly.

The light curves lack the white dwarf egress feature characteristic of quiescent eclipses, but have well-defined minima allowing timings to within a few seconds. Lack of an up-to-date ephemeris prevents us from interpreting the timings absolutely, but there seems to be a small advance in the time of minimum light relative to the published ephemeris through the August outburst. In three cases there were sufficient observations to show that the light level was about 0.2 mag brighter before the eclipse than after it. We attribute this to the 'orbital hump' effect due to the changing aspect of the hot spot [1]; in quiescence the difference is about 1 mag in the B–band [5, 6].

Three of the August/September eclipses have been covered sufficiently well (eclipse numbers 25 170, 25 195 and 25 214) for estimates of the total duration of the eclipses to be made. In terms of orbital phase interval (time interval between first and last contact divided by orbital period) these eclipses were 0.214, 0.247 and 0.198 wide respectively. The third value is probably too short as the beginning of the eclipse may have been missed, but the other two suggest a possible increase in the disc diameter between JD 2 449 597 and 2 449 601 (days 7 and 11 of the outburst). Definition of the points of first and last contact is intrinsically difficult, so this result should be treated with caution.

Finally we note that the eclipses were all different in detail. There were varying degrees of asymmetry implying time-dependent azimuthal structure in the disc and/or hot spot. However there was no obvious pattern of development of the detailed eclipse structure through an outburst.

5. Conclusions and recommendations

(i) Eclipses of IP Peg occurring during its 1994 August/September and 1994 December outbursts have been observed using both visual and CCD techniques.
(ii) The largest source of error in the visual estimates is an inadequate comparison sequence.
(iii) Allowing for problems due to the comparison sequence, the eclipse depth, the time of minimum light and the beginning and end of an eclipse can be determined to a useful degree of accuracy from visual data.
(iv) The visual data represent the details of the ingress and egress light curves to an erratic degree of accuracy.
(v) Unfiltered CCD results may not be comparable with results obtained in standard photometric wavebands.
(vi) In the 1994 August outburst the main outburst light source, which

must initially have been very compact, spread out and stabilised within about 2 days to give eclipses about 1.5 mag deep in the V−band for the plateau phase of the outburst.
(vii) Modelling work suggests that, if we require a steady state disc, to reproduce the eclipse data it is necessary to postulate a flared disc with a semi-opening angle of about 14°.
(viii) In the plateau phase of the 1994 August outburst the orbital hump had an amplitude of about 0.2 mag in the V−band.
(ix) The white dwarf egress feature is not visisble in the eclipse light curves obtained during outburst.
(x) Although a recent orbital phase calibration is not available, eclipse minima were approximately 0.002 d earlier than predicted for most of the 1994 August outburst. However this time difference was smaller earlier (day 5) in the same outburst and on the only occasion (December 13) in the later outburst when it could be measured.
(xi) Eclipses were observed with various degrees of asymmetry and with various changes of slope on the ingress and egress light curves, but there was no discernible pattern of development through an outburst.

In future visual observers should aim to make estimates about once per minute to improve coverage of subtle features of the eclipse light curves. Observations are encouraged for some time both before and after eclipses to check the size of the orbital hump in outburst and its possible relationship with eclipse asymmetries. Accurate white dwarf egress timings made in quiescence are needed to permit full interpretion of minimum light timings made during outburst. More work evidently needs to be done and the IP Peg observing project is continuing.

Acknowledgements. Contributing observers: S. J. Robinson, T. Naylor, T. C. Platt, E. Broens, G. Coady, A. Diepvens, F. Van Loo, G. Poyner, T. Vanmunster, W. Worraker, J. Pietz and M. Westlund.

References

1. Frank, J., King, A.R., Raine, D.J., 1992, "Accretion Power in Astrophysics" (2nd Edition), Cambridge University Press
2. Skiff, B.A., 1994, VSNET message 'Skiff on IP Peg, issued on 9th October
3. Somers, M. W., Mukai, K., Naylor, T., 1996, MNRAS, in press
4. VSNET postings, August 1994
5. Wood, J. H., Marsh, T. R., Robinson, E. L., et al., 1989, MNRAS, **239**, 809
6. Wood, J. H., Crawford, C. S., 1986, MNRAS, **222**, 645

SLINGSHOT PROMINENCES DURING DWARF NOVA OUTBURSTS?

D. STEEGHS[1], K. HORNE[2], T. R. MARSH[3]
1. Astronomical Institute, Utrecht University, The Netherlands
2. Physics & Astronomy, University of St.Andrews, UK
3. Astronomy Group, University of Southampton, UK

Abstract. We present phase-resolved spectroscopy of two dwarf novae, IP Peg and SS Cyg, observed in outburst with the WHT/ISIS spectropolarimeter. Doppler tomograms of several emission lines show peculiar narrow, stationary emission components.

1. Introduction

IP Peg is an eclipsing dwarf nova with an orbital period of 3.8 h, and an outburst cycle of about three months. IP Peg was observed during outburst on 1993 August 27 (0.37 Å pixel^{-1}). Secondary star emission was very strong during the outburst, and a peculiar emission component, with a very low velocity, was observed in the Hα line, visible as a spot in the Doppler map (Fig. 1, top left). SS Cyg is a U Gem-type dwarf nova with an orbital period of 6.6 h, and an outburst cycle of about 50 d. SS Cyg was observed for about 6 h on 1993 August 28 during an anomalous (slow) outburst, just after maximum light. The wavelength ranges 6390 Å to 6850 Å (0.37 Å pixel^{-1}) and 4200 Å to 5000 Å (0.78 Å pixel^{-1}) were covered. Again low-velocity emission components were observed in Hα, He I (6678) and He II (4686) (see Fig. 1). Secondary star emission is strong in most lines, except in the He II (4686) line which is only visible during outburst, but which shows a very clear stationary emission component.

2. Stationary emission components

Both dwarf novae observed show low velocity components in their emission lines. The low velocities make it incompatible with secondary star emission or emission from the white dwarf although secondary star emission is observed to be strong during outburst.

Although there is no obvious part of the binary system moving with the observed velocity, a possible interpretation may be 'slingshot prominences';

A. Evans and J. H. Wood (eds.), Cataclysmic Variables and Related Objects, 71–72.

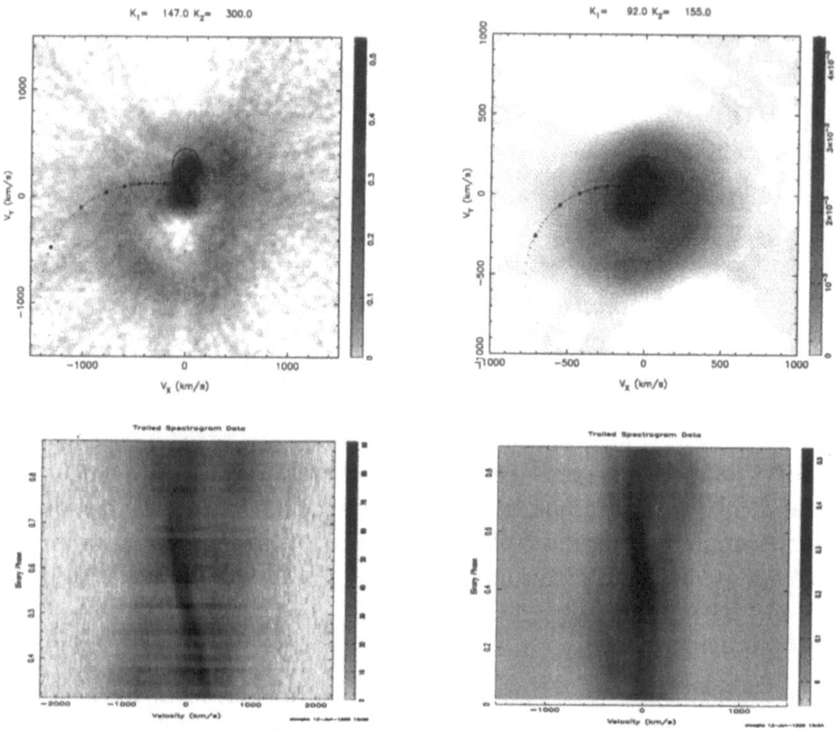

Figure 1. Top panels: Doppler tomograms constructed using the filtered back-projection method (Horne 1991). Left shows the Hα map for IP Peg, right Hα emission from SS Cyg. Apart from a ring of disk emission and emission from the secondary star, both systems display low velocity emission, outside the secondary star's Roche lobe. Bottom panels: Observed data in the form of trailed spectrograms.

material trapped in magnetic loops and co-rotating with the secondary star (similar to the prominences observed from single, rotating dwarf stars). In IP Peg the supposed prominence extends about a stellar radius beyond the L_1 point into the Roche lobe of the white dwarf. Here gravity pulls material toward the top of the loop, where it can collect.

At this point we do not know whether our interpretation is correct, or how widespread such emission features may be during dwarf nova outbursts. Are large prominences always present, but just lit up during the outburst, or could they somehow be a cause or effect of the outburst? Work is in progress to constrain possible prominence parameters from the observed fluxes. In the case of IP Peg, eclipse observations can also be used to constrain the emission geometry.

References

Horne, K., 1991, in "Fundamental Properties of CVs" ed. A.W. Shafter (U. San Diego)

RESULTS OF OPTICAL MONITORING OF THE DWARF NOVA SS CYG DURING THE 1993 CAMPAIGN OF CO-ORDINATED OBSERVATIONS WITH THE ASCA SATELLITE

IRINA VOLOSHINA
Sternberg Astronomical Institute, 119899 Moscow, Russia
e-mail:volosh@sai.msk.su

SS Cyg is the brightest known and therefore best studied classical dwarf nova and it was the first one from which X-ray radiation was discovered. SS Cyg is unique because it has been detected at very wide range of energies, from a few eV up to $\sim$ 10 keV (Jones & Watson 1992). It was chosen as the first cataclysmic variable for X-ray observations with the Japanese satellite ASCA. Simultaneous optical photometric and spectroscopic observations were also provided for completeness of the study.

SS Cyg was observed by ASCA in 1993 from 20:47 UT May 26 to 19:05 May 27. Our photometric monitoring of SS Cyg was started two days before, on May 24, and lasted two weeks. The observations were made with 60 cm telescope of the Sternberg Astronomical Institute in Crimea. The star BD+42°4186 served as the local standard. The measurement errors do not exceed 1%. The results of our *UBV* photometry are given in Table 1. Using these data and the measurements communicated by the AFOEV the light curve of SS Cyg was constructed. The analysis of this light curve shows that the ASCA observations were made while the system was at outburst maximum and on the first steps of decline. The outburst started on May 17 and ended on June 3 when SS Cyg returned to quiescence. According to the criteria described in Howarth (1978), such as the form of the light curve at maximum, the magnitude at the maximum, the outburst width at 9.5 mag, the outburst width at 10.5 mag and the rate of decline, we may conclude that this outburst is an anomalous one.

References

Howarth, I.D., 1978, J.British Astron.Assoc., **88**, 458
Jones, M.H., Watson, M.G., 1992, MNRAS, **257**, 633

A. Evans and J. H. Wood (eds.), Cataclysmic Variables and Related Objects, 73–74.

TABLE 1. Photoelectric Photometry of SS Cyg

Date	$JD_\odot$	V	$B-V$	$U-B$
24 May	2449132.525	$9^m.003$	$-0^m.056$	$-0^m.940$
	529	8.998	−0.052	−0.940
25 May	2449133.478	9.070	−0.005	−0.855
	482	9.039	−0.057	−0.846
26 May	2449134.417	9.345	−0.047	−0.916
	422	9.390	−0.004	−0.931
	428	9.376	−0.012	−0.946
	433	9.390	−0.028	−0.950
	439	9.402	−0.014	−0.964
	448	9.543	−0.019	−0.969
30 May	2449138.453	10.847	−0.087	−0.996
	459	10.884	−0.061	−0.894
31 May	2449139.424	11.149	0.150	−0.823
	431	11.329	0.165	−0.822
	436	11.243	0.081	−0.906
	442	11.207	0.127	−0.940
	447	11.183	0.157	−0.849
2 June	2449141.450	11.718	0.455	−0.894
	455	11.915	0.524	−0.867
	460	11.987	0.529	−0.873
	465	12.006	0.548	−1.024
3 June	2449142.512	12.137	0.513	−0.952
	517	12.165	0.601	−1.067
	522	12.144	0.465	−1.017
7 June	2449146.356	12.011	0.395	−1.028
	363	12.103	0.430	−0.983
	369	12.039	0.471	−1.056
	374	11.996	0.484	−0.956
	381	11.953	0.452	−1.013
	387	12.086	0.523	−0.968
	393	12.076	0.552	−0.942
	400	11.873	0.356	−1.020
	406	11.868	0.488	−0.988
	411	11.963	0.463	−1.050
	417	11.929	0.454	−0.992
	424	11.970	0.528	−1.051
	431	12.141	0.577	−0.937
	438	12.099	0.763	−1.121

THE LONG PERIOD DWARF NOVA UY PUP IN OUTBURST

J. J. LOCKLEY[1], JANET H. WOOD[1], D. H. P. JONES[2],
S. MINESHIGE[3]
1. Physics Department, Keele University, Keele, ST5 5BG, UK
2. Royal Greenwhich Observatory, Madingley Road, Cambridge, CB3 0EZ, UK
3. Dept. of Astronomy, Kyoto University, Sakyo-Ku, Kyoto 606-01, Japan

1. Introduction

UY Pup is a little studied, frequently outbursting dwarf nova whose brightness varies from $V \sim 13.5$ mag in outburst to $V \sim 15.8$ mag in quiescence, which typically lasts ~15...25 d. We observed it spectroscopically at the INT in La Palma on 1989 Febuary 15 and 16 during outburst maximum. In this outburst, the rise lasted at least 3 d and there was a long decline of about 6...9 d. The whole outburst lasted about 18 d (Mattei, private communication).

2. Results

Hα, Hβ and He I (6678Å) lines are clearly present. The Hα emission line is superimposed on a very weak absorption component. In Hβ, however, the emission is very weak and the absorption component stronger. The emission components of the Hα and Hβ lines both have FWZI $\sim$ 925 km s^{-1}, whereas the absorption around Hβ has FWZI $\sim$ 4600 km s^{-1}. These lines are typical of Balmer lines seen in low inclination, high mass transfer rate systems, both nova-like variables and other dwarf novae at the same stage of outburst. Theoretical models with mass transfer rates of $\dot{M} = (1\ldots3)\,10^{16}$ g s^{-1} and orbital inclination $i = 15°$ have very similar profiles to the Hα and Hβ profiles in UY Pup [see Fig. 5i–j in Cheng & Lin (1989)]. Therefore Balmer lines of the type seen in UY Pup can be produced by discs alone. We find no evidence for a second component to the line, from, for example, irradiation of the secondary star. There is variation by a factor of 2 in the equivalent width of Hα, though this does not appear to be related to the orbital cycle or the evolution through the outburst.

A. Evans and J. H. Wood (eds.), Cataclysmic Variables and Related Objects, 75–76.

Radial velocities were measured from Hα using the double Gaussian method (Shafter 1983). The resulting radial velocities gave a best fit orbital period of 0.423 ± 0.003 d, however a period of 0.297 ± 0.001 d was also found. From Monte Carlo simulations of radial velocity data sets generated at both periods we concluded that, while the longer of the two periods remains the most likely, our sampling would generate an alias at the longer period if the shorter one were real. Our best fit parameters are $P = 0.423$ d, $K_{em} = 24.5 \pm 8.5$ km s^{-1}, $\gamma = 54.8 \pm 3.5$ km s^{-1} and $T_0 = 2447574.072 \pm 0.006$ d. If this is the true period, then UY Pup has one of the longest periods of all dwarf novae. With $P = 0.423$ d, the relations given by Szkody & Mattei (1984) predict an outburst duration of 17.6 d and a decline of 7.9 d, in good agreement with the outburst of UY Pup that we observed.

3. Estimation of the system parameters

Assuming the longer period is correct and that the secondary is a main sequence star, given a value for the mass ratio, q, we can calculate the primary mass, the primary radius, the orbital separation and the radius of the secondary. We can then place limits on q by comparing the FWZI of the emission lines with the Keplerian velocity of material at the primary surface and using the upper mass limit of the primary star. We also consider that the secondary may be evolved from the main sequence (Friend et al. 1990 and references therein) and re-calculated the parameters assuming a secondary mass only half that of the main sequence mass. Both sets of parameters are listed in Table 1.

TABLE 1. Estimated Binary Parameters

	Main Sequence Secondary	Non Main Sequence Secondary
Secondary Mass	$1.12\,M_\odot$	$0.56\,M_\odot$
Primary Mass	$0.79\,M_\odot \ldots 1.44\,M_\odot$	$0.53\,M_\odot \ldots 1.44\,M_\odot$
Seconday Radius	$1.2\,R_\odot \ldots 1.1\,R_\odot$	$0.92\,R_\odot \ldots 0.88\,R_\odot$
Orbital Separation	$2.9\,R_\odot \ldots 3.2\,R_\odot$	$2.4\,R_\odot \ldots 2.9\,R_\odot$
Inclination	$6.8^\circ \ldots 8.3^\circ$	$9.4^\circ \ldots 14.1^\circ$

References

Cheng, F.H., Lin, D.N.C., 1989, Ap. J., **337**, 432
Friend, M.T., Martin, J.S., Smith, R.C., Jones, D.H.P., 1990, MNRAS, **246**, 654
Shafter, A.W., 1983, Ap. J., **267**, 222
Szkody, P., Mattei, J.A., 1984, PASP, **96**, 988

OBSERVATION OF THE WZ SGE-TYPE DWARF NOVA AL COM

T. KATO[1], D. NOGAMI[1], H. BABA[1], K. MATSUMOTO[2], J. ARIMOTO[2], K. TANABE[2], K. ISHIKAWA[2]

1. Department of Astronomy, Kyoto University, Sakyo-ku, Kyoto 606-01, Japan
2. Astronomical Institute, Osaka Kyoiku University, Asahigaoka, Kashiwara, Osaka 582, Japan

Abstract. Although presently classified as a SU UMa-type dwarf nova, WZ Sge is well known as one of the most peculiar objects in that it shows only superoutbursts with exceptional duration and amplitude, and no normal outbursts. Furthermore, on its decline from the 1978 outburst, WZ Sge showed a deep temporal dip. All of these characteristics have puzzled both theoreticians and observers.

The dwarf nova AL Com was photometrically observed during the outburst in 1995 April, which occurred for the first time since 1975. The striking similarity of AL Com to WZ Sge, as demonstrated by the present observation (Fig. 1), provides plenty of material in interpreting the enigmatic nature of WZ Sge-type dwarf novae.

1. The early stage: superhumps and 'orbital' superhumps

In the early stage of the outburst, two distinct types of periodic variation were observed: (i) a 0.05666 d variation with decaying amplitude in the earliest stage of the outburst, and (ii) 0.0572 d superhumps which grew in ~ 11 d after the onset of the outburst. We attributed the former period to the orbital period of this cataclysmic binary, which seems to be later confirmed by a period analysis of quiescent light variation. The double-humped profile of this variation seems to preclude the explanation of a hot spot enhanced by the mass-transfer burst from the secondary. This has been believed to be the best observational evidence of a mass-transfer burst in WZ Sge, obtained during its 1978 outburst.

A. Evans and J. H. Wood (eds.), Cataclysmic Variables and Related Objects, 77–78.

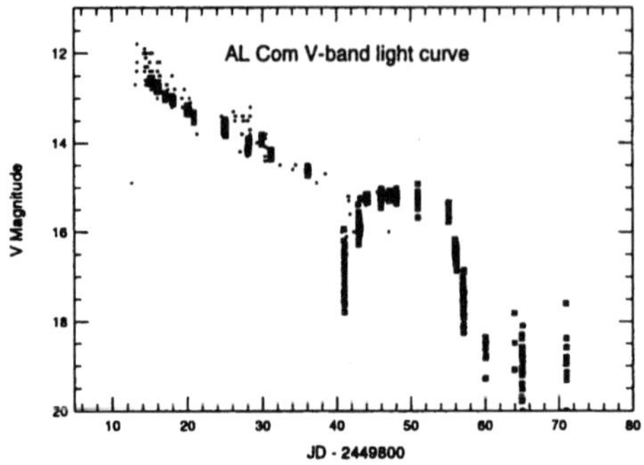

Figure 1. The overall *V*-band light curve of the 1995 outburst of AL Com. Note the exceptionally long duration of the outburst interrupted by a deep dip, both of which are clearly reminiscent of the 1978 outburst of WZ Sge.

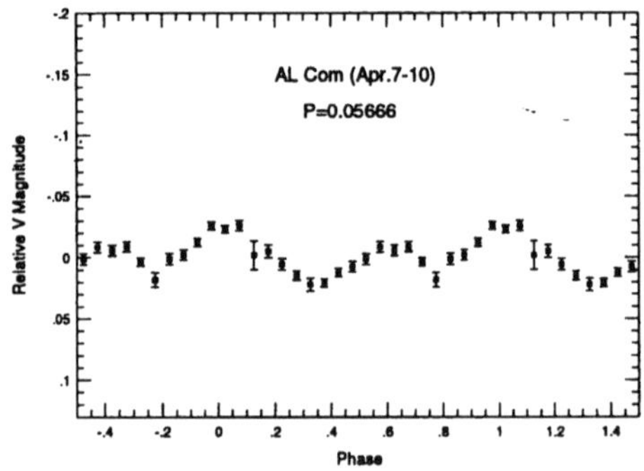

Figure 2. 'Orbital' superhumps observed in the earliest stage of the outburst. The double-humped profile of this variation seems to preclude explanation by the enhanced hot spot.

2. After the 'dip' behavior

The present outburst of AL Com showed a 'dip' very similar to the one observed in the 1978 outburst of WZ Sge. Non-existence of orbital humps during the recovery from the dip strongly supports the interpretation that the re-brightening was not powered by a mass-transfer event, but by the thermal instability of the accretion disk. After the dip, AL Com showed a short-lived maximum followed by a rapid decline. After this 'secondary dip' the star entered a new stage; the 'plateau phase'. The observation has shown that the superhumps again grew during this phase. We can therefore identify this stage as the start of a new superoutburst. The short-lived maximum after the main dip may represent a normal outburst which triggered this superoutburst. The peculiar outburst light curve common to WZ Sge and AL Com now seems to be naturally understood by the occurrence of two successive superoutbursts, which are powered by the usual combination of thermal and tidal disk instabilities.

SIMULTANEOUS MULTIWAVELENGTH OBSERVATIONS OF DWARF NOVAE SYSTEMS AT QUIESCENCE AND IN OUTBURST

G. TOVMASSIAN[1], J. ECHEVARRIA[1], E. PEREZ[1], D. H. P. JONES, M. SHARA, M. TAPIA[1], J. BOHIGAS[1], R. GILMOZZI, R. COSTERO, J. A. LOPEZ[1], M. ROTH[1], M. ALVAREZ[1], L. F. RODRIGUEZ, E. DE LARA, R. STOVER, C. MARTINEZ-ROGER, F. GARZON, N. ASTRIAN, E. ZOLDOS, J. MATTEI, F. BATESON

1. Instituto de Astronomía, Universidad Nacional Autónoma de México, Ensenada, México

The results of photometric and spectroscopic observations of dwarf novae are presented. The data were obtained during an international program of multiwavelength observations, held in 1986 February at several observatories, of dwarf novae during the first and subsequent days of outburst. During the campaign numerous dwarf novae were monitored in order to catch them in outburst. Preliminary results and analysis of some objects are reported elsewhere. A total of 30 dwarf novae were observed in the northern and southern hemispheres. Among them 37% were caught in outburst, including 10% on the rise to outburst and 17% in decline. Photometric observations were carried out in the *UBVRI* system and colour indexes were calculated. Colour-colour diagrams and light curves of the particular stars were constructed. $U - B$, $B - V$ and $V - R$, $B - V$ diagrams are presented in Figs. 1 and 2 respectively. The data shown in Figs. 1 and 2 confirm many results obtained previously, and shows clustering of the objects, depending on their orbital periods, above the main sequence at minimum (Echevarria 1984; Echevarria, Costero & Michel 1993). This region is a very large one, reflecting differences in types of dwarf novae. SU UMa type stars and other short period systems are considerably bluer in both $U - B$ and $B - V$ colours. During an outburst cycle, the systems trace a loop on the colour-colour diagram, with the general pattern shown in the $U - B$, $B - V$ diagram. However the colour changes are very individual and rather characteristic for each star. The large part of the loop covers the rise to outburst. The systems at maximum occupy a very distinct region, slightly above main sequence B9–A0 stars. However several systems

A. Evans and J. H. Wood (eds.), Cataclysmic Variables and Related Objects, 79–80.

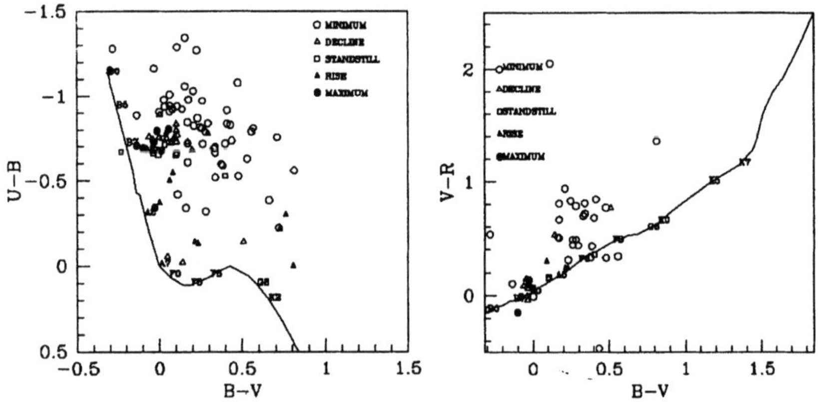

Figure 1. Left: Colour-colour diagram of dwarf novae at several stages of outburst and quiesence. The solid line corresponds to the main sequence. Right: $B-V$ versus $V-R$ diagram for the same objects shown in the left plot. Note the clustering of points at outburst maximum near the main sequence.

displayed peculiarities, which are also presented. A clear delay of UV flux was observed in the case of SS Aur in outburst, as opposed to MU Cen, which displays increasing U flux during the slow rise to outburst. DO Dra is also peculiar for its extremely red colours in the near-IR, and it exhibits a huge contribution from the secondary despite its relatively short orbital period.

References

Echevarria, J., 1984, Rev. Mex. Astr., **9**, 99

Echevarria, J., Costero, R., Michel, R., 1993, A&A, **275**, 201

IS THERE OBSERVATIONAL EVIDENCE FOR THE EVAPORATION OF THE INNER ACCRETION DISK IN DWARF NOVAE AT QUIESCENCE?

C. LA DOUS[1], E. MEYER-HOFMEISTER[2], F. MEYER[2]
1.ESA/IUE Observatory, Villafranca del Castillo, Apartado 50727, 28080 Madrid, Spain
2.Max-Planck-Institut für Astrophysik, Karl-Schwarzschild-Str. 1, 85740 Garching, Germany

Abstract. We present the preliminary results of a comparison of observed general patterns of quiescent dwarf novae and theoretical predictions of the disk evaporation model. We demonstrate that, even though details still have to be clarified, on the whole there is strong evidence that observations and theory are in agreement.

1. Introduction

The observations of dwarf novae in quiescence and during rise to outburst provide some features which seem to be in contradiction to the standard theoretical modelling of dwarf nova outbursts, where mass is accumulated during quiescence until the instability causes the next outburst.

(i) The angle dependence of the equivalent widths of UV emission lines: at low inclinations all lines are strongly in emission; as the inclination increases, the equivalent width decreases and at angles larger than some 60° or 70° only an essentially featureless continuum can be seen[1] (la Dous 1991). From this it can be concluded that there is an optically thin emission source radiating at UV wavelengths and that this source is predominantly confined to the orbital plane of the system.

(ii) The white dwarfs: it has been demonstrated that on the basis of the SWP spectrum alone it is possible to roughly determine the temperature of the white dwarf in quiescent dwarf novae (la Dous 1996a,

[1]Except for the Lyα absorption line which is strong in some objects; we will deal with this in the next section.

A. Evans and J. H. Wood (eds.), Cataclysmic Variables and Related Objects, 81–84.

1996b; Hassall & la Dous 1996). The white dwarf spectrum then can be subtracted from the observed integral spectrum. What remains is interpreted as the spectrum of the quiescent accretion disk. To zeroth approximation this can be fitted with a single blackbody having a temperature between 9 000 and 15 000 K. In all objects for which more than one quiescent spectrum is available, the flux level is not constant with time, but changes are consistent with the assumption that they are entirely due to a temperature change in the central object.

(iii) In many dwarf novae a sometimes pronounced delay in the rise to outburst between optical and UV wavelengths radiation can be observed. It is too long to be explained by the spreading of the hot state in the disk. It seems that the inner disk first has to be filled with material before it can go into outburst (Meyer & Meyer-Hofmeister 1989).

(iv) In a cool accretion disk mass flows with a rate of $10^{-13} \ldots 10^{-12}\, M_\odot\, yr^{-1}$ towards the white dwarf. This is at least an order of magnitude too low for explaining the actually observed X-ray and UV fluxes at quiescence, which require rates on the order of $10^{-11}\, M_\odot\, yr^{-1}$ (e.g. van der Woerd & Heise 1987; Szkody et al. 1991).

2. The evaporation model

All these features can be understood within the framework of the disk evaporation model (Meyer & Meyer-Hofmeister 1994; Liu, Meyer & Meyer-Hofmeister 1995). The interaction of a hot corona with the underlying cool disk in dwarf nova systems leads to an evaporation of the inner disk around the white dwarf. In contact with the corona, material evaporates at higher gravitational potential and accretes on the compact star at lower gravitational potential. Due to angular momentum this mass flow takes the form of a coronal accretion disk, which is heated by friction. The process works best close to the white dwarf, thus evaporating the innermost part of the disk first. As a consequence this hole has to be filled up on the diffusion time-scale during the next outburst causing a delay of radiation from the innermost disk, that is, of UV flux.

The evaporation model gives typical mass flow rates of about 10^{-11} to $3\,10^{-11}\, M_\odot\, yr^{-1}$. Part of it leaves the system as a very hot wind, the remaining part is accreted onto the white dwarf. X-rays are produced mainly in the optically thin thermal conductive boundary layers above the accretion disk (a small fraction) and the white dwarf (main part of the radiation). The less hot regions below are heated, so that the UV radiation originates from both the disk and the white dwarf, whereby the latter varies its surface temperature depending on the accretion rate.

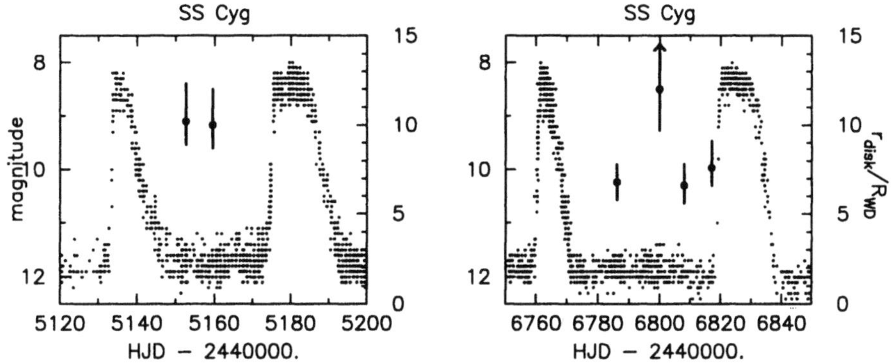

Figure 1. **Variations of the inner disk radius during the course of two quiescent intervals of SS Cyg. The error bars shown are those resulting from uncertainties of ±1000 K in the temperature of the optically thick accretion disk. The photometric data are courtesy of the AAVSO.**

3. Interpretation of observations

The ratio of fluxes from the white dwarf and the disk provide information on the size of the disk area where most of the radiation is emitted. Taking blackbody radiation as a first approximation, the ratio of the fluxes is related to the size of the emitting areas and, therefore, to the ratio of the square of the radii. For assumed inclination, and mass and radius of white dwarf we derive the disk radius near the inner edge where most radiation comes from. If we further assume that the cool underlying disk region is heated by the coronal layers above, we can derive the evaporation rate (= mass flow rate in the corona). From the observational side the necessary inputs are the fluxes and temperatures of white dwarf and accretion disk, which can be derived directly from UV spectra. The most delicate point, and at the same time the most difficult one, is the determination of the temperature of the optically thick accretion disk.

Neglecting the effect of mass accumulation in the disk, theory would predict a progressive evaporation of the accretion disk during a quiescent state and, consequently, a monotonic increase of the inner disk radius. Analyzing sequences of spectra taken during the same quiescent intervals of SS Cyg, we find that this is not the case, that the inner disk radius varies non-monotonically (Fig. 1). This behaviour might be due to the heating of the disk by the corona above and the material flowing inward from outer regions. The derived radii, of order 5 to 10 white dwarf radii and evaluated coronal mass flow rates of about $10^{-11}\,M_{\odot}\,yr^{-1}$ confirm the basic predictions of the evaporation model. For details we refer to a forthcoming publication (la Dous, Meyer-Hofmeister & Meyer 1996).

References

Hassall, B.J.M., la Dous, C., 1996, these proceedings, p85
la Dous, C., 1991, A&A, **252**, 100
la Dous, C., 1996a, in preparation
la Dous, C., 1996b, A&A, in preparation
la Dous, C., Meyer-Hofmeister, E., Meyer, F., 1996, A&A, in preparation
Liu, F.K., Meyer, F., Meyer-Hofmeister, E., 1995, A&A, **300**, 823
Meyer, F., Meyer-Hofmeister, E., 1989, A&A, **221**, 36
Meyer, F., Meyer-Hofmeister, E., 1994, A&A, **288**, 175
Szkody, P., Mattei, J.A., Waagen, E.O., Staeblein, C., 1991, Ap. J. Supp., **76**, 359
van der Woerd, H., Heise, J., 1987, MNRAS, **225**, 141

THE TEMPERATURES OF WHITE DWARFS IN DWARF NOVAE

B. J. M. HASSALL[1], C. LA DOUS[2]
1. Centre for Astrophysics, University of Central Lancashire, Preston, PR1 2HE, UK
2. ESA/IUE Observatory, Villafranca del Castillo, Apartado 50727, 28080 Madrid, Spain

Abstract. We attempt to measure the temperature of all the quiescent dwarf novae that have been observed with IUE, by fitting the observed flux distribution of white dwarfs (WDs) with known temperatures. This approach gives results similar to those using much more sophisticated modelling, but allows us to investigate the relationship with system parameters and subclasses.

1. Introduction

In recent years, there has been an upsurge in interest in measuring the temperature of the white dwarf in dwarf novae as it is fundamental in understanding various aspects of the WD's binary environment and the evolutionary status of CVs. A large number of dwarf novae and other CVs have been observed with IUE, allowing us to determine the temperatures of their white dwarfs. While the results for any one system are not as reliable as the detailed WD atmosphere models applied to HST data, their number allows a statistical approach that will not be possible with HST results for some years to come.

2. The method

In quiescence, the flux in the SWP range shortward of about 1600 Å is dominated by emission from the white dwarf (la Dous 1996). In our simple approach, we fit WD spectra to this region of the spectrum, using both the Lyman alpha profile and the continuum shape. Considering that we have not used detailed modelling, it is remarkable what good agreement

A. Evans and J. H. Wood (eds.), Cataclysmic Variables and Related Objects, 85–86.

is achieved with the more sophisticated techniques. For example, for OY Car, we find $T_{WD} = 15\,200$ K while Horne et al. (1994) find the WD is best fitted by a 16 500 K WD with solar metallicity shrouded by a curtain of disk material at 10 000 K. From this we deduce that we probably systematically underestimate T_{WD}. Although this difference is not negligible, it is not sufficiently large to invalidate the overall trends seen in our study.

3. Discussion

The complete list of the temperatures of those WDs best fitting the dwarf novae spectra can be found in la Dous (1996). When arranged by orbital period, the sample reveals a tendency for the short period systems to have cooler white dwarfs. This could be due to a combination of two evolutionary effects, namely age and accretion rate. When the WD temperatures are looked at as a function of the the orbital inclination, there is no obvious correlation for intermediate values; whereas the highest inclination systems inexplicably tend to have the lowest temperatures and the shortest orbital periods. The absence of data for inclinations below about 20° probably reflects the difficulty of determining system parameters in these cases. Thus it is important to investigate any possible selection effects.

For a good fraction of the systems we cannot fit a DA WD and the spectrum does not resemble an optically thick accretion disc either. But at least much of the short wavelength spectrum, if not all the IUE range, can be fitted well with a DB WD, despite the fact that there is no theoretical explanation for this, given that the WD is accreting hydrogen-rich material from the secondary star. Clearly, this aspect requires further investigation.

4. Conclusions

- The IUE archive provides a unique opportunity to perform statistical studies of the white dwarfs in quiescent dwarf novae.
- The temperatures obtained give reasonable agreement with detailed models using recent HST data of a few systems so far observed.
- In about half of the systems, the best (or even only) fit is obtained with a DB WD.
- There is a correlation of T_{WD} with orbital period.
- Most of the coolest white dwarfs appear in the short period, high inclination systems.

References

Horne, K., Marsh, T.R., Cheng, F.H., et al., 1994, Ap. J., **426**, 294
la Dous, C., 1996, in preparation.

LINE FORMATION IN U GEM AND T LEO

W. HUMMEL[1], K. HORNE[2], T.R. MARSH[3], JANET H. WOOD[4]
1. Astrofysisch Instituut, Vrije Universiteit Brussel, Pleinlaan 2, B-1050 Brussel, Belgium
2. University of St. Andrews, School of Physics & Astronomy, North Haugh, St. Andrews, Fife KY16 9SS, Scotland
3. University of Southampton, Department of Physics, Highfield, Southampton SO17 1BJ, England
4. Keele University, Department of Physics, Keele ST5 5BG, England

We present 3-D LTE radiative transfer calculations [1] for H, He and Ca in accretion disks (AD) of dwarf novae in quiescence. The model disk is assumed to be in hydrostatic equilibrium vertically, and to rotate with Keplerian velocities. Calculated emission lines are fitted to phase-averaged, continuum-subtracted spectra of U Gem (Fig. 1) and T Leo (Fig. 2). Up to four parameters of the AD have been fitted: distance D, baryonic number density N, isotropic turbulence $V_{\rm tu}$, and disk temperature T; the latter two are assumed to be constant throughout the disk. Geometrical parameters are from [2] and [3].

Our preliminary results for U Gem show agreement between observation and theory for the line intensities and line ratios, indicating that both Ca II and He I emission lines can show up in an isothermal AD. However, the line profile shapes are not matched correctly by the simple model disk, giving rise to a surface density and/or a temperature gradient.

Two models are shown for T Leo demonstrating the sensitivity of model spectra on temperature.

Acknowledgement. WH thanks the *Belgian Federal Scientific Services (DWTC/SSTC)*, project number SC/005, for funding.

References

1. Hummel, W., 1994, A&A, **289**, 458
2. Marsh, T.R., Horne, K., Schlegel, E.M., et al., 1990, Ap. J., **364**, 637
3. Shafter, A.W., Szkody, P., 1984, Ap. J., **276**, 305

A. Evans and J. H. Wood (eds.), Cataclysmic Variables and Related Objects, 87–88.

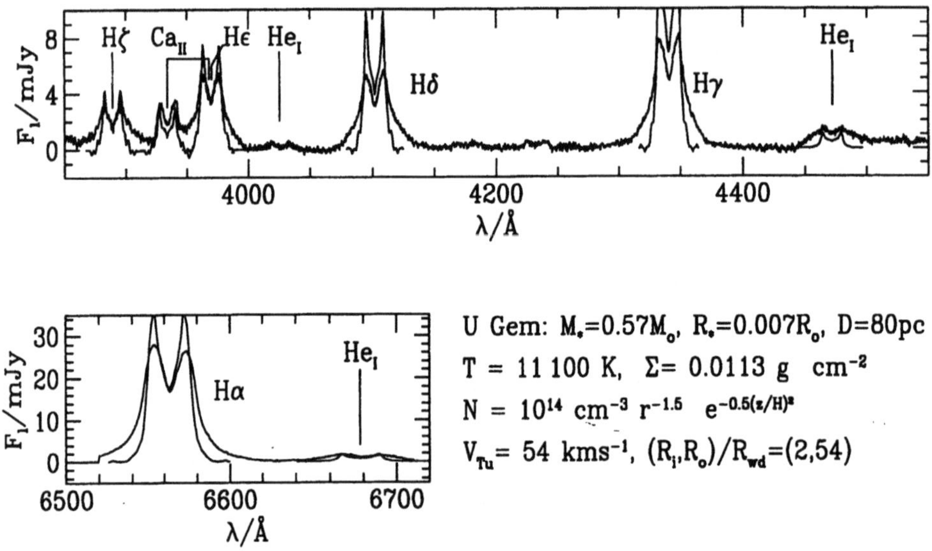

Figure 1. Calculated emission lines of the model disk compared with the continuum subtracted spectrum of U Gem in quiescence.

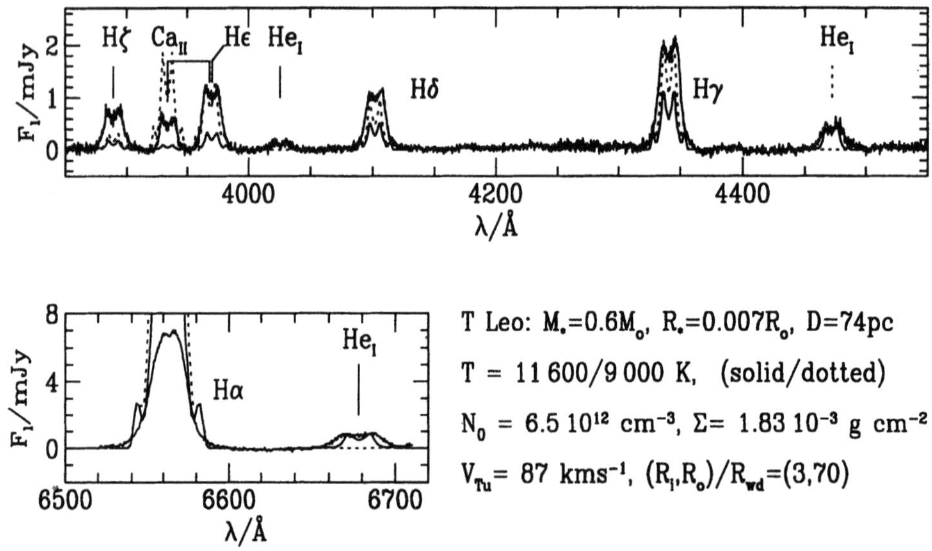

Figure 2. Calculated emission lines of two model disks with T=11 600 K (solid) and T=9 000 K (dotted) compared with the continuum subtracted spectrum of T Leo in quiescence.

POPULATION STUDIES OF CATACLYSMIC VARIABLES

F. A. RINGWALD
Dept. of Physics, Keele University, Keele, ST5 5BG, England
Present address: Planetary Science Institute,
620 North Sixth Avenue, Tucson, AZ 85705-8331, USA

Cataclysmic variables (CVs) are idiosyncratic objects, but progress can be made by studying groups of them. To do this, one must identify samples that are representative of their true properties. Because of the wide variety of physics that occurs in CVs, they have a complex phenomenology in which effects are still being discovered (e.g. permanent superhumps: Patterson & Richman 1991). CV outbursts have their own complex phenomenology (but see Osaki 1995a,b and this volume, for ideas on unification for dwarf novae), so it is unwise to rely on outburst properties to give unbiased samples.

One wants to identify CVs by some property common to all of them. One such property is flickering; another is color excess. CVs are very blue, especially in the ultraviolet, with $U - B < -0.46$ for 95% of CVs (Bruch & Engel 1994). This color index also defines inclusion in the Palomar-Green (PG) survey (Green, Schmidt & Liebert 1986), and a preliminary list of PG CVs was given by Green et al. (1982). My Ph.D. thesis (Ringwald 1993) completed this work, being the first optically selected complete sample of CVs at high latitude. At the same time, Andy Silber was writing his Ph.D. thesis on an X-ray-selected sample of CVs, from the HEAO-A1 MC-LASS survey (Silber 1992). One might say I was Andy's optical counterpart.

The Palomar-Green catalog includes 1715 objects of all types (QSOs, sdBs, WDs) in its complete sample. Of these, 70 were listed by Green et al. (1986) as candidate CVs, having emission lines at Galactic velocities, or energy distributions with $f_\nu \gtrsim \nu^0$. Follow-up spectra were obtained of all 70 candidates. Of these, 30 were found to be genuine CVs. Five more CVs were found, but were excluded from the complete sample, since they were seen by the PG survey only because they were in outburst when Green was exposing his films on the 18-inch Palomar Schmidt telescope. The rest of the candidates, some 35 objects, were found not to be CVs. Many were misidentified sdB and other hot, high-gravity stars, with several binaries (including, but not limited to, PG 1114+187, PG 1119+147, BE UMa, NN Ser, KQ Peg = PG 2240+193, and PG 2300+166), two com-

A. Evans and J. H. Wood (eds.), Cataclysmic Variables and Related Objects, 89–92.

pact emission-line galaxies (PG 1136+581 = Mrk 1450 and PG 0947+462 = Mrk 125), and one planetary nebula nucleus (PG 1712+493).

Because it selects by color and not by outburst properties, the PG survey found new classes of CVs not showing conspicuous outbursts. These include the dreaded SW Sex stars (Thorstensen et al. 1991; Dhillon, this volume), faint nova-likes with mysterious but consistent behavior. Of 30 PG CVs, five are SW Sex stars. Nine are dwarf novae and three are magnetic: the SW Sex stars are not 'peculiar', as claimed by most of the discovery papers. All but WX Ari (Hellier, Ringwald & Robinson 1994) are eclipsing.

Another new class of stars from the PG survey are RZ LMi and ER UMa (PG 0948+344 and PG 0943+521, respectively: Robertson, Honeycutt & Turner 1994, 1995; Kato & Kunjaya 1995). These are low-amplitude dwarf novae with cycles that repeat unusually coherently, for dwarf novae. HS Vir = PG 1341−079 is another low-amplitude dwarf nova, with outbursts rising only 1.5 mag brighter than quiescence (Kato et al. 1995).

Permanent superhumps have been found in several nova-likes, including BK Lyn = PG 0917+342 (Skillman & Patterson 1993), V795 Her = PG 1717+413 (Patterson & Skillman 1994), and PG 1633+115 (Misselt & Shafter 1995). That these nova-likes are those in the PG catalog with the shortest orbital periods (1.80, 2.60, and 3.94 h, respectively) is consistent with the theory that superhumps are tidally induced (Whitehurst 1988), resulting from the high mass ratios short-period CVs should have.

An unbiased sample allows ensemble properties to be determined. The complete sample from the PG survey gives a space density of $6\,10^{-6}$ pc^{-3}. This is about ten times higher than previous estimates of $7 \pm 3.5\,10^{-7}$ pc^{-3} (Della Valle & Duerbeck 1993), $8\,10^{-7}$ pc^{-3} (Downes 1986), $7\,10^{-7}$ pc^{-3} (Patterson 1984), and $1\,10^{-6}$ pc^{-3} (Warner 1974). Nearly all the excess space density from the PG survey is from the least luminous CVs.

This space density suggests a recurrence time between novae of 15 000 yr, or about ten times longer than for the low-latitude sample of Downes (1986), assuming $D_{\rm N}/T = 3.8\,10^{-10}$ pc^{-3} yr^{-1}, where $D_{\rm N}$ is the average space density and T is the average recurrence time (Duerbeck 1984), assuming that all CVs have classical nova outbursts at some time. This may not be so if CVs are in two populations, one in the Galactic disk and one in the bulge (Della Valle et al. 1992). Because of the PG survey's relatively bright average limiting magnitude, the PG CVs extend only about one scale height away from the Galactic plane, so nearly all are in the disk.

The fraction of magnetic PG CVs is 3/30 (10%); the AM Her systems MR Ser and AN UMa, and the DQ Her star YY Dra. This contrasts with 28% from the X-ray-selected sample of Silber (1992), and 2% for the ratio of magnetic white dwarfs to field white dwarfs (Schmidt & Liebert 1987).

In response to a poster shown at the San Diego meeting (Ringwald 1994), several observers, including Hellier et al. (1994) and Misselt & Shafter (1995), obtained time-resolved photometry of many PG CVs. All systems in the complete sample have now had sufficient phase coverage to determine the fraction of eclipsing CVs: 7/30 (23%) eclipse. Silber (1992) also obtained a fraction of 23% for his X-ray-selected sample; an isotropic distribution should have 14%. Silber suggested this to be evidence of massive white dwarfs, although the fraction for the PG CVs can be turned into an isotropic distribution if one excludes the SW Sex stars.

The cumulative surface density of the PG CVs is $d \log n(B)/dB = 0.35$ mag^{-1} over $14.4 \leq B \leq 16.2$, where $n(B)$ is the number of CVs per square degree per B magnitude. This is similar to the surface density of sdO and sdB stars (Green et al. 1986). The orbital period distribution is like that of the HEAO A-1 CVs, except that the 'AM Her spike' is less pronounced; it differs significantly from the distribution for the AM Her stars found by ROSAT (Kolb & de Kool 1993).

Other population studies of CVs include the Kitt Peak-Downes survey (Downes 1986), which was optically selected near the Galactic plane. Deep high-latitude UV-excess surveys are in progress, including the Edinburgh-Cape survey (Stobie et al. 1992), which reaches to $B \sim 18$, and the Montréal-Cambridge survey (Demers et al. 1987), which reaches to $B \sim 17.5$. These surveys will be statistically cleaner than the PG survey, since their cutoff can be held at $B = 16.5$; the limiting magnitude in the PG survey changes from field to field, between 15.49 and 16.67. They may also avoid problems with incompleteness suspected of the PG survey, since most objects were near the detection limit of Green's Schmidt films.

A sample of high-latitude CVs, selected by variability, was compiled from the GCVS by Howell & Szkody (1990), intending to explore CVs in the Galactic halo. Despite selection effects and incompleteness, it has found many of the faintest, nearest CVs (Howell, Szkody & Cannizzo 1995).

Shafter (this volume) reviews studies of CVs in M31 and other galaxies. Searches for CVs in globular clusters are ongoing with HST (Paresce & de Marchi 1994). Perhaps too few CVs are present (Shara, Bergeron & Moffat 1994; Shara et al. 1995), but the case is not yet clear (Cool et al. 1995).

Another, different sort of survey has been carried out by Ringwald, Naylor & Mukai (1996), to explore the behavior through time in novae that had outbursts between 1783 and 1986. Essentially, they don't do anything: there is no sign of the secondary stars, as one might expect if novae went into hibernation. High-excitation lines such as He II λ4686 Å have comparable strengths in centuries-old novae and in novae just out of the nebular phase, suggesting that the white dwarfs or disks stay hot for centuries.

There has been work on detached post common-envelope binaries, often wrongly called pre-cataclysmic binaries (Ritter 1986a), which have turned up in the EUV surveys of ROSAT/WFC (Pounds et al. 1993) and EUVE (Bowyer et al. 1994). The first Ph.D. thesis on these objects will soon appear (Catalán 1995), surely establishing these objects as a part of the astrophysical landscape. Theory includes work on selection effects, begun by Ritter (1986b), and on population modeling (Kolb 1993; also this volume).

References

Bruch, A., Engel, A., 1994, A&AS, **104**, 79
Bowyer, S., Lieu, R., Lampton, M., et al., 1994, Ap. J. Supp., **93**, 569
Catalán, M.S., 1995, Ph.D. thesis, University of Sussex
Cool, A.M., Grindlay, J.E., Cohn, H.N., et al., 1995, Ap. J., **439**, 695
Della Valle, M., Bianchini, A., Livio, M., Orio, M., 1992, A&A, **266**, 232
Della Valle, M., Duerbeck, H., 1993, A&A, **271**, 175
Demers, S., Fontaine, G., Wesemael, F., et al., 1987, in "The Second Conference on Faint Blue Stars", Davis P.A.G., et al. (ed.s) L. Davis Press, Schenectady, New York, p497
Downes, R.A., 1986, Ap. J., **307**, 170
Duerbeck, H.W., 1984, A&SS, **99**, 363
Green, R.F., Ferguson, D.H., Liebert, J., Schmidt, M., 1982, PASP, **94**, 560
Green, R.F., Schmidt, M., Liebert, J., 1986, Ap. J. Supp., **61**, 305
Hellier, C., Ringwald, F.A., Robinson, E.L., 1994, A&A, **289**, 148
Howell, S.B., Szkody, P., 1990, Ap. J., **356**, 623
Howell, S.B., Szkody, P., Cannizzo, J.K., 1995, Ap. J., **439**, 337
Kato, T., Kunjaya, C., 1995, PASJ, **47**, 163
Kato, T., Nogami, D., Masuda, S., Hirata, R., 1995, IBVS, 4193
Kolb, U., 1993, A&A, **271**, 149
Kolb, U., de Kool, M., 1993, A&A, **279**, L5
Misselt, K.A., Shafter, A.W., 1995, AJ, **109**, 1757
Osaki, Y., 1995a, PASJ, **47**, L11
Osaki, Y., 1995b, PASJ, **47**, L25
Paresce, F., de Marchi, G., 1994, Ap. J., **427**, L33
Patterson, J., 1984, Ap. J. Supp., **54**, 443
Patterson, J., Richman, H., 1991, PASP, **103**, 735
Patterson, J., Skillman, D.R., 1994, PASP, **106**, 1141
Pounds, K.A., Allan, D.J., Barber, C., et al., 1993, MNRAS, **260**, 77
Ringwald, F.A., 1993, Ph.D. thesis, Dartmouth College
Ringwald, F.A., 1994, in "Interacting Binary Stars", ed. A. Shafter, ASP Conf. Ser. **56**, 294
Ringwald, F.A., Naylor, T., Mukai, K., 1996, MNRAS, submitted
Ritter, H., 1986a, A&A, **168**, 139
Ritter, H., 1986b, A&A, **168**, 105
Robertson, J.W., Honeycutt, R.K., Turner, G.W., 1994, in "Interacting Binary Stars", ed. A. Shafter, ASP Conf. Ser. **56**, 298
Robertson, J.W., Honeycutt, R.K., Turner, G.W., 1995, PASP, **107**, 443
Schmidt, G.D., Liebert, J., 1987, A&SS, **131**, 549
Shara, M.M., Bergeron, L.E., Moffat, A.F.J., 1994, Ap. J., **429**, 767
Shara, M.M., Drissen, L., Bergeron, L.E., Paresce, F., 1995, Ap. J., **441**, 617
Silber, A.D., 1992, Ph.D. thesis, Massachusetts Institute of Technology
Skillman, D.R., Patterson, J., 1993, Ap. J., **417**, 298
Stobie, R.S., Chen, A., O'Donoghue, D., Kilkenny, D., 1992, in "Variable Stars and Galaxies", ed. B. Warner, ASP, San Francisco, p87
Thorstensen, J.R., Ringwald, F.A., Wade, R.A., et al., 1991, AJ, **102**, 272
Warner, B., 1974, MNASSA, **33**, 21
Whitehurst, R., 1988, MNRAS, **232**, 35

SPECTROPHOTOMETRY OF 106 TO–BE–CONFIRMED CVS

TOMAŽ ZWITTER[1], ULISSE MUNARI[2],
ANGELA BRAGAGLIA[3]
1. Dept. of Physics, University of Ljubljana, Slovenia
2. Astronomical Observatory of Padova, Italy
3. Astronomical Observatory of Bologna, Italy

Abstract. We present a status report on our ongoing spectrophotometric program to survey as many as possible of the 359 objects in the 751 entry Downes & Shara (1993) catalogue of CVs that are reported as having no published spectroscopy.

1. Introduction

The Downes & Shara (1993; hereafter DS93) catalogue and atlas of CVs lists 751 objects of which 359 have no published spectrum to confirm their CV classification. In 1993 we began to secure optical CCD spectra of as many as possible of them, with a two-fold aim: (i) to prune mis-entries from the excellent DS93 catalogue and (ii) to provide high quality, flux calibrated spectroscopic data for a large sample of faint CVs. The emission line and continuum fluxes will serve as the basis for statistical investigations of the properties of these intrinsically faint and/or distant CVs.

2. Results

We are collecting CCD spectra over the wavelength range 3200...9150 Å with the ESO 1.5 m (at 7 Å resolution), Asiago 1.8 m (18 Å resolution) and Loiano 1.5 m (10 Å resolution) telescopes. 106 targets have been successfully observed so far, while an additional 38 objects were too faint to be put on the spectrograph slit. Their spectra were published in Bragaglia et al. (1995), Munari, Zwitter & Mikuž (1995) and Zwitter & Munari (1994, 1995, 1996) Among the 106 surveyed systems, 59 show spectra confirming their CV nature (see Table 1), whilst 47 do not show a CV-like spectrum (Table 2). Mis-classified objects include: planetary nebulae, starburst galaxies,

A. Evans and J. H. Wood (eds.), Cataclysmic Variables and Related Objects, 93–94.

Mira variables, VV Cep and symbiotic stars, OB subdwarfs, white dwarfs and detached WD + M dwarf binaries.

The rate of mis-classification is particularly high for objects from the Palomar-Green survey. We have observed 22 candidates from the PG list: only 5 of them show CV-like spectra (23%).

TABLE 1. Objects showing a CV-like spectrum

AG Aps	NN Cen	UV Gem	GS Pav	HZ Pup	CU Vel
WX Ari	V373 Cen	GY Hya	HX Peg	UY Pup	1H 0204−023
AT Cnc	RR Cha	TT Ind	FY Per	VZ Pyx	1H 0459+248
WZ CMa	BP CrA	TU Ind	PY Per	V478 Sco	1H 0616−818
AQ CMi	EY Cyg	AD Men	TZ Per	UZ Ser	PG 0859+415
SV CMi	V476 Cyg	TU Men	AY Psc	RW Sex	PG 0911−066
V365 Car	V516 Cyg	HQ Mon	BV Pup	KK Tel	PG 0943+521
AM Cas	V747 Cyg	KQ Mon	BX Pup	TW Tri	PG 1114+187
V592 Cas	V751 Cyg	V972 Oph	DY Pup	BB Vel	PG 1524+622
MU Cen	HR Del	BD Pav	HS Pup	CN Vel	

TABLE 2. Objects not showing a CV-like spectrum

V1327 Aql	MV Gem	V3914 Sgr	PG 0248+056	PG 1445+583
KY Ara	SY Gem	V4019 Sgr	PG 0947+036	PG 1459−026
V422 Ara	AN Gru	LQ Sgr	PG 1104+022	PG 1520−050
CT Boo	CG Mus	OV Tau	PG 1116+349	PG 1712+493
CG CMa	V699 Oph	WW Tel	PG 1119+147	PG 2300+166
V411 Car	DP Pav	YY Tel	PG 1128+098	PG 2315+071
BC Cas	V1089 Sgr	EG Uma	PG 1136+581	PG 0240+066
BM Cha	V2038 Sgr	GD 1401	PG 1146+228	
CP Eri	V2493 Sgr	GD 1555	PG 1314+041	
KT Gem	V3909 Sgr	NSV 11561	PG 1403−111	

References

Bragaglia, A., Duerbeck, H.W., Munari, U., Zwitter, T., 1995, A&A, **297**, 759
Downes, R.A., Shara, M.M., 1993, PASP, **105**, 127
Munari, U., Zwitter, T., Mikuž, H., 1995 A&A, **296**, 310
Zwitter, T., Munari, U., 1994, A&AS, **107**, 503
Zwitter, T., Munari, U., 1995, A&AS, **114**, 575
Zwitter, T., Munari, U., 1996, A&AS, submitted

ACCRETION DISCS

ACCRETION DISC VISCOSITY

CHRISTOPHER A. TOUT
Institute of Astronomy, The Observatories, Madingley Road, Cambridge CB3 0HA, UK

Abstract. We review the various physical processes that could lead to viscosity in accretion discs. A local magnetic dynamo offers the most plausible mechanism and we discuss a simple model in some detail. The dynamo operates even in partially and very weakly ionized discs without much modification.

1. Introduction

Accretion discs occur where high angular momentum material is falling on to a central object. Material that cannot be accreted directly forms a disc rotating around the object. Accretion discs (of about $1\,R_\odot$ in radius) form in cataclysmic variables. Material overflowing from the low-mass star filling its Roche lobe cannot accrete directly on to the white dwarf, collides with itself and forms an accretion disc around the star. Accretion discs (of about $100\text{AU} \approx 20\,000\,R_\odot$) are inferred to be an integral part of the process of star formation. Once a denser, self-gravitating core has formed within a rotating cloud it cannot directly accrete further material. This material first collapses and then accretes on to a disc perpendicular to the rotation axis. Accretion discs (of about $1\,\text{kpc} \approx 4\,10^{10}\,R_\odot$) are probably the supply route to black holes at the centre of active galactic nuclei.

If material is to fall inwards through the accretion disc its angular momentum must be transferred outwards. If the material in the disc is orbiting in circular Keplerian orbits with angular velocity $\Omega \propto r^{-3/2}$, where r is the radius in the disc, such a transfer is energetically favourable because the lowest energy state for any rotating object of given angular momentum is one of complete corotation. The transfer of angular momentum outwards serves to slow down the inner, rapidly rotating parts of the disc while spinning up the outer, more slowly rotating parts. This can be achieved if a simple shear viscosity acts within the disc fluid. The stress between adja-

A. Evans and J. H. Wood (eds.), Cataclysmic Variables and Related Objects, 97–106.

cent annuli will then be

$$\sigma_{r\phi} = \eta_{\rm v}\frac{dV}{dr}, \tag{1}$$

where the suffices $r\phi$ indicate the force per unit area in the azimuthal direction owing to the velocity gradient dV/dr in the radial direction. The dynamic viscosity is $\eta_{\rm v} = \rho\nu$, where ρ is the density of the fluid and ν its kinematic viscosity. Gravitational forces will generally dominate over viscous forces ensuring that material in the disc follows near Keplerian orbits for a non-self-gravitating disc and tidal effects will ensure that these orbits are circular. If ν is known we can write a diffusion equation for the disc surface density (Lynden-Bell & Pringle 1974). The time-scale for viscous processes to influence the disc will be $\tau_\nu = R^2/\nu$, where R is the radius of the disc. For the disc to remain in a steady accreting state this must be much less than the accretion time-scale, which is about 10^9 yr for typical cataclysmic variables. Taking the radius of the disc to be about 10^{10} cm we find $\nu \gg 3 \times 10^3\,{\rm cm^2 s^{-1}}$.

2. Molecular viscosity

For normal material, molecular viscosity is several orders of magnitude too small. Even honey has a kinematic viscosity of about $5\,{\rm cm^2\,s^{-1}}$ and astrophysical discs are made of much more slippery stuff. On the other hand if a disc becomes degenerate then its electrons have very long mean free paths. In just the same way that degenerate matter is highly conducting it is also very viscous (Paczyński & Jaroszyński 1978).

3. Turbulent viscosity

Turbulent motions within the fluid can transport angular momentum. First consider a radial turbulent cell of length l, average velocity $c_{\rm t}$ and cross section σ. There will be a velocity difference δV between the ambient material at the two ends of the cell such that $\delta V = Sl$, where S is the shear in the medium. Now in a time $\Delta t = l/c_{\rm t}$ the cell transports momentum $\Delta p = \rho\sigma l.Sl$ from one end to the other. This provides an effective force

$$F = \frac{\Delta p}{\Delta t} = \rho\nu S\sigma \tag{2}$$

from which we deduce that $\nu \approx lc_{\rm t}$. We expect $l \le H$, the disc thickness, and $c_{\rm t} \le c_{\rm s}$, the sound speed, so that

$$\nu \approx \alpha H c_{\rm s}, \tag{3}$$

with $\alpha \leq 1$ (Shakura & Sunyaev 1973). A calculation of vertical structure within the disc gives $H \approx c_s/\Omega$ so that

$$\nu \approx \frac{\alpha c_s^2}{\Omega} \approx 10^{14}\alpha \text{ cm}^2\text{s}^{-1}. \tag{4}$$

However, this simple argument breaks down because we have exchanged higher angular momentum material from the outer end of the cell for lower from the inner end. This is neither what we want nor is it energetically favourable. Had we consulted Rayleigh's criterion for stability to axisymmetric perturbations we would have found our disc to be stable to such turbulence. Consider an incompressible uniform fluid rotating in cylindrical shells at $\Omega(r) = V(r)/r$. Interchange two cylinders of fluid each of mass m at r_1 and r_2 with $r_2 > r_1$. If the specific angular momentum $h = rV = r^2\Omega$ is conserved then the energy change on interchange is

$$\Delta E = \frac{1}{2}m\left\{\left(\frac{h_2}{r_1}\right)^2 + \left(\frac{h_1}{r_2}\right)^2 - \left(\frac{h_1}{r_1}\right)^2 - \left(\frac{h_2}{r_2}\right)^2\right\} \tag{5}$$

$$= \frac{1}{2}m\left(\frac{1}{r_1^2} - \frac{1}{r_2^2}\right)\left(h_2^2 - h_1^2\right). \tag{6}$$

If the interchange is energetically favourable and thence unstable $\Delta E < 0$ and hence $h_1^2 > h_2^2$ or specific angular momentum must decrease outwards. Alternatively if

$$\frac{d}{dr}(r^2\Omega)^2 > 0 \tag{7}$$

the disc is stable to axisymmetric perturbations. This is indeed the case with Keplerian discs.

On the other hand, should the disc be unstable to turbulent motions predominantly in an azimuthal direction, the situation is different. Consider a small region of fluid moving at an azimuthal velocity c_t faster than the ambient flow. This region has excess angular momentum and experiences a coriolis force in the outward radial direction. In this way higher angular momentum material can be transported outwards while lower angular momentum material falls inwards.

We can now ask when such turbulence might set in. For fluid of known viscosity we can construct the Reynolds number

$$\Re_e = \frac{LV}{\nu}, \tag{8}$$

where L is a typical length scale over which the fluid velocity, typically V, varies. In the disc $L \approx R$ and $V \approx R\Omega$. If we write $\nu \approx \alpha H^2\Omega$ then

$$\Re_e \approx \frac{1}{\alpha}\left(\frac{R}{H}\right)^2 \gg 1. \tag{9}$$

In the laboratory, turbulence sets in when $\Re_e \geq 10^3$. Lynden-Bell & Pringle (1974) argued that the fluid in the disc might be self-regulating in the sense that the viscosity is such that the fluid is just turbulent and

$$\alpha \approx 10^3 \left(\frac{H}{R}\right)^2 . \tag{10}$$

If the viscosity were smaller $\Re_e$ would be larger and turbulence would increase. If the viscosity were larger then turbulence would be suppressed. Although a reasonable numerical value is obtained there is no real physical justification for this approach. It has proved popular because it allows α to take much smaller values in thin discs with $H/R \ll 1$ than in thicker discs, a phenomenon that might explain dwarf novae outbursts.

4. Convection

Convective turnover would indeed give rise to turbulence that can cause viscosity. Whether or not a disc is convectively unstable depends on how and where the gravitational energy of the material flowing through the disc is liberated. Although parts of some cataclysmic variable discs may be convective, we can conclude from Livio & Shaviv (1977) that they are generally stable and convection is not the source of viscosity. However in protostellar discs Lin & Papaloizou (1980) have claimed that convection will be important simply because, as the disc contracts to its mid plane, the released gravitational energy sets up a temperature gradient sufficient to drive convection. Convective cells, although essentially vertical, must close on themselves to avoid accumulation of matter. Suppose a cell is able to move a distance l before being disrupted. If disruption takes place when the two extremes of the cell have been sheared around the disc, $l \approx v_c/\Omega$, where v_c is the convective velocity to be calculated using mixing length theory. The viscosity is then $\nu \approx v_c^2/\Omega$.

Kley, Papaloizou & Lin (1993) were able to create two-dimensional axisymmetric fluid simulations in which they found convective cells transporting angular momentum. However they had to introduce an additional artificial viscosity to get the model to work and then found that the cells themselves carry angular momentum inwards because their axisymmetry forces them to close in a radial direction. The situation might be improved if convective cells can close in an azimuthal direction but, since top and bottom must move in opposite directions, each end of the cell will experience oppositely directed coriolis forces and it is difficult to see how such a cell can close on itself at all.

5. Dynamical processes

Paczyński (1978) pointed out that, if it is left to cool, an accretion disc will contract until its own self-gravity becomes important. The density at which this occurs is given by

$$\rho_{\mathrm{disc}} > \frac{M_{\mathrm{WD}}}{r^3} < 1\,\mathrm{g\,cm^{-3}}, \tag{11}$$

where the second condition avoids degenerate support. A similar situation to that seen in galactic discs results, with the formation of denser clumps throughout the disc. Torques between these clumps will act so as to enforce corotation which directly transfers angular momentum outwards in Keplerian discs. Laughlin & Bodenheimer (1994) constructed numerical models of a cool disc which becomes unstable and gives rise to viscosity equivalent to $0.01 < \alpha < 0.03$. Cataclysmic variable discs, however, are generally hot and far from self-gravitating.

6. Magnetic fields

Magnetic fields are the most promising source of viscosity. Magnetohydrodynamics in astrophysics is based on Maxwell's equations with two simplifying assumptions. The first that

$$|\nabla \wedge \mathbf{B}| \gg \frac{1}{c^2}\left|\dot{\mathbf{E}}\right|, \tag{12}$$

that electromagnetic waves are unimportant, holds well in all quantifiable situations. This leads to the induction equation

$$\dot{\mathbf{B}} = \nabla \wedge (\mathbf{u} \wedge \mathbf{B}) + \eta \nabla^2 \mathbf{B}, \tag{13}$$

where $\eta = 1/4\pi\sigma$ is the magnetic diffusivity (σ being the electrical conductivity). The second assumption commonly made is that η is small in astrophysical plasmas so that, except where $|\nabla^2 \mathbf{B}|$ is large, field lines are linked to the fluid or fluid can flow freely only along the field lines.

A particular consequence of this is that if there are radial field lines in a disc the shear flow will tend to wrap them around the disc generating azimuthal field. In so doing work must be done on the field and mechanical energy is converted to magnetic energy. This conversion is the basis of a magnetic dynamo. At the same time the field lines will be bent. Curved field lines will attempt to straighten themselves and in so doing enforce corotation of the fluid and hence angular momentum transport outwards in the disc. Lynden-Bell (1969) suggested that magnetic fields are in this way responsible for viscosity in accretion discs. The magnetic torque can be calculated from the Lorentz force

$$\begin{aligned} \mathbf{F} &= \mathbf{j} \wedge \mathbf{B} \\ &= \frac{1}{4\pi}(\nabla \wedge \mathbf{B}) \wedge \mathbf{B} \\ &= \frac{1}{4\pi}[(\mathbf{B}.\nabla)\mathbf{B} - \nabla(\tfrac{1}{2}|\mathbf{B}|^2)]. \end{aligned} \tag{14}$$

The second term provides an additional contribution to isotropic pressure while the first is the magnetic curvature force. By evaluating the mean of the component of this force in the azimuthal direction,

$$\overline{F}_{\mathrm{curv},\phi} = \frac{1}{4\pi}\overline{[(\mathbf{B}.\nabla)\mathbf{B}]_\phi} \approx \frac{B_R B_\phi}{4\pi r}, \tag{15}$$

we obtain an effective viscosity [from equation (1)] of

$$\nu \approx \frac{B_R B_\phi}{4\pi\rho\Omega} \tag{16}$$

or

$$\alpha \approx \frac{B_R B_\phi}{4\pi\rho c_\mathrm{s}^2}. \tag{17}$$

Now in order to have viscosity all we need do is generate radial magnetic field in the disc. Radial turbulent motions can generate radial field from azimuthal but we have already shown that the disc is stable to such turbulence and isotropic turbulence might in itself generate the necessary viscosity. A major break-through was made when Balbus & Hawley (1991) recognised the importance of an instability first recorded by Velikhov (1959).

6.1. THE BALBUS–HAWLEY INSTABILITY

Discs with weak vertical fields are unstable if

$$\frac{d(\Omega)^2}{dr} < 0, \tag{18}$$

if $|\Omega|$ decreases outwards. The instability can be understood by considering a vertical field line. Suppose a perturbation moves material together with the field line in the azimuthal direction of the ambient flow. In the absence of the field this would be a neutral perturbation. In this case the field line is stretched and bent so that the curvature force slows down the displaced material relative to the surrounding fluid. Losing angular momentum it falls radially inwards where it is caught up in yet faster flowing material.

If the shear is stronger than the magnetic tension the material will be dragged yet further from its equilibrium increasing the magnetic tension as it goes. More angular momentum is carried outwards and the material falls in further. Alternatively, if the field is strong enough the tension will win over the shear straightening the field line and stabilizing against the inflow. Growth occurs only on wavelengths $\lambda > \lambda_{\rm crit} \propto B_z$ so that if $\lambda_{\rm crit} > 2H$, the total disc thickness, it will be stable.

We now have an instability that will generate viscosity if there is a weak vertical magnetic field in the disc. Various sources of vertical field can be envisaged. If the whole system is embedded in a region of magnetised space, field might be advected in with the material flowing through the disc. Alternatively field anchored on the central accreting object may thread through the disc and interact with it. This is the case in DQ Her systems or intermediate polars in which the central-object field disrupts the inner parts of the disc. However it is not clear that weaker fields can thread a differentially rotating disc at all. The disc may appear super-conducting to external field which can then be excluded entirely. A third, more promising mechanism is the regeneration of vertical field within the disc itself by some dynamo process.

6.2. PARKER INSTABILITY

Magnetic buoyancy can provide the source of vertical field. Consider a tube of magnetic flux **B** embedded in non-magnetic fluid of density $\rho_{\rm e}$ and pressure $p_{\rm e}$. If the tube is in pressure balance with its surroundings its own thermal pressure $p_{\rm i}$ will be supplemented by its magnetic pressure.

$$p_{\rm e} = p_{\rm i} + \frac{B^2}{8\pi} > p_{\rm i}. \tag{19}$$

Thus its internal density $\rho_{\rm i} < \rho_{\rm e}$ and the tube will float to the surface. Once a section of a flux tube begins rising material can flow down the tube towards the mid plane leaving the rising part of the tube less dense still, so that it rises yet faster. This is the Parker instability. Its fastest growing mode has a wavelength some eight times the disc scale height (Horiuchi et al. 1988) and a growth rate $\tau_{\rm P}^{-1}$ two to five times slower than the Alfvén crossing rate. In the disc it is the azimuthal and radial field in the plane of the disc that buoys up generating new vertical field.

6.3. RECONNECTION

Unless the vertical field B_z decays it will build up until the Balbus–Hawley instability is stabilized, when the generation of radial and azimuthal field ceases. The vertical field generated through the two instabilities will not be uniform but will change direction on a length scale $\lambda_{\rm rec} \approx H$, determined by the wavelengths of the fastest growing modes of the Balbus–Hawley and Parker instabilities and the action of shear in the disc which will tend to reduce $\lambda_{\rm rec}$ (see Tout & Pringle 1992 for details). Between regions of oppositely directed field reconnection can take place rapidly. Once reconnection begins at one point reconnected loops attempt to straighten pulling material away from the reconnection region. Excess pressure outside the region pushes more field in close enough to continue reconnecting. The reconnection time-scale is then $\tau_{\rm rec} \approx \lambda_{\rm rec}/0.1[V_{\rm A}]_z$, where the vertical Alfvén speed, $[V_{\rm A}]_z = B_z/\sqrt{4\pi\rho}$.

6.4. OPERATIONAL MODEL

Tout & Pringle (1992) put together these processes in a description of a magnetic dynamo that leads directly to disc viscosity without the need for any externally imposed turbulence. The equilibrium $\mathbf{B} = \mathbf{0}$ is unstable. Initially both B_R and B_ϕ grow on a time-scale Ω^{-1} but decay only on a time-scale $\max(\tau_{\rm P}, \tau_{\rm rec}) \gg \Omega^{-1}$; a second equilibrium is reached only when B_z is close to its maximum for instability when the ratios of the Alfvén speeds to the ambient sound speed for the three field components are

$$\frac{[V_{\rm A}]_z}{c_{\rm s}} \approx \frac{[V_{\rm A}]_\phi}{c_{\rm s}} \approx 0.8, \quad \frac{[V_{\rm A}]_R}{c_{\rm s}} \approx 0.1. \tag{20}$$

The equilibrium value of α is about 0.1. This equilibrium is also unstable and Fig. 1 shows how the fields oscillate around their equilibrium values. In each cycle B_R and B_ϕ build up until B_z is sufficient to shut off the Balbus–Hawley instability. While B_R and B_ϕ begin to decay B_z increases further. Eventually B_z decays faster than it is replenished until the Balbus–Hawley instability begins to operate again and B_R and B_ϕ rise once more. Fig. 2 illustrates the variation of α [from equation (17)] for the same system.

6.5. NUMERICAL SIMULATIONS

The above model simplifies all the processes involved to make the solution tractable. Numerically we can try to model more of the details but a number of specific problems beset such attempts. First, because of limited resolution, an artificial magnetic diffusivity is always present so that the magnetic Reynolds number $\Re_{\rm m} = RV/\eta$ is several orders of magnitude too large, seed magnetic fields must already be large in order to grow and reconnection cannot be modelled properly. Second, only a small region of the disc can be followed and boundary conditions limit the size of large scale growth and enforce certain field structures.

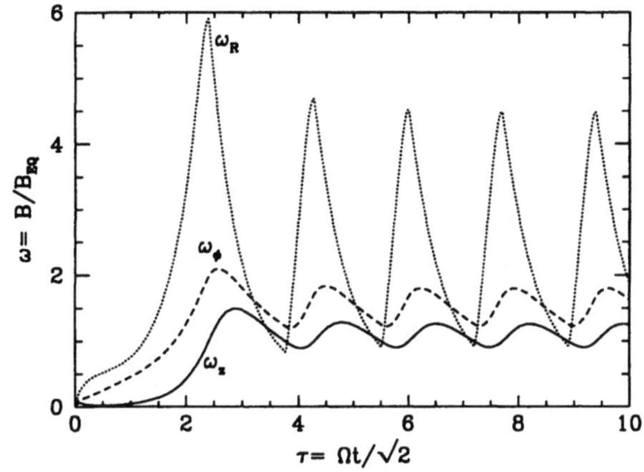

Figure 1. Variation of component field strengths relative to their equilibrium values.

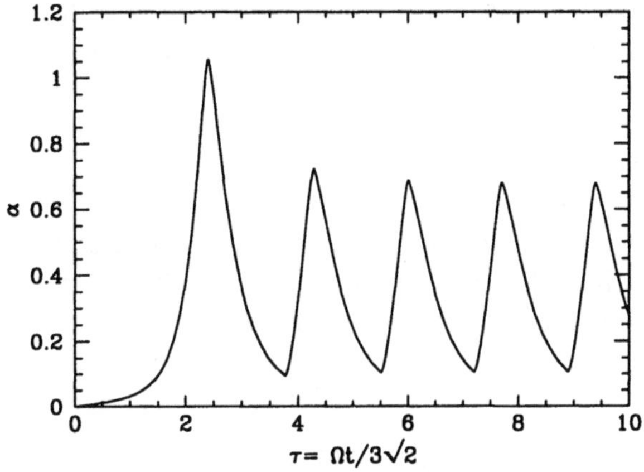

Figure 2. The variation of α with time.

Not withstanding these problems Hawley, Gammie & Balbus (1995) showed that a self excited dynamo can be maintained even with zero gravity. Brandenburg et al. (1995) have included gravity but their surface boundary condition still limits the growth of Parker's instability. They found $\alpha \approx 0.004$ from magnetic stress.

6.6. PARTIAL IONIZATION

We have described processes operating in fully ionized media. Regős (1996) has asked the question of what happens when some material is neutral. Neutral particles feel the magnetic fields only through collisions with ions and two-component magnetohydrodynamic equations are needed. Cataclysmic variable discs in quiescence are partially ionized with $0.1 \leq \rho_i/\rho_n \leq 0.9$. Regős has recalculated the rates for the various instabilities and finds that α does not differ much from the fully ionized case.

In protostellar discs the ionization fraction may be as little as $\rho_i/\rho_n \approx 10^{-10}$. In this case the velocities of neutral particles and ions differ widely and the induction equation becomes a diffusion equation. As a result reconnection is faster and α increases because B_R and B_ϕ can build up to larger values before the Balbus–Hawley instability switches off. Brandenburg et al. find their models in agreement with these conclusions.

7. Conclusions

Magnetic fields are probably the source of viscosity in cataclysmic variable discs and are a viable source in protostellar and AGN discs too. A radial component of magnetic field is needed and this can be generated by the Balbus–Hawley instability which overcomes the Rayleigh stability criterion. The Parker instability or ensuing turbulence can regenerate vertical field by dynamo activity. Theoretically acceptable values of $\alpha \approx 0.1$ are predicted. Limited numerical models give smaller but still reasonable values, $\alpha \approx 0.004$. The gravitational energy from the inward flowing material will not all be dissipated where the viscosity acts. It is released by reconnection of vertical field which may occur in the disc corona or it may escape altogether in a disc wind or jet. The inability of a disc dynamo to find a stable equilibrium may account for disc flickering.

References

Balbus, S. A., Hawley, J. F., 1991, Ap. J., **376**, 214
Brandenburg, A., Nordlund, Å., Stein, R. F., Torkelsson, U., 1995, Ap. J., **446**, 741
Hawley, J. F., Gammie, C. F., Balbus, S. A., 1995, Ap. J., **440**, 742
Horiuchi, T., Matsumoto, R., Hanawa, T., Shibata, K., 1988, PASJ, **40**, 147
Kley, W., Papaloizou, J. C. B., Lin, D. N. C., 1993, Ap. J., **416**, 679
Laughlin G., Bodenheimer, P., 1994, Ap. J., **436**, 335
Lin, D. N. C., Papaloizou, J. C. B., 1980, MNRAS, **191**, 37
Livio, M., Shaviv, G., 1977, A&A, **55**, 95
Lynden-Bell, D., 1969, Nature, **223**, 690
Lynden-Bell, D., Pringle, J. E., 1974, MNRAS, **168**, 603
Paczyński, B., 1978, Acta Astr., **28**, 91
Paczyński, B., Jaroszyński, M., 1978, Acta Astr., **28**, 111
Regős, E., 1996, MNRAS, submitted
Shakura, N. I., Sunyaev, R. A., 1973, A&A, **24**, 337
Tout, C. A., Pringle, J. E., 1992, MNRAS, **259**, 604
Velikhov, E. P., 1959. Sov. Phys. JETP, **36**, 1398

DYNAMO DRIVEN ACCRETION DISCS AND DWARF NOVA ERUPTIONS

P. J. ARMITAGE[1], M. LIVIO[2], J. E. PRINGLE[1,2]
1. Institute of Astronomy, Madingley Road, Cambridge, CB3 0HA, UK
2. Space Telescope Science Institute, 3700 San Martin Drive, Baltimore, MD 21218, USA

1. Introduction

A magnetic dynamo whose crucial ingredient is the Balbus-Hawley instability (Balbus & Hawley 1991) is a promising candidate for generating accretion disc viscosity. Both semi-quantitative analyses (Tout & Pringle 1992) and simulations (e.g. Brandenberg et al. 1995) suggest that such dynamos can operate in a disc environment. Here we aim to apply simple dynamo ideas to models of dwarf novae, and in particular investigate how dynamo driven discs can produce different values for α in quiescence and in outburst.

2. Generating outbursts

The Balbus-Hawley instability operates with an arbitrarily weak vertical field, so a dynamo requires no pre-existing strong fields or disc turbulence. However a *strong* field,

$$B_Z^2 > \frac{24}{\pi}\rho c_S^2, \tag{1}$$

where B_Z is the vertical field, ρ the density, and c_S the sound speed – shuts off the instability. Therefore rapid cooling at the end of a dwarf nova outburst may act to depress the thermal energy relative to the magnetic energy, shut off the instability, and quench dynamo action. The turbulence generated by the dynamo in the disc could then die away, leaving a quiescent disc with a relatively strong vertical field and very low viscosity. The dynamo could then be restarted (and a new outburst begin), when sufficient mass had been added at the outer edge to violate inequality (1).

A. Evans and J. H. Wood (eds.), Cataclysmic Variables and Related Objects, 107–108.

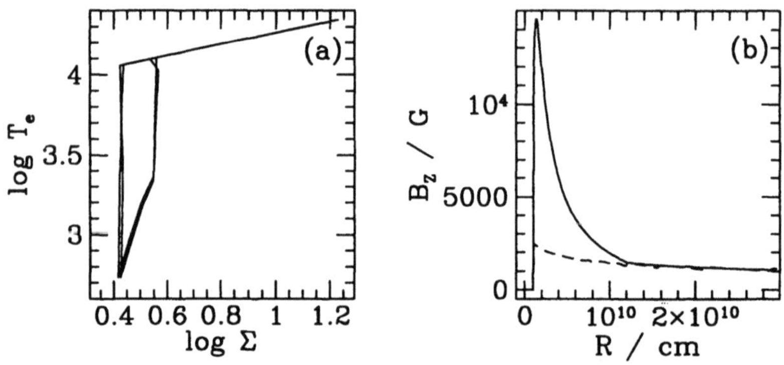

Figure 1. (a) Path traced out in the effective temperature-surface density plane by an annulus at the inner edge of the disc. (b) Vertical magnetic field strengths during outburst (solid line) and quiescence (dashed line).

3. Results

We have modified a time-dependent disc code similar to that described by Cannizzo (1993) to model outburst cycles generated by the above mechanism. The details of the code are described elsewhere (Armitage, Livio & Pringle 1996). We can produce well-separated outbursts provided that the working dynamo generates an α of $\sim 0.2 \ldots 0.3$. Our model differs from the normal thermal instability in not requiring a double valued 'S-curve' in the surface density-temperature plane, as the transition to the hot state is controlled by a magnetic trigger condition [inequality (1)]. Fig. 1a shows the path followed by an annulus in our model, with a prolonged cooling phase at constant surface density in quiescence. Observational consequences include:

- An active disc corona, as dissipation is concentrated in low density regions of the disc (leading also to a different vertical thermal structure).
- Magnetic fields (on scales of order the disc scale height) of a few kG in quiescence and reaching up to $\sim$ 10 kG in outburst [see Fig. 1b].
- Flickering due to variability in dynamo 'cells' across the disc.
- A cold quiescent disc, with a surface density profile strongly weighted towards the outer edge as the next outburst approaches. The temperature profile in the quiescent disc is flat.

References

Armitage, P.J., Livio, M., Pringle J.E., 1996, Ap. J., in press
Balbus, S.A., Hawley, J.F., 1991, Ap. J., **376**, 214
Brandenburg, A., Nordlund A., Stein, R.F., Torkelsson U., 1995, Ap. J., **446**, 741
Cannizzo, J.K., 1993, Ap. J., **419**, 318
Tout, C.A., Pringle J.E., 1992, MNRAS, **259**, 604

ON THE TURBULENT VISCOSITY PRESCRIPTION IN ACCRETION DISCS

PATRICK GODON
NASA/Jet Propulsion Laboratory, 4800 Oak Grove Dr., MS 238-332, Pasadena, CA 91109, USA

Abstract. The validity of the standard α viscosity prescription in turbulent accretion discs is analyzed. It is found that the α viscosity prescription is no longer valid when the disc is not Keplerian. The analysis is based on simple scaling arguments for fully developed turbulences in accretion discs. An analytical treatment is carried out for geometrically thin, non-Keplerian discs in vertical hydrostatic equilibrium, and is compared with the standard values of the viscosity in the boundary layer region and with the viscosity prescription derived from a general causal equation for the evolution of the stress tensor.

1. Introduction

Instability analyses and numerical simulations have associated the origin of instabilities in accretion discs to the α viscosity prescription. Three recent works (Narayan, Loeb & Kumar 1994; Papaloizou & Szuszkiewicz 1994; Kato & Inagaki 1994) have dealt with the problem of the viscosity in rotating flows in detail. Their results show that the viscosity law for a non-Keplerian disc cannot be represented by the standard α viscosity prescription. In this work, we develop a turbulent viscosity prescription for non-Keplerian discs, using simple scaling arguments. We find very good agreement with the results of the three recent works mentioned above.

2. The standard α viscosity prescription

In the theory of fully developed turbulences (Landau & Lifshitz 1987), the viscosity is given by $\nu = a\Delta u\lambda$, where a is a constant of order unity, Δu is the change of the flow mean velocity u over a distance λ – the scale length of the largest eddies, responsible for the transport of angular momentum.

A. Evans and J. H. Wood (eds.), Cataclysmic Variables and Related Objects, 109–110.

For a geometrically thin, Keplerian disc in vertical hydrostatic equilibrium one obtains: $\lambda = H << r$, $u = v_{\rm K}$, $\Delta u = v_{\rm K}H/2r$, $H/r = v_{\rm s}/v_{\rm K}$ and

$$\nu = a v_{\rm s} H/2.$$

3. Turbulent viscosity prescription for non-Keplerian discs

In this case one expects $\lambda = \bar{H} = P/|\nabla P|$ (Papaloizou & Stanley 1986) and $\Delta u(\lambda) = r|\Omega'|\lambda$. Assuming $\Omega' \neq 0$ leads to

$$\nu = a v_{\rm s} H \frac{|\Omega'|}{|\Omega'_{\rm K}|}(\bar{H}/H)^2 = a v_{\rm s} H \mathcal{M}_{\rm t}^2 \frac{|\Omega'|}{|\Omega'_{\rm K}|},$$

where constant terms of order unity have been included in a and $\mathcal{M}_{\rm t}$ is the turbulent Mach number. For convenience we have written the viscosity in the form $\nu = a v_{\rm s} H F$. The stress tensor is then written

$$\tau_{r\phi} = -aP\frac{|\Omega'|}{|\Omega'_{\rm K}|}F,$$

where $F = \nu/(a v_{\rm s} H)$. The stress tensor vanishes whenever $\Omega' = 0$. This prescription of the viscosity is completely consistent with the treatment of Shakura & Sunyaev (1988) for the boundary layer region, and is in good agreement with the causality formalism, in which one expects the viscosity to be reduced when the radial infall velocity increases such that $v_{\rm r} \approx \max(v_{\rm s}, v_{\rm t})$, where $v_{\rm t}$ is the turbulent velocity [see Godon (1995) for more details]. Recent observations (Horne 1995) show supersonic turbulent velocities with $\mathcal{M}_{\rm t} \approx 10$ in accretion discs. In this case the supersonic turbulences might contribute only partially to the cascade effect, (since they also dissipate in part through shocks), and the present treatment would need improvement to take this effect into consideration.

References

Godon, P., 1995, MNRAS, **177**, 157
Horne, K., 1995, poster presentation, IAU Coll. 158
Kato, S., Inagaki, S., 1994, PASJ, **46**, 289
Landau, L., Lifshitz, E., 1987, Fluid Mechanics, Pergamon Press (2nd Edition), Oxford
Narayan, R., Loeb, A., Kumar, P., 1994, Ap. J., **431**, 359
Papaloizou, J.C.B., Stanley, G.Q.G., 1986, MNRAS, **220**, 253
Papaloizou, J.C.B., Szuszkiewicz, E., 1994 MNRAS, **268**, 29
Shakura, N.I., Sunyaev, R.A., 1988, Adv. Sp. Res., **8**, 135

THE RISE AND FALL OF α-MODEL VISCOSITY

F. V. HESSMAN, C. OBACH
Universitäts-Sternwarte Göttingen, Germany

Abstract. For nearly two decades, our only useful model for the viscosity in accretion disks has been the so-called 'α-model'. However, it has become clear that the simplest models – in which α is constant – are inadequate to explain the range of behaviours seen in real disks. We show that the properties of steady-state, vertically-averaged models can be determined without any assumptions other than that the disks obey the classical Navier-Stokes equation. These solutions have derived values of α which vary with radius by many orders of magnitude even in small CV disks.

1. A short history of α-viscosity

The α-ansatz was first formulated by Shakura & Sunyaev (1973), both as a plausible physical model (the stresses are proportional to pressure in, for example, turbulent viscosity) as well as an attempt to close the system of algebraic equations determining the structure of vertically-averaged disks. This remarkably simple model was immediately applied to a host of situations and the properties of so-called α-disks were studied in great detail. α-disks ceased to be a mere theoretical construct when Hoshi (1979) realized that α-disks could suffer a limit-cycle instability. If the disks could cycle between periods of high and low accretion rates, one might explain the dwarf nova eruptions. Detailed calculations needed for the study of the dynamical evolution of such disks were then made (e.g. explicit vertical structure calculations and the inclusion of convection). The many and varied additions and improvements to the α-disk models are described by Cannizzo (1993). As the other difficult physical and numerical details behind the limit-cycle models were worked out, detailed comparisons between observed dwarf novae light curves and theoretical models began to place constraints on the value of α. It became clear that there could not be a *universal* and *single* value of α, since α-disk models needed roughly a factor of

A. Evans and J. H. Wood (eds.), Cataclysmic Variables and Related Objects, 111–114.

10 difference between the values of α in quiescent (i.e. cool) and outburst (i.e. hot) states. As the same numerical technology began to be applied to other systems, the signs that α must be a rather difficult function of something important like radius or temperature became even more obvious: one needs a value of α which is many orders of magnitude smaller in FU Ori variables (e.g. Meyer 1990); and in symbiotic stars, the value of α is apparently affected by the size of the outer disk (Duschl 1986).

2. Positive things to say about α-viscosity

Despite the blank transparency I showed at the conference on this topic (which did not elicit the desired comic relief but only an "Aw come now!" from Joe Smak), I must admit that the α prescription has taken us a long way. Although we have learned practically nothing about viscosity, we have learned a tremendous amount about accretion physics in general and cataclysmic variables in particular. All of the theoretical studies, the observational campaigns, and the detailed comparisons between theory and reality would not have been possible had we not had the vague feeling that we might have a good physical idea of what is really going on.

3. Negative things to say about α-viscosity

Though one is tempted to sing paeons, we should admit to ourselves that our reliance on α has been dangerous. Imagine having to understand stars using only the assumption that the (unknown) energy production rate in stars is a simple function of pressure: we might be able to construct a qualitative model for stellar evolution and would now be calculating 3-D numerical models of rapidly rotating stars with magnetic fields but we would have no idea about how most stars really work. The values of α in CV's disks have been 'constrained', but no real increase in physical insight has occurred. Certainly, the generally accepted values of 0.01–0.1 are too high to be explained by physical processes like spiral shocks or non-local waves. Another danger of taking α-models too seriously is the natural tendency to ignore other possibilities because of a 'fixed mind-set'. For example, observations of quiescent disks suggest that they are optically thin near the central star, whereas the presence of a hot spot suggests that there is lots of mass in the outer disk. However, constant values of α produce surface densities which decrease rather than increase with radius. Thus, clear differences between empirical evidence and theoretical predictions exist: should we add new features or change the paradigm?

4. Further constraints on viscosity?

In an effort to see if we cannot better constrain the viscous mechanisms *without the help of an α-ansatz*, let us attempt to construct a 'detailed balance' argument very similar in spirit to that used with such success in atomic physics. Assume that we have a geometrically thin, axisymmetric and steady disk in which an unknown but *local* viscous process, dependent only upon the shear in a (nearly) Keplerian velocity field, is at work: assuming $\frac{\partial}{\partial t} = \frac{\partial}{\partial \phi} = v_Z = 0$, $v_R \ll c_S$, and $\sigma_{\rm visc} = \sigma_{R\phi}$, the radial component of the Navier-Stokes equation reduces to

$$\frac{1}{\rho}\frac{\partial P}{\partial R} \approx \frac{v_\phi^2}{R} - \frac{GM}{R^2} \tag{1}$$

(the missing advection term is negligible as long as the radial velocities are subsonic). This is just the equation for what are normally called 'slim disks', in which the radial pressure gradients are taken into account without actually calculating the full 2-D disk structure. The time-scale for radial pressure equilibrium is very short compared with the viscous time-scale

$$\frac{\tau_{\rm visc}}{\tau_{\rm R}} \approx \frac{2\pi R^2 \Sigma}{\dot{\rm M}}\frac{c_S}{R} \approx 600\ r_{10}\ \dot{m}_{17}^{-1}\ \Sigma_3\ T_4^{-1/2} \tag{2}$$

where, e.g. $r_{10} \equiv R/10^{10}\,{\rm cm}$. Reasonably steady accretion disks will have had *lots* of time to come into radial pressure equilibrium.

The implication of equation (1) is that the large body of all Keplerian disks must have constant pressures. Traditional α-disks, however, do not, unless one is willing to let α be a strong function of radius. This new constraint on the structure of steady accretion disks tells us ***absolutely nothing*** about viscosity. Indeed, *this is exactly the point!* If there is a viscous process conforming to the few requirements we have made at work in real disks, they will produce disks with this type of structure.

5. Analytic models

Since the deviations from Keplerian motion are small and are highly localized in the dynamical boundary layer, and since the deviations from constant pressure are felt largely within the thermal boundary layer, one can construct simple analytical models for the large body of vertically-averaged accretion disks simply by assuming that the disk has a constant pressure. For the high accretion rates typical of dwarf novae in eruption, $\kappa \propto \rho T^{-5/2}$, one can solve for the properties of the disk:

$$\begin{array}{lcllllr} T & \approx & 5.7\,10^5\,{\rm K} & m_0^{+0.06} & r_9^{-0.19} & \dot{m}_{17}^{+0.13} & p_9^{+0.25} \\ \Sigma & \approx & 620\,{\rm g\,cm^{-2}} & m_0^{-0.53} & r_9^{+3.19} & \dot{m}_{17}^{-0.06} & p_9^{+0.88} \\ \alpha & \approx & 7.9\,10^{-2} & m_0^{+0.97} & r_9^{-2.91} & \dot{m}_{17}^{+0.94} & p_9^{-1.13} \end{array} \tag{3}$$

For comparison: the standard model in which α is implicity assumed to be constant yields:

$$\begin{aligned} T &\approx 1.5\,10^5\,K \quad m_0^{0.25} \quad r_9^{-0.75} \quad \dot{m}_{17}^{0.30} \quad \alpha^{-0.20} \\ \Sigma &\approx 150\,\mathrm{g\,cm^{-2}} \quad m_0^{0.25} \quad r_9^{-0.75} \quad \dot{m}_{17}^{0.70} \quad \alpha^{-0.80} \end{aligned} \tag{4}$$

Note the drastic difference in the properties of the disks, particularly in the dependence of the surface density Σ: this difference is reconciled by the steep dependence of the viscosity with radius in the constant pressure models.

6. Realistic models

We have constructed more self-consistent models for non-α-disks, including radial energy transport and full radial pressure equilibrium with a boundary layer. The thermal width of the boundary layer is always much larger than its dynamical width. Its structure depends upon the viscosity parameter, the stellar rotation rate, and the white dwarf boundary pressure. Away from the boundary layer, the solutions are in good agreement with the constant pressure analytic solutions. We have also derived analytic disk solutions for optically thin, constant pressure disks and fully wavelength-dependent solutions using realisitic opacities and cooling. These solutions have the nice property that the surface density *increases* rather than *decreases* with radius – just what one needs to explain the structure of quiescent dwarf novae disks.

7. Conclusions

It is clear that *simple*, *local*, and *steady* viscous processes in disks require that α be a strongly decreasing function of radius – at least! Since standard α-models are inconsistent with this requirement, non-physical α-models have ceased to be useful. This means that much more work must be expended in the production of truly physical models for the viscosity [see the contribution by Tout (1996)].

References

Cannizzo, J.K., 1993, in "Accretion Disks in Compact Stellar Systems", ed. J.C. Wheeler (World Scientific), p6
Duschl, W.J., 1986, A&A, **163**, 56
Glatzel, W., 1992, MNRAS, **257**, 572
Hoshi, R., 1979, Progr. Theor. Phys., **61**, 1307
Meyer, F., 1990, Rev. Mod. Astron., **3**, 1
Shakura, N.I., Sunyaev, R.A., 1973, A&A, **24**, 337
Tout, C. A., 1996, these proceedings, p97

SPH SIMULATIONS OF TIDALLY UNSTABLE ACCRETION DISKS

JAMES MURRAY
Canadian Institute for Theoretical Astrophysics
University of Toronto
Ontario M5S 1A7, Canada
jmurray@cita.utoronto.ca

We use two dimensional Smoothed Particle Hydrodynamics simulations to investigate the the Tidal Resonance Instability (TRI) model for superhumps in SU UMa stars.

In the TRI model, superhumps are the signature of an eccentric disk being periodically stressed by the binary system's rotating tidal field. A slow prograde motion of the disk in the inertial frame is then responsible for the superhump period slightly exceeding the binary orbital period. The disk eccentricity is caused by a 3:1 resonance between orbits in the disk and the orbit of the binary itself. The development of the TRI model is due mainly to Whitehurst (1988), Hirose & Osaki (1990) and Lubow (1991a). Lubow in particular carried out a nonlinear mode analysis of a fluid disk's response to a weak tidal field and identified the feedback mechanism responsible for disk eccentricity growth.

The tidal field due to the secondary star can be decomposed into a set of functions $\phi_m(r)\cos[m(\theta-\Omega_b t)]$, where m (an integer) is the mode number, and Ω_b is the angular speed of the secondary about the primary. The mth component of the tidal field generates a sinusoidal response with argument $m(\theta-\Omega_b t)$. At the 3:1 eccentric inner Lindblad resonance the disk's $m=3$ tidal response couples with any existing eccentricity in the disk to produce a travelling wave (argument $2\theta-3\Omega_b t$). In turn, the interaction of the travelling wave with the tidal response reinforces the disk eccentricity. Lubow (1991a) calculated the initial exponential rate of growth of the eccentricity.

SPH simulations performed to test the analysis (Lubow 1991b, 1992) showed that the disk's tidal response to the $m=2$ component of the binary potential interfered with the eccentricity growth mechanism by clearing material out of the Lindblad resonance. Lubow observed a much reduced eccentricity growth rate in simulations that included the $m=2$ term.

A. Evans and J. H. Wood (eds.), Cataclysmic Variables and Related Objects, 115–116.

In this poster, the results of two SPH accretion disk simulations are described. System parameters ($q = \frac{3}{17}$, $\dot{M} = 10^{-9}\,M_\odot\,yr^{-1}$, $T = 2\pi\,\Omega_b^{-1} = 0.063121$ d) were chosen to be appropriate for OY Car in (super)outburst. For an introduction to SPH the reader is referred to Monaghan (1992). For details of these simulations and tests of the code see Murray (1996).

In the first calculation, a disk was built up via a steady mass transfer stream originating at the L_1 point. In the second simulation, mass was added to the disk in Keplerian orbit at the disk circularisation radius r_c. Both simulations were isothermal with sound speed $c = 0.02\,d\,\Omega_b$, where d is the inter-star separation. The kinematic viscosity $\nu = 2.5\,10^{-4}\,d^2\,\Omega_b$. The calculations were followed for over 160 binary periods. In that time the disks reached an eccentric, precessing equilibrium state. Mode strengths in the disk were determined using a Fourier analysis similar to that in Lubow (1991b).

For this particular system, theory predicts an eccentricity growth rate $\lambda/\Omega_b = 0.10 \pm 0.01$. In both simulations, the growth of the eccentricity was well fit by an exponential curve. With the L_1 stream the growth rate was $\lambda_1/\Omega_b = 0.046 \pm 0.002$. In the second simulation $\lambda_2/\Omega_b = 0.058 \pm 0.002$. These values are approximately half the predicted value. Adding mass to the disk from L_1 further reduced the rate. However our simulations did not show the much diminished eccentricity growth described in Lubow (1992). In fact we observed that as our disks became more eccentric, the tidal response to the $m = 2$ component of the binary potential became weaker.

The eccentricity grew sufficiently rapidly to agree with superhump observations. In particular, luminosity variations associated with the disk precession developed in both calculations within a day or so of the disk encountering the Lindblad resonance. Superhumps appear approximately a day into superoutburst.

References

Hirose, M., Osaki, Y., 1990, PASJ, **42**, 135
Lubow, S.H., 1991a, Ap. J., **381**, 259
Lubow, S.H., 1991b, Ap. J., **381**, 268
Lubow, S.H., 1992, Ap. J., **401**, 317
Monaghan, J.J., 1992, ARAA, **30**, 543
Murray, J.R., 1996, MNRAS, in press
Whitehurst, R., 1988, MNRAS, **232**, 35

INNER EDGE DRAG BY AN ASYNCHRONOUS PRIMARY AND ACCRETION DISC STRUCTURE IN CLOSE BINARIES

G. LANZAFAME[1], G. BELVEDERE[2], D. MOLTENI[3]
1. Osservatorio Astrofisico di Catania, Italy
2. Istituto di Astronomia, Università di Catania, Viale A. Doria 6, I-95125 Catania, Italy
3. Istituto di Fisica, Università di Palermo, Via Archirafi 36, I-90139 Palermo, Italy

1. Model and results

In this work a 3–D 'Smoothed Particle Hydrodynamics' ([1]; [4]; [5]) accretion disc is simulated where particles at its inner edge are dragged by a fast spinning compact central star, as in the case of the intermediate polars. The angular velocity of the central star is twice the orbital angular velocity ω_o. This drag can be attributed mainly to viscous interaction in the dense compact star atmosphere, although magnetic coupling may also play a role.

Our simulations have been performed for a system with a primary (a white dwarf) mass $M_1 = 1.3\,\mathrm{M}_\odot$, a secondary mass $M_2 = 2.2\,\mathrm{M}_\odot$, a separation between the components (centre to centre) $d_{12} = 6.33\,10^{11}\,\mathrm{cm}$ and an orbital period $= 1.7\,\mathrm{d}$, the integration time being as long as $\simeq 75\,\mathrm{d}$. As in [2] and [3], we have considered quasi-polytropic structures with $\gamma = 1.01$. We adopt the term quasi-polytropic since we have in fact solved also an energy equation.

The results of our simulations indicate that the drag due to a fast spinning compact primary has a fundamental influence on accretion disc formation, structure, dynamics and energetics, in a way largely different from the co-rotation case. The disc mass becomes stationary after about 25 orbital periods.

In Fig. 1 four XY plots (orbital plane) of the accretion disc are shown for four different times: $t = 268, 270, 273$ and 275, where the time is measured in units of $P_{\mathrm{orb}}/2\pi$. The corresponding numbers of particles are: $N = 10\,137, 10\,182, 10\,169, 10\,138$. At all these times the disc seems to be surrounded by an expanding cloud of hot particles, released by its outer

A. Evans and J. H. Wood (eds.), Cataclysmic Variables and Related Objects, 117–118.

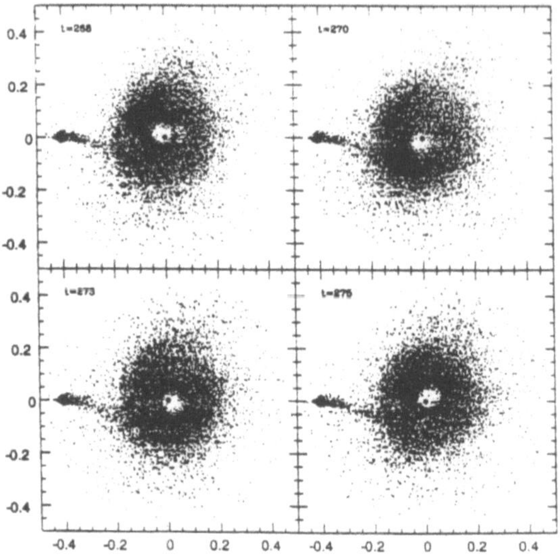

Figure 1. XY plots of the accretion disc at $t = 268$, 270, 273 and 275 (see text).

edge and interpretable as an anisotropic hot wind. The most evident characteristic of the disc dynamical evolution is an inner edge asymmetric azimuthal oscillation with a frequency equal to ω_o, as the four XY plots of Fig. 1 clearly show. Therefore, the centre of the inner edge of the accretion disc and the centre of the compact star, which is also the centre of the rotating frame, do not coincide. This oscillation produces a particle compression wave, localized in inner edge regions closer to the primary, which causes a local thickening of the disc itself in regions more distant from the primary. In other words, during the disc rotation, particles at the inner edge, being at a certain time closer to the primary, are scattered toward the opposite side of the edge by a 'sling' effect due to the rotational drag. Owing to collisions among particles, this results in a local bump in the disc thickness, with an azimuthal shift of about 180° and at an increased distance from the primary.

The radial temperature distribution shows a disc heating up and a flat profile as a consequence of the additional energy source provided by the rotational drag, which is converted into heat owing to the high collisional rate among particles.

References

1. Gingold, R.A., Monaghan, J.J., 1983, MNRAS, **204**, 715
2. Lanzafame, G., Belvedere, G., Molteni, D., 1992, MNRAS, **258**, 152
3. Molteni, D., Belvedere, G., Lanzafame, G., 1991, MNRAS, **249**, 748
4. Monaghan, J.J., 1985, Comp. Phys. Reports, **3**, 71
5. Monaghan, J.J., 1992, ARAA, **30**, 543

SYNTHETIC SPECTRA FOR ACCRETION DISCS: THE UV ABSORPTION SPECTRUM

R. A. WADE
Pennsylvania State University,
Department of Astronomy & Astrophysics,
525 Davey Lab, University Park PA, 16802, USA

1. Spectrum synthesis on a large grid of disc models

With my collaborators (Ivan Hubeny, NASA/Goddard Space Flight Center; Marcos Diaz, Catherine Grant, and Darren Williams, all of Penn State), I am computing the structure of discs and their corresponding UV spectra for a large grid of choices of the white dwarf mass $M_{\rm wd}$ and the mass transfer rate $\dot{M}$. In our approach, we divide the problem into a number of parts, each corresponding to a particular FORTRAN-implemented computer program. Program **TLUSDISK** computes the vertical structure for individual rings in the disc (T, ρ, $n_{\rm e}$, τ *versus* z). The rings are treated independently, as plane-parallel entities without irradiation, obeying the standard expression for $T_{\rm eff}$ as a function of $M_{\rm wd}$, $R_{\rm wd}$, and $\dot{M}$. With this vertical structure, program **SYNSPEC** computes the detailed local (rest frame) spectrum for each ring. Both *specific intensities* and angle-averaged *fluxes* can be computed. Program **DISKSYN** is used to combine the spectra of individual rings to obtain the integrated or whole-disc spectrum, taking into account Doppler shifts (dependent on the chosen disc inclination) and limb darkening. Finally, program **ROTINS** is used to smooth the spectrum to an appropriate instrumental resolution. Intermediate results, e.g., from **TLUSDISK** or **SYNSPEC**, are stored for repeated use.

Specific modeling choices have been made. These include the use of the Reynolds number to specify the disc viscosity, a vertical dissipation law parameterized by ζ as in equation (6) of Long et al. (1994), the assumption of LTE, and the neglect of irradiation. Winds and wind albedos have also been neglected. Steady-state disc structure has been assumed, and boundary layer emission has been neglected. Simple opacities have been used in computing the vertical structure, including H and He line and continuum opacities and scattering by free particles. Full line lists of many elements

A. Evans and J. H. Wood (eds.), Cataclysmic Variables and Related Objects, 119–122.

have been used in computing the spectra from the models. Further details of the modeling can be found in Long et al. (1994) and references therein.

For the models discussed here, a Reynolds number $\Re_e = 5000$ has been chosen. The α viscosity parameter varies inversely with $\Re_e$. The ζ parameter has been chosen to be 2/3. Reasonable changes in these values are explored below.

The computer codes are flexible enough to handle certain non-LTE computations, to incorporate prescribed irradiation or wind albedos, and additional opacity sources can be included. The steady-state assumption can be dropped when appropriate. A long-term goal is to explore the effect of the choices on the computed spectra. Nevertheless, the present grid of models and spectra can usefully be taken as benchmarks and as initial guides to the interpretation of observations.

We have computed models over most of a rectangular grid in (M_{wd}, $\dot{M}$) space, with $\log \dot{M}$ (in $M_\odot\,yr^{-1}$) varying in steps of 0.50 from −10.5 to −8.50 and M_{wd} taking values of 0.350, 0.550, 0.800, 1.030, and 1.210 $M_\odot$. The corresponding values of R_{wd} are 11.424, 9.049, 6.985, 5.177, and 3.783 ($\times 10^8$ cm). Except for the largest value of M_{wd}, this makes a grid whose diagonals correspond to the same run of T_{eff} with dimensionless radius $x = R/R_{wd}$. Ring radii increase roughly logarithmically from $x = 1.05$ until $T_{eff}(x)$ falls below roughly 10 000 K. The maximum disc temperature occurs at $x = 1.36$ and varies from 16 500 K to 69 500 K in the present models. Fluxes and specific intensities (at several emergent angles) have been computed on the wavelength interval 1150...1900 Å for the rings and full discs of these models. Over this interval, the restriction to $T_{eff} > 10\,000$ K is not harmful.

2. Results

Only a few points can be highlighted here. A fuller discussion will be published in the serial literature.

2.1. OPTICAL DEPTH

In this class of models (with $T_{eff} > 10\,000$ K) the Rosseland mean optical depth τ, measured vertically from the disc midplane, is least at the inner edge of the disc. At $x = 1.05$, it is larger for low M_{wd} and for high $\dot{M}$. Values of $\tau < 4$ occur at $x = 1.05$ in some models with $\dot{M} \leq 10^{-10}\,M_\odot\,yr^{-1}$. For a broad range of temperatures and densities, the continuous absorption coefficient declines substantially with decreasing wavelength toward the H I Lyman edge, suggesting that transparency effects will likely show up first at the inner edge and at far UV wavelengths.

2.2. HEIGHT OF DISC PHOTOSPHERE

The height of the photosphere $z(\tau = 1)$ measured from the midplane will be governed by the temperature scale height at the midplane, which in turn varies with $T_{\rm eff}$ and the midplane optical depth. Since τ depends on the surface density, one expects that increasing the Reynolds number will increase the scale height at fixed R for a given $(M_{\rm wd}, \dot{M})$ model. This is verified in the detailed computations, with $z(\tau = 1) \propto \Re_{\rm e}^{1/4}$ for intermediate models. The standard result that $z(\tau = 1)$ increases more rapidly than R (e.g. Pringle 1981) is recovered also. For $M_{\rm wd} = 1.15\,{\rm M}_\odot$ and $\dot{M} = 4.5\,10^{-9}\,{\rm M}_\odot\,{\rm yr}^{-1}$(not on the regular grid), the outer disc thickness corresponds to an opening angle of about 3° for $\Re_{\rm e} = 5000$. The vertical component of the gravitational acceleration falls off somewhat more slowly than $g \propto R^{-2}$.

For disc rings that are quite optically thick, the choice of the viscosity distribution parameter, ζ, has very little effect on $z(\tau = 1)$ for $0.05 < \zeta < 1$. Varying ζ from 0.05 to 1 varies the midplane temperature by less than 10% in optically thick rings.

2.3. LIMB DARKENING AND LINE BLENDING

Figure 1 of Wade, Hubeny & Polidan (1994) illustrates several features of the UV spectrum for a particular disc with $M_{\rm wd} = 0.8\ {\rm M}_\odot$ and $\dot{M} = 10^{-10}\ {\rm M}_\odot\ {\rm yr}^{-1}$. These include the complex blending of many absorption lines and the very strong limb darkening in the UV. The line blends change dramatically with disc inclination in details of strength and pattern because of the Doppler effect. This will be of diagnostic value for UV spectra of high resolution and S/N ratio.

Limb darkening (LD) will occur in any photosphere with a non-inverted temperature gradient. LD is qualitatively similar in (optically thick) discs and stars, but is more important for discs than for stars because the entire disc is viewed at (nearly) the same emergent angle, so the effect occurs *without dilution.* LD is stronger in the UV (describable with linear LD coefficients close to unity) as a consequence of the wavelength dependence of the source function. It is also stronger for cooler photospheres as a consequence of the temperature dependence of the source function at fixed wavelength. This directly affects the interpretation of 'brightness temperature' maps and of UV–R 'colour–magnitude' diagrams that have been made for eclipsing cataclysmic variables. The locus of disc surface elements in a colour–magnitude plane may be distorted because of the λ–dependence of LD. The shape of the $T_{\rm br} - \log R$ plot (made at a specific wavelength) may be distorted because of the T–dependence of LD. Statements regarding the distance of a CV or whether it is in steady state need to take LD into account. The contribution by Diaz, Wade & Hubeny (1996) in the present volume discusses disc LD further.

The finite opening angle of a disc means that the front and back parts of a disc are viewed at different angles, as noted by Robinson et al. (1995). In this case *differential* LD will need to be taken into account in eclipse-mapping studies. It is important to note that the amount of disc flaring is not a free parameter, to be determined independently of the $T_{\rm br}$–R relation; rather, they must be determined in a mutually consistent way, since they are both manifestations of the underlying disc structure.

2.4. DISC SPECTRA AND WHITE DWARF SPECTRA

As discussed in somewhat more detail in Wade, Hubeny & Polidan (1994), discs at low mass transfer rates have UV spectra that are similar to solar abundance white dwarfs. The kinematic line smearing in discs mimics the pressure broadening of lines in a white dwarf photosphere. With typical *IUE* spectra and the usual uncertainty in the distances of CVs, this results in a major interpretive problem. The ambiguity can be broken given sufficient wavelength coverage; sufficient wavelength resolution and S/N; or independent information about the distance or disc inclination of the CV to remove the excess degrees of freedom in fitting the spectrum.

3. Conclusions

LTE spectra of model rings differ little from stellar atmosphere spectra computed in the same way with the same opacities. This is because radiative transfer and hydrostatic equilibrium in stars and discs are basically similar. Intuition about stellar atmospheres pays off if carefully applied to discs. In spite of their approximations and omissions at the present stage, disc models are sufficiently far advanced that one need not resort to blackbody approximations or angle-averaged fluxes when interpreting observations of cataclysmic variable discs.

Acknowledgements. This work is supported in part by NASA grants NAGW-3171 and NAG 5-2125.

References

Diaz, M. P., Wade, R. A., Hubeny, I., 1996, these proceedings, p123
Long, K.S., Wade, R.A., Blair, W.P., et al., 1994, Ap. J., **426**, 704
Pringle, J.E., 1981, Ann. Rev. Astron. Astrophys., **19**, 137
Robinson, E.L., Wood, J.H., Bless, R.C., et al., 1995, Ap. J., **443**, 295
Wade, R.A., Hubeny, I., Polidan, R.S., 1994, in "Interacting Binary Stars", ed. A. Shafter, ASP Conf. Ser. **56**, 319

A QUANTITATIVE STUDY OF LIMB DARKENING IN ACCRETION DISKS

M. P. DIAZ[1], R. A. WADE[1], I. HUBENY[2]
1. Department of Astronomy, Pennsylvania State University, University Park, PA 16802, USA
2. Goddard Space Flight Center, Greenbelt, MD 20771, USA

Abstract. The UV emission of accretion disks was synthesized for a grid of optically thick models that cover the high-$\dot{M}$ nova-like regime, aiming to quantify the effect of limb darkening in the integrated and radially resolved spectrum of the disk.

1. Introduction

The method described by Hubeny (1990) is employed to compute the disk properties. The steady-state disk is assumed geometrically thin and its structure is solved for a set of concentric rings where a plane-parallel atmosphere calculation is performed. Such an atmosphere is in hydrostatic equilibrium with the depth-dependent gravity. The mass surface density is given by the standard model and $\bar{w}$, the depth-averaged viscosity, is parameterized in terms of the Reynolds number of the flow. The viscosity $w(z)$ varies as a power law of the mass column density above depth z. The 'disk' equations and the radiative transfer equation are solved iteratively in LTE by the complete linearization method using the code TLUSDISK (Kriz & Hubeny 1986). Once the temperature and density are self-consistently modeled, the radiative transfer equation is solved by the general spectral synthesis program SYNSPEC (Hubeny, Lanz & Jefferys 1994). Several continuum opacity sources are included and the lines of elements up to $Z = 30$ (Kurucz 1990) were considered. The local line profile has the form of a Voigt function which includes Stark and thermal broadening. The integrated disk spectrum is obtained by coadding the specific intensities in the observed wavelength scale defined by the inclination and the local Keplerian velocity. In addition, the partial occultation of the inner disk by the white dwarf is considered.

A. Evans and J. H. Wood (eds.), Cataclysmic Variables and Related Objects, 123–124.

2. Results

A comparison between the limb darkening effect in disk models with 'gravity equivalent' model atmospheres indicates that the local behavior of limb darkening in disks is not too different from the same process in stellar atmospheres (cf. Wade & Rucinski 1985). The continuum limb darkening effect as a function of the disk radius was measured by defining a 'line-free' passband at 1448 Å. The limb darkening correction factors for the integrated disk were then interpolated in $\dot{M}$, M_1 coordinates. It can be seen from our results that the limb darkening effect is larger at outer disk radii and for low white dwarf masses, reflecting the temperature and gravity variations in the disk. It also increases at low mass transfer rates up to a typical value of 2.5 mag for $i = 83°$. However, at high mass transfer rates onto massive white dwarfs, negligible effects are expected over *most* of the disk, even for an inclination of 72°. The disk emissivity at 1448 Å is largest close to the canonical maximum temperature radius and its radial decline is flatter for low-mass white dwarfs, as expected from the standard theory with a simple blackbody emissivity. Gross systematic trends in the Lyman line profiles can be seen. The line width obviously increases with i while the core depth increases as the white dwarf mass decreases. Another sequence is seen in the Lyman line width as a function of the mass transfer rate; the larger, the narrower the lines. The continuum shape appears steeper as the disk temperature ($\dot{M}$ or M_1) increases. In general, the level of the continuum adjacent to the blue wing of Lyγ is a good indicator of the mass transfer rate for the whole mass range. This simulation shows that the limb darkening is an important effect, especially in the UV, and should be taken into account when the overall or local emission of accretion disks in CV's is modeled and compared with observations. Limb darkening corrections such as those computed in this work should be considered in the luminosity and/or distance estimations made on the basis of observed color–magnitude diagrams and radial continuum emissivity distributions. These quantities have been measured for several nova-like systems using eclipse mapping techniques. A detailed discussion of the results summarized here will be published elsewhere.

Acknowledgement. Supported in part by NASA grant NAGW-3171.

References

Hubeny, I., 1990, Ap. J., **351**, 632
Hubeny, I., Lanz, T., Jefferys, C.S., 1994, Collaborative Computational Project 7, **20**, 30
Kriz, S., Hubeny, I., 1986, Bull. Ast. Inst. Czech. **37**, 129
Kurucz, R.L., 1990, Trans IAU 20B, 169
Wade, R.A., Rucinski, S.M., 1985, A&AS, **60**, 471

PHYSICAL PARAMETER MAPPING OF ACCRETION DISKS

S. VRIELMANN[1], F. V. HESSMAN[1], K. HORNE[2],
R. BAPTISTA[2]
1. Universitäts-Sternwarte Göttingen, Germany
2. University of St. Andrews, Scotland, UK

Abstract. We present a further development of the classical eclipse mapping method, which reconstructs the spatial distribution of physical parameters in the accretion disks of cataclysmic variables.

1. Limitations of classical eclipse mapping

The classical 'eclipse-mapping' method (Horne 1985; Baptista & Steiner 1991) employs a maximum-entropy approach (MEM) (Skilling & Bryan 1984) to derive the spatial distribution of the continuum or line intensities of accretion disks in cataclysmic variables. This widely used method has revealed circular disks with more or less pronounced hot spots and radially decreasing brightness temperature distributions (e.g. Wood, Horne & Vennes 1992; Baptista et al. 1995).

The classical method, however, suffers from the following limitations: (i) a single data set is mapped into a single quantity (intensity); (ii) light curves at different wavelengths are assumed to be independent; (iii) the requirement of phase independence is contradicted by anisotropic emission, e.g. from the hot spot or from optically thick lines due to shear effects and turbulence (Horne & Marsh 1986; Horne 1995) and (iv) the information in the line profiles (radial velocities) is ignored.

2. A more physical approach

We have extended the classical method by mapping physical parameters (e.g. temperature, density and height of the optically thin layer) instead of continuum or line intensities. Using data at several different wavelengths, one can use wisely chosen physical models to deduce a (hopefully) small number of physical parameter maps. Depending upon the success of the fit, different models with increasing complexity can be attempted: optically thick blackbody emission; optically thin LTE disk spectra; 'stellar

A. Evans and J. H. Wood (eds.), Cataclysmic Variables and Related Objects, 125–126.

atmosphere' spectra or the inclusion of NLTE effects. Furthermore, special effects can be easily included: anisotropic emission; velocity profiles; UV illumination by the central source.

3. A simple test case

As a first test of our method, we constructed *UBV* light curves (S/N = 50) for a blackbody accretion disk. The simultaneous fits to the data are good (Fig. 1 left). The reconstructed temperature distribution (Fig. 1 right) qualitatively reproduces the general behaviour of the test distribution, but shows typical MEM effects: steep inner and outer gradients are smoothed and to compensate this, the smooth intermediate gradient is steepened. The large extension of the disk at temperatures of about 4000 K is also artificial: as the observed *UBV* light curves are rather insensitive to regions as cold as 4000 K MEM produces a disk which is as flat as possible and still in agreement with the data. Since the effects of MEM are known, Physical Parameter Mapping can be considered as a promising tool to interpret the physical properties of accretion disks.

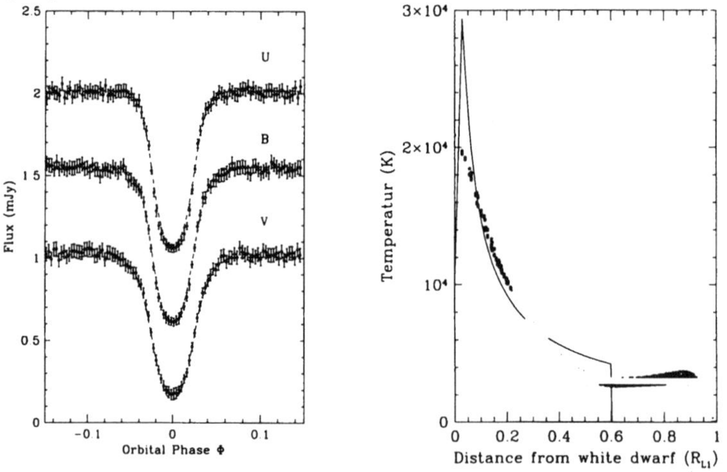

Figure 1. **Left**: Test *UBV* eclipse light curves (dots with errorbars; *U*, *B* shifted by 0.5*n* mJy) and fitted eclipse light curves of the reconstructed disk image (dashed lines). **Right**: Temperature distributions of the test image (solid line) and of the reconstructed image (dots) derived by fitting all 3 light curves simultaneously.

References

Baptista, R., Steiner, J.E., 1991, A&A,, **249**, 284
Baptista, R., Horne, K., Hilditch, R.W., et al., 1995, Ap. J., **448**, 395
Horne, K., 1985, MNRAS, **213**, 129
Horne, K., 1995, A&A, **297**, 273
Horne, K., Marsh, T.R., 1986, MNRAS, **218**, 761
Skilling, J., Bryan, R.K., 1984, MNRAS, **211**, 111
Wood, J.H., Horne, K., Vennes, S., 1992, Ap. J., **385**, 294

THERMAL-TIDAL INSTABILITY MODEL OF DWARF NOVAE BELOW THE PERIOD GAP: A UNIFICATION THEORY

Y. OSAKI
Department of Astronomy, School of Science, University of Tokyo, Bunkyo-ku, Tokyo, 113, Japan

Abstract. A unification theory is proposed in which almost all of outburst behavior in dwarf novae may be understood in a unified way within the basic framework of the disk instability model. In particular, a rich variety of outburst behavior of cataclysmic variables below the period gap may be understood by the thermal-tidal instability model in which the coupling of the two intrinsic instabilities in the accretion disk plays a unique role.

1. A unification model of dwarf nova outbursts

It is fairly widely accepted that the outbursts of dwarf novae (DNe) can basically be explained by the disk instability (DI) model, at least for those above the well-known period gap. In this model, the mass transfer rate from the secondary star is assumed to be constant but accretion from the disk to the central white dwarf is not steady but intermittent because of the intrinsic instability in accretion disks. The instability itself is a kind of thermal instability associated with ionization of hydrogen.

Another intrinsic instability that operates in accretion disks – called the 'tidal instability' – was later discovered by Whitehurst (1988). In this instability the accretion disk is deformed to an eccentric form and its eccentric pattern slowly precesses in the prograde direction in the inertial frame of reference. The so-called 'superhump' phenomenon observed during superoutbursts of SU UMa stars is now believed to be explained by this tidal instability. The tidal instability is found to occur when the disk's outer edge reaches the 3:1 resonance radius (Whitehurst 1988; Hirose & Osaki 1990; Lubow 1991), a condition that is possible only in binary systems with extremely low mass ratio, q, with $q < 0.25$, where $q = M_2/M_1$ and M_1 and M_2

A. Evans and J. H. Wood (eds.), Cataclysmic Variables and Related Objects, 127–130.

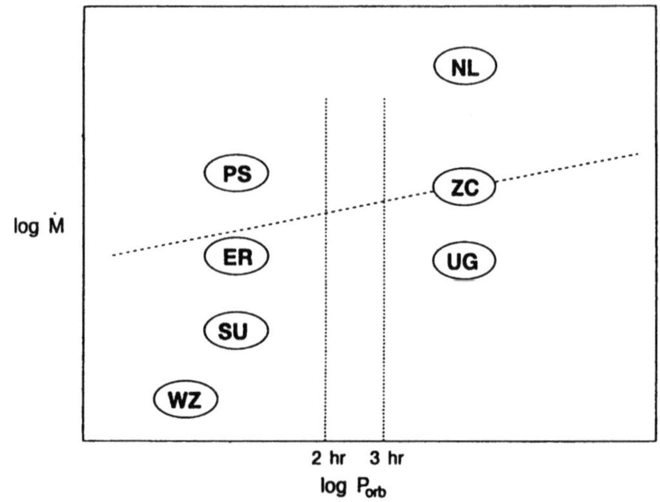

Figure 1. $P_{\rm orb} - \dot{M}$ diagram showing different outburst behaviors of non-magnetic CVs. The region surrounded by dotted vertical lines shows the CV's 2...3 h period gap. The dashed line shows the border line between the thermally stable and unstable disks. Symbols in the figures are; NL: nova like stars, ZC: Z Cam stars, UG: U Gem stars, PS: 'permanent superhumpers', ER: ER UMa stars, SU: SU UMa stars, and WZ: WZ Sge stars. See text for more details.

are the masses of the mass-accreting primary star and the secondary star. The latter condition is in turn satisfied exclusively in cataclysmic variables (CVs) below the period gap.

The present author (Osaki 1989) has proposed the so-called 'thermal-tidal instability model' to explain the superoutburst phenomenon of SU UMa stars, in which the two intrinsic instabilities, the thermal instability and tidal instability, are properly coupled. This model is basically within the general framework of the DI model.

An intriguing possibility has now opened in that almost all varieties of outburst light curves of dwarf novae may be explained by a single principle: the DI model. In this unification model, different outburst behaviors among non-magnetic CVs are basically classified by two parameters characterizing accretion disks in these systems; that is the orbital period of the system and mass transfer rate from the secondary star.

Fig. 1 schematically illustrates different classes of non-magnetic CVs in this unification model. They are basically divided into four regions depending on different combinations of the two intrinsic instabilities.

The CVs above the period gap can be classified into nova-like, Z Cam-type DNe, and U Gem-type DNe depending on the thermal stability nature, which is in turn determined by the mass transfer rate. It has turned out that a similar classification of non-magnetic CVs below the period gap is possible.

2. Dwarf novae below the period gap

Dwarf novae below the period gap are basically classified in the SU UMa sub-type that exhibit two distinct types of outbursts: more frequent, short 'normal outbursts' and less frequent, long 'superoutbursts'; a cycle from one superoutburst to the next one is called a 'supercycle'. The typical supercycle length is about a few hundred days.

Observations show a wide variety of activity within SU UMa stars. In fact, Vogt (1993) has classified SU UMa stars into three groups by their activity: group A, 'active' stars showing frequent outbursts (e.g. VW Hyi), group B, 'intermediate' in activity between two extremes (e.g. OY Car), and group C, very inactive stars or 'WZ Sge' stars that exhibit large outbursts on very rare occasions. Furthermore, a new type of extreme SU UMa stars called either 'ER UMa stars' or 'RZ LMi stars' has recently been discovered (Kato & Kunjaya 1995; Robertson, Honeycutt & Turner 1995), which exhibit extremely short supercycles of 19 d to 45 d. We may thus add to the Vogt activity sequence two more groups: permanent superhumpers and ER UMa stars, and these two new groups should be put above Vogt's group A (see Fig. 1). Here permanent superhumpers are nova-like stars and thus thought to have hot thermally stable disks but still exhibit permanent superhumps (e.g. Skillman & Patterson 1993).

We first discuss the case with $\dot{M} < 10^{16}\,\mathrm{M_\odot\,yr^{-1}}$. The present author (Osaki 1995a) has already suggested at the Garching conference and the Padova conference (Osaki 1995d) that Vogt's (1993) activity sequence may be explained as a sequence of decreasing mass transfer rate in the thermal-tidal instability model. It is found in this model that the supercycle length, T_S, is inversely proportional to mass transfer rate while the recurrence time, T_N, of normal outbursts is inversely proportional to the *square* of mass transfer rate. Thus if the mass transfer rate is decreased along this sequence, the supercycle length is increased in proportion to the inverse of mass transfer rate while the number of normal outbursts in a supercycle decreases in proportion to the inverse of the supercycle length.

However, in order to explain the extremely long recurrence time of WZ Sge itself, 30 yr, it is not enough to decrease the mass transfer rate but also we need to decrease the viscosity parameter in the cold state, $\alpha_{\rm cold}$.

Let us now discuss the case of high mass transfer rate with $\dot{M} > 10^{16}\,\mathrm{M_\odot\,yr^{-1}} \equiv \dot{M}_{16}$. As already noted, there is a critical mass transfer rate above which an accretion disk is always in a hot and thermally stable state. If the mass transfer rate is higher than the critical one, the system remains in a hot state. These CVs below the period gap are identified as permanent superhumpers, nova-like stars exhibiting permanent superhumps because they are thermally stable but tidally unstable.

Osaki (1995b,c) has studied outburst light curves of the ER UMa stars. He has examined the supercycle length as a function of the mass transfer rate under fixed model parameters. It is found that the supercycle length is inversely proportional to the mass transfer rate if $\dot{M}_{16} < 1$, as expected. However, if the mass transfer rate is further increased above $\dot{M}_{16} \sim 1$, the supercycle length has a broad minimum and it finally goes to infinity when the mass transfer rate becomes larger than the critical value for the thermal instability. ER UMa itself is found to be simulated very well as a system having a mass transfer rate corresponding to the broad minimum mentioned above (see Fig. 2 of Osaki 1995b).

For model parameters used to simulate ER UMa, the minimum supercycle length was about 45 d and thus a supercycle length as short as 19 d observed for RZ LMi could not be reproduced by these model parameters. It was found (Osaki 1995c) that the minimum supercycle length was sensitive to one model parameter that describes the strength of tidal torques during the supermaximum when the disk becomes eccentric. It has turned out that the supercycle length as short as 19 d of RZ LMi can be reproduced if the tidal torques during the superoutburst in RZ LMi are significantly weaker than those of ER UMa and ordinary SU UMa stars. The light curve of RZ LMi is also found to be simulated very well in the case of weaker tidal torques (see Fig. 2 of Osaki 1995c).

The wide variety in activity of CVs below the period gap indicates that the mass transfer rates from the secondary stars should have much wider range than that expected from the standard scenario for CV evolution based on gravitational-wave radiation. A cyclic variation in mass transfer rate in time scale of $10^3 \ldots 10^4$ yr is thus inferred.

References

Hirose, M., Osaki, Y., 1990, PASJ, **42**, 135
Kato, T., Kunjaya, C., 1995, PASJ, **47**, 163
Lubow, S.H., 1991, Ap. J., **381**, 259
Osaki, Y., 1989, PASJ, **41**, 1005
Osaki, Y., 1995a, PASJ, **47**, 47
Osaki, Y., 1995b, PASJ, **47**, L11
Osaki, Y., 1995c, PASJ, **47**, L25
Osaki, Y. 1995d, in "Cataclysmic Variables", eds A. Bianchini et al., Kluwer, p307
Robertson, J.W., Honeycutt, R. K., Turner, G. W. 1995, PASP, **107**, 443
Skillman, D.R., Patterson, J., 1993, Ap. J., **417**, 298
Vogt, N., 1993, in "Cataclysmic Variables and Related Physics", eds O. Regev, G. Shaviv, Ann. Israel Phys. Soc., **10**, 63
Whitehurst, R., 1988, MNRAS, **232**, 35

NON CIRCULAR DISKS IN AM CVN SYSTEMS?

J.-E. SOLHEIM
Institutt for Matematiske Realfag, Universitetet i Tromsø,
N-9037 Tromsø, Norway

1. Introduction

The AM CVn stars are mass transferring, interacting binary, white dwarf systems with orbital periods of 15...45 minutes. Hydrogen is completely lost from these systems, and we observe small helium disks which may show thermal and tidal instabilities if the mass transfer rate is large enough (Osaki 1995). A tidal instability brings the disk into a superoutburst state, and in the light curve we may observe superhumps. Based on the observed periods and mass transfer rates we can divide the AM CVn stars into three groups:

A: In permanent superoutburst: AM CVn and EC 15330−1403

B: Regular superoutbursts: CR Boo, V803 Cen and CP Eri

C: Not yet observed superoutburst: GP Com

Group A consists of systems with disks which are too hot to decline into a low state. These disks are in a constant superoutburst state, analogous to nova-like CVs which show no outbursts, but still exhibit permanent superhumps (Skillman & Patterson 1993). Group B shows normal outbursts which can trigger superoutbursts, analogous to the VY Scl dwarf novae. The group C object may also be a superhumper, but with very infrequent outbursts analogous to the SU UMa stars (Warner 1995). In the following we will discuss evidence for superoutbursts in these systems, and the likelihood for these systems to develop elliptical disks.

2. Observations

AM CVn shows variable absorption line profiles, which have been analysed by Patterson, Halpern & Shambrook (1993). They found a line profile skewness modulation with a period of 13.4 h, which was interpreted as due to the variable aspect of a precessing elliptical disk. This modulation is also observed as high frequency sidebands of 20.8 μHz to a set of harmonic fre-

A. Evans and J. H. Wood (eds.), Cataclysmic Variables and Related Objects, 131–132.

quencies of the superhump modulation frequency in the Fourier transform of the light curve (Solheim et al. 1991, Warner 1995). From these observations we can determine an orbital period of 1028 s and a superhump period of 1051.2 s.

For EC 15330 a photometric period of 1119 s is observed. Its spectrum and general properties are almost like those of AM CVn (O'Donoghue et al. 1994). A search for periodic line profile variations is so far inconclusive.

For CR Boo a stable period of 1471 s and an unstable period of 1493 s have been determined (Provencal et al. 1991). The unstable period was only observed during high state observations, and is interpreted as the superhump period, while the 1471 s period is interpreted as the orbital period (Warner 1995).

For V803 Cen and CP Eri no superhump periods have been detected, but the mass ratios are comfortably lower than the ratio $q = 0.22$, which is the limit for development of elliptical disks.

For GP Com the situation is not settled. No outbursts have been detected so far, and the disk may be in a permanent cold state without outbursts. However, even with a low mass transfer rate, the disk may become hot enough for a normal outburst; this may trigger a superoutburst which may last considerably longer than a normal superoutburst because much more mass has been accumulated in the disk during the long dormant phase. This is analogous to the model for WZ Sge proposed by Osaki (1995).

3. Conclusions

All the AM CVn systems have low mass ratios and their disks can, with suitable mass transfer rates, expand beyond the tidal radius and become elliptical. AM CVn and EC 15330 are in a constant superoutburst state and show superhump periods. For the 3 objects which cycle between low and high states, a superhump period is observed only in one case (CR Boo). The object in a (quasi) stable low state (GP Com) will probably show a superoutburst during the next few years.

References

O'Donoghue, D., Kilkenny, D., Chen, A.L., et al., 1994, MNRAS, **271**, 910
Osaki, Y., 1995, PASJ, **47**, 47
Patterson, J., Halpern, J., Shambrook, A., 1993, Ap. J., **419**, 83
Provencal, J.L., Winget, D.E., Nather, R.E., et al., 1991, in "White Dwarfs", eds. G. Vauclair, E. Sion, Kluwer, Dordrect, p449
Skillman, D.R., Patterson, J., 1993 Ap. J., **417**, 298
Solheim, J.-E., Emanuelson, P.-I., Vauclair, G., et al., 1991, in "White Dwarfs", eds. G. Vauclair, E. Sion, Kluwer, Dordrect, p431
Warner, B., 1995, A&SS, **225**, 240

MODEL SPECTRA FOR THE HELIUM CATACLYSMIC VARIABLE: AM CVN

S. BARD

The Auroral Observatory, University of Tromsø, 9037 Tromsø, Norway. E-mail: bards@astro.uio.no

Abstract. Accretion through Roche lobe overflow in the helium cataclysmic binary AM CVn is considered and its emitted spectrum is computed using model stellar atmospheres. The accretion disc in AM CVn is helium-rich and such discs have hitherto not been constructed. We have computed the models by numerically solving the radiative transfer equation for helium opacities. The method adopted is that of treating the accretion disc as a set of optically thick and geometrically thin concentric annuli. Each annulus is treated individually as a blackbody or as a stellar atmosphere. We find that our description of AM CVn is consistent with both photometric and spectroscopic measurements of the system which have been made using the *Nordic Optical Telescope* (NOT) and the *International Ultraviolet Explorer* (IUE). We find that the stellar model synthetic spectrum provides a better fit to the observed spectrum than the Planckian disc spectrum. Also, for a given accretion rate, the synthetic spectrum composed of the sum of a series of stellar atmospheres is flatter than the corresponding blackbody sum.

1. Introduction

Since its discovery some sixty years ago (Malmquist 1936) the number of publications devoted wholly or partly to AM CVn has risen to ~200. Despite the attentions of the scientific community this stellar system remains an enigma – is AM CVn a single white dwarf or does it have a mass donor companion?

A. Evans and J. H. Wood (eds.), Cataclysmic Variables and Related Objects, 133–134.

2. Our results

Even although the exact nature of AM CVn is still the subject of controversy (Bard 1995) there are compelling arguments in favor of a binary system in which mass transfer takes place (Patterson, Halpern & Shambrook 1993). Our calculations of the emitted spectrum are in good agreement with the binary model and the IUE data, as can be seen in Fig. 1.

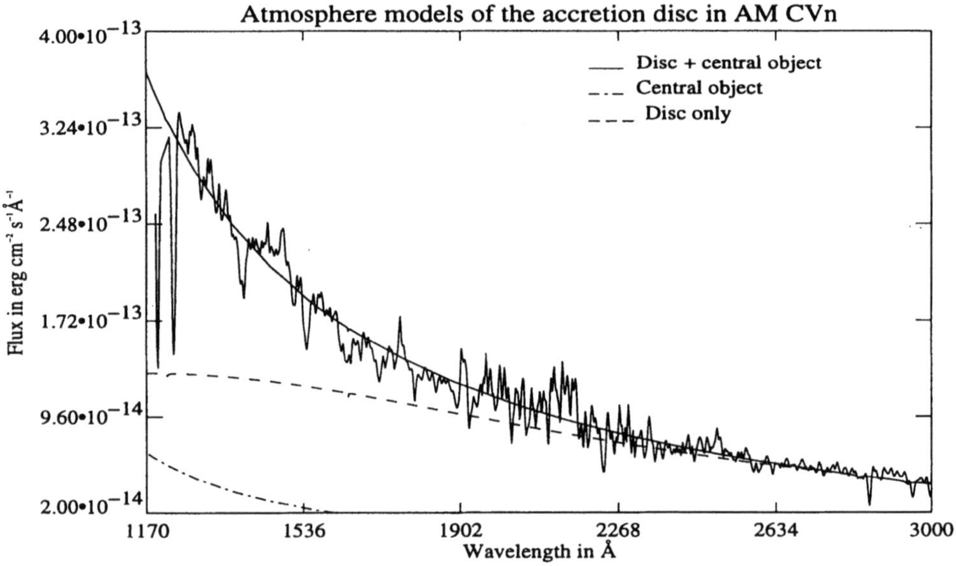

Figure 1. The stellar atmosphere disc approximation of the spectrum of AM CVn. Here the solid line is the sum of the DB white dwarf atmosphere and the central object, the dashed line is the disc alone, and the dotted-dashed line is the central object. The models are calculated in the LTE regime and are fully blanketed. $T_{\rm eff}$ and $\log g_{\rm eff}$ range from 30 000 K and 8.00 $\rm ms^{-2}$ to 11 000 K and 6.16 $\rm ms^{-2}$ respectively.

3. Discussion

The primary in AM CVn is likely to be a one solar mass white dwarf with an effective temperature of ~150 000 K. At these temperatures and gravities the models ought to be calculated in the NLTE regime. This will be the subject of future work.

References

Bard S., 1995, Masters thesis, p38
Malmquist K.G., 1936, Stockholm Obs. Ann., **12**, 7
Patterson J., Halpern J., Shambrook A., 1993, AJ, **419**, 803

DISK INSTABILITIES IN THE BLACK HOLE BINARIES

J. K. CANNIZZO
Universities Space Research Association
NASA/GSFC/Code 666, Greenbelt, MD 20771, USA

1. Introduction

I will briefly review recent progress in our understanding of the global propagation of accretion disk instabilities operating in interacting binary systems in which the accreting star is a $\sim 10\,M_\odot$ black hole. A detailed discussion of this work can be found in Cannizzo, Chen & Livio (1995).

The X-ray light curves of the brightest and best studied X-ray novae – systems such as A0620–00, GS2000+25, GS1124–68, and GRO J0422+32 – are notable in two respects: the rise times are fast (a few days) and the decays have an exponential character, where the e–folding time constant is about 30 to 40 d (Mineshige, Yamasaki, & Ishizaka 1993, hereafter MYI). It is noteworthy that outbursts in dwarf novae also decay exponentially, with the time constant being about a factor of 10 shorter. In this contribution I will discuss the constraints on the physics of accretion disks which the (observed) exponential decay enables us to make.

2. Discussion

Why does the X-ray flux fade as the X-ray nova progresses? X-ray novae are thought to be powered by accretion onto a massive compact object, most probably a black hole (BH), and therefore the fading must be due to a diminution of the rate of mass flow onto the BH. On the assumption that the mass flow rate is determined by the reservoir of material available in the accretion disk, one might imagine that the rate of mass flux would become smaller as the mass of the disk decreases. Furthermore, from the exponential nature of the decay, we know that the rate at which the accretion disk mass is becoming smaller must vary with the disk mass: $\dot{M}_{\rm disk} = -M_{\rm disk}/\tau$ (MYI). The solution to this differential equation has the desired form $M_{\rm disk} = M_0 e^{-t/\tau}$, or $L_{\rm disk} \propto \dot{M}_{\rm disk} \propto e^{-t/\tau}$.

A. Evans and J. H. Wood (eds.), Cataclysmic Variables and Related Objects, 135–138.

How can the mass of the accretion disk be made to decrease? The most obvious way is simply to allow accretion onto the BH to proceed at a faster rate than material is being resupplied into the outer disk from the mass losing star. Unfortunately, it turns out to be quite difficult with this simple mechanism to obtain the law $\dot{M}_{\rm disk} \propto M_{\rm disk}$. Many investigators have examined the problem of the nature of the decay of the light curve produced by loss of disk matter due to accretion, and, for a variety of parameterizations of the viscous dissipation, one finds that $\dot{M}_{\rm disk} \propto M_{\rm disk}^p$ where $p \neq 1$ (Mantle & Bath 1983; Lyubarskii & Shakura 1987; Cannizzo, Lee & Goodman 1990; MYI). Disk depletion through accretion inevitably leads to power law decays of the accretion luminosity rather than exponential decays.

What about removing matter from the outer edge of the disk rather than the inner edge? Could this provide an additional degree of freedom to allow theorists to concoct a plausible physical way to get $\dot{M}_{\rm disk} \propto M_{\rm disk}$? The primary failing with the idea of taking matter out from the inner edge was that the rate of mass loss depends on the local viscosity. By removing matter from the outer edge we are not necessarily limited by this restriction. For example, in the limit cycle model for dwarf nova outbursts, there exists a strong outflow of matter from the inner disk to the outer disk due to the action of a cooling transition front. One finds the outflow in time dependent modeling of the limit cycle, and the outflow is caused by the existence of an interface between the material in the high viscosity state (in the inner disk) and the material in the low viscosity state (in the outer disk).

We do not know *a priori* whether the action of the cooling front can lead to an exponential decay of the accretion luminosity. No one has attempted a numerical study in which some control parameter for the accretion disk physics is varied in a systematic way to see what conditions (if any) must exist so that $L_{\rm disk} \propto e^{-t/\tau}$ during the decay. Our uncertainty about the physics of the viscosity in accretion disks – a topic which was covered so well by Chris Tout in his review talk – is customarily lumped into the 'α–parameter' (Shakura & Sunyaev 1973) wherein one sets the $\phi - r$ component of the stress tensor equal to α times the gas pressure. In order to obtain well separated outbursts for dwarf novae using the limit cycle model, most workers have found it necessary to increase α in going from the low to high viscosity state in the model. One of the popular functional forms for α has been $\alpha = \alpha_0(h/r)^n$, where h is the local semi-thickness of the disk and r is the local radius. In an attempt to test this law for α, we ran numerical models in which different values for n were tried. The results were quite interesting. For $n = 1$ the decay in the accretion luminosity has a faster-than-exponential form. That is, when plotted as $\log L_{\rm disk}$ versus time, the resulting light curve is concave down. For $n = 2$ the decay has a slower-than-exponential form – i.e., the light curve is concave up. For $n = 1.5$ the decay is exponential, just as observed. Our first finding, therefore, is that, based on the observed exponential decay, the law $\alpha = \alpha_0(h/r)^{1.5}$ seems to be the preferred one. Further tests reveal that the e–folding time constant varies directly with the mass of the accreting object M_1, and inversely with the constant α_0 which appears in the α scaling.

By following the evolution of surface density and disk temperature during the decay of an outburst, one sees that the physical process ultimately responsible for the nature of the decay is the speed of the cooling front. The law $\dot{M}_{\rm disk} \propto M_{\rm disk}$ occurs only for $n = 1.5$ because, for that n value, $\dot{r}_{\rm F} \propto r_{\rm F}$, where $r_{\rm F}$ is the radius of the front. In other words, an exponential decrease in the mass of the disk (or, more properly, in the mass of the hot portion of the disk), is caused by an exponential constriction of its outer radius $r_{\rm F} = r_0 e^{-t/\tau'}$.

By examining the width of the transition front, $w_{\rm F}$, in the time dependent computations, we find $w_{\rm F} = \sqrt{h_{\rm F} r_{\rm F}}$, where $h_{\rm F}$ and $r_{\rm F}$ are the disk semi-thickness and radius evaluated in the disk annulus lying just interior to the cooling front. This appears to be true regardless of how the viscosity is parameterized, and therefore seems to be quite general. The discovery of this law was made possible through the use of our high resolution code in which we divide the disk into 1000 radial elements. Most previous studies have taken $\sim 20 \ldots 40$ radial elements. Although we are confident that the relation $w_{\rm F} = \sqrt{h_{\rm F} r_{\rm F}}$ represents a fundamental property of the cooling front, we have yet to derive this relation from first principles.

The speed of the cooling transition front is given by

$$v_{\rm F} = \dot{r}_{\rm F} = -\frac{\nu_{\rm F}}{w_{\rm F}} = -\frac{\alpha_{\rm F} h_{\rm F}^2 \Omega_{\rm F}}{w_{\rm F}}, \tag{1}$$

where ν is the viscosity coefficient and again, the subscript F denotes values in the high viscosity disk just interior to the cooling front. If we adopt $\alpha = \alpha_0(h/r)^n$ and use $h = c_s/\Omega$ (from hydrostatic equilibrium), we get

$$\dot{r}_{\rm F} = -\alpha_0 c_s^{(n+3/2)} \left(\frac{r_{\rm F}}{GM_1}\right)^{(2n+1)/4}. \tag{2}$$

For $n = 1.5$ this reduces to

$$\dot{r}_{\rm F} = -\frac{\alpha_0 c_s^3}{GM_1} r_{\rm F} \tag{3}$$

or

$$r_{\rm F} = r_0 e^{-t/\tau'}, \tag{4}$$

where $\tau' = (GM_1/\alpha_0)c_s^{-3}$. Note that the sound speed c_s was taken to be constant. This is because $T_{\rm F} \simeq 30\,000$ K just interior to the cooling front, independent of radius, giving $c_s \simeq 16$ km s^{-1}.

The law $v_F = \sqrt{h_F r_F}$ has turned out to be crucial for understanding why $n = 1.5$ is special. Equation (2) shows that $\dot{r}_F \propto r_F$ only for $n = 1.5$. Furthermore, for $n = 1.5$ we also see that $\tau' \propto M_1/\alpha_0$, which explains the numerical results regarding M_1 and α_0 which were quoted earlier. By taking $M_1 = 10\,M_\odot$, we find $\alpha_0 \simeq 50$ is needed to get $\tau \simeq 30$ d as is observed. Meyer & Meyer-Hofmeister (1984) used the exact same law we do, $\alpha = 50(h/r)^{1.5}$, in modeling dwarf novae. These systems also possess outbursts with exponential decays, but with $\tau \sim 3$ d. This is expected if α_0 were the same in both cases, because the primary masses are $\sim 1\,M_\odot$ in cataclysmic variables – of order a factor 10 smaller than those estimated for the X-ray novae (Shahbaz, Naylor & Charles 1994; Shahbaz et al. 1994).

Acknowledgement. This research was supported through the visiting scientist program under the Universities Space Research Association (USRA contract NAS5-32484) in the Laboratory for High Energy Astrophysics at Goddard Space Flight Center.

References

Cannizzo, J. K., Lee, H. M., Goodman, J., 1990, Ap. J., **351**, 38
Cannizzo, J. K., Chen, W., Livio, M., 1995, Ap. J., **454**, 880
Lyubarskii, Yu. É., Shakura, N. I., 1987, Soviet Astronomy Letters, **13**, 386
Mantle, V. J., Bath, G. T., 1983, MNRAS, **202**, 151
Meyer, F., Meyer-Hofmeister, E., 1984, A&A, **132**, 143
Mineshige, S., Yamasaki, T., Ishizaka, C., 1993, PASJ, **45**, 707
Shahbaz, T., Naylor, T., Charles, P. A., 1994, MNRAS, **268**, 756
Shahbaz, T., Ringwald, F. A., Bunn, J. C., et al., 1994, MNRAS, **271**, L10
Shakura, N. I., Sunyaev, R. A., 1973, A&A, **24**, 337

THE ACCRETION DISK MODEL FOR THE FIRST HUNDRED DAYS OF THE OUTBURST EVOLUTION IN THE BLACK HOLE X-RAY NOVAE

S.-W. KIM[1], J. C. WHEELER[1], S. MINESHIGE[2]

1. Department of Astronomy, University of Texas at Austin, RLM 15.308, Austin, TX 78712-1083, USA
2. Department of Astronomy, Faculty of Science, Kyoto University, Sakyo-ku, Kyoto 606-01, Japan

We present time-dependent, irradiated, accretion disk models for the black hole X-ray novae in the first hundred days of the dwarf nova-like outbursts, including the rise, precursor, maximum and the secondary re-flare. This work is based on the disk instability model (Kim, Mineshige & Wheeler 1996, Kim, Wheeler & Mineshige 1996). The model is reasonably consistent with the observed optical light curves. The irradiators are the central hot region around the black hole, and the corona or chromosphere above the accretion disk. In addition, we include the time-dependent shadowing effect and consequent blocking of the outer portions of the disk from the central irradiator. We find the stagnation phenomenon whereby the disk stays in the intermediate temperature stage between the hot and cool state. This can explain the recently discovered optical precursor rise prior to the maximum light in Nova Sco 1994 (Bailyn et al. 1995: see Fig. 1). We suggest the secondary re-flare after the maximum is due to the coupled effects of the irradiation and stagnation. In the model, the stagnation phenomenon during the rise results from the partial ionization and molecular opacity. In addition, we find irradiation-induced stagnation during the decay phase, which is consistent with the observed secondary re-flare in X-ray novae (see Fig. 1). In the overall evolution of model outbursts in the first hundred days, the outer disk is blocked from the irradiation and, in turn, the companion star may not be strongly irradiated. This suggests that there is no appreciable increase of mass transfer rate during the decay prior to the secondary re-flare, unlike the behaviour in the mass transfer burst models.

A. Evans and J. H. Wood (eds.), Cataclysmic Variables and Related Objects, 139–140.

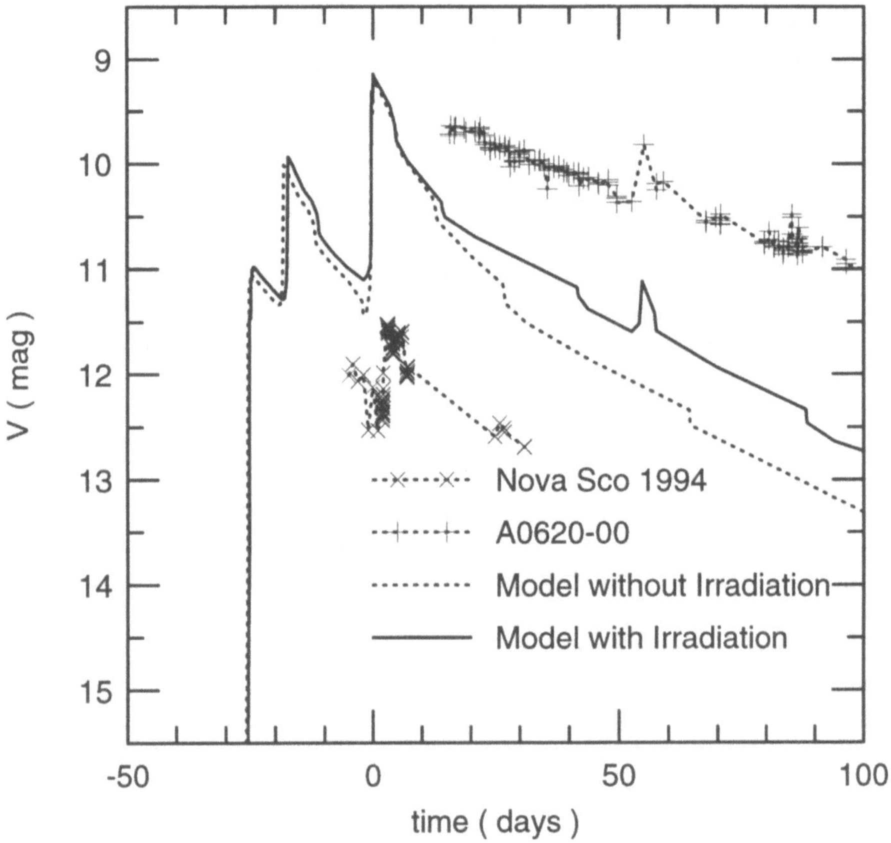

Figure 1. Model light curves for A0620−00 are presented for the case with no irradiation (dash line) and irradiation (solid line) along with the optical observations of A0620−00 (Webbink 1978) and Nova Sco 1994 (Bailyn et al. 1995). The data for A0620−00 and Nova Sco 1994 are plotted 1.5 mag brighter and 2.5 mag dimmer than observations, respectively. Note that Nova Sco 1994 revealed a secondary re-flare much sooner than that in A0620−00.

References

Bailyn, C. D., Orosz, J. A., Girard, T. M., et al., 1995, Nature, **374**, 701
Kim, S.-W., Mineshige, S., Wheeler, J. C., 1996, to be submitted to ApJ
Kim, S.-W., Wheeler, J. C., Mineshige, S. 1996, to be submitted to ApJ
Webbink, R. F., 1978, University of Illinois, and private communication

INTERMEDIATE POLARS

THE INTERMEDIATE POLARS

COEL HELLIER
Department of Physics, Keele University, Keele, Staffordshire, ST5 5BG, UK

Abstract. Recently discovered intermediate polars extend the limits of known behaviour. They now fill the range between, and overlap with, non-magnetic CVs and AM Hers. They have a wide range of asynchronism, a full range of field strengths, and several different accretion modes and pulsation mechanisms. I highlight (i) a puzzle over the symmetry between magnetic poles, (ii) the X-ray spectra, (iii) the similarity with polars, and (iv) the mass of EX Hya.

1. The known intermediate polars

Patterson (1994) provided a comprehensive review of intermediate polars (IPs or DQ Her stars), while my Capetown review (Hellier 1995) focussed on the accretion geometry. Here I do not try to be comprehensive, but instead highlight issues which need updating or further attention.

The significant change since Patterson's parade of the 'cast list' has been the addition of new IPs discovered by ROSAT. In addition to the WFC source PQ Gem (RE 0751+14; Mason et al. 1992) we have the XRT sources RX 0558+53 (Haberl et al. 1994), RX 1712–24 (Buckley et al. 1995), RX 1238 –38 (Buckley 1995) and probably RX 0028+59, RX 0153+74 and RX 1914+24 (Haberl & Motch 1995), with further identifications likely.

With these additions I present the known IPs in Table 1. The list is conservative, including only certain or almost certain members, placing most weight on a convincing X-ray periodicity, since this is the hardest IP signature to fake.

It has become traditional to plot IPs on a spin-orbit diagram (Fig. 1) and then draw conclusions from it (e.g. King & Lasota 1991 and references therein). The previously touted $P_{spin}=0.1P_{orb}$ relation is not apparent in up-to-date IP lists — indeed least-square fits produce a roughly orthogonal line. The safest interpretation is as a scatter diagram with three outliers.

A. Evans and J. H. Wood (eds.), Cataclysmic Variables and Related Objects, 143–152.

TABLE 1. The known intermediate polars.

Star	P_{spin} (s)	P_{orb} (h)	M_V	Mag Field (MG)
EX Hya	4022	1.638	13	~1
RX 1238–38	1850:	3.7:		
TX Col	1911	5.72	15	"
TV Col	1910	5.49	14	"
1E 0830.9–2238	1558		17	
RX 0153+74	1414			
FO Aqr	1254	4.85	13.5	"
RX 1712–24	927	3.4:	14	8
BG CMi	913	3.23	14.5	4
PQ Gem	833	5.19	14.5	12
AO Psc	805	3.59	13.5	~1
V1223 Sgr	746	3.37	13	"
RX 1914+24	568		>19	
RX 0558+53	545	4.15	14.6	"
YY Dra	529	3.96	11	"
GK Per	351	47.9	14	"
RX 0028+59	313	5:	14.6	
XY Ari	206	6.06	>23	"
DQ Her	142	4.65	15	0.14
AE Aqr	33	9.88	11	0.3

Notes: References in Sect. 1, with Allan et al. 1996 (RX 0558+53) and O'Donoghue et al. 1996 (1E 0830.9–2238). Following Zhang et al. (1995) we quote DQ Her's period as 142 s, however 71 s is also possible. Magnetic field estimates are reliable only where direct evidence of the field is seen (BG CMi, PQ Gem, RX1712–24). Values are from Piirola et al. 1993 (PQ Gem); Buckley et al. 1995 (RX 1712–24); Chanmugam et al. 1990 (BG CMi); Zhang et al. 1995 (DQ Her). Other systems are likely to have magnetic fields of ~1 MG (see Patterson 1994 for further details).

However, interpreting the P_{spin}=0.1P_{orb} line as an upper limit for disc-fed accretors is also tenable: the only systems certainly in the zone between this line and the AM Her synchronism line are EX Hya, which is spinning up rapidly towards the other IPs, and the asynchronous polars such

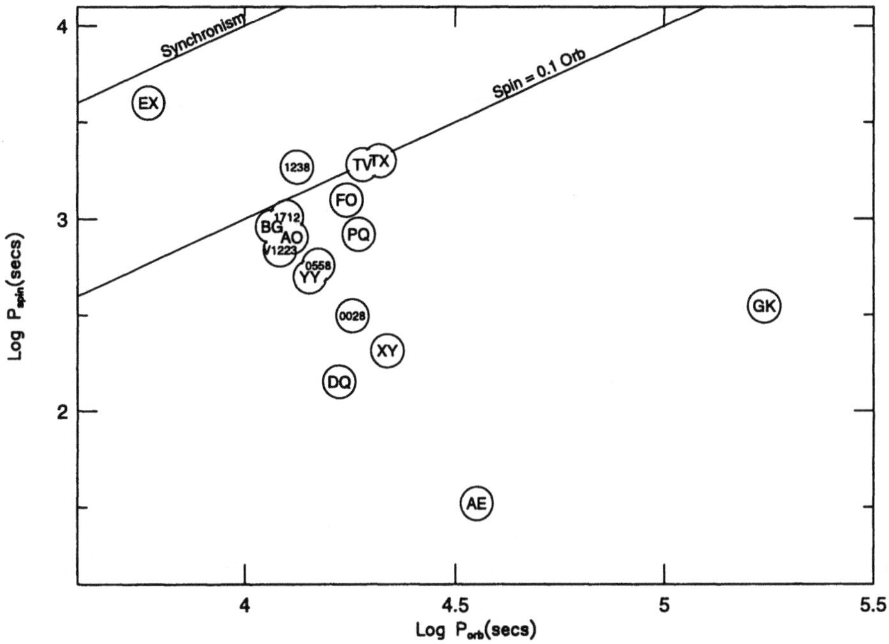

Figure 1. An update of the spin-orbit diagram from Patterson (1994).

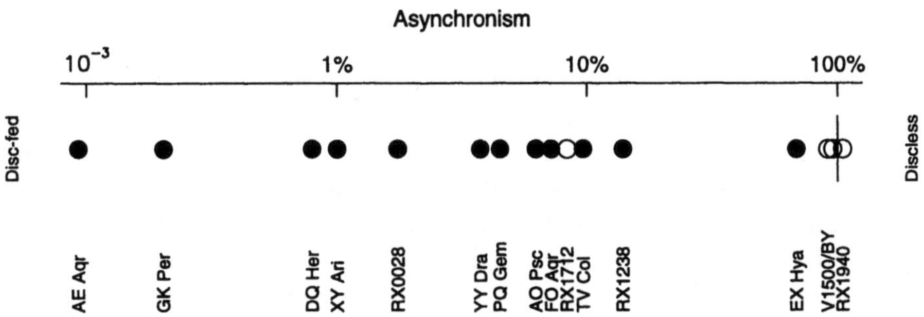

Figure 2. The degree of asynchronism in magnetic CVs. Disc-fed systems are filled symbols, discless systems are open symbols.

as V1500 Cyg, which is heading rapidly for synchronism. In contrast the zone below the $P_{spin}=0.1P_{orb}$ line is well populated. This pattern fits the model by Wu & Wickramasinghe (1991) well. The range of synchronism for magnetic CVs is illustrated in Fig. 2.

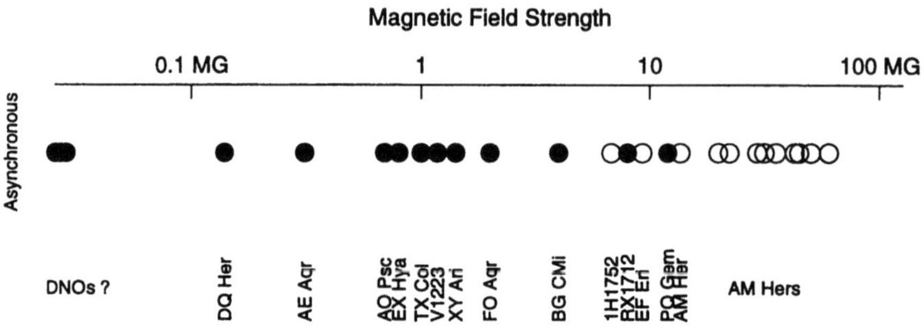

Figure 3. Field strength of magnetic cataclysmic variables (open circles are AM Hers).

2. A disparate bunch

The newly discovered systems make it clearer that IPs are not a homogeneous set, but have properties ranging from the non-magnetic CVs to the AM Hers, and even overlapping to some extent. Thus a 'typical IP' model will not apply to many members. Further, magnetic CVs can change modes over time, and, as emphasized by Wynn, King & Horne (1995), history can be as important as basic characteristics such as field strength in determining the current appearance of a system.

2.1. RANGE OF FIELD STRENGTHS

In addition to the range of synchronism, over 3 orders of magnitude, there is a great range of field strengths. BG CMi's status as the most polarized IP has been superseded by PQ Gem and RX 1712–24, both showing polarisation variations over the spin cycle. The implied fields are $\sim 10\,\mathrm{MG}$, overlapping with the lower end of the AM Her field distribution (Fig. 3). DQ Her and AE Aqr probably have fields two orders of magnitude weaker. As discussed by Warner (1995), the dwarf nova oscillations indicate a field strength another order of magnitude lower still in 'non-magnetic' CVs. V533 Her could be the 'overlap' between magnetic and other CVs.

2.2. RANGE OF ACCRETION MODES

IPs have traditionally been thought to have accretion discs truncated by the magnetic field in the inner regions. While this probably holds for the majority of known IPs, it cannot be the whole story. Buckley et al.'s (1995) discovery of RX 1712–24 presents us with a system showing all the hallmarks of an IP with a spin period of 1003 s — except for polarization pulsed at 927 s, implying that 927 s must be the spin period and that 1003 s is therefore the beat period. Variations at the beat period, caused by the orbitally-locked accretion stream flipping between poles, had been proposed as the signature of discless IPs (e.g. Hellier 1991), suggesting that RX 1712–24 is the first of the discless IPs predicted by Hameury, King & Lasota (1986).

FO Aqr shows a dominant X-ray pulse at the spin period, and also a weaker, variable, beat period pulsation (Norton et al. 1992). Thus it might be accreting via a hybrid 'disc-overflow' mode, proposed for IPs by Hellier et al. (1989) and Hellier (1993).

As if three different modes weren't sufficient, there is a fourth possibility. Wynn & King (1995) have suggested that AE Aqr is an 'ejector' in which most of the mass transfer stream is ejected from the system, and only a minority accreted.

Future work will determine the relative incidence of different accretion modes in IPs. Although current evidence points to a truncated disc as most common (e.g. Hellier 1991), this is not necessarily the final word. Notice that if the field in RX 1712–24 were lower, so that no polarization were observed, we would have no inkling of the 927-s period, and might suppose that it was a normal IP with a spin period of 1003 s. Thus other systems could be similarly mis-identified. Careful comparison of RX 1712–24 with other IPs might show whether or not this is the case.

2.3. RANGE OF PULSE MECHANISMS

In AE Aqr the high temperature of the optical/UV spin-pulsation implies that it originates from a region only $\sim 3\%$ of the white dwarf area, presumably the polecaps. However, in other IPs the evidence instead points towards the 'accretion curtain' model of Hellier et al. (1987), where the pulsations arise in the curtains of material between the inner disc and the white dwarf. Both the temperature of the pulsed emission in FO Aqr and the phase wandering of the optical eclipse in EX Hya are strong evidence for the accretion curtain model in these stars (see Hellier 1995 for further details). DQ Her is different again, with the observed pulsation arising from the entire disc (Patterson, Robinson & Nather 1978). For a thorough investigation of pulse temperatures see Welsh & Martell (1996).

The explanation for the differences is likely to be the relative size of the accretion curtains — a slower rotator will have a larger magnetosphere (assuming equilibrium rotation) and so have large curtains, cool at their outer edges. A faster rotator such as AE Aqr would have smaller curtains, so the pulsed flux would be dominated by hotter regions nearer to or on the white dwarf. The same mechanism probably explains the correlation between pulse amplitude and spin period noted by Patterson (1994): all IPs have the hot pulse from the inner regions, but in slower rotators the larger curtains add in a large, cooler pulse. This curtain model implies that pulses generally have a range of temperature and can only be characterised with a single temperature in cases such as AE Aqr, where the pulsation comes from one small region.

In DQ Her we are probably not seeing these regions, due to the high inclination, so we only see a much weaker component reflected off the disc.

3. Single or double humped pulses?

An issue which has recently come to prominence is whether an IP produces single or doubled peaked pulses every white dwarf rotation. The old ideas in which X-ray/optical maxima occur when a pole points towards us suggest that a system with one dominant pole would produce a roughly sinusoidal profile, while a system with two bright poles would produce a double peaked pulse. In contrast the accretion curtain model (with pulse maximum when the upper pole points away) predicts a sinusoidal pulse from both one and two pole systems, since in this model the poles act in phase.

This is almost certainly the case for EX Hya, where both the partial X-ray eclipse and the width of the optical emission lines insist that we are seeing roughly equal emission from two poles (see e.g. Hellier 1995 for more details) yet the pulsation is single peaked at the 67-m spin period (unless one wants to postulate that the spin period is actually 134 m, far longer than the orbital period).

However, Zhang et al. (1995) have revisited the phase-shift through eclipse of the 71-s pulse in DQ Her. Using a new technique to boost the pulse's signal-to-noise ratio they show convincingly that the disc must be illuminated by two beams with a rotation speed of 142 s (rather than 1 beam rotating at 71 s). However, since there is no direct sign of a 142-s pulse, the two beams must be equally intense to high accuracy. Thus we have the opposite of EX Hya — a spin-pulse with all the power in the first harmonic rather than in the fundamental.

The only likely way two beams could illuminate the disc equally is if the dipole were tilted at 90° and is spinning in the orbital plane. However, the phenomenon is more widespread than DQ Her: both RX 0558+53 (Allan et al. 1996) and YY Dra (Haswell & Patterson 1996) show strong power at the first harmonic, although the fundamental is also present in these stars. Thus, further investigation of the pulse shape from an IP is required to understand why this is seen in some systems and not in others.

The basic ideas of the accretion curtain model (that spin maximum occurs when the upper pole points away; that optical and X-ray maxima are in phase; that opacity in the curtain and its changing projected area are the cause of the pulse; Hellier et al. 1987; Rosen, Mason & Cordova 1988; Hellier, Cropper & Mason 1991) have recently received strong support from the modelling by Kim & Beuermann (1995, 1996). They reproduce all these features, confirming that the two-pole accretion curtain geometry gives rise to a single-peaked pulse, at maximum when the upper pole points away. Thus we now need to explain the double humps in DQ Her, YY Dra and RX 0558+53. One clue is that systems furthest from synchronism are likeliest to show double-peaked pulses. In addition to these three stars, AE Aqr, PQ Gem and XY Ari also show double humps to some degree, and all six are in the same region of Fig. 1. Double humps arise if the optical depth in the accretion polecap is greatest horizontally (poles acting anti-phased) while single pulses arise if the optical depth is greatest vertically (as in the accretion curtain model, with poles acting in phase). Possibly, rapid rotators, having relatively smaller accretion curtains, are more likely to have light curves dominated by the accretion polecap, producing double humps.

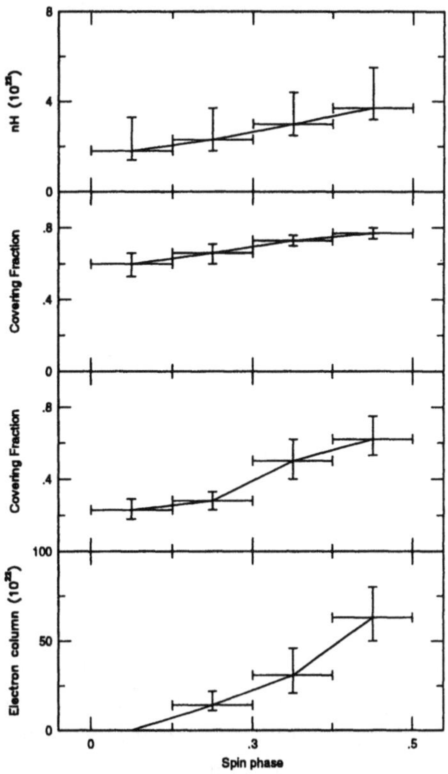

Figure 4: Variation in absorption in AO Psc to reproduce the spin pulse. One absorbing component varies in both density (panel 1; top) and covering fraction (panel 2). A second absorber of density $2\,10^{23}$ cm^{-2} varies in covering fraction (panel 3), while a 'grey absorber' varies in density (panel 4).

4. The X-ray spectra

Low resolution data from the previous generation of X-ray satellites showed that IP spectra were 10–30 keV bremsstrahlungs with heavy but patchy absorption (e.g. Ishida 1991). Typical spin pulses are sinusoidal, with minima caused by increased absorption (although stars such as PQ Gem and FO Aqr also show other effects).

So far, ASCA results confirm this picture, their main impact being to increase the complexity of the required absorption. For instance Mukai, Ishida & Osborne (1994) model the phase-averaged spectrum of FO Aqr with a bremsstrahlung absorbed by three different columns. Hellier et al. (1996) attempted phase-resolved analysis of ASCA data on AO Psc, a star showing a particularly strong and simple, sinusoidal pulse. Given an underlying bremsstrahlung and an interstellar column of 10^{21} cm^{-2}, the pulsation is largely reproduced by a partially covered $\sim 3\,10^{22}$ cm^{-2} absorber, increasing in both column and covering fraction during minima (Fig. 4). In addition, though, they require a denser absorber to reduce the high energy flux at minimum. Given the limited energy range of ASCA, the parameterisation is not unique — one possibility combines a $2\,10^{23}$ cm^{-2} absorber, with varying covering fraction, and a 'grey' electron scattering column, denser at spin minimum. Such divisions between components are probably artificial; improved data will likely require increasingly complicated distributions of absorbers. Although the need for larger columns and covering fractions during minima is robust, the exact parameterisation is (so far) ambiguous.

The interpretation is still uncertain: it is probable that the accretion columns are tall with a small footprint, producing most absorption along their length, so that flux minimum occurs when the upper pole points towards us. To reproduce the leaky absorbers the material in the flow must be clumpy. However, so far we cannot determine whether the grey-ish absorber is highly ionized material beneath the accretion shock, producing opacity by electron scattering, or dense, opaque blobs obscuring a significant fraction of the lines of sight. Similarly, we don't yet know whether all significant absorption occurs near to the accretion shock, or whether material much further out, where material rises from the plane along field lines, also plays a role. Progress in resolving such issues is likely to come from comparing ASCA data on several systems.

5. Comparison with AM Hers

Recent discoveries make IPs look increasingly similar to AM Hers: we now have asynchronous polars and IPs with polar-like field strengths, while the hard X-ray spectrum of both IPs and polars is a bremsstrahlung, showing greater absorption when the pole points towards us. Further, ROSAT has discovered a soft X-ray component in some IPs (Haberl & Motch 1995). Probably this component is present in all IPs, but at too low an energy to be seen in many systems.

This emission is the counterpart of the soft component ubiquitous in AM Hers, where it is blackbody emission from the white dwarf polecap, heated both by reflection of hard X-rays and by thermalising of blobs which bury into the surface. Simple interpretation of the ROSAT IP data yields heated areas only $\sim 10^{-5}$ of the white dwarf surface (e.g. Haberl et al. 1994) — small even by the standards of small polecap models. If this is the size of the accretion footprint then electron scattering in the accretion column must be substantial.

Given these similarities there are two significant differences between polars and IPs. The first is that no IP shows an 'off-phase', as occurs every cycle in AM Hers, if the dominant accreting pole is occulted by the white dwarf. The interpretation has to be that accretion is roughly even between the two poles, with little dependence on orbital phase, so that the pole facing us is always accreting.

Further, the extreme flickering characteristic of polars, caused by individual blobs of material hitting the white dwarf, is not seen in IPs. A possible explanation is that the higher accretion rate in IPs causes more blobs to be present at any one time, effectively smoothing the flickering.

Alternatively, the accretion could be much less blobby in an IP (although some blobbiness is required to explain the partial covering in the X-ray spectra). If accreting material were stripped from a disc, rather than coming from a blobby stream as in polars, it could account for the difference. This might further explain why the soft component is much weaker in IPs (no blobby accretion, therefore no soft excess), and, if smoother accretion were more effectively channelled by the magnetic field, why the estimate of the accretion area is so small. A snag with such an argument is that the discless IP RX 1712–24, which is presumably more AM Her-like, does not, so far, have a detected soft component.

6. The mass of EX Hya

On a different topic to finish, the high resolution X-ray spectra from ASCA provide a new way of measuring the mass of a white dwarf. The emission line ratios give the plasma temperature of the accretion shock, which, through accretion shock theory and a mass-radius relationship, yields the white dwarf mass (e.g. Ishida & Fujimoto 1995). For EX Hya, which has the strongest emission lines and so the most reliable result, this method gives a white dwarf mass of 0.47 ± 0.04 $M_\odot$ (Fujimoto & Ishida 1995).

EX Hya is also the best IP for measuring the mass by traditional means, being an eclipsing double-line binary. Thus we have $i = 77° \pm 1°$, $K_1 = 69 \pm 9$ km s^{-1} (Hellier et al. 1987) and $K_2 = 356 \pm 4$ km s^{-1} (Smith, Collier-Cameron & Tucknott 1993). Combining these gives a white dwarf mass of 0.49 ± 0.03 $M_\odot$ and a secondary mass of 0.095 ± 0.013 $M_\odot$. The agreement with the ASCA result is remarkable given the tight error bars on both determinations.

I make three remarks on the masses. Firstly, the secondary mass is lower than the 0.128 $M_\odot$ predicted by Patterson's (1984) period-mass relation. Secondly, the secondary mass is very much lower than suggested by recent infrared work by Dhillon et al. (1996). They find spectral features consistent with an M3 or M4 dwarf, which normally has a mass of ~ 0.3 $M_\odot$. Lastly, the white dwarf mass implies an escape velocity of 3600 km s^{-1}. The optical emission line wing extends to almost this velocity (for example Hellier et al. 1987 quoted widths of ± 3500 km s^{-1}). Thus the mass can only be correct if we are seeing emission from infalling material which is (i) extremely close to the white dwarf, and (ii) has not been significantly slowed in its journey.

References

Allan, A., Horne, K., Hellier, C., et al., 1996, MNRAS, in press
Buckley, D.A.H., 1995, private communication
Buckley, D.A.H., Sekiguchi, K., Motch, C., et al., 1995, MNRAS, **275**, 1028
Chanmugam, G., Frank, J., King, A.R., Lasota, J.P., 1990, Ap. J., **350**, L13
Dhillon, V.S. et al., 1996, in preparation
Fujimoto, R., Ishida, M., 1995, ASP Conf series, **85**, 136
Haberl, F., Thorstensen, J.R., Motch, C., et al., 1994, A&A, **291**, 171
Haberl, F., Motch, C., 1995, A&A, **297**, L37
Hameury, J.-M., King, A.R., Lasota, J.-P., 1986, MNRAS, **218**, 695
Haswell, C.A., Patterson, J., 1996, these proceedings, p169
Hellier, C., 1991, MNRAS, **251**, 693
Hellier, C., 1993, MNRAS, **265**, L35
Hellier, C., 1995, ASP Conf series, **85**, 185
Hellier, C., Mason, K.O., Rosen, S.R., Córdova, F.A., 1987, MNRAS, **228**, 463
Hellier, C., Mason, K.O., Smale, A.P., et al., 1989, MNRAS, **238**, 1107
Hellier, C., Cropper, M.S., Mason, K.O., 1991, MNRAS, **248**, 233
Hellier, C., Mukai, K., Ishida, M., Fujimoto, R., 1996, MNRAS, in press
Ishida, M., 1991, Ph.D. thesis, University of Tokyo
Ishida, M., Fujimoto, R., 1995, ASP Conf series, **85**, 132
Kim, Y., Beuermann, K., 1995, A&A, **298**, 165
Kim, Y., Beuermann, K., 1996, A&A, in press
King, A.R., Lasota, J.P., 1991, Ap. J., **378**, 674
Mason, K.O., Watson, M.G., Ponman, T.J., et al., 1992, MNRAS, **258**, 749
Mukai, K., Ishida, M., Osborne, J.P., 1994, PASJ, **46**, L87
Norton, A.J., Watson, M.G., King, A.R., et al., 1992, MNRAS, **254**, 705
O'Donoghue D., Koen C., Kilkenny D., 1996, MNRAS, **278**, 1075
Patterson, J., 1984, Ap. J. Supp., **54**, 443
Patterson, J., 1994, PASP, **106**, 209
Patterson, J., Robinson, E.L., Nather, R.E., 1978, Ap. J., **224**, 570
Piirola, V., Hakala, P., Coyne, S.J., 1993, Ap. J., **410**, L10
Rosen, S.R., Mason, K.O. Córdova, F.A., 1988, MNRAS, **231**, 549
Smith, R.C., Collier-Cameron, A., Tucknott, D.S., 1993, in "Cataclysmic Variables and Related Physics", eds O. Regev, G. Shaviv, Ann. Israel Phys. Soc., **10**, 70
Warner, B., 1995, ASP Conf series, **85**, 343
Welsh, W.F., Martell, P., 1996, MNRAS, submitted
Wu, K., Wickramasinghe, D.T., 1991, MNRAS, **252**, 386
Wynn, G., King, A.R., Horne, K. 1995, ASP Conf series, **85**, 196
Zhang, E., Robinson, E.L., Stiening, R.F., Horne, K., 1995, Ap. J., **454**, 447

MASS AND ANGULAR MOMENTUM FLOWS IN MAGNETIC CVS

A. R. KING
Astronomy Group, University of Leicester,
Leicester LE1 7RH, UK

Abstract. I review recent progress in understanding the accretion geometry and angular momentum flows in magnetic CVs. The recent discovery that AE Aqr is an ejector demonstrates that the criterion for disc formation depends on the spin rate of the white dwarf, in addition to the strength of its magnetic field. It seems likely that AE Aqr represents one phase of a cyclic evolution followed by magnetic CVs. An important consequence is that an observed white dwarf spin rate is not uniquely determined by current properties of the system (e.g. magnetic field and instantaneous accretion rate), but also by its history.

1. Introduction

Among all CVs, magnetic systems have a particular interest because one can directly measure the white dwarf spin and thus make statements about angular momentum transport within the binary. Accordingly, one of the main goals in studying these systems is to explain their observed spin rates in terms of other properties. The fact that no convincing explanation currently exists shows that present theories of the angular momentum flows in these binaries are deficient. This is hardly surprising, given the many idealizations that these theories are forced to adopt in order to give tractable results. These simplifying assumptions may include those of infinite conductivity, flow along field lines, or other ingredients. At bottom however, it appears that the most fundamental of these is the (often implicit) assumption that the accretion flow is very smooth, and thus homogeneously penetrated by the magnetic field. In fact there is ample observational evidence to the contrary; I shall mention here just three examples of such evidence, but there are many more.

A. Evans and J. H. Wood (eds.), Cataclysmic Variables and Related Objects, 153–160.

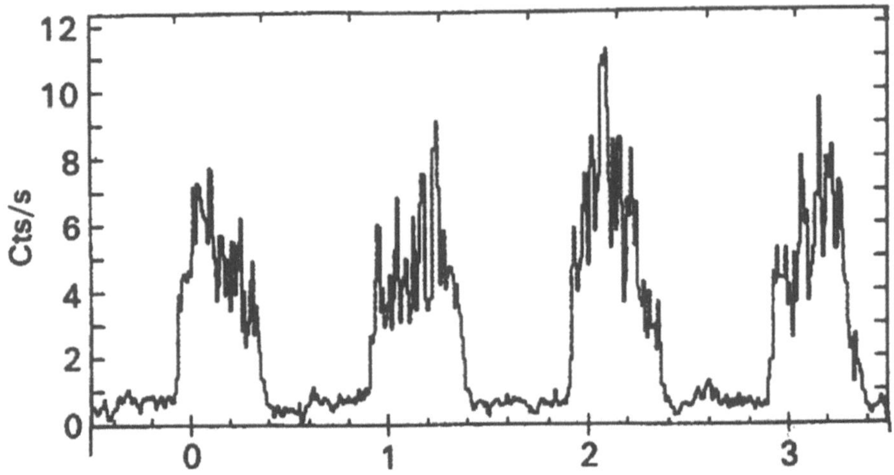

Figure 1. Soft X-ray light curve of AM Her in an anomalous state (Heise et al. 1985).

X–ray light curves: these are strikingly noisy in both non-synchronous and synchronous systems (cf. Fig. 1). In the so-called anomalous states of the latter one can directly simulate the light curves as the arrival of rather small numbers ($\sim$ 15) of blobs of gas at the accretion region (Hameury & King 1988).

X-ray/IR dips: Fig. 2 shows an observation of the synchronously-rotating system EF Eri through an X-ray dip. These dips have long been understood (King & Williams 1985) as photoelectric absorption in the accretion stream as it crosses the line of sight to the accretion region, once per orbital period in synchronous systems (similar dips are seen in the *spin* cycle of the non-synchronous system PQ Gem). The accompanying IR dips were supposed by King & Williams (1985) to result from free-free absorption in the stream, as a homogeneous stream capable of absorbing the X-rays should give such an effect. Yet detailed observation in the J and K infrared bands does not show the expected λ^2 wavelength dependence of this absorption: the process appears to be 'grey'. The explanation (Watson et al. 1989) is that the gas stream is not the simple homogeneous structure envisaged by King & Williams (1985) and subsequent authors, but contains dense filaments or blobs whose optical depths are very large, so that they are effectively opaque. The material between these blobs is diffuse and thus

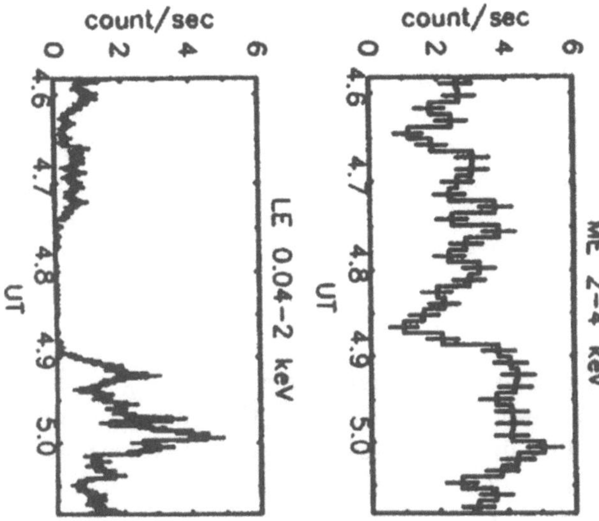

Figure 2. Profiles of an absorption dip in EF Eri in the soft X-ray (LE: 0.04...2 keV) and hard X-ray (ME: 2...4 keV) bands (Watson et al. 1989).

transparent, so that the opacity of the stream is essentially geometrical, i.e. given by the relative area filled by the dense blobs as seen in projection.

The soft X-ray excess: It is now well established that most of the luminosity of the synchronous systems emerges at soft X-ray energies; a similar picture may well be emerging for the non-synchronous systems. The commonly-accepted explanation is that the accretion is mainly in the form of separated blobs of gas, which are able to penetrate the white dwarf photosphere before releasing their infall energy to radiation.

All this evidence suggests that the accretion flow in magnetic CVs is very inhomogeneous. This is of fundamental importance in understanding the gas-field interaction, because the length-scale, l, of a gas element controls the time-scale for the magnetic field to diffuse into it as

$$t_{\rm diff} = l^2 \sigma, \qquad (1)$$

where σ is the conductivity. The gas leaving the secondary star at the inner Lagrange point is probably not threaded by the white dwarf magnetic field, so if the flow is in the form of large gas blobs it is likely that these will behave diamagnetically. This is in direct contrast to many theoretical treatments which assume that the gas will flow along field lines as soon as the magnetic stresses ($\sim B^2/8\pi$) exceed material ones ($\sim \rho v^2$).

2. Inhomogeneous diamagnetic accretion.

Of course an inhomogeneous flow is more complicated than a smooth one; we might then think that the description of accretion flows in magnetic CVs should be even more complex than current MHD treatments. However, it appears that a rather simple phenomenological treatment of the inhomogeneous case is possible. Arons & Lea (1976) already suggested that one should treat the interactions of diamagnetic blobs with the field through a surface drag term. Drell, Foley & Ruderman (1965) showed that the drag time-scale is

$$t_{\text{drag}} = \frac{c_{\text{A}} m}{B^2 l^2} \tag{2}$$

for a blob of mass m and length-scale l. Here c_{A} is the Alfvén speed in the plasma between the blobs (in practice this is usually very diffuse, so that $c_{\text{A}} \sim c$). This drag will predominate over viscous interactions between blobs if t_{drag} is smaller than the viscous time-scale, t_{visc}. For the neutron star systems studied by Arons & Lea (1976) the drag force is probably never important, as the gas passes rapidly from a regime with $t_{\text{visc}} < t_{\text{drag}}$ to one in which the blobs are quickly penetrated by the field and flow along it. For typical conditions in magnetic CVs however, King (1993) showed that

$$t_{\text{dyn}} \lesssim t_{\text{drag}} < t_{\text{visc}} \tag{3}$$

where t_{dyn} is the dynamical ($\sim$ free-fall, Keplerian) time-scale of a gas blob. Under these conditions the gas blobs will behave as independent particles influenced by gravity and the drag force. Per unit mass, this can be written as

$$\mathbf{f}_{\text{drag}} = k[\mathbf{v} - \mathbf{v}(\text{field})]_{\perp}, \tag{4}$$

where $k = t_{\text{drag}}^{-1}$, $\mathbf{v}, \mathbf{v}(\text{field})$ are the blob and field velocities, and the suffix on the square bracket denotes that only the component of this velocity difference perpendicular to field lines is taken into account. The form of (4) immediately shows that if k becomes large, i.e. t_{drag} becomes small compared with t_{dyn}, the blobs will try to reduce the square bracket to zero. Thus (4) naturally describes the tendency of gas to flow along field lines in cases where the field becomes very strong. Of course we do not in general have easy ways of discovering the values of the quantities entering (2), so in practice k is usually regarded as a parameter rather like the viscosity parameter α used in accretion disc theory. On quite general grounds we expect k to fall off with radial distance, r, from the white dwarf, probably as $k \sim r^{-2}$. However the normalization inevitably differs for different blobs [cf. (2)], so it is important in simulations to allow a range of normalizations.

The simple form of (4) makes it extremely easy to add it to particle codes. However, because the blobs are almost independent of each other, a simple analytic treatment of the effects of the force (4) can be given (King 1993). Clearly the blobs can exchange their orbital energies and angular momenta with the spinning white dwarf by interacting with its field through (4). The most important result is the existence of a critical specific orbital energy

$$E_{\rm E} \simeq -\frac{\omega J}{2} \tag{5}$$

for the blobs, where J is the specific angular momentum and ω the white dwarf spin rate. Blobs with $E > E_{\rm E}$ gain energy by interacting with the field, while those with $E < E_{\rm E}$ lose energy. For fixed ω, low-energy blobs lose more energy to the white dwarf and ultimately accrete to it, while high-energy blobs ultimately gain enough energy from the white dwarf that their E becomes positive and they are expelled. Numerical simulations using (4) in a particle code (Wynn & King 1995) qualitatively show precisely this behaviour. Often the expelled blobs are simply caught by the secondary star's gravity and re-accreted rather than being expelled to infinity.

We can use (5) to understand the equilibrium spin rate of a white dwarf accreting in this way. Clearly if ω is close to zero $E_{\rm E}$ will be near zero and thus much larger than the typical value E_1 for blobs leaving the inner Lagrange point L_1 with specific angular momentum J_1. Thus all blobs will be accreted, spinning up the white dwarf, raising ω and lowering $E_{\rm E}$. The process continues until $E_{\rm E}$ becomes of order the typical specific orbital energy of a blob at L_1: clearly ω cannot increase significantly beyond the value $\omega_{\rm eq} \simeq -2E_1/J_1$ since no accretion would be possible. This defines an equilibrium spin rate which is proportional to the orbital rotation rate, the constant of proportionality depending somewhat on the binary mass ratio. For typical parameters one finds the connection

$$P_{\rm spin} \simeq 0.1 P_{\rm orb} \tag{6}$$

between spin and orbital periods (King 1993; Wynn & King 1995).

As the white dwarf is at least potentially a sink of orbital angular momentum, thus tending to shrink the binary, there is the real possibility of dynamical instability, i.e. that the Roche lobe could shrink faster than the stellar radius, causing very rapid mass transfer. Wynn & King (1995) show that spin-up to the 'discless' or 'intermediate polar' (IP) equilibrium (6) always involves dynamical instability, and is therefore likely to be very rapid. Systems with spin periods longer than given by (6), such as EX Hya, thus presumably have accretion discs (spin-up via a disc is always stable, and therefore slow). By contrast, spin-*down* to (6) is always stable: further, once (6) is achieved it is stable, provided that the mass ratio $M_2/M_1 \lesssim 0.7$. Because of the connection $M_2 \sim 0.1 P_{\rm orb}({\rm h})\ {\rm M}_\odot$ this limits the orbital periods of systems accreting in this way to $P_{\rm orb} \lesssim 5$ h. It is also possible to have a perfectly stable system in which all of the mass transferred from the secondary is expelled and returned to it by the white dwarf, particularly if the drag coefficient, k, is large.

3. AE Aqr: an ejector system.

The ideas discussed above have recently received strong support from observations of AE Aqr. This magnetic CV has $P_{\rm orb} = 9.88\,{\rm h}$, $P_{\rm spin} = 33\,{\rm s}$. For many years it was straightforwardly interpreted in terms of a magnetic white dwarf accreting from an extensive accretion disc which was disrupted close to the white dwarf (e.g. Patterson 1994). Difficulties with this idea began with the discovery that the white dwarf is spinning down on a timescale of a few $\times 10^6$ yr (de Jager et al. 1994). The rapid spin rate means that the implied spin-down luminosity is $\sim 6\,10^{33}$ erg s^{-1}, far larger than the observed radiative output of the system. The true nature of AE Aqr is revealed by Doppler tomography performed by Welsh, Horne & Gomer (1996). This shows that, far from possessing a fully developed disc, AE Aqr is actually ejecting most of the matter being transferred from the secondary star. The tomograms show the unmistakable signature of matter being violently accelerated near the white dwarf and flung out of the system by centrifugal force. This is of course where most of the spin-down luminosity is going.

AE Aqr is thus a clear example of a magnetic CV spinning down and ejecting most of the mass transferred to it, just as envisaged in Sect. 2. Using the drag prescription (4) it is straightforward to reproduce the Doppler tomogram from particle simulations of the flow (Wynn & King 1995; Wynn, King & Horne 1995, 1996). These simulations also reproduce the observed spin-down rate. One can use the simulations to produce a pictorial representation of the system (Fig. 3).

The success of the simulations should not obscure the fact that it is extremely surprising that AE Aqr does not have a disc. Indeed in many ways one would regard it as the best *a priori* candidate for such a system. The short spin period must mean that the magnetic field is fairly weak ($\lesssim 10^6$ G), as otherwise the system could never have accreted and spun up to this period. Combined with this, the very long (for a CV) orbital period should make disc formation very easy, as the circularization radius is large, and one would imagine the magnetosphere to be very small, thus presenting no obstacle to the self-intersection of the gas stream and the

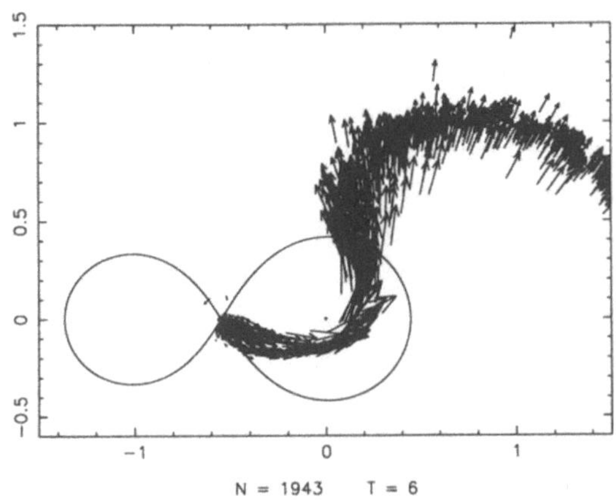

Figure 3. Simulated ejection stream in AE Aqr. The plot shows the Roche lobes of the primary and secondary, with orbital motion being in a clockwise direction. Inertial velocity vectors are plotted for each particle.

consequent disc formation. Yet this argument is clearly wrong. The reason is that the criterion for disc formation evidently depends not just on the magnetic field strength and the accretion rate, but also on the spin rate of the white dwarf. The form (4) of the drag force embodies precisely this connection, and the example of AE Aqr is powerful evidence for such a formulation.

4. What determines the spin rates of magnetic CVs?

The obvious question to ask is how AE Aqr ever got into its present state. The discussion above shows that it would have had no difficulty forming a disc if the spin rate were low. If this was true in the past, the disc would have spun the white dwarf up to $P_{spin} \leq 33$ s in about 10^7 yr. Presumably as long as the disc remained stable, accretion could proceed quite normally. But if accretion was interrupted for a few viscous times, i.e. long enough for the disc to disappear (a few weeks at most), it is evident that it could not re-form. The system would thus have reached something like its present state. There are two possibilities for the future spin behaviour. One is that the white dwarf will spin all the way down to the equilibrium (6), i.e. with $P_{spin} \sim 1$ h. A more likely eventuality is that the spin-down will proceed until the drag time-scale at the circularization radius becomes longer than the viscous time there, allowing a disc to re-form with P_{spin} much smaller than this. Disc accretion will spin the white dwarf up again, possibly to an equilibrium value ≤ 33 s, but any further interruption in mass transfer will allow the system to become an ejector once more. AE Aqr thus seems destined to spend its lifetime moving between disc accretion, ejection, and possibly the IP equilibrium (6), at essentially random intervals determined by the vagaries of the mass transfer rate. Something similar is clearly true for all magnetic CVs, depending on their precise parameters. The important result is that the current observed value of P_{spin} is not simply fixed by short-term conditions in the system, such as the relationship of the accretion rate to the white dwarf magnetic moment μ; on the contrary, we have seen that it must depend on the previous history of the system. The implicit belief in a simple relationship between P_{spin}, P_{orb} and μ, which has characterized research in this field up to now, must be abandoned.

Acknowledgements. Much of the work reported here was performed with Graham Wynn. I thank him for many discussions, and much help in the preparation of this article.

References

Arons J., Lea S. M., 1976, Ap. J., **207**, 914

de Jager, O. C., Meintjes, P. J., O'Donoghue, D., Robinson, E. L., 1994, MNRAS, **267**, 577

Drell, S. D., Foley, H. M., Ruderman, M. A., 1965, J. Geophys. Res., **70**, 3131

Hameury, J.M., King, A.R., 1988, MNRAS, **235**, 433

Heise, J., Brinkman, A. C., Gronenschild, E., et al., 1985, A&A, **148**, L14

King, A. R., 1993, MNRAS, **261**, 144

King, A. R., Williams, G. A., 1985, MNRAS, **215**, 1P

Patterson J., 1994, PASP, **106**, 209

Watson, M. G., King A. R., Jones, M. H., Motch, C., 1989, MNRAS, **237**, 299

Welsh, W.F., Horne, K., Gomer, R., 1996, in preparation

Wynn, G. A., King, A. R., 1995, MNRAS, **275**, 9

Wynn, G. A., King, A. R., Horne, K., 1995, in "Magnetic Cataclysmic Variables", eds D. Buckley, B. Warner, ASP Conf. Ser. **85**, 196

Wynn, G. A., King, A. R., Horne, K., 1996, in preparation

ACCRETION DISC FORMATION IN INTERMEDIATE POLARS

G. A. WYNN, A. R. KING
Astronomy Group, University of Leicester,
Leicester LE1 7RH, UK

1. Introduction

The large-scale accretion flow in the intermediate polars (IPs) is still a matter of vigorous debate. It is known that the magnetic field of the white dwarf (WD) controls the accretion flow close to the surface, channeling the plasma onto the polecaps and giving rise to X-ray emission modulated at the WD spin period ($P_{\rm spin}$). After their discovery it was assumed that IPs were the WD analogues of the pulsing X-ray binaries, where a magnetic neutron star accretes from a disrupted accretion disc. However, a number of authors have pointed out that the criteria for disc formation in IPs are less certain than those for the X-ray binaries.

The simplest possible criterion for disc formation in a binary is that the accretion flow should be able to orbit freely about the primary star (see Frank, King & Raine 1991 for a review). In non-magnetic systems this is merely the condition that the minimum approach distance of the free stream ($R_{\rm min}$) should exceed the radius of the primary. The situation in magnetic systems is more complex, as the magnetic field of the primary presents an obstacle to the infalling accretion stream. In many treatments of IPs it is assumed that the plasma stream is able to orbit freely about the WD until the ram pressure of the stream is of the same order as the magnetic pressure $\rho v^2 \sim B^2/8\pi$, where ρ is the stream density, v the stream velocity and B the local magnetic field strength. This condition fixes the magnetospheric radius, $R_{\rm mag}$, inside which the magnetic field is assumed to thread the stream material and direct the accretion flow along the fieldlines. The criterion for disc formation in these treatments is then

$$R_{\rm min} > R_{\rm mag}. \tag{1}$$

A. Evans and J. H. Wood (eds.), Cataclysmic Variables and Related Objects, 161–164.

The radius $R_{\rm mag}$ is a difficult quantity to estimate in general, but is usually written as a fraction of the spherical Alfvén radius (eg. Frank, King & Raine 1991) such that $R_{\rm mag} \propto \mu^{4/7}\dot{M}^{-2/7}$, where μ is the magnetic moment of the white dwarf and $\dot{M}$ is the accretion rate.

If a disc forms, the WD will tend to spin up as it accretes the Keplerian angular momentum of material at the inner edge of the disc. An equilibrium spin period ($P_{\rm eq}$) will be reached at which the WD is able to centrifugally expel as much angular momentum as it accretes. This equilibrium spin period depends on the complex global interaction between the field and the accretion disc and, as yet, there remains no consensus on the actual spin evolution of the primary in the disc case. However a number of models (eg. Ghosh & Lamb 1979) predict that this equilibrium will occur when the co-rotation radius $R_{\rm co} = (GM_1P_{\rm spin}^2/4\pi^2)^{1/3}$, where the field lines move with the local Keplerian speed, is of order $R_{\rm mag}$. This gives $P_{\rm eq} \propto R_{\rm mag}^{3/2} \propto \mu^{6/7}\dot{M}^{-3/7}$, and leads to an equilibrium with

$$P_{\rm eq} \ll P_{\rm orb}, \tag{2}$$

in general, where $P_{\rm orb}$ is the orbital period of the binary.

2. Inhomogeneous, diamagnetic accretion in IPs

In contrast to the picture presented above, a recent treatment (King 1993; Wynn & King 1995; see also King 1996 for a review) considered the accretion flow in IPs to consist of a number of diamagnetic blobs. Applying the work of Arons & Lea (1980) and Drell, Foley & Ruderman (1965) to conditions typical of IPs, shows that the blobs orbit the WD and interact with the magnetic field via a surface drag force, which can be written, per unit mass, as

$$\mathbf{f}_{\rm drag} = -k\mathbf{v}_{\rm r}. \tag{3}$$

Here $\mathbf{v}_{\rm r}$ is the relative velocity between the field and the blob across the field lines, and $k \propto \mu^2(\rho_{\rm b}l_{\rm b})^{-1}$ is the drag coefficient, where $\rho_{\rm b}$ is the blob density and $l_{\rm b}$ is the blob length-scale. The blobs are able to exchange orbital energy and angular momentum with the field, allowing a particular blob to be accreted or ejected depending upon the WD spin rate (King 1993). So, in contrast to the previous treatment, $P_{\rm spin}$ has a direct influence on the dynamics of the plasma flow, and therefore must affect disc formation. The diamagnetic blob model predicts a longer $P_{\rm eq}$ than the disc equilibrium (2), with $P_{\rm eq} \sim 0.1P_{\rm orb}$ (King 1993; Wynn & King 1995), a position at which a number of IPs have been noted to cluster (eg. Barrett, O'Donoghue & Warner 1988).

A recent application of the diamagnetic blob model to the long period ($P_{\rm orb} = 9.88\,{\rm h}$) cataclysmic variable AE Aqr, has shown this system to be in an ejector state (Wynn, King & Horne 1995). The observation that the rapidly spinning WD ($P_{spin} = 33$ s) is spinning down with an implied spin down power exceeding the X-ray and UV luminosities (de Jager et al. 1994), as well as the unusual Doppler tomogram (Welsh, Horne & Gomer 1996) can be explained if the system is ejecting $\sim 99\%$ of the mass overflowing the L_1 point. By independently fitting the simulated flow pattern with the tomogram and the observed and predicted spin down rates, it is possible to estimate μ and the mass overflow rate ($\dot{M}_{ov}$) as

$$\begin{aligned} \mu &\sim 10^{32}(l_9\,\rho_{-8})^{1/2}\ \mathrm{G\,cm^3} \\ \dot{M}_{ov} &\sim 5\,10^{17}\ \mathrm{g\,s^{-1}}, \end{aligned} \tag{4}$$

where ρ_{-8} is ρ_b in units of 10^{-8} g cm^{-3}, and l_9 is l_b in units of 10^9 cm. Note that the more recent $\dot{M}_{ov}$ estimate presented here is higher than that of Wynn, King & Horne (1995).

3. Discussion: a possible accretion history for AE Aqr

A dramatic example of the differences between the treatments of the accretion flow in IPs outlined in Sect. 1 and 2, when considering the question of disc formation, can be illustrated by the following possible accretion history of AE Aqr.

If, at the onset of mass transfer, we assume AE Aqr to have an initial orbital period $\gtrsim 10$ h, and a μ and $\dot{M}$ in line with the estimates (4), it is possible to consider the disc formation criteria in both treatments discussed above. In the case of a homogeneous stream, ram and magnetic pressure balance leads us to condition (1). It is possible to estimate these radii as $R_{mag} \sim 10^9$ cm, and $R_{min} \sim 0.5 R_{circ} \sim 10^{10}$ cm where R_{mag} is taken as half the spherical Alfvén radius, and R_{min} is calculated from the stream circularization radius R_{circ}. From this we can see that condition (1) is satisfied and an accretion disc is expected to form. For the case of diamagnetic blob accretion we must further take into account the spin rate of the WD. It is probable that the initial P_{spin} of the WD was much longer than the current value of 33 s; if we consider $P_{spin} \sim$ hours initially the combination of a low μ and $\mathbf{v}_r$ in (3) (at $r \sim R_{min}$) allows the blobs to orbit freely about the white dwarf. So again we expect the formation of an accretion disc at the onset of mass transfer.

Once the accretion disc has formed the WD will spin up to the disc P_{eq} as in (2). It is interesting to note that $P_{eq} \simeq 33$ s gives an estimate of μ (cf. Ghosh & Lamb 1979, with a fastness parameter of unity) as $5\,10^{31} \lesssim \mu$ (in G cm^3) $\lesssim 3\,10^{32}$ for accretion rates in the range $10^{17} \lesssim \dot{M}$ (in g s^{-1}) $\lesssim 10^{18}$. This independent estimate is encouraging in that it is in agreement with the diamagnetic blob model result (4).

Once the WD has been spun up, the evolution of the system depends on short-term fluctuations of $\dot{M}$. If we assume that $\dot{M}$ drops to a value low enough to allow the accretion disc to be eroded (typically only a few weeks), then on the resumption of mass transfer the accretion flow will now interact with a rapidly rotating magnetosphere ($P_{\rm spin} \sim 33$ s). As far as the pressure balance condition is concerned the situation is unchanged if $\dot{M}$ increases to its previous value, and an accretion disc reforms (in fact we would expect accretion disc formation even for much lower $\dot{M} \lesssim 10^{16}$ g s^{-1}). However, in the framework of the diamagnetic model, $\mathbf{f}_{\rm drag}$ has increased by 2...3 orders of magnitude because of the increased value of $\mathbf{v}_r$, which is now governed by $P_{\rm spin}$. In this case we expect the blobs to gain angular momentum and to be rapidly expelled from the system as they approach $R_{\rm min}$, as confirmed by numerical simulations (cf. Wynn, King & Horne 1995).

So it can be seen that the two prescriptions for the disc formation criteria in IPs give very different results for the same μ and $\dot{M}$; the difference being that $P_{\rm spin}$ plays a central role in determining the dynamics of the plasma flow when the flow is inhomogeneous and diamagnetic. Under the homogeneous stream treatment outlined in Sect. 1, AE Aqr would be an ideal system for an accretion disc to form, having long $P_{\rm orb}$ (and thus large $R_{\rm min}$) and low μ. However, once a disc has formed and spun up the WD, any interruption of mass transfer causes the system to enter an ejector state, as currently observed. So it is probable that the IPs are a mixed bag of systems, which show disc accretion, blob accretion and ejection phases determined by μ and the evolution of $\dot{M}$ and $P_{\rm spin}$ in the system. This fact may well account for the failure of any single model to explain the observational properties of IPs (cf. Hellier 1996).

References

Arons, J., Lea, S.M., 1980, Ap. J., **235**, 1016
Barrett, P., O'Donoghue, D., Warner, B., 1988, MNRAS, **233**, 759
de Jager, O.C., Meintjes, P.J., O'Donoghue, D., Robinson, E.L., 1994, MNRAS, **267**, 577
Drell, S.D., Foley, H.M., Ruderman, M.A., 1965, J. Geophys. Res., **70**, 3131
Frank, J., King, A.R., Raine, D.J., 1991, Accretion Power in Astrophysics (Cambridge University Press)
Ghosh, P., Lamb, F.K., 1979, Ap. J., **232**, 259
Hellier, C., 1996, these proceedings, p143
King, A.R., 1993, MNRAS, **261**, 144
King, A.R., 1996, these proceedings, p153
Welsh, W.F., Horne, K., Gomer, R., 1996, in preparation
Wynn, G.A., King, A.R., 1995, MNRAS, **275**, 9
Wynn, G.A., King, A.R., Horne, K., 1995, in "Magnetic Cataclysmic Variables", eds D. Buckley, B. Warner, ASP Conf. Ser. **85**, 196

THE FLARES OF AE AQR

M. ERACLEOUS[1], K. HORNE[2], M. LIVIO[1], W. F. WELSH[3]
1 Space Telescope Science Institute, 3700 San Martin Drive, Baltimore, MD 21218, USA
2 Physics and Astronomy, The University of St. Andrews, North Haugh, St. Andrews, Fife, KY16 9AJ, UK
3 Physics Department, Keele University, Keele, ST5 5BG, UK

Abstract. The enigmatic flaring activity of AE Aqr, which extends from the radio to the X-ray bands, may be the exotic outcome of a relatively feeble accretion flow onto the rapidly spinning white dwarf. We summarize and interpret the results of recent observations by HST and ROSAT of the aperiodic flares. Optical and UV flares are tightly correlated and are clearly involved in powering the emission lines. The spectrum of a flare consists of a Balmer continuum, and a plethora of emission lines, which vary synchronously. The large optical/UV flares are *not* accompanied by *large* X-ray outbursts. The radial velocity curves of the UV lines, and of Hα, suggest that they originate in the accretion stream. The observational results dis-favor scenarios invoking coronal activity on the secondary star or instabilities at the disk/magnetosphere interface. Rather, they support a recently proposed scenario in which the accretion flow consists of blobs which are shocked upon encountering the white dwarf magnetosphere.

1. Observations and results

AE Aqr was observed twice in the UV with the HST: in 1992 November and in 1993 October (see the accounts by Eracleous et al. 1994, 1995). The second observation was made during the *World Astronomy Days* campaign, and coincided with a ROSAT observation (described by Osborne et al. 1995), yielding about 35 m of truly simultaneous X-ray and UV data.

The time-resolved UV spectroscopy carried out with the HST shows that the flare spectrum consists of a strong Balmer recombination continuum ($T \sim 10^4$ K), and a plethora of emission lines from highly ionized species (ionization energies: 50...100 eV). *All* the lines were observed to

A. Evans and J. H. Wood (eds.), Cataclysmic Variables and Related Objects, 165–166.

vary together through the flares, with no discernible time lag (down to a resolution of 2.5 m). The phase and amplitude of the radial velocity curves of the He II λ1640 and Si III] λ1892 lines suggest very strongly that the lines originate in the accretion stream from the companion star, much like the radial velocity curves of the Balmer lines (Welsh, Horne & Gomer 1993; Reinch & Beuermann 1994). Moreover, the simultaneous UV and X-ray observations show that the large optical/UV flares are not accompanied by *large* X-ray outbursts.

2. Discussion and conclusions

The observational results, and in particular the emission-line radial velocity curves, dis-favor an origin of the flares in material following the orbit of the white dwarf, or on the companion star. In view of these results, scenarios attributing the flares to instabilities at the white dwarf magnetosphere (van Paradijs, Kraakman & van Amerongen 1989; Spruit & Taam 1993), and suggestions invoking coronal mass ejections from the companion star (Chincarini & Walker 1981) appear unlikely. Rather, the observations favor a recently proposed scenario, in which the flares are the results of the impact of a fragmented ('blobby') accretion stream on the white dwarf magnetosphere (Wynn, King & Horne 1995; Wynn 1996).

In the context of the latter scenario, what is observed as a flare is the arrival of a new blob at the magnetosphere of the white dwarf, where it is shock-ionized and/or photo-ionized. Some of the matter is eventually accreted onto the white dwarf, but most of it is ejected from the system by a propeller-type mechanism. As the matter moves away from the system, it cools and recombines and the flare declines. All of the activity occurs near the white dwarf magnetosphere where the dynamical time and the recombination time for hydrogen ($n \sim 10^{10}$ cm^{-3}) are comparable to the rise time of the flares ($\lesssim$ 30 m).

References

Chincarini, G., Walker, M.F., 1981, A&A, **104**, 24
Eracleous, M., Horne, K., Robinson, E.L., et al., 1994, Ap. J., **433**, 313
Eracleous, M., Horne, K., Osborne, J.P., et al., 1995, in "Magnetic Cataclysmic Variables", eds D. Buckley, B. Warner, ASP Conf. Ser. **85**, 392
Osborne, J.P., Clayton, K.L., O'Donoghue, D., et al., 1995, in "Magnetic Cataclysmic Variables", eds D. Buckley, B. Warner, ASP Conf. Ser. **85**, 368
van Paradijs, J., Kraakman, H., van Amerongen, S., 1989, A&AS, **79**, 205
Reinch, K., Beuermann, K., 1994, A&A, **282**, 493
Spruit, H. C., Taam, R.E., 1993, Ap. J., **402**, 593
Welsh, W.F., Horne, K., Gomer, R., 1993, Ap. J., **410**, L39
Wynn, G., King, A.R., Horne, K., 1995, in "Magnetic Cataclysmic Variables", eds D. Buckley, B. Warner, ASP Conf. Ser. **85**, 196
Wynn, G., 1996, these proceedings, p161

TRIGGERING THE RADIO EMISSION FROM AE AQR

D. STEEGHS[1], J. KUIJPERS[1], L. FLETCHER[1]
M. ABADA-SIMON[1], K. HORNE[2], G. RAMSAY[1]
1. Astronomical Institute, Utrecht University, The Netherlands
2. Physics & Astronomy, University of St.Andrews, UK

Abstract. We propose a model for the acceleration of particles by magnetic pumping to drive the radio emission from AE Aqr.

1. Introduction

The peculiar CV AE Aqr displays irregular, flare-like radio emission which varies in intensity and duration (minutes to hours). Even during the inter–flare periods there is usually some quiescent radio flux (Abada-Simon et al. 1993). The flare-like emission is attributed to synchrotron emission from high energy electrons in expanding plasma clouds (Bastian, Dulk & Chanmugam 1988). We propose that magnetic pumping (betatron acceleration) accelerates these electrons.

As the white dwarf rotates, the magnetic field strength is periodically varied as it opens and closes around the accreting diamagnetic 'blobs' and the secondary star, which lies within the light cylinder (Fig. 1). As the magnetic pressure experienced by the accreting blobs changes by a factor of ~ 4 every rotation, internal shocks are generated and surface plasma instabilities arising from the relative velocity of the blob and the field disrupt and strip off the surface layers. This results in the loading onto the oscillating field of particles which can be energized by magnetic pumping.

2. Magnetic pumping

In a changing magnetic field $p_\perp^2/B$ is conserved, so that the particle perpendicular momentum, $p_\perp$, increases as the field increases. If some momentum can be transferred, by particle scattering, to the parallel direction (not affected by the field change) the particle energy increases with time, t, as

A. Evans and J. H. Wood (eds.), Cataclysmic Variables and Related Objects, 167–168.

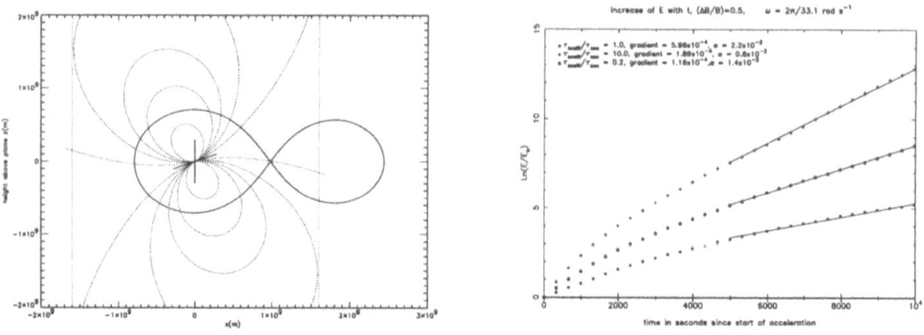

Figure 1. Left panel: side view of the binary system, showing the tilted dipole field and the position of the light-cylinder. Right panel: a particle simulation result, showing the energy increase due to magnetic pumping. (Kuijpers et al. 1996)

$(\Delta B/B)^2 \exp(\alpha t)$ (at relativistic energies), where α is a parameter which maximizes when the particle collision time equals the field oscillation time (white dwarf rotation period). Particle simulations of pumping (Kuijpers et al. 1996) show that standing oscillations accelerate more efficiently than traveling Alfvén waves. We calculate values for α and find that, under realistic conditions, enough particles to account for the flare energy can be accelerated up to the required energies in a few thousand seconds, even when energy losses (e.g. synchrotron) are included.

3. Radio emission

The particles are trapped by magnetic mirroring in the converging white dwarf magnetic field lines. Fresh material is continually supplied by accretion and, as acceleration continues, the loading of the field increases. When the total kinetic energy of the particles becomes comparable with the total energy of the magnetic field, the field no longer contains them and a MHD instability is triggered resulting in a catastrophic loss of particle confinement and the ejection of a plasmoid, which produces the enhanced radio emission. The magnetosphere can then be energized again for the next flare. The quiescent radio emission is attributed to the synchrotron emission of the fast particles as they are accelerated between eruptions.

References

Abada-Simon, M., Lecacheaux, A., Bastian, T.S., et al., 1993, Ap. J., **406**, 692

Bastian, T.S., Dulk, G.S., Chanmugam, G., 1988, Ap. J., **324**, 431

Kuijpers J., Fletcher L., Abada-Simon M., et al., 1996, in preparation

HST FOS OBSERVATIONS OF YY DRA

CAROLE A. HASWELL, JOSEPH PATTERSON
Columbia University, 538 W. 120th St., NY, NY 10027, USA

Abstract. The UV continuum of YY Dra exhibits a sinusoidal modulation with 15.5% semi-amplitude and period 264.7 s, commensurate with one half the white dwarf spin period. The C IV line profile varies with spin phase, implying accretion onto, and visibility of, both magnetic poles. The line width is greatest at the phases of continuum pulsation maximum, when the wings extend out to $\pm 3000\,\mathrm{km\,s^{-1}}$. This implies a minimum white dwarf mass of $0.62\,\mathrm{M_{\odot}}$, and constrains the angle between the line of sight and the magnetic axis.

1. Introduction

Coherent X-ray pulsations (Beuermann & Thomas 1993; Patterson & Szkody 1993) have confirmed YY Dra as a DQ Her star/intermediate polar (Patterson et al. 1992; Patterson 1994; Hellier 1996). The X-ray and optical data imply a white dwarf spin period, $\mathrm{P_{spin}}$, of 529 s, and variable accretion geometry: sometimes modulation occurs predominantly at P_{spin}, while at other times $P_{\mathrm{spin}}/2$ dominates.

We present a brief discussion of high time resolution UV observations of YY Dra[1]; a full presentation will be given in Haswell et al. (1996).

2. Observations

HST observed YY Dra on 1994 June 8...9. FOS/RAPID mode with the G130 grating was used, yielding a time resolution of 5.6 s, a duty cycle of 87%, and coverage of 1147...1605 Å. The spectrum reveals strong C IV $\lambda\lambda 1548.1, 1550.8$ emission and several other line emission features.

[1]Based on observations with the NASA/ESA Hubble Space Telescope obtained at the Space Telescope Science Institute, which is operated by the Association of Universities for Research in Astronomy, Incorporated, under NASA contract NAS5-26555

A. Evans and J. H. Wood (eds.), Cataclysmic Variables and Related Objects, 169–172.

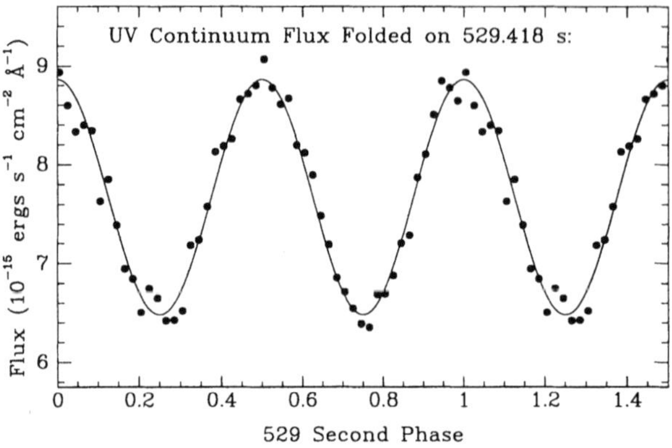

Figure 1. The UV continuum light folded on the inferred white dwarf spin period, 529.4 s; the modulation occurs purely at one half the spin period: 264.7 s.

2.1. THE UV CONTINUUM PULSATIONS

A UV continuum light curve was generated by computing the mean flux in the region 1420...1525 Å for each of the 3873 individual exposures. Strong pulsations were detected. The best-fitting sine wave has period 264.709 ± 0.05 s, with mean flux $7.67\,10^{-15}\,\mathrm{erg\,s^{-1}\,cm^{-2}\,Å^{-1}}$ and a semi-amplitude of 15.5%. Fig. 1 shows the light curve folded on a period of 529.418 s: there are no significant deviations from a sinusoidal modulation with period 264.7 s. If the emission is due to two-pole accretion with $P_{\mathrm{spin}} = 529.4$ s, then we must conclude that the accretion geometry produced extremely symmetric pulses at the epoch of our HST observations.

2.2. PULSATIONS IN THE UV LINES

To search for line profile modulations, our spectra were folded on the inferred spin period: 529.4 s. Fig. 2 shows the trailed spectrogram of the C IV line after subtracting the (spin-phase dependent) continuum. The line profile variations are subtle, but the line centre flux is weakly pulsed (4.4% semi-amplitude) in antiphase with the underlying continuum pulsations, and broad wings appear at the phases of continuum maxima. These wings appear simultaneously on both sides of the line and extend out to $\pm \sim 3000\,\mathrm{km\,s^{-1}}$.

3. Discussion and conclusions

The most plausible origin for the high velocity emission is radially infalling material in the accretion columns. Assuming free-fall from infinity, a radial velocity of 3000 km s^{-1} is attained at $r = 2.5\,10^9 \cos^2(|\alpha - i|)$ cm; here

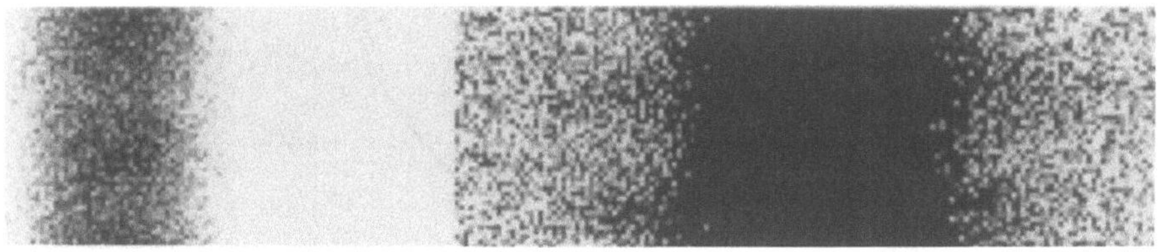

Figure 2. The trailed spectrogram of the C IV line emission after folding the data on the white dwarf spin period. The same data appear twice: on the left-hand side with grayscale adjusted to show the line center modulations; on the right-hand side to show the modulations in the wings. Spin phase increases (through one 529-s cycle) vertically upwards, and wavelength increases to the right.

α is the angle between the magnetic axis and the spin axis of the white dwarf, i is the orbital inclination, and a white dwarf mass of $M_{wd} = 0.85\,M_\odot$ has been used (Mateo, Szkody & Garnavich 1991, henceforth MSG). This corresponds to about four white dwarf radii (r_{wd}) from the center of the white dwarf. Since we see both red *and* blue wings, we are seeing two columns simultaneously. Hence, we can definitely rule out the hypothesis of there being only one accreting column from which the radiation beaming pattern produces two flux maxima per white dwarf spin period. Maximum radial velocity occurs when the accretion columns are most closely aligned with our line of sight. Thus we deduce that the maximum continuum flux also occurs when the accretion columns are aligned with our line of sight.

For infalling material in the far column with radial velocity K to be visible, rather than being occulted by the white dwarf, we require:

$$\sin(|\alpha - i|)\cos^2(|\alpha - i|) \geq \frac{r_{wd}K^2}{2GM_{wd}}. \tag{1}$$

The Hamada-Salpeter mass-radius relation: $r_{wd} \approx 0.56(M_{wd}/0.85\,M_\odot)^{-0.8}$ then gives a minimum allowed value of M_{wd} for any $|\alpha - i|$. This relationship for $K = 3000\,\mathrm{km\,s^{-1}}$ is the curve shown in Fig. 3; the region of parameter space below the curve is disallowed. Hence the C IV line profile implies a minimum white dwarf mass of $0.62\,M_\odot$. This is in good agreement with the completely independent mass determination of MSG.

MSG also determined the orbital inclination of the system: $i = 42° \pm 5°$. Since both α and i are $< 90°$, this implies $|\alpha - i| < 53°$; which yields the vertical dotted line in Fig. 3. Putting all these constraints together requires the parameters to lie within the region bounded by the MSG $\pm 1\sigma$ mass limits at the top and bottom; on the left by our curve, and on the right

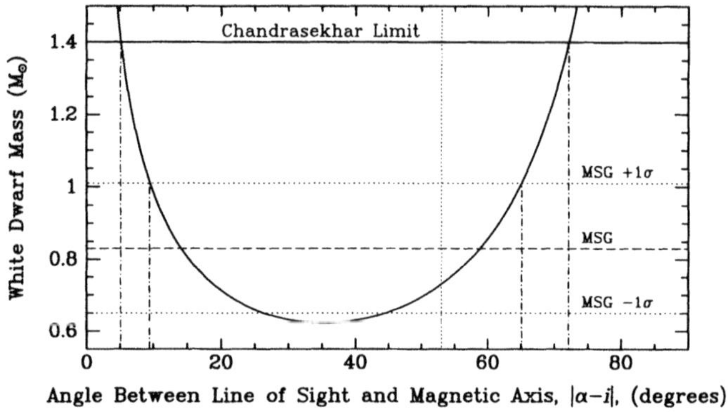

Figure 3. Constraints on white dwarf mass: the region of parameter space below the curve is disallowed (see text). Hence a minimum white dwarf mass of 0.62 $M_\odot$ is derived. MSG's independent mass determination (lower three horizontal lines) agrees well. The Chandrasekhar upper limit on the white dwarf mass leads to the outer pair of dot-dashed vertical lines delimiting the allowed angle. The inner pair of dot-dashed vertical lines delimit the allowed range of angles arising from the combination of the MSG mass and the curve resulting from the C IV line width. The dotted vertical line at 53° is the constraint which arises from MSG's orbital inclination: the region to the right of this line is disallowed.

by our curve and $|\alpha - i| < 53°$. We can add a further weak qualitative constraint by noting (Fig. 1 and Fig. 2) that the two accretion columns appear remarkably similar to us. This is easiest to understand if $\alpha \sim 90°$, and $|\alpha - i|$ is towards the right hand side of the allowed region. A similar requirement on the orientation of the magnetic axis in DQ Her was recently argued by Robinson et al. (1996) and Zhang et al. (1996).

Acknowledgments. Support for this work was provided by NASA through grant number GO-5500-01-93A from the Space Telescope Science Institute, which is operated by the Association of Universities for Research in Astronomy, Incorporated, under NASA contract NAS5-26555.

References

Beuermann, K., Thomas, H.-C., 1993, Adv. Space Res., **13(12)**, 115
Haswell, C.A., Patterson, J., Hellier, C., et al., 1996, in preparation.
Hellier, C., 1996, these proceedings, p143
Mateo, M., Szkody, P., Garnavich, P., 1991, Ap. J., **370**, 370
Patterson, J., 1994, PASP, **106**, 209
Patterson, J., Schwartz, D.A., Pye, J.P., et al. 1992, Ap. J., **392**, 233
Patterson, J., Szkody, P., 1993, PASP, **105**, 1116
Robinson, E.L., Zhang, E., Bless, R.C., et al., 1996, these proceedings, p173
Zhang, E., Robinson, E.L., Stiening, R.F., Horne, K., 1996, Ap. J., in press

THE 71-SECOND OSCILLATION OF DQ HER AT 2180 Å

E. L. ROBINSON[1], E. ZHANG[1], R. C. BLESS[2], J. F. DOLAN[3], J. L. ELLIOT[4], G. W. VAN CITTERS[5]

1. Department of Astronomy, University of Texas, Austin, TX 78712, USA

2. Department of Astronomy and Space Astronomy Laboratory, University of Wisconsin, Madison, WI 53706, USA

3. Laboratory for Astronomy and Solar Physics, NASA Goddard Space Flight Center, Greenbelt, MD 20771, USA

4. Department of Earth, Atmospheric, and Planetary Sciences and Department of Physics, Massachusetts Institute of Technology, Cambridge, MA 02139, USA

5. Division of Astronomical Sciences, National Science Foundation, 4201 Wilson Blvd. Arlington, VA 22230, USA

Abstract. The 71-s modulation in the light curve of the old nova DQ Her has a mean semi-amplitude of 1.8% at 2180 Å, although the amplitude varies by at least a factor of two from orbit to orbit. The disturbance in the amplitude and phase of the modulation during eclipses is similar to that observed at visual wavelengths. We conclude that at 2180 Å, as at visual wavelengths, the modulated flux is reprocessed and comes from the accretion disk.

1. Introduction

The visual light curve of the old nova DQ Her has a highly-coherent, sinusoidal modulation with a period of 71.06 s and an amplitude of 1% induced by rotation of the magnetic, accreting white dwarf in the system [1]. The light curve also shows eclipses, which recur every 4 h 39 m. The visual flux from DQ Her is dominated by the accretion disk around the white dwarf and, consequently, the eclipses, which are occultations of the white dwarf and its accretion disk by their cool companion star, are wide, lasting ~ 0.2 of the orbital period. The phase of the oscillation is disturbed during the eclipses: it increases by at least 100° during eclipse ingress, then jumps

A. Evans and J. H. Wood (eds.), Cataclysmic Variables and Related Objects, 173–176.

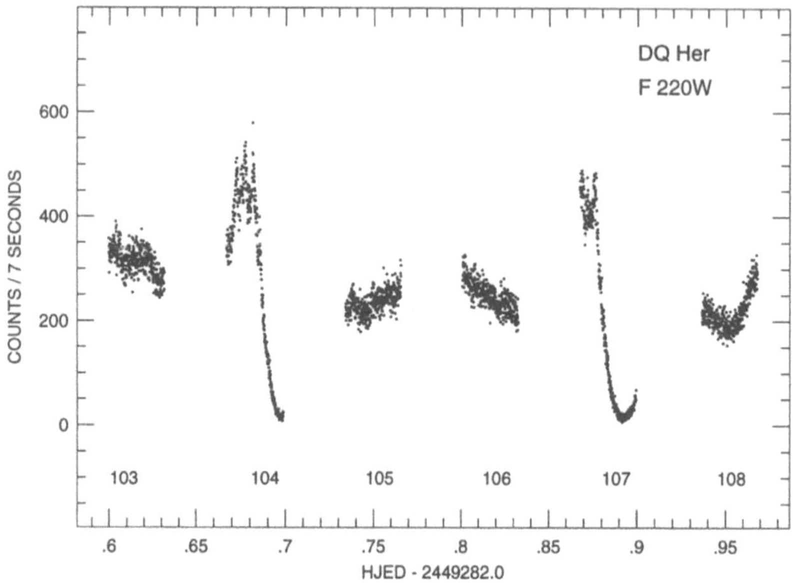

Figure 1. The ultraviolet light curve of DQ Her. The F220W filter has a full width at half maximum of 480 Å and an effective wavelength of 2180 Å. The light curve has been compressed to 7 s per sample in the figure.

by $\sim 120°$ at mid eclipse, and finally climbs smoothly by another $\sim 140°$ during eclipse egress, returning to its phase outside of eclipse [3]. The amplitude is also disturbed: it is greater than its uneclipsed value near the beginning and end of eclipse, but fades to less than 10% of its uneclipsed average at mid-eclipse. Since the amplitude and phase disturbances take place throughout the entire eclipse, most of the flux in the 71-s modulation is reprocessed light coming from the accretion disk. Quantitative modeling shows that little or no modulated flux is received directly from the white dwarf [2, 3].

2. The observations

We observed DQ Her for six successive HST orbits in 1993 October with the HST high-speed photometer and the F220W filter, which has an effective wavelength near 2180 Å. The light curve is shown in Fig. 1, where the sections of light curve from each HST orbit are labeled by a number from 103 to 108; the gaps between sections are caused by earth occultations. The light curve covers two orbits of DQ Her. In addition to the deep eclipse, it has an unexpectedly prominent hump with strong flickering just before the eclipse. The hump is similar to the humps in the light curves of many

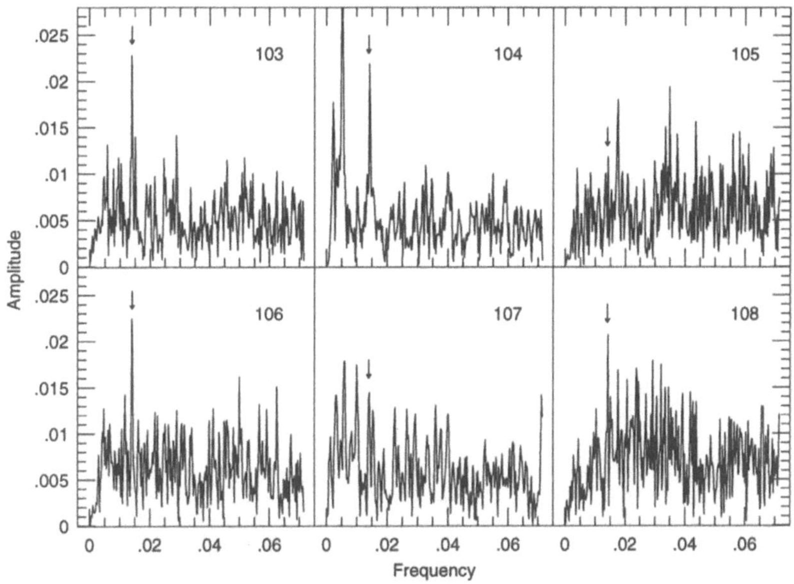

Figure 2. Power spectra of each section of the ultraviolet light curve of DQ Her. Only the portions of the power spectra near 71 s are shown; the expected positions of the 71-s modulation are marked by arrows.

dwarf novae leading us to presume that, as in the dwarf novae, the hump is caused by a bright spot on the disk.

3. The 71-second modulation

We calculated the power spectrum of each section of the light curve to test for the presence of the 71-s modulation. The pertinent portions of the power spectra are shown in Fig. 2; the expected positions of the 71-s modulation are shown by arrows. We detected the modulation in four sections of the ultraviolet light curve – 103, 104, 106, and 108 – but not in section 105 and probably not in section 107. The measured semi-amplitude varied from 2.3% in section 103 to less than 1.2% in section 105. The variation in amplitude is mostly uncorrelated with orbital phase since the amplitude of the modulation at a given orbital phase can be different at different times (compare the amplitudes in sections 105 and 108, both of which cover orbital phases near $\phi = 0.3$). The mean semi-amplitude during our observations was 1.8%.

Two sections of the ultraviolet light curve cover eclipse ingresses, sections 104 and 107. The 71-s modulation was too weak to detect during section 107, leaving section 104 for measuring the phase and amplitude of

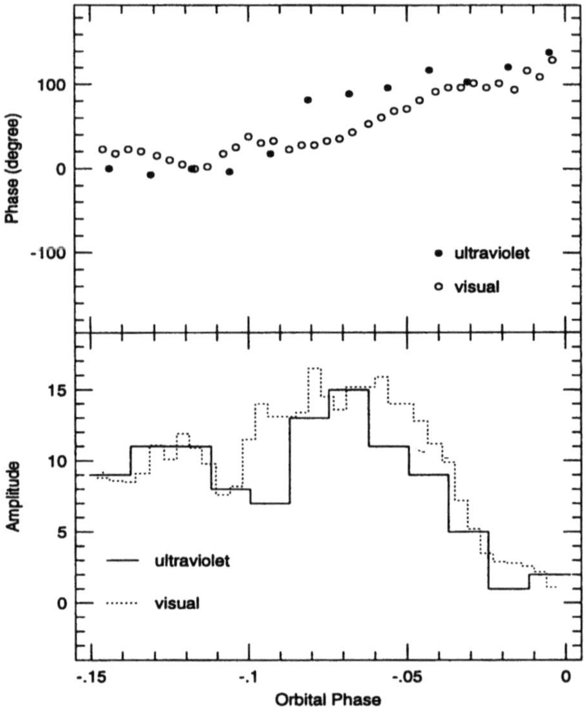

Figure 3. The amplitude (lower panel) and phase (upper panel) of the 71-s oscillation as a function of orbital phase during section 104 of the ultraviolet light curve. The phase and amplitude at optical wavelengths are shown for comparison [1].

the 71-s oscillation during eclipse. We extracted the phase and amplitude from section 104 in the usual way, by fitting a sine curve with a period of 71.06 s to the light curve. The measured amplitude and phase are shown in Fig. 3, where we have fitted a sine curve with 6 cycles, shifting 3 cycles between fits. The amplitude and phase variations at visual wavelengths are shown for comparison. The variations at ultraviolet wavelengths mimic those at visual wavelengths; the differences, if any, are too small to measure.

We conclude that at 2180 Å, as at visual wavelengths, the flux in the 71-s oscillation is reprocessed and comes from the accretion disk, not directly from the white dwarf.

References

1. Patterson, J., Robinson, E. L., Nather, R. E., 1978, Ap. J., **224**, 570
2. Petterson, J., 1980, Ap. J., **241**, 247
3. Zhang, E., Robinson, E. L., Stiening, R. F., Horne, K., 1996, Ap. J., in press

SPIN-RESOLVED Hα SPECTROSCOPY AND PHOTOMETRY OF THE INTERMEDIATE POLAR RXJ0558+5353

S. R. DUCK[1], M. D. STILL[2], A. ALLAN[2], K. HORNE[2]
R. W. HILDITCH[2]
1. Oxford University, Astrophysics, Keble Road, Oxford, UK
2. St. Andrews University, N. Haugh, St. Andrews, Fife, UK

Abstract. RXJ0558+5353 was discovered and classified as an intermediate polar during the ROSAT all-sky survey and subsequent optical follow up programme (Haberl et al. 1994). A further pointed ROSAT observation revealed a spin periodicity of 272.74 s and a soft X-ray component well represented by an absorbed 57 eV blackbody, the spin modulation being due principally to intensity variations of the component (rather than variable absorption). Optical spectroscopy showed the orbital period to be 4.15 h. In this paper we show the true spin period to be 545.4555(8) s (Allan, Horne & Hilditch 1995), twice the published X-ray period, and also discuss preliminary results of the first spin-resolved spectroscopy of RXJ0558+5353.

1. Observations

The photometry presented here, the results of which were first published in Allan, Horne & Hilditch (1995), uses V and R band light curves obtained at the James Gregory 1 m telescope, sited at St. Andrews University, between 1994 November 24 and 1995 February 22.

The spin-resolved spectroscopy was obtained in 1995 February on the INT. The set-up used provided a resolution of approximately $80\,\mathrm{km\,s^{-1}}$ and coverage enabling the simultaneous monitoring of Hα and He I 5875. A total of 173, 30...45 s, flux calibrated spectra were obtained from about two hours on each of three successive nights in sub-arcsecond seeing.

2. Results

A periodogram of the photometric data shows significant power at twice the spin period derived from the ROSAT data. Fig. 1 shows the phase folded light curves from several nights observations, the morphology of which appears to change between the observations.

A. Evans and J. H. Wood (eds.), Cataclysmic Variables and Related Objects, 177–178.

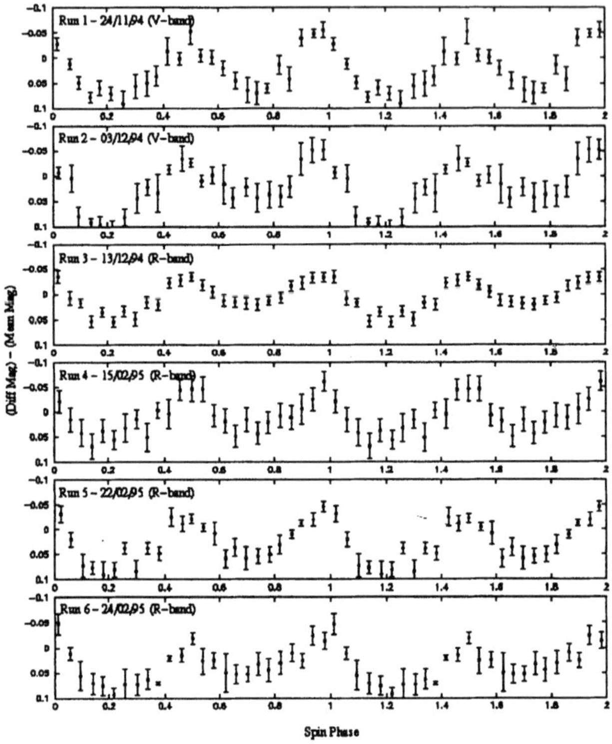

Figure 1. V and R band photometry of RXJ0558+5353 taken on several nights and folded on the new spin period of $\sim$ 545 s.

The continuum data from the spectroscopic observations also show the same double peaked spin light curve, however the Hα flux spin light curve does not obviously show this, although the orbital scatter may hide a lower level of modulation. Similarly the phase binned Hα line profile shows no clear changes over the spin period.

A Doppler map of the Hα emission presents problems in interpretation, at least partially due to the lack of a published orbital ephemeris. If the bright asymmetrical emission is due to emission from the secondary or hot spot, then although the implied phasing agrees with that of the orbital light curves, the disc-like ring in the tomogram is then not centered anywhere near to the expected location of the primary. This may indicate that this system contains a very distorted disc or that some direct stream accretion is occurring.

References

Allan, A., Horne, K., Hilditch, R.W., 1995, IAUC 6154
Haberl, F., Thorstensen J. R., Schwarzenburg-Czerny, A. et al., 1994, A&A, **291**, 171

ROTATIONAL DISTURBANCE IN BG CMI

M. A. GARLICK
Astronomy Centre, University of Sussex,
Brighton BN1 9QH, UK

Abstract. I report on the independent discovery of a feature in BG CMi which resembles a rotational disturbance (RD). A similar effect has already been reported in the He II λ 4686 line, whereas here it is seen in the *Balmer* lines, 0.25 orbital cycles prior to that reported previously. These results suggest that BG CMi is at a relatively high inclination.

1. The data

These data have been published before (see Garlick et al. 1994 for details), but that paper was limited to a spin-resolved analysis. Here, I present the results of an orbital analysis. The left panels of Fig. 1 show the V/R ratio (top) and the intensity in arbitrary units (bottom) of the Balmer lines of BG CMi, rebinned and plotted against orbital phase (Patterson & Thomas 1993). Data from the *wings only* (400...1400 km s^{-1}) of Hδ–Hβ are included. There is a strong feature near phase 0.7...0.8 which is very much like a classic RD. The feature was fully covered only on the first night, but there is clear evidence for a redward excursion in the Balmer line V/R ratio just before the coverage stops, at phase 0.75, on the second night. There is also evidence for an increase in the V/R ratio at phase zero—at the same phase as an RD reported previously but in the He II line (de Martino et al. 1995). In the He II line here, however (Fig. 1, right), no activity is seen in the V/R ratio, whilst the intensity shows a dip at phase 0.05 and marginal evidence for another at 0.75. These observations suggest that we are seeing evidence for two line eclipses in this system, near phases 0.75 and zero.

2. Interpretation?

A possible interpretation is that the accretion stream hits the edge of a disc, a ring, or the magnetosphere, throwing material out of the orbital

A. Evans and J. H. Wood (eds.), Cataclysmic Variables and Related Objects, 179–180.

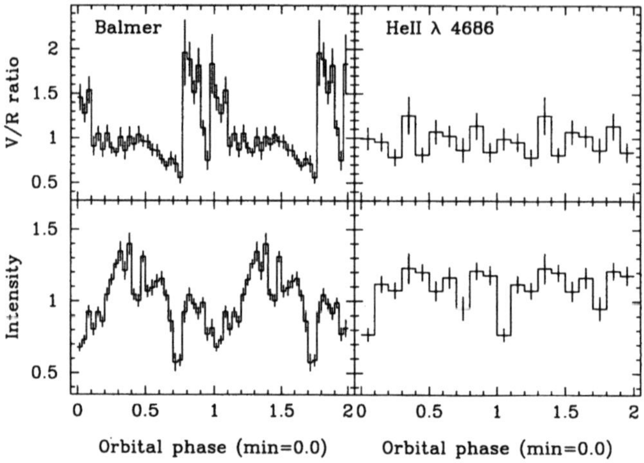

Figure 1. Left: The average V/R ratio (top) and arbitrary intensity (bottom) in the wings of the Balmer lines of BG CMi as a function of phase. Right: the same for the He II λ 4686 data. Fewer phase bins were used because of the weaker signal in He II.

plane. This material, on the observer's side of the white dwarf at phases 0.75...0.90, then obscures first the red- then the blue-shifted regions of the inner edges of the accretion disc, creates an RD in the wings of the Balmer lines, and may also be responsible for the X-ray dips in this system as has already been proposed (Hellier, Garlick & Mason 1994). An eclipse by the secondary star follows ~ 0.25 cycles later.

This simple picture is capable of explaining the phases of the events seen in the data, but raises questions. Most notably: how is it that the Balmer line region suffers RDs at phases 0.75 and possibly phase zero, whilst the He II line merely undergoes a dip in intensity at these phases? A possibility is that the He II region is localised—not in a ring around the white dwarf—and is small enough to be wholly obscured by the thrown up material so that the RD behavior is not seen. However, as an RD in He II at phase zero *was* seen by de Martino et al. (1995), this indicates that significant changes in the whereabouts of the line emission regions in the system have taken place.

In conclusion, these results strongly suggest that BG CMi is at a high inclination and possesses at least a truncated disc.

References

Garlick, M.A., Rosen, S.R., Mittaz, J.P.D., et al., 1994, MNRAS, **267**, 1095
Hellier, C., Garlick, M.A., Mason, K.O., 1993, MNRAS, **260**, 299
de Martino, D., Mouchet, M., Bonnet-Bidaud, J.M., et al., 1995, A&A, **298**, 849
Patterson, J., Thomas, G., 1993, PASP, **105**, 59

SIMULTANEOUS UBVRIJK PHOTOMETRIC AND POLARIMETRIC OBSERVATIONS OF PQ GEM

STEPHEN POTTER[1], K. O. MASON[1], M. S. CROPPER[1],
J. A. BAILEY[2], J. H. HOUGH[3]

1. Mullard Space Science Laboratory, Holmbury St. Mary, Dorking, Surrey, RH5 6NT, England
2. Anglo-Australian Observatory, PO Box 296, Epping, NSW 2121, Australia
3. Division of Physical Sciences, University of Hertfordshire, Herts, AL10 9AB, England

We present simultaneous multiwavelength observations of the intermediate polar PQ Gem (Mason et al. 1992) obtained on 1993 February 18 and 19 using the Hatfield Polarimeter on UKIRT.

The data are folded on the 13.9 m spin period in Fig. 1. The light curves are double peaked at long wavelengths, with dips at phase $\sim$ 0.15 and phase $\sim$0.65, but almost sinusoidal in the U and B bands where the phase $\sim$0.65 dip is absent. The percentage of circular polarisation also varies with the spin cycle, most notably in the I band, with both positive and negative excursions. The peaks in the positive and negative polarisation occur at phase $\sim$ 0.15 and phase $\sim$ 0.65 respectively, approximately coincident with the two intensity dips.

We have modelled these light curves using arc-shaped emission regions. The spectrum of each point along the arc is calculated using the data of Meggitt & Wickramasinghe (1982) and the spectrum summed as a function of spin phase for various viewing geometries. We have used $\Lambda \sim 10^5$ (where Λ is the optical depth parameter), $kT \sim 10$ keV models with a field of $\sim$18 MG at the magnetic pole. The best fit to the observations was obtained for a system inclination of 60° and magnetic dipole offset of 30°, with two extended arcs located at opposite ends of the magnetic dipole (Fig. 2).

Our model explains the dip in the negative polarisation at phase $\sim$0.65 as absorption or scattering of the cyclotron radiation by the accretion column as the accretion region traverses the line of sight. Half a cycle later the main emission region is occulted by the white dwarf resulting in the wavelength independent dip at phase $\sim$ 0.15.

A. Evans and J. H. Wood (eds.), Cataclysmic Variables and Related Objects, 181–182.

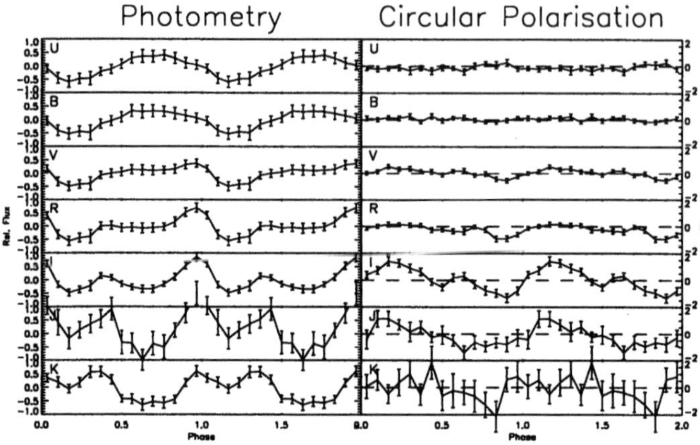

Figure 1. Photometry and circular polarimetry of PQ Gem plotted against the phase of the 13.9 m spin period with the ephemeris $2449297.973 + 0.009645966N$.

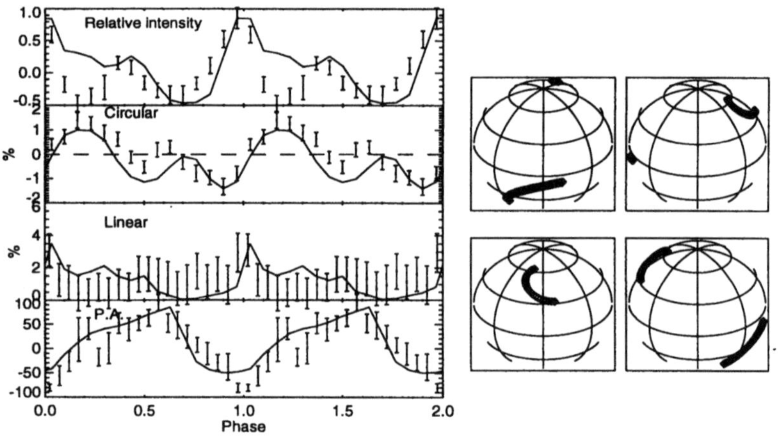

Figure 2. Model fit to the I band photometry and circular polarisation. The linear and position angle observations taken from Piirola, Hakala & Coyne (1993). Best fit obtained for an inclination of 60° and with the magnetic dipole offset at an angle of 30°.

References

Mason, K.O., Watson, M.G., Ponman T.J., et al., 1992, MNRAS, **258**, 794
Meggitt, S.M., Wickramasinghe, D.T., 1982, MNRAS, **198**, 71
Piirola, V., Hakala, P., Coyne, G.V., 1993, Ap. J., **410**, L107

THE MAGNETIC FIELD OF THE INTERMEDIATE POLAR RE 0751+14

H. VÄTH
Institut für Astronomie und Astrophysik der Universität, 24098 Kiel, Germany (supas097@astrophysik.uni-kiel.d400.de)

Piirola, Hakala & Coyne (1993) modeled the optical/IR light curve of RE 0751+14 assuming a uniform shock structure and neglecting the hard X-ray emission. In this paper, we model the light curves at optical/IR and hard X-ray wavelengths and include the effects of the shock structure.

We base our model on accretion onto a white dwarf with a displaced magnetic dipole for a range of likely white dwarf masses. We find that the observed intensity variations of X-rays and in the I band over one spin period largely determine the position of the emission regions. Furthermore, the observed maximum X-ray flux constrains the specific accretion rate. We deduce that the magnetic field at the pole is likely to be in the range 9 ... 21 MG, which is consistent with the estimates of Piirola et al. (1993). It had been proposed previously that there must exist asynchronous rotators with sufficiently strong magnetic fields such that the binaries will evolve into AM Her binaries (Chanmugam & Ray 1984; King, Frank & Ritter 1985). With this deduced high magnetic field RE 0751+14 is the most likely example of such a system known to date.

For more details see Väth, Chanmugam & Frank (1996).

Acknowledgements. The author thanks Ganesh Chanmugam and Juhan Frank for their contribution to this work. This work was supported by NASA grant NAGW 2447 to LSU. The calculations were performed on the MasPar computer of the Department of Physics and Astronomy, Baton Rouge, USA and on the MasPar computer of the University of Karlsruhe, Germany.

References

Chanmugam, G., Ray, A., 1984, Ap. J., **285**, 252
King, A.R., Frank, J., Ritter, H., 1985, MNRAS, **213**, 181
Piirola, V., Hakala, P., Coyne, G.V., 1993, Ap. J., **410**, L107
Väth, H., Chanmugam, G., Frank, J., 1996, Ap. J., in press

A. Evans and J. H. Wood (eds.), Cataclysmic Variables and Related Objects, 183.

ON THE LONG–TERM LIGHT CURVE BEHAVIOUR OF THE INTERMEDIATE POLAR TX COL

D. A. H. BUCKLEY
South African Astronomical Observatory
PO Box 9
Observatory 7935
Cape Town
South Africa

1. Introduction

TX Col (1H0547–407; $P_{\rm orb}$ = 5.72 h) is an intermediate polar (IP) with spin and synodic (beat) periods of 1911 s and 2106 s respectively (Buckley & Tuohy 1989). EXOSAT observations (Tuohy et al. 1986) showed the beat period dominated at higher energies (ME; 1...9 keV), while the spin period was stronger in softer X-rays (LE1; 0.1...1 keV). The high flux level, and pulse fraction (>70%), of the soft X-rays were unique in an IP at that time. Now three other IPs, recently discovered from the ROSAT survey, have been found to show a true separate soft X-ray component, similar to the polars (Motch & Haberl 1994).

The strength of the synodic X-ray period in TX Col was, until the recent discovery of the variable polarised IP, RX J1712.6–2414 (Buckley et al. 1995), unique. This was taken as evidence for direct accretion onto the magnetosphere (e.g. Mason, Rosen & Hellier 1988), through a component of the accretion stream overflowing the disc (Hellier 1992).

2. Long-term monitoring

High speed, white-light photometry of TX Col was conducted over the last ~9 yr, in annual runs lasting typically 7...10 d. Most of the data come from the SAAO, supplemented with occasional multi-site data from CTIO and MJUO (New Zealand). While a detailed analysis is still in progress, a number of interesting results have already arisen:

- the amplitude of the 2106 s beat period has decreased. It is now only ~3% in the 1994 observations, a factor of 2 less than in 1985.

A. Evans and J. H. Wood (eds.), Cataclysmic Variables and Related Objects, 185–186.

- the 1989 data were dominated by the harmonic of the synodic period, at 1053 s (Buckley & Sullivan 1992). This was attributed to two-pole reprocessing, or emission from both sides of the accretion curtain.
- since 1990 the periodograms have been dominated by low frequency QPO-like behaviour, with typical periods of 80...100 m.

The light curves of TX Col, which used to be dominated by the 2106 s beat period, now show gross aperiodic variations, typical of flickering, but at much longer time-scales than usual. This is exemplified in the 1994 data (from SAAO, CTIO and MJUO), which show a strong 102 m period, and other significant low frequency variations. These changes in the light curve over time have never been seen to this extent in other IPs, although most have probably not been as closely scrutinised as TX Col.

Most intriguing is the development of large amplitude, low frequency 'QPO-like' variations, and possibly even more coherent variability at ~100 m. The origin of these low frequency variations is unknown, although they may be related to disc instabilities and superhump phenomena. These strong QPOs are close to 1/3 of the orbital period.

The beat amplitude changes seen in both TX Col and RX J1712.6−2414 may be related to stochastic changes in the accretion stream, and in particular the amount of mass transferred directly into the magnetosphere. Whereas a disc mediates the accretion, a discless component is more susceptible to short-term changes in the mass transfer rate from the secondary. It is intriguing that both objects are strong candidates for discless accretion, or at least a component which is discless (e.g. from stream overflow).

It remains to be seen whether TX Col shows similar changes in its X-ray light curves. To date there is only one published X-ray observation, by EXOSAT (Tuohy et al. 1986), at a time when it exhibited a strong optical synodic periodicity, and lower frequency QPOs were absent.

References

Buckley, D.A.H., Tuohy, I.R., 1989, Ap. J., **344**, 376

Buckley, D.A.H., Sullivan, D.J., 1992, in "Proceedings if the Viña del Mar Workshop on Cataclysmic Variables", ASP Conference Series (N. Vogt, ed.), **29**, 387

Buckley, D.A.H., et al., 1995, MNRAS, in press

Hellier, C., 1992, MNRAS, **258**, 578

Mason, K.O, Rosen, S.R., Hellier, C,. 1988, in "The Physics of Compact Objects", eds. White, N.S, Filipov, L.G., Pergamon Press, Adv. Space Res. **8** (2), 293

Motch, C., Haberl, F., 1995, in "Magnetic Cataclysmic Variables", eds D. Buckley, B. Warner, ASP Conf. Ser. **85**, 109

Tuohy, I.R., et al., 1986, Ap. J., **311**, 275

POLAR SYSTEMS

MAGNETIC FIELDS AND ACCRETION STREAMS IN POLARS

AXEL D. SCHWOPE
Astrophysical Institute Potsdam, An der Sternwarte 16, 14482 Potsdam, Germany. (ASchwope@aip.de)

Abstract. The flow of matter, the accretion geometry and the energy release in polars (AM Her binaries) are to a large extent dominated by the strong magnetic field of the accreting white dwarf. In this review I summarize the main tools used to investigate the magnetic fields and the accretion streams, using optical and X-ray spectroscopy and photometry.

1. Magnetic fields in polars

Cyclotron Radiation. The optical continua of polars in their high accretion states are fed by a strong cyclotron source at the footpoints of accreting field lines. These regions, of size $f \sim 10^{-4}$ (fraction of the white dwarf surface), are heated by hydrodynamic shocks or by particle bombardment to temperatures of a few keV. The ionized plasma is forced to emit electromagnetic radiation on the cyclotron fundamental and in the overtones. Depending on plasma temperature, magnetic field strength and field orientation, resolvable cyclotron harmonics lie in the optical window between 3 500 and 10 000 Å and can be used in order to infer the field.

Historically, the first direct observational evidence for the presence of a magnetic compact star in AM Her was derived from strong optical polarization (Tapia 1977) and a field strength in excess of 10^8 G (100 MG) was estimated on the assumption that the observed radiation corresponds to the cyclotron fundamental. The first genuine measurement of the field strength in a polar was possible after identification of cyclotron harmonics in spectra of VV Pup (Visvanathan & Wickramasinghe 1979; Wickramasinghe & Meggitt 1982). Meanwhile, this kind of measurement has been done for a number of systems and the scientific trade-off is rather large. The necessary prerequisite in order to do cyclotron spectroscopy is access to a medium- or large-sized telescope equipped with a low-resolution op-

A. Evans and J. H. Wood (eds.), Cataclysmic Variables and Related Objects, 189–198.

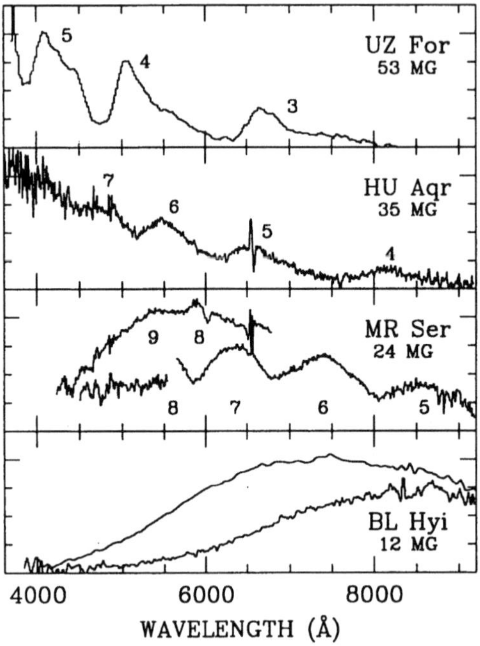

Figure 1. Collection of cyclotron spectra of AM Her binaries (polars) with different field strengths. The spectra in different panels were scaled arbitrarily, the low- and high-state spectra shown for BL Hyi and MR Ser, respectively, were scaled by factors of 20 and 2 with respect to each other. Integer numbers designate identified cyclotron harmonics.

tical and/or infrared spectrograph, coverage of preferably at least one orbital period (80...480 m) of the object concerned, and some co-operation of the target regarding e.g. brightness. Although in some cases cyclotron humps are obvious even in the raw spectra, thus allowing a field measurement, it is necessary to correct for other sources of light (stellar continua, recombination radiation, second accretion region) in order to uncover genuine cyclotron spectra. Results of this procedure for a number of systems sorted according to field strength are shown in Fig. 1. The current census of field measurements is compiled in Table 1. All the spectra shown were extracted at binary phases which correspond to large angles, 70°...90°, between the observer and the local magnetic field in the accretion spot. The main features of the observed cyclotron spectra are compatible with basic cyclotron theory (Wickramasinghe & Meggitt 1985; Schwope 1990). The occurrence of resolvable cyclotron humps is restricted to low harmonic numbers $m_{\rm max} \leq 10$, large viewing angles $\gtrsim 45°$, and relatively low temperatures 5...10 keV (compared with the adiabatic shock temperature). If the field is low, as in the case of BL Hyi (Schwope, Beuermann & Jordan 1995a), the harmonic range covered by optical observations is between $m_{\rm H} = 10$ and 22. The harmonics are then spaced so closely and the overlap is so large that they are smeared to a smooth quasi-continuum; the field measurement is then possible only using either Zeeman lines or by cyclotron spectroscopy in the infrared. This has been exercised for the cases of AM Her and ST LMi by Bailey, Ferrario & Wickramasinge (1991) and Ferrario, Bailey & Wickramasinge (1993). The absence of optical cyclotron lines, the red optical continua and the occasional detection of halo Zeeman lines in a low field make BL Hyi, WW Hor, EF Eri and RXJ1957–57 suitable future targets for cyclotron spectroscopy in the infrared.

If the field is sufficiently high, and overlap of harmonics not too large, their detection is considerably facilitated if a specific system is observed at reduced accretion rate (e.g. MR Ser, second panel in Fig. 1). Then neither a high temperature nor the spread in angle, temperature, particle density and/or magnetic field are large enough to dilute the cyclotron line pattern. For high-field polars (as HU Aqr and UZ For, upper panels in Fig. 1) the detection of cyclotron lines becomes obviously possible, independent of accretion rate (brightness).

The overall shape of the cyclotron spectra suggest that the bulk of emission is shifted from long to short wavelengths with increasing field strength (an additional but smaller blue-shift within one system results from enhanced accretion). This may change our perception of ultraviolet spectra of high-field polars. It appears possible that a major fraction of the cyclotron flux is released in the ultraviolet regime (a never-confirmed prediction of the basic model of Lamb & Masters 1979) and it seems promising to detect cyclotron lines in low-resolution UV spectra of e.g. UZ For, VV Pup (second pole), RXJ0515 and QS Tel (RXJ1938). It also appears likely that at long last the pieces of the energetic puzzle in AM Hers can be put together and that the role of the magnetic field will be understood. Recent steps ahead on the observational side have been undertaken e.g. by Beuermann & Schwope (1994) who have shown that the hard X-ray bremsstrahlung in ROSAT spectra is systematically suppressed in high-field polars (Fig. 2). Woelk & Beuermann (1996) (see also Beuermann & Woelk 1996) have investigated theoretically the relation between maximum electron temperature, thickness of the emission region, magnetic field strength and specific mass accretion rate, by solving the stationary radiation hydrodynamics including cyclotron absorption. The behaviour of the observed cyclotron spectra can be described by simply assuming a homogeneous emission region which is characterized for a given field strength and viewing angle by a single size parameter $\Lambda = n_e s/B$ (n_e = electron density, s = geometrical path length, B = field strength). It essentially determines the switchover frequency from optically thick to thin radiation (Wickramasinghe & Meggitt 1985). Its dependence on field strength for a couple of systems is displayed in Fig. 2. From there it seems to be evident that a strong correlation exists between Λ and B in a sense predicted by Woelk & Beuermann (1996: collapse of the size of the emission region with increased cyclotron efficiency). The size

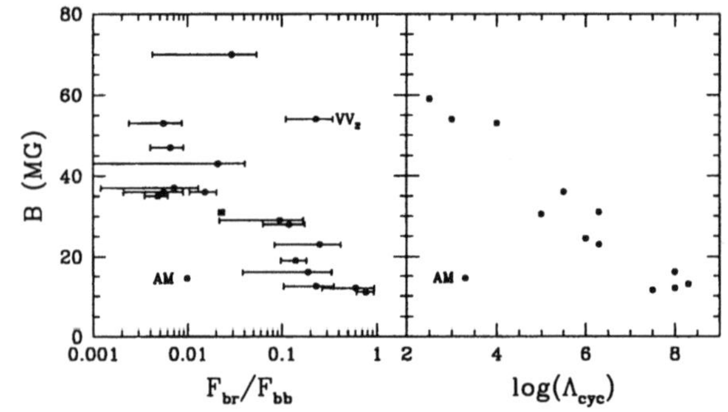

Figure 2. Soft X-ray excess in the ROSAT band (left panel, adapted from Beuermann & Schwope 1994) and size parameter of cyclotron spectra (right panel) for emission regions in polars as a function of field strength.

of this effect, however, is much larger (~5 decades) than the theoretical prediction (~2 decades). This is a hint of a biased determination of Λ in the presence of pronounced cyclotron lines. An extreme example supporting this view is AM Her, whose optical radiation peaks at ~8000 Å, suggesting a high $\log\Lambda \sim 7$ (no optical cyclotron lines), whereas a fit to the resolved low harmonics in the infrared yields the surprisingly low value of $2\,10^3$. The same may apply to the high-field regions in UZ For (main pole), VV Pup, and DP Leo (second poles), where only the low-harmonic optical tail was observed. Summing up, recent observational and theoretical progress are in qualitative agreement. A detailed quantitative comparison is difficult at present due to the lack of sufficient coverage of the cyclotron spectra and to the one-dimensionality of the observational approach (white dwarf mass and specific mass accretion rate $\dot{M}$ must be included).

Another facet of cyclotron spectroscopy is its potential to determine the accretion geometry from the observed motion of cyclotron harmonics. Once the field strength and temperature are fixed, the wavelength of a cyclotron line is completely determined by the viewing angle Θ, which itself is a combination of i, δ and ϕ (orbital inclination, field line inclination, and phase angle). Thus cyclotron spectroscopy can substitute for photoelectric polarimetry (although presently with less time resolution) and recent examples for full-phase tracing of the moving cyclotron pattern can be found in Schwope et al. (1995b) and Burwitz et al. (1996) for QS Tel and RXJ0453.

Zeeman Spectroscopy. In Fig. 3 normalized spectra of MR Ser obtained in high and low accretion states (adapted from Schwope et al. 1993b) are shown. They highlight two different situations where Zeeman-shifted absorption can potentially be observed in polars, here as a very rare case in one system. In the high state (top panel) narrow Zeeman absorption lines are seen against the bright cyclotron background, hence they are clearly of non-photospheric origin. The field strength deduced from the observed

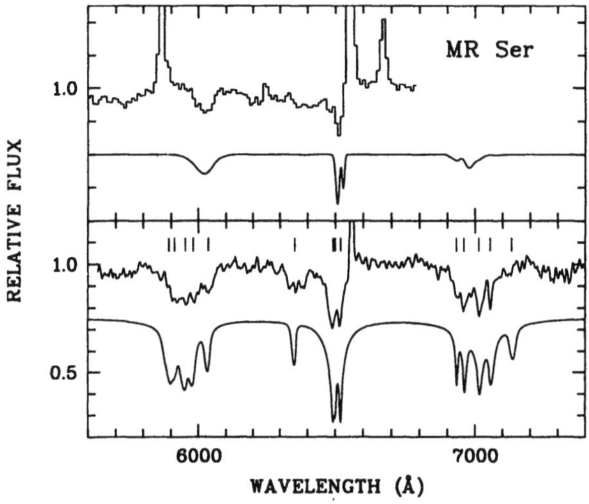

Figure 3. Normalized Zeeman spectra of MR Ser detected in high (top) and low states (bottom), respectively. In the high state the Hα Zeeman triplet is seen against the bright cyclotron background (halo absorption, cyclotron humps at 5950Å and 6700Å), in the low state the Zeeman spectrum originates in the white dwarfs photosphere. Model curves below the spectra are explained in the text.

Zeeman split is 24.5 MG with a spread $\sigma = 1\,\mathrm{MG}$ (Gaussian distribution assumed; fit below the observed spectrum). The field strength is the same as that deduced from the cyclotron humps. Obviously, cool absorbing matter can coexist with the hot emission plasma. In the low state, when cyclotron radiation from the spot and recombination radiation from the stream are both negligible, the observed spectra are dominated by the stellar photospheres. The white-dwarf's Zeeman-spectrum, which appears after subtraction of a template M-star spectrum, is the flux-weighted average over the visible hemisphere of the white dwarf. Attempts to fit the observed spectra of polars in general, and that of MR Ser in particular, with centered dipolar models regularly fail, either because the motion of Zeeman lines at different magnetic phase is not reflected by the simple models (see e.g. Ferrario et al. 1992, Schwope et al. 1995a for the cases of V834 Cen and BL Hyi) or because the observed Zeeman lines are narrower than predicted (the most striking case is that of MR Ser, where the complete set of non-degenerate Zeeman lines of Hα could be resolved). Models applied so far therefore assume either shifted dipoles, shifted mostly along the z-axis, or dipole-quadrupole combinations ('D + Q' in Table 1). The effect of both refinements is that the field strength on one hemisphere becomes more homogeneous with respect to the centered dipole, which reduces both the motion as well the width of the lines (the curve below the low-state Zeeman spectrum of MR Ser is a white-dwarf model calculated for $T_{\rm eff} = 10\,000\,\mathrm{K}$, $B_{\rm pole} = 59\,\mathrm{MG}$, $z_{\rm off} = 0.3\,\mathrm{R_{wd}}$), where $z_{\rm off}$ is the off-centering of the dipole along the symmetry axis.

TABLE 1. Field strengths of white dwarfs in polars. The field strengths refer to cyclotron emission line regions at the main (B_{cyc1}) and the secondary (B_{cyc2}) accretion region, the accretion halo (B_h), and the mean or effective, i.e. flux-weighted, photospheric field (B_{ph}). This is an update of Schwope (1995), where a complete reference list is given. Here only references which were not included in the previous edition, or which have changed in the meantime, are given. The systems are sorted according to decreasing field strength of the centered-dipole equivalent.

Object	B_{cyc1}	B_{cyc2}	B_h	B_{ph}	Ref./Comment
RXJ0515+01	62				Ref. 1
QS Tel (RXJ1938–46)	47	70 (80)			Ref. 2, $z_{off} = 0.1$
UZ For	53				
MR Ser	24...25		24.5	27.3	$z_{off} = 0.3$
RXJ1149+28	43				
DP Leo	30...32	59	29		
VV Pup	30...31.5	54...56			
EU Cnc	42:				
RXJ0453–42	36				Ref. 3
HU Aqr (RXJ2107–05)	35				
RE 1307+535	35:				Ref. 4
EK UMa	35:/47:				
QQ Vul	31:/36:				
V834 Cen	23		23	22	$z_{off} = 0.1$
BL Hyi			12	22	Ref. 5, D + Q
AN UMa	29				
BY Cam	28				
AM Her	14.5			13	$z_{off} = 0.17$
RXJ0531–46	19				
ST LMi	12			19	
RXJ1957–57			16		Ref. 6
EF Eri			13		
V2301 Oph (1H1752+08)				7	Ref. 7

REFERENCES: (1) Shafter et al. 1995, ApJ 443, 419; (2) Schwope et al. 1995b; (3) Burwitz et al. 1996, A&A; (4) Osborne et al. 1994, MNRAS 270, 650; (5) Schwope et al. 1995a; (6) Thomas et al. 1996, A&A, in press; (7) Ferrario et al. 1995, MNRAS 273, 17

The Census. In Table 1 I have collected all available measurements of magnetic fields in polars. It contains 23 systems, hence a fraction of ~50% of all known systems. It is therefore probably only a question of telescope time before more field measurements will be available in binary white dwarfs than in single ones [currently 41 known (Schmidt 1995)]. The latter cover three decades in field strength, the polar WDs only one (7...70 MG). At low fields they lose synchronism and the AM Her phenomenon vanishes, whereas the ultimate reason for the high-field cutoff is still a matter of debate. One hypothesis is that high-field systems evolve so fast that their detection probability becomes very low (Hameury, King & Lasota 1988). Wickramasinghe & Wu (1994) on the other hand suspected that high fields prevent the evolution of these systems in general.

Cyclotron spectroscopy is by far the most succesful method of measuring field strengths in polars; it provides about 2/3 of all entries in Table 1. The numbers given for BY Cam and AN UMa are revised compared to the original publications (Cropper et al. 1989). This became necessary through the application of full spectral models (Schwope 1991) instead of an analytic formula describing the position of individual cyclotron humps.

In all three cases where both cyclotron emission and halo Zeeman absorption lines have been observed the field strengths agree, thus qualifying halo Zeeman lines in general as indicators of the field in the accretion region, even in those cases where no cyclotron lines are detected. Three confirmed polars exhibit two systems of cyclotron lines originating from a main (X-ray bright) and a secondary accretion region with, in all cases, very different field strengths (Table 1), which is a strong hint of the presence of non-dipolar field structures on the white dwarf. This is supported also by Zeeman spectroscopy, which has been applied in detail to six AM Her binaries. In no case was a successful fit achieved using centered dipole models, significantly decentered dipoles or dipole-quadrupole configurations were always required to reflect the data.

2. Accretion streams

Being the vehicles transporting matter from the red companion star to the accreting white dwarf, the accretion streams did not attract so much attention, either theoretically or observationally. Their physics is quite complex and their state subject to several plasma-physical instabilities, most of which are discussed by Hameury, King & Lasota (1986). Observationally they manifest themselves in each polar by strong, asymmetric and variable emission lines (mainly H, He I and He II), and by so-called absorption dips observed at IR, optical and X-ray wavelengths in systems with a high enough inclination (provided observer and accretion spot are on the same hemisphere). In these cases the out-of-plane part of the accretion stream (vertical stream) crosses the line of sight towards the accretion spot and causes absorption features (photoelectric absorption at X-ray wavelengths, free-free or grey absorption in the optical). An overview of relevant data and analysis of the dips was given recently by Watson (1995). A third possibility to investigate the properties of the stream are the light curves of eclipsing systems where the stream emerges as a distinct source of light

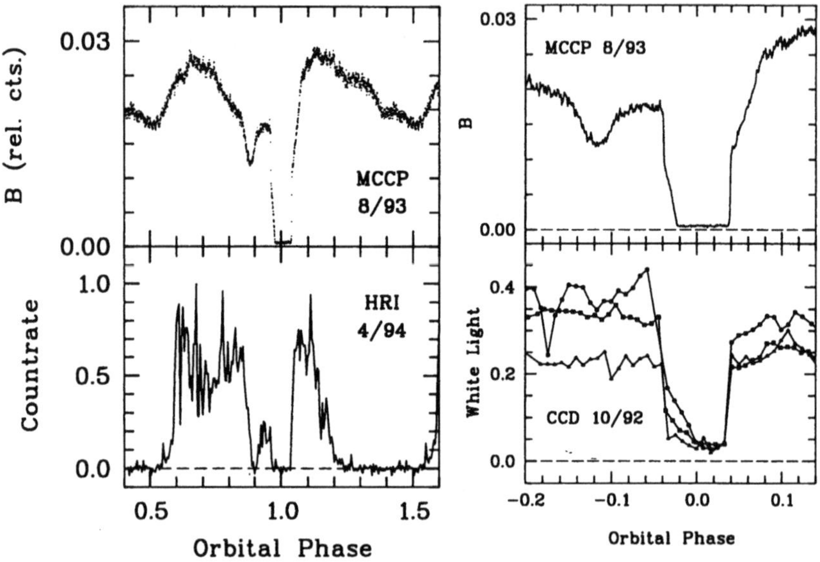

Figure 4. Optical and X-ray light curves of HU Aqr optained at high and intermediate (lower left) accretion states. Left: phase-folded ROSAT HRI and optical B−band light curves in overview; right: optical B−band (phase-folded) and white-light measurements (three original time sequences) on an expanded scale around eclipse.

during ingress and egress. A good example for the case of UZ For using HST data can be found in Schmidt et al. (1993).

The analysis of emission line data to yield definite statements about the stream has been hampered so far, mainly by lack of data at sufficient time resolution. For the remainder of this Section I will discuss a unique example among the AM Her binaries, HU Aqr (= RX J2107.9–0518), which displays all the features mentioned: an extended eclipse with a bright accretion stream, a pronounced pre-eclipse dip at optical and X-ray wavelengths and a distinct fingerprint of the accretion stream in trailed high-resolution spectrograms. It may serve as a template object to reach a deeper understanding of many of the spectroscopic data brought together in the literature for the different objects. For further reading on HU Aqr see Schwope et al. (1993a, 1995c), Glenn et al. (1994), Hakala (1995), Schwope (1995).

In Fig. 4 I show optical and X-ray light curves of HU Aqr obtained on different occasions in states of high and intermediate accretion. In Fig. 5 a trailed spectrogram of the spectral region around the He II 4686 line with high time and spectral resolution is shown (phase resolution 0.01 of the 125 m binary) as well as its transformation to velocity space (Doppler tomogram). The high-speed photometry and spectroscopy of Figs. 4 and 5

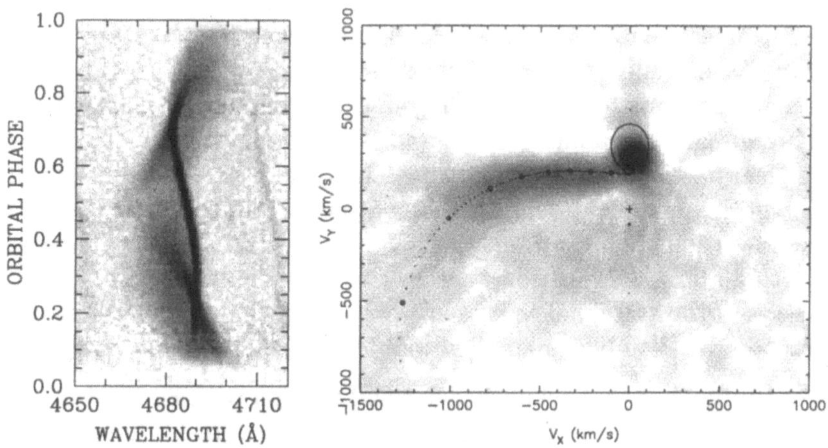

Figure 5. **HU Aqr. Left: phase-folded continuum-subtracted spectrogram of He II 4686. Right: Doppler map of He II 4686 calculated by filtered back-projection. The overlay shows the secondary star and the horizontal stream for an assumed velocities $K_1 = 85\,\mathrm{km\,s^{-1}}$, $K_2 = 340\,\mathrm{km\,s^{-1}}$.**

were obtained simultaneously. Since HU Aqr is a high-inclination system, $i \simeq 85°$, the phase of the absorption dip prior to eclipse indicates the location of the stagnation region and its width gives the size of the stagnation region in the orbital plane. The phase of mid-dip is somewhat variable, 43° in 1993 August and 35° in 1994 April, respectively. Its shape in the optical is well described by a Gaussian curve. The X-ray eclipse at dip-phase is total, the X-ray flux between the dip and the genuine eclipse around phase $\phi = 0.0$ is reduced with respect to pre-dip or post-eclipse phases, indicating the presence of some absorbing matter above the whole horizontal stream between the coupling region and the L_1 point. At optical wavelengths only the cyclotron source is obscured at dip-phase, the remaining light is the bright accretion stream. It appears as a prominent and distinct source of light also at eclipse ingress and egress. The optical eclipse reveals the presence of two sources of light, the small accretion spot (cyclotron radiation) which causes the steep initial ingress and egress ($\Delta t_{\mathrm{i,e}} \simeq 3\,\mathrm{s}$) and the accretion stream which is much more extended. Apart from the back-and-forth motion of the coupling region which derives from the variable phase of dip-center, HU Aqr displays episodes with completely different appearance (Fig. 4, bottom right). Then no distinct and well-shaped (macro-) dips can be detected (but perhaps there are several micro-dips at arbitrary phase prior to eclipse) and ingress/egress times of the stream are much longer than in a mode of normal accretion (top right). It is not clear at all whether the stream in this anomalous mode of accretion is bound to a ballistic trajectory in its horizontal part and to e.g. dipolar field lines in its vertical part, whereas this can be assumed with high confidence for the state of regular accretion. The trailed spectrogram can be deconvolved into three components, a narrow emission line (NEL) from the secondary star, a high-velocity and a broad base component (HVC and BBC) originating in the horizontal and vertical parts of the stream. The NEL transforms to a point-like structure in the Doppler map (inside the contour of the secondary), the BBC fills the lower left quadrant and the HVC finally transforms to a cometary tail starting at L_1. Its location and curvature in the tomogram depends on the masses of the stars, hence it seems promising that Doppler tomography can be elaborated to become a new tool for mass determinations in polars.

Acknowledgements. I acknowledge collaboration with K.-H. Mantel (Munich), K. Horne (St. Andrews), S. Jordan (Kiel), H.-C. Thomas (MPA Garching) and K. Beuermann (Göttingen). This work has been supported by the BMFT under grant 50 OR 9403 5.

References

Bailey, J., Ferrario, L., Wickramasinghe, D.T., 1991, MNRAS, **251**, 37P
Beuermann, K., Schwope, A.D., 1994, in "Interacting Binary Stars", ed. A. Shafter, ASP Conf. Ser. **56**, 119
Beuermann, K., Woelk, U., 1996, these proceedings, p199
Burwitz, V., Reinsch, K., Schwope, A.D., et al., 1996, A&A, **305**, 507
Cropper, M., Mason, K.O., Branduardi-Raymont, G., et al., 1989, MNRAS, **236**, 29P
Ferrario, L., Bailey, J., Wickramasinghe, D.T., 1993, MNRAS, **262**, 285
Ferrario, L., Wickramasinghe, D.T., Bailey, J., et al., 1992, MNRAS, **256**, 252
Glenn, J., Howell, S.B., Schmidt, G.D., et al., 1994, Ap. J., **424**, 967
Hakala, P.J., 1995, A&A, **296**, 164
Hameury, J.-M., King, A.R., Lasota, J.-P., 1986, MNRAS, **218**, 695
Hameury, J.-M., King, A.R., Lasota, J.-P., 1988, MNRAS, **237**, 848
Lamb, D. Q., Masters, A.R., 1979, Ap. J., **234**, L117
Schmidt, G., 1995, Rev. in Mod. Astr., **8**, 147
Schmidt, G., Liebert, J., Stockman, H., Holberg, J., 1993, Ann. Isr. Phys. Soc., **10**, 13
Schwope, A.D., 1990, Rev. in Mod. Astr., **3**, 44
Schwope, A.D., 1991, PhD thesis, TU Berlin
Schwope, A.D., 1995, Rev. in Mod. Astr., **8**, 125
Schwope, A.D., Beuermann, K., Jordan, S., 1995a, A&A, **301**, 447
Schwope, A.D., Thomas, H.-C., Beuermann, K., 1993a, A&A, **271**, L25
Schwope, A.D., Beuermann, K., Jordan, S., Thomas, H.-C., 1993b, A&A, **278**, 498
Schwope, A.D., Schwarz, R., Mantel, K.-H., et al., 1995c, in "Magnetic Cataclysmic Variables", eds D. Buckley, B. Warner, ASP Conf. Ser. **85**, 166
Schwope, A.D. et al., 1995b, A&A, **293**, 764
Tapia, S., 1977, Ap. J., **212**, L125
Visvanathan, N., Wickramasinghe, D.T., 1979, Nature, **281**, 47
Watson, M.G., 1995, in "Magnetic Cataclysmic Variables", eds D. Buckley, B. Warner, ASP Conf. Ser. **85**, 179
Wickramasinghe, D.T., Meggitt, S.M.A., 1982, MNRAS, **198**, 975
Wickramasinghe, D.T., Meggitt, S.M.A., 1985, MNRAS, **214**, 605
Wickramasinghe, D.T., Wu, K., 1994, MNRAS, **266**, L1
Woelk, U., Beuermann, K., 1996, A&A, in press

RADIATION HYDRODYNAMICS OF ACCRETING MAGNETIC WHITE DWARFS

K. BEUERMANN, U. WOELK
Univ.-Sternwarte, Geismarlandstr. 11, 37083 Göttingen, FRG

Abstract. We solved the stationary one-dimensional two-fluid radiation hydrodynamic equations including cyclotron radiation for a wide range of mass flow rates. Here, we discuss the implications for accretion phenomena on the white dwarfs in AM Her binaries.

1. Introduction

In accreting magnetic white dwarfs, the atmosphere is heated by a stream of free falling protons. If the mass flow rate is low, $\dot{m} \leq 10^{-1}\,\mathrm{g\,cm^{-2}s^{-1}}$, the protons lose their energy by Coulomb encounters with atmospheric electrons and the energy is radiated within the stopping length (bombardment case, Kuijpers & Pringle 1982). For high mass flow rates, a hydrodynamic shock forms and the energy is radiated from the extended post-shock cooling region. The radiative transfer problems for the bombardment (non-hydrodynamic) case and for the hydrodynamic case were solved by Woelk & Beuermann (1992, 1993, hereafter WB92, WB93) and by Woelk & Beuermann (1996, hereafter WB96), respectively. We discuss the implications of these results for the study of AM Her binaries.

2. Computational method and results

WB96 solved the stationary one-dimensional two-fluid viscous hydrodynamic equations simultaneously with the fully frequency and angle dependent radiative transfer, including anisotropic cyclotron absorption. The connecting link between the hydrodynamic and the radiative transfer equations is the radiative flux; its gradient balances the energy loss of the settling matter. We obtained solutions for wide ranges of the mass flow rates and field strengths, $\dot{m} = 10^{-3} \ldots 10^{2}\,\mathrm{g\,cm^{-2}s^{-1}}$ and $B = 5 \ldots 70\,\mathrm{MG}$. The

A. Evans and J. H. Wood (eds.), Cataclysmic Variables and Related Objects, 199–202.

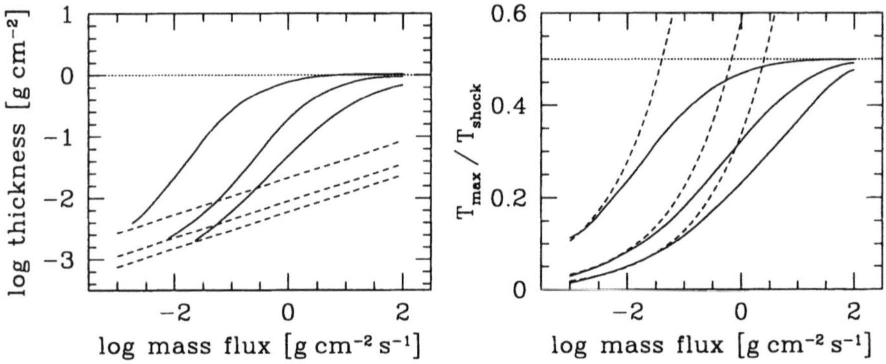

Figure 1. **Maximum post-shock electron temperature and thickness of the cooling region for a white dwarf mass $M_{wd} = 0.7\,M_\odot$ and magnetic field strengths $B = 10, 30$, and 50 MG (from top); zero-field case (dotted line); bombardment case (dashed curves).**

electron temperature behind the shock is determined by the competition between Coulomb collisions with the ions and radiative losses; the maximum temperature achieved decreases with decreasing $\dot{m}$ and increasing B. At the same time, the thickness (column density) of the flow collapses by up to 3 orders of magnitude below that of the zero-field case due to the increasing cyclotron losses. Fig. 1 depicts some results from WB96 for a white dwarf mass of $0.7\,M_\odot$ and field strengths of 10, 30, and 50 MG, along with the zero-field case. For low $\dot{m}$ and high B, the temperature distribution and cooling lengths approach those of the bombardment case (dashed curves; WB92). The transition occurs between 10^{-3} and $10^{-1}\,\mathrm{g\,cm^{-2}s^{-1}}$ for field strengths of 10 to 70 MG. For higher flow rates, the bombardment solution becomes invalid and the hydrodynamics must be taken into account.

For comparison with observations, the geometric height, h, of the cooling region is also of interest (Fig. 2). This is the stand-off distance of the shock above the bottom of the cooling flow. As noted by Beuermann (1988) and Stockman (1988), the bottom of the flow at h_{ram} (i.e. $p = p_{ram}$, where p_{ram} is the ram pressure) is increasingly pushed into the atmosphere and the formation of trenches will affect the escape of the radiation. We determine the trench depth, d, relative to some reference level, h_N, in the atmosphere which we took to correspond to a residual atmospheric column density, $N_H = 10^{23}$ H-nuclei cm^{-2}. If the shock drops below this level, bremsstrahlung may be considered effectively trapped, because it can neither escape sideways nor through the high column density of the pre-shock flow. For an isothermal atmosphere with effective temperature T_{eff} given by $\sigma T_{eff}^4 = GM_{wd}\dot{m}/R_{wd}$, the atmospheric scale height is $H = kTR_{wd}^2/m_H GM_{wd}$ and $d = h_N - h_{ram} = H\ln(p_{ram}/p_N)$, where $p_N = GM_{wd}m_H N_H/R_{wd}^2$. Fig. 2 shows that d exceeds h for $\dot{m} > 30\,\mathrm{g\,cm^{-2}s^{-1}} \simeq \frac{1}{3}\dot{m}_{Edd}$. Finally, we comment on the geometry of the cool-

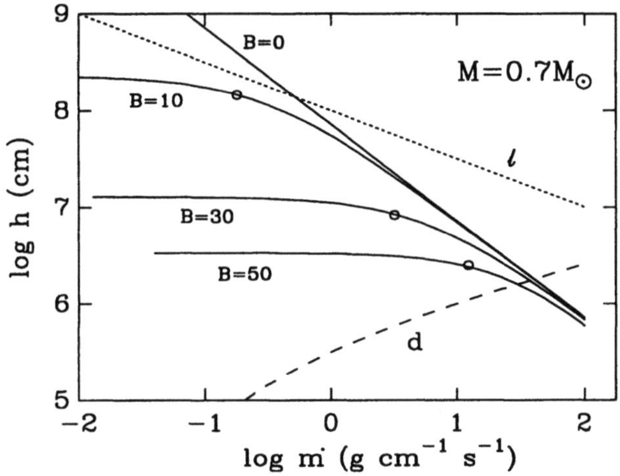

Figure 2. Shock height (cooling length) as a function of the mass flow rate for $M_{\rm wd} = 0.7\,{\rm M}_\odot$ and field strength of 0, 10, 30, and 50 MG. Also shown is the depth, d, of the pressure-induced trench and a typical linear extent, l, of the accretion spot for $\dot{M} = 10^{16}$ g s^{-1}. All curves terminate at $\dot{m}_{\rm Edd} = 75\,R_9^{-1}$ g cm^{-2}s^{-1} $\simeq 100$ g cm^{-2}s^{-1}.

ing region by comparing its linear extent, l, with the stand-off distance, h. For a typical accretion rate $\dot{M} = A\dot{m} = 10^{16}$ g s^{-1} and $A = l^2$, we have $l = (\dot{M}/\dot{m})^{1/2}$ (Fig. 2, dotted line). In the zero-field case, $h > l$ for low $\dot{m}$ and the emission region becomes tall. For the larger B–values, however, $h \ll l$ and the region as a whole is rather pill-box shaped.

3. Discussion and conclusion

Our present results are important for an understanding of the cyclotron and bremsstrahlung emission from magnetic CVs. The observed accretion spots in AM Her binaries display internal structure with a large range of $\dot{m}$ (e.g. Beuermann, Stella & Patterson 1987), often referred to as core-halo structure. The low-$\dot{m}$ halo produces part of the observed cyclotron emission including the optically thin cyclotron lines while the core(s) produce(s) the bremsstrahlung and an underlying cyclotron continuum. The theory allows us to interpret a given cyclotron spectrum in terms of B *and* of the range of peak temperatures or of the flow rates which contribute to this spectrum. For AM Her binaries with pronounced cyclotron lines (as e.g. UZ For; Beuermann, Thomas & Schwope 1988) a fit by theoretical cyclotron spectra implies that a range of $\dot{m}$ values is present, e.g. from 10^{-3} to 10^{-1} g cm^{-2}s^{-1} for UZ For (Rousseau et al. 1996). The fraction of the accretion energy released as bremsstrahlung is very low at these low $\dot{m}$-values and values of the order of $\dot{m} = 10$ g cm^{-2}s^{-1} are needed to account for the hard X-ray emission seen in some systems in their high states.

Beuermann & Schwope (1994) have demonstrated that bremsstrahlung is systematically suppressed at high B. Fig. 2 provides a hint why this may be so. Following Beuermann et al. (1987) we assume that the accretion stream stagnates in the outer magnetosphere at a point where $B^2/4\pi = m_{\rm H} n\, v^2$. To a first approximation, n will equal the particle density in the line-emitting region. Following the dipole lines, this matter arrives at the surface of the white dwarf with mass flow rate $\dot{m} = B(m_{\rm H}n/4\pi)^{1/2} = 36 B_7 n_{14}^{1/2}$, where B_7 and n_{14} are the surface field strength and the particle density in the stagnation region in units of 10^7 G and $10^{14}\,{\rm cm}^{-3}$, respectively. For $n = 10^{13} \ldots 2\,10^{14}$ ${\rm cm}^{-3}$ (Liebert & Stockman 1985), $\dot{m}$ ranges from $10 \ldots 50$ and from $50 \ldots 250\,{\rm g\,cm^{-2}s^{-1}}$ for $B_7 = 1$ and 5, respectively. This is the material which produces the dense accretion cores. With increasing B, this material is driven further into the region where shocks tend to be covered up. In fact, for $B_7 = 3$, where bremsstrahlung starts to be significantly suppressed, the above $\dot{m}$-range has just moved entirely to the right of the intersection of solid and dashed line in Fig. 2. At the same time the output in soft X-rays will increase. To be sure, our radiation hydrodynamic calculations do not bear directly on the open problem of how the intense soft X-ray emission from AM Her binaries is produced. A formal solution of the so-called 'soft X-ray puzzle' requires the time-dependent treatment of the hydrodynamics simultaneously with an appropriate two-dimensional radiative transfer (because edge effects are important). It will also be important to include the radiation pressure in the hydrodynamics and heat conduction in the energy transport, both of which were neglected in the calculations of WB96. Such calculations are not presently available. The above example suggests, however, that even in the stationary case increasing field strength implies a suppression of the bremsstrahlung flux which is likely to be reprocessed into soft X-rays.

References

Beuermann, K., Stella, L., Patterson, J., 1987, Ap. J., **316**, 360

Beuermann, K., 1988, in "Polarized Radiation of Circumstellar Origin", eds. G.V. Coyne et al., University of Arizona Press, p125

Beuermann, K., Thomas, H.-C., Schwope, A.D., 1988, A&A, **198**, L15

Beuermann, K., Schwope, A.D., 1994, in "Interacting Binary Stars", ed. A. Shafter, ASP Conf. Ser. **56**, 119

Kuijpers, J., Pringle, J.E., 1982, A&A, **114**, L4

Liebert, J., Stockman, H.S., 1985, in "Cataclysmic Variables and Low-mass X-ray Binaries", eds. D.Q. Lamb, J. Patterson, Reidel Publ. Co.

Rousseau, T., Fischer, A., Beuermann, K., Woelk, U., 1996, A&A, in press

Stockman, H.S., 1988, in "Polarized Radiation of Circumstellar Origin", eds., G.V. Coyne et al., University of Arizona Press, p237

Woelk, U., Beuermann K., 1992, A&A, **256**, 498 (WB92)

Woelk, U., Beuermann K., 1993, A&A, **280**, 169 (WB93)

Woelk, U., Beuermann K., 1996, A&A, **306**, 232 (WB96)

MAGNETIC FIELDS IN AM HER BINARIES

KINWAH WU[1], PAUL A. MASON[2]
1.RCfTA, The University of Sydney, NSW 2006, Australia
2. Department of Astronomy, Case Western Reserve University, Cleveland, OH 44106, USA

1. Introduction

There is growing evidence of complex white dwarf magnetic fields in CVs (e.g. Meggitt & Wickramasinghe 1989; Robinson & Córdova 1994). The properties of systems with non-dipolar white dwarf magnetic fields have been investigated, and it was found that a non-negligible quadrupole component is essential for the magnetic locking of AM Her binaries (AM Hers) (Wu & Wickramasinghe 1993). In a recent study the ratio of the quadrupole to dipole component of the surface field of white dwarfs was shown to grow by a factor of 10 during the white dwarf cooling lifetime (Muslimov, van Horn & Wood 1995), thus the quadrupole field component is long lived and significant.

2. Synchronous versus asynchronous systems

For synchronous AM Hers, it is generally not easy to distinguish between an offset dipole field and a dipole-quadrupole field. Nevertheless if a system has more than two accretion poles the evidence for a multipole field is strong. Systems with a multipole field also have other observational properties that distinguish them from the systems with a pure dipole field. If a system has a dipole field, the polar field strengths deduced from the cyclotron spectra are about a factor of 1.5 higher than the photospheric magnetic fields deduced from the Zeeman lines. This, however, does not hold for dipole-quadrupole fields. In fact, the suggestion that BL Hyi has a strong quadrupole component for its white dwarf magnetic field is based on this argument (Schwope 1995).

For asynchronous AM Hers, the accretion stream impacts different parts of the white dwarf magnetosphere throughout the spin-orbit beat cycle. For a dipole-quadrupole field (Fig. 1), the two main accretion regions are located near the intersection of the magnetic equator and the orbital plane,

A. Evans and J. H. Wood (eds.), Cataclysmic Variables and Related Objects, 203–204.

Figure 1. A white dwarf with comparable quadrupole and dipole field strengths

and accretion switches back and forth between the two poles with the spin-orbit beat period. The field strength of the two poles are generally similar and have the same polarity, thus the polarization of the cyclotron emission they emit have same sign. At certain beat phases accretion onto a third (or even a fourth) accretion pole may also occur.

BY Cam is a typical example of an asynchronous AM Her system (e.g. Ramsay & Mason 1996). In the model proposed by Mason et al. (1995) the main two accretion regions appear to be diametrically opposed but show the same sign of polarization (see Mason, Liebert & Schmidt 1989). As a note of caution we point out that a competing model for BY Cam exists, in which the dominant spike in the periodogram of Ramsay & Mason's figure 1 is the spin period rather than a side-band period and the spin-orbit beat period is 7.07 d rather than 14.15 d. However, the periods shown in Ramsay & Mason's figure 1 are not readily explained in terms of the 7.07 d beat period model. In any case there is strong evidence for a complex field for the white dwarf in BY Cam due to extremely complex and variable polarization curves.

3. Conclusions

AM Her binaries generally have complex white dwarf magnetic fields. The field structure can be easier determined if the systems are asynchronous. Observations indicate that the magnetic field of BY Cam likely has a strong quadrupole component.

References

Mason, P.A., Andronov, I.L., Kolesnikov, S.V., et al., 1995, in "Magnetic Cataclysmic Variables", eds D. Buckley, B. Warner, ASP Conf. Ser. **85**, 493
Mason, P.A., Liebert, J., Schmidt, G.D., 1989, Ap. J., **346**, 941
Meggitt, S.M.A., Wickramasinghe, D.T., 1989, MNRAS, **236**, 31
Muslimov, A.G., van Horn, H.M., Wood, M.A., 1995, Ap. J., **442**, 758
Ramsay, G., Mason, P.A., 1996, these proceedings, p213
Robinson, C.R., Córdova, F.A., 1994, Ap. J., **437**, 436
Schwope, A., 1995, Rev. Mod. Astron., in press
Wu K., Wickramasinghe D.T., 1993, MNRAS, **260**, 141

THE HARD X-RAY SPECTRA OF EF ERI AND OTHER CVS

C. DONE[1], J. P. OSBORNE[2], A. P. BEARDMORE[3]
1. Department of Physics, University of Durham, Durham DH1 3LE, UK
2. Department of Physics and Astronomy, University of Leicester, Leicester LE1 7RH, UK
3. Department of Physics, The Open University, Milton Keynes, MK7 6AA, UK

Abstract. Reflection of the X-ray emission from the surface of the white dwarf should be present in all hard X-ray emitting CV systems. It is clearly identified in the Ginga 2...20 keV spectra of the polars AM Her and EF Eri, and also in the quiescent spectra of the dwarf novae SS Cyg. The inclusion of the hard reflection spectrum lowers the derived continuum temperature, so resolving a long standing problem of the mismatch of the iron line and continuum properties, and, with more realistic multi–temperature continuua, the derived shock temperatures are close to those expected from theoretical models of a strong shock.

1. The hard X-ray spectra of CVs

The AM Her systems, or polars (Cropper 1990), of which EF Eri is a long established example, are binary systems in which a low mass star overflows its Roche lobe onto a magnetised white dwarf companion. The flow is entrained by the magnetic field before it reaches the circularization radius, and thus falls freely through the entire gravitational potential without forming an accretion disk. If the accretion stream shocks above the surface of the white dwarf then all the gravitational energy is released as bremsstrahlung emission (optically thin thermal radiation), with a predicted shock temperature of $kT = 31\, M_{\rm wd}/\, {\rm M}_{\odot}$ keV, i.e. producing hard X-rays (see e.g. Frank, King & Raine 1992). Since the plasma contains heavy elements, this hot gas produces X-ray line emission as well as continuum. At high temperatures the most important lines are from iron, either at 6.7 keV (from Fe^{+24}, He–like) and/or 6.9 keV (Fe^{+25}, H–like).

A. Evans and J. H. Wood (eds.), Cataclysmic Variables and Related Objects, 205–208.

The polars are generally faint, hard X-ray sources so detailed spectral studies are generally not possible. However, the compilation of polar spectra from the Ginga X-ray satellite (2...20 keV) showed that, while the spectra are generally well fit by a single temperature thermal bremsstrahlung continuum with $kT \sim 10 - 30$ keV, the iron line emission is at systematically lower energies and higher intensity than predicted by the continuum temperature (Ishida 1991).

Since the hard X-ray emission is expected to bc isotropic, this spectrum *must* illuminate the white dwarf surface, and some fraction is reflected back towards the observer. The reflection probability is energy dependent, as at low energies there are many elements on the white dwarf surface that can photoelectrically absorb the incident photon. For example, photons below 1 keV can be absorbed by the K shell of carbon, nitrogen, oxygen etc., thus the reflection probability is small at these energies. However, at higher energies, these low atomic number elements have very little cross section and only high Z elements can contribute to the photoelectric cross section. But these are generally not very abundant, so the total absorption probability decreases and hence the reflected fraction increases. Iron is the last abundant high Z element, so leading to a pronounced feature at the iron K edge, energy in the reflected spectrum and an associated Fe $K\alpha$ fluorescence line at 6.4 keV from the cool reflecting material, i.e. distinct from the more highly ionised lines from the hot plasma. Thus the reflected continuum spectrum is harder than the incident spectrum in the 2...20 keV Ginga range, with an iron Kα fluorescence line superimposed on it. The dependence on photoelectric absorption also means that the reflected spectrum is a function of the ionisation state of the gas (the more the reflector is ionised, the fewer bound electrons are left to provide absorption opacity and so the larger the reflected fraction), its elemental abundance (higher abundance gives more absorption opacity and hence less reflection); see e.g Lightman & White (1988), George & Fabian (1991), Matt, Perola & Piro (1991). The amount of reflection also depends strongly on geometry, both on the solid angle subtended by the reflecting material to the hard X-ray source (the higher the solid angle, the higher the relative normalisation of the reflected continuum with respect to the incident spectrum) and also the inclination of the reflecting region with respect to the observer (the higher the inclination, the longer the path length through the white dwarf photosphere the reflected photon has to travel before escaping, and so the higher its absorption probability and hence the lower its reflected fraction); see e.g. George & Fabian (1991), Matt, Perola & Piro (1991).

The signature of reflection can be clearly seen in the Ginga 2...20 keV hard X-ray spectra of AM Her and EF Eri (Beardmore et al. 1995; Done, Osborne & Beardmore 1995). With a single temperature plasma and com-

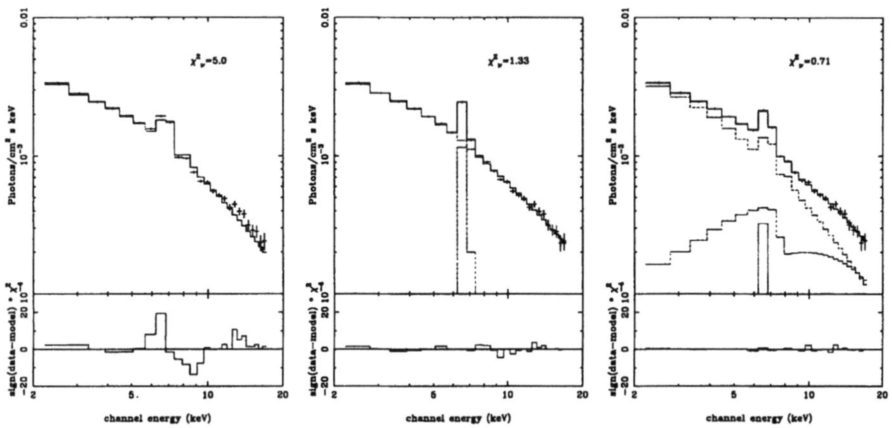

Figure 1. **Spectra and residuals to data from AM Her at X-ray maximum. From left to right the model is (a) an optically thin thermal plasma (includes continuum and self consistent line emission); (b) optically thin thermal bremsstrahlung and arbitrary line emission; (c) an optically thin thermal plasma and its reflection spectrum.**

plex absorption the resulting fit is unacceptable, as shown in Fig. 1a, with a $\chi^2_\nu = 5$ (for a temperature of 14.5 keV and best fit abundance of solar). The problem derives again from a mismatch between the line properties and continuum temperature. Fig. 1b shows the results from fitting a bremsstrahlung continuum, together with arbitrary Gaussian line emission and complex absorption. This is statistically (just) acceptable, with $\chi^2_\nu = 1.3$ for a temperature of 21 keV, line energy of 6.7 keV and equivalent width 640 eV. However, in a plasma at 21 keV, much of the iron will be completely ionised, and the little that remains will have just 1 electron (H–like), which gives a line energy of 6.9 keV, and with solar abundances this should have an equivalent width of 420 eV. Fig. 1c shows the significant improvement in the fit ($\chi^2_\nu = 0.71$) from a single temperature plasma model that results from inclusion of the reflected spectrum and its associated 6.4 keV fluoresence line emission. As reflection is a hard spectral component, its inclusion means that not all the high energy photons have to be produced from the intrinsic continuum spectrum, leading to a marked decrease in the derived plasma temperature, $\sim$ 12 keV, compared to the $\sim$ 21 keV required in the bremsstrahlung fits. This in turn leads to a prediction of a lower iron line energy and a higher iron line intensity, matching the observed trend in these objects.

A similar set of plots for EF Eri again show that reflection is significantly detected in the spectrum (see Done, Osborne & Beardmore 1995), but that here the iron line both from the hot plasma and that associated with the reflection continuum are smaller by a factor 2 than those seen in AM Her. The plasma codes used to model the spectra are produced in the coronal approximation, which almost certainly does not hold for these objects (Done, Osborne & Beardmore 1995). In particular, the iron lines from the hot plasma are probably optically thick and so will be scattered many times before escaping. However, as there seems to be no viable alternative decay path for the Lyα transition which produces the lines then it is difficult to see how the lines can be suppressed. Also, the 6.4 keV fluoresence line from the reflection spectrum cannot be affected by the hot plasma line (6.7 and 6.9 keV) optical depth effects, so the only explanation seems to be a true underabundance of iron in this system.

While the inclusion of reflection is extremely successful in obtaining a good fit to the spectrum, the derived continuum temperature of $\sim$ 10 keV is much lower than the 31 $M_{\rm wd}/\,{\rm M}_\odot$ keV predicted by the simple strong shock models. However, we also do not expect the continuum to be described by a single temperature. The material must cool as it settles from the shock to the white dwarf photosphere, so the expected spectrum is a density weighted sum of multiple temperature hot plasma spectra. This can be modeled simply by assuming that the bremsstrahlung cooling is dominant and that the pressure and gravitational potential remain constant throughout the cooling structure (Done, Osborne & Beardmore 1995). These cooling spectra give very similar line and continuum emission to a single temperature 10 keV plasma when the maximum shock temperature is of order 25 keV, matching well with that expected from a solar mass white dwarf.

Of course, reflection is not limited to one particular type of object. The dwarf nova SS Cyg produces hard X-rays in quiescence. Fitting this spectrum with a single temperature plasma results in a very poor fit (χ^2_ν = 2.7 for kT = 15 keV). Including a second temperature component, as is often suggested (see e.g. Nousek et al. 1994), does not help substantially, with χ^2_ν = 2.0 for 7.5 keV and 60 keV components, but a single temperature plasma and its reflection give an excellent fit to the data (χ^2_ν = 1.0 for kT = 10 keV).

References

Beardmore, A.P., Done, C., Osborne, J.P., Ishida, M., 1995, MNRAS, **272**, 749
Cropper, M., 1990, Sp. Sci. Rev., **54**, 195
Done, C., Osborne, J.P., Beardmore, A.P., 1995, MNRAS, **276**, 483
Frank, J., King, A.R., Raine, D., 1992, Accretion power in Astrophysics, 2nd edition, Cambridge, UK: Cambridge University Press
George, I., Fabian, A.C., 1991, MNRAS, **249**, 352
Ishida, M., 1991, PhD thesis, University of Tokyo
Lightman, A.P., White, T.R., 1988, Ap. J., **335**, 57
Matt, G., Perola, G.C., Piro, L., 1991, A&A, **247**, 25
Nousek, J.A., Baluta, C.J., Corbet, R.H., et al., 1994, Ap. J., **436**, L19

DIFFERENT TYPES OF PHOTOMETRIC AND POLARIMETRIC BEHAVIOUR OF AM HER IN 1989-1994

N. M. SHAKHOVSKOY[1], I. L. ANDRONOV[2], S. V. KOLESNIKOV[2]
1. Crimean Astrophysical Observatory
334413 Nauchny, Crimea, Ukraine
2. Department of Astronomy, Odessa State University
T.G. Shevchenko Park, 270014 Odessa, Ukraine

Abstract. Variations in *UBVRI* are classified as: luminosity– and wave–length– dependent phase variations with differences from cycle–to–cycle; flares caused by accretion events; UV Cet–type flares from the secondary.

More than 60 nights of observations in various luminosity states allow us to distinguish the following types of photometric activity.

(i) Regular variations with the orbital phase in R and I, which are less pronounced in V and practically invisible in B and U.

(ii) Irregular flares with different time-scales. Usually they are best seen in R and I, but sometimes 'white' flares occurred with similar amplitudes in all bands. Both types of variability are observed mainly in the bright state of the star and are due to cyclotron emission in the accretion column(s).

(iii) Smooth variations in U and B (at the 1st harmonic of the orbital period). These are often opposite to variations in R and I. They are most prominent in the 'intermediate' and 'low' states and may be explained by the inhomogeneity of the surface brightness of the red component (reflection effect or spot).

(iv) Fast 'blue' flares, most prominent in U and B. These were similar to a giant flare which occurred on 1992 August 29 and are probably arising on the red dwarf. They resemble the flares of UV Cet-type stars.

References

1. Andronov, I.L., Kolesnikov, S.V., Shakhovskoy, N.M., 1992, in "Stellar Magnetism", ed. Yu.V. Glagolevsky, (S.-Peterburg), p148
2. Shakhovskoy, N.M., Alexeev, I.Yu., Andronov, I.L., Kolesnikov, S.V., 1993, in "Cataclysmic Variables and Related Physics", eds O. Regev, G. Shaviv, Ann. Israel Phys. Soc., **10**, 237

A. Evans and J. H. Wood (eds.), Cataclysmic Variables and Related Objects, 209–210.

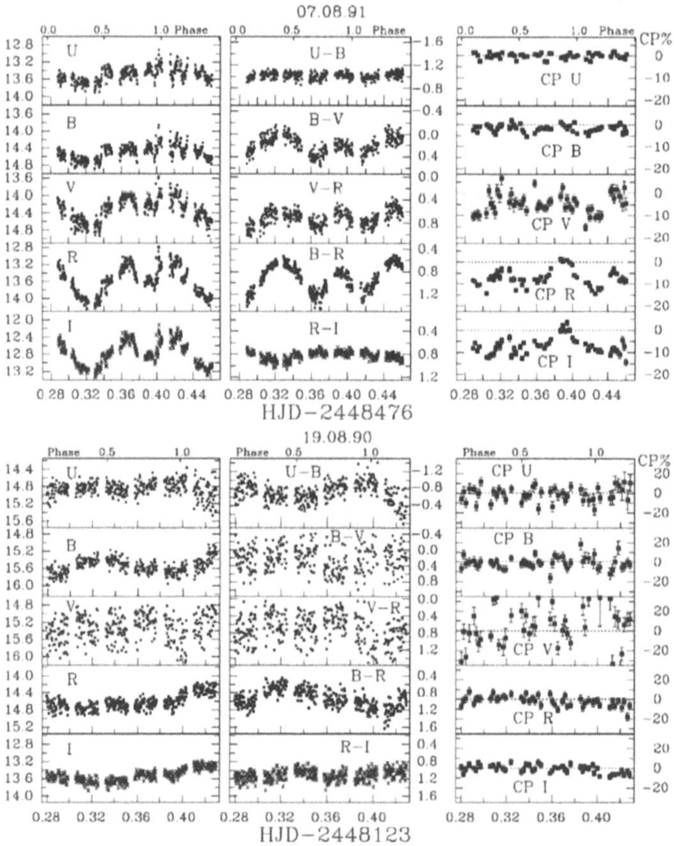

Figure 1. *UBVRI* brightness, color and circular polarization variations of AM Her in the high (top; 1991 August 7, JD 2 448 476) and low (bottom; 1990 August 19, JD 2 448 123) luminosity states. Bottom abscissae show the decimal part of HJD, upper ones, the polarimetric phase. The flares in the light curves (top) are not seen in all colours. During long states of low activity, the circular and linear polarization and the fast variability sometimes completely disappear. At these times AM Her loses all characteristics of the type and formally can not be classified as a 'polar' [1]. The unprecedented UV Cet-type flare on 1992 August 29 was also observed during the low state [2].

SOFT X-RAY FLICKERING OF AM HER

PETER J. WHEATLEY
Astronomical Institute, Utrecht University, Postbus 80 000, 3508 TA Utrecht, The Netherlands

1. Introduction

I present preliminary results from an analysis of the flickering in the only off-axis, high-state observation of AM Her with the ROSAT PSPC. At high $\dot{M}$, the ROSAT count rates are dominated by optically-thick emission from the white-dwarf surface, which is heated directly by the accretion flow. This emission is very soft, predominantly below 0.4 keV, and exhibits strong flickering.

2. Flickering

The brightness of AM Her and the unprecedented sensitivity of ROSAT combine to extend sensitivity to flickering to time-scales less than 1 s. Fig. 1 shows the light curve of one data section, binned at 1 s. The power spectrum of this light curve is plotted in Fig. 2, which shows a red noise spectrum, steepening to a slope of $1/f^2$ at high frequencies.

This is characteristic of shot noise, which is a reasonable interpretation considering the blobby accretion proposed to explain the direct heating of

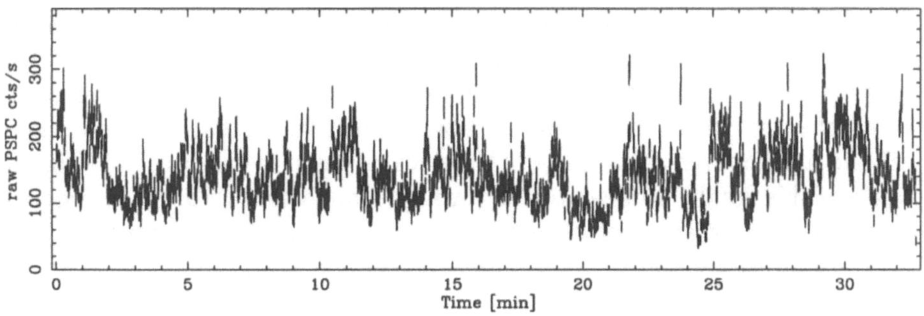

Figure 1. ROSAT PSPC light curve of AM Her

A. Evans and J. H. Wood (eds.), Cataclysmic Variables and Related Objects, 211–212.

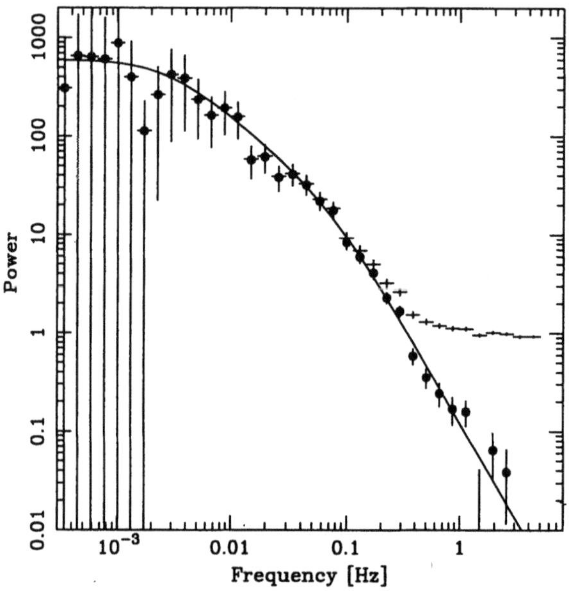

Figure 2. Power spectrum of the light curve of Fig. 1. The filled circles show the spectrum after subtraction of the white noise. The curve shows an example fit with a shot-noise model (see text).

the white-dwarf photosphere (Kuijpers & Pringle 1982). The solid curve in Fig. 2 represents a fit to the shot-noise model of Lehto (1989). In this case, exponentially-decaying shots are evenly distributed over time-scales in the range 5 to 300 s.

The steepening to a slope of $1/f^2$ is crucial, since it constrains the short time-scale limit. In this interpretation (possibly not unique) the flickering is dominated by time-scales longer than ~ 5 s. This may reflect the accretion time for the shortest structures, or alternatively the fastest time-scales on which the white dwarf surface can react to perturbation by the accretion flow: either its cooling or dynamical time-scale.

3. Colour of flickering

The flickering appears suprisingly grey. The variability of the PSPC hardness ratio is in the range expected from counting noise alone ($\chi^2_\nu = 1.18$). However, cross-correlating the hardness ratio with the light curve does reveal spectral variations. These are always positive (i.e. harder when brighter) and are apparent at a range of time-scales (as short as 10 s). The significance of this result has been tested through simulations. Work is in progress to quantify these spectral variations in terms of physical quantities.

References

Kuijpers, J., Pringle, J.E., 1982, A&A, **114**, L4
Lehto, H.J., 1989, in 23rd ESLAB Symp. on Two Topics in X-ray Binaries, p499

X-RAY AND OPTICAL OBSERVATIONS OF BY CAM

GAVIN RAMSAY[1], PAUL A. MASON[2]

1. Astronomical Institute, The University of Utrecht, Postbus 80000, Utrecht, NL 3508 TA, The Netherlands
2. Astronomy Department, Case Western Reserve University, Cleveland, OH 44106, USA

1. Introduction

We present preliminary results of an analysis of X-ray and optical data of the asynchronous AM Her star BY Cam [1]. We use X-ray data from EXOSAT (0.1...50 keV), Ginga (1.5...50 keV) and ROSAT (0.1...2.0 keV) and optical data obtained during a 45-day campaign in 1994.

2. The periods of BY Cam

There are 4 known periods: the orbital period (201.30 m), the spin period of the white dwarf (199.3303 m), a spin-orbit beat period (14.15 d) and a side-band period (197.4 m). The detection of this side-band period lends credence to the theory of [2], who suggest that for a stream accreting, diskless, magnetic CV a frequency, $f = 2\omega_{\rm spin} - \Omega_{\rm orb}$ (=197.399 m), will appear as a strong spike in the power spectrum for certain systems. Wu & Mason (this volume) discuss a competing model where $P_{\rm spin} = 197.4$ m.

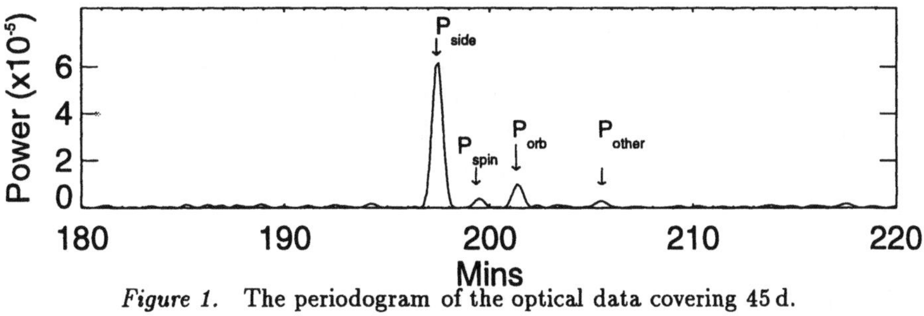

Figure 1. The periodogram of the optical data covering 45 d.

A. Evans and J. H. Wood (eds.), Cataclysmic Variables and Related Objects, 213–214.

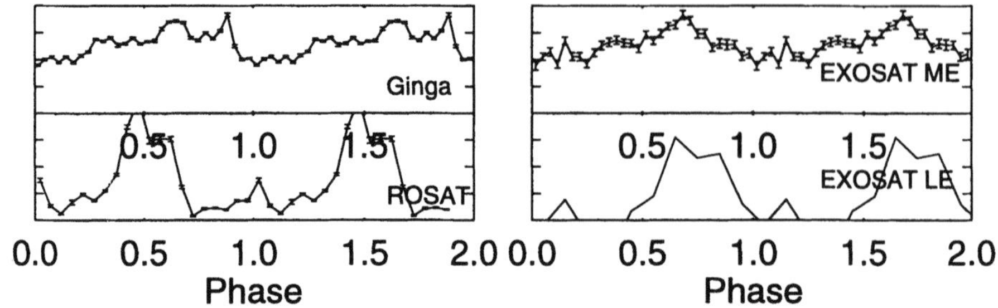

Figure 2. X-ray and optical data folded on the side-band period P=197.399 m as a function of phase. A proposed period derivative of 0.264 s yr^{-1} is included.

3. The optical data

Fig. 1 shows the periodogram obtained from data covering a baseline of 45 d (see Silber, this volume). A very strong peak is seen at the side-band period, a weaker secondary peak at the orbital period and a very weak peak near the spin period. Data from a few days during which the accretion geometry is stable show the spin period, while data with a baseline of more than a few days show the side-band period with negligible power at the spin period. Another frequency, $3\omega_{\rm spin} - 2\Omega_{\rm orb}$, predicted by [2] is seen near 205.5 m.

4. X-ray data

Periodograms of the ROSAT data show a peak at exactly twice the side-band period. Since the ROSAT baseline covers ~1/2 of a beat cycle we would expect a pole switch to take place during this observation and hence the spin peak would not be seen. It is likely that twice the side-band period is seen rather than the side-band period, due to the very sparse sampling inherent in the ROSAT observations. The Ginga data shows a very broad peak, but favours the 199.33 m spin period.

5. The folded light curves

To obtain phase folded light curves which can be suitably compared we need to derive the quadratic term (which is almost certainly present) with some accuracy. At this stage we tentatively propose $\dot{P}_{\rm side} = 0.264\,{\rm s\,yr^{-1}}$ based on what is expected from the ephemeris of [1]. Fig. 2 suggests that the EXOSAT and Ginga data are well phased, but the ROSAT data is possibly out of phase by ~0.2 cycles. Work is in progress to refine both $P_{\rm side}$ and $\dot{P}_{\rm side}$ from optical and X-ray data.

References

1. Mason, P.A., Andronov, I.L., Kolesnikov, S.V., et al. 1995, in "Magnetic Cataclysmic Variables", eds D. Buckley, B. Warner, ASP Conf. Ser. **85**, 493
2. Wynn G., King, A.R., 1992, MNRAS, **255**, 83

ASYNCHRONOUS ROTATION IN BY CAM: IT'S GOT A GOOD BEAT AND YOU CAN DANCE TO IT (VERY SLOWLY)

A. D. SILBER
University of Washington, Dept. of Astronomy, Box 351580, Seattle, WA 95810-1580, USA

1. Introduction

Since its discovery, BY Cam was known to be an atypical AM Her type CV (Remillard et al. 1986). Observations showed the orbital and white-dwarf spin periods differ by about 1% (Silber et al. 1992). As the white dwarf rotates relative to the secondary, the accretion stream encounters different magnetic geometries, and hence follows different accretion paths to the white dwarf. The magnetic geometry seen by the accretion stream repeats at the beat period ($1/P_{\rm beat} = 1/P_{\rm spin} - 1/P_{\rm orb}$), which is 7 to 20 d (the range is due to the uncertainty in the measured periods). The shape of the light curve changes from night to night, which can be explained by the changing geometry of the highly-beamed cyclotron emission. Even if the asynchronous rotation is the underlying cause of the changing light curve, the light curve could differ at similar beat phases if the threading process is chaotic. Previous attempts to study the beat period using IUE (Zucker et al. 1995) and optical data (Mason & Chanmugam 1992) have been inconclusive.

2. Observations

Our goal was to observe BY Cam every night for 40 consecutive nights to see if we could detect any sign of the beat period. Observations were carried out at the University of Washington's Manastash Ridge Observatory, Case Western Reserve's Burrell Schmidt, several telescopes at the Crimean Observatory, and the University of Keele's 0.6 m telescope. We succeeded in obtaining at least one spin period of data on 43 nights over a 65 night span in the fall of 1994.

A. Evans and J. H. Wood (eds.), Cataclysmic Variables and Related Objects, 215–216.

3. Summary

We collected a large and well-sampled light curve to study the changes through the beat period. We found a photometric period (0.137123 d) that is slightly shorter than the spin period determined through polarization studies. This shorter period bolsters the model of BY Cam as an asynchronous rotator (Wynn & King 1992). We also found a clear, though subtle, sign of the beat period. Once a week the light curve became more complex and the phase shifts relative to the photometric ephemeris. We believe that the accretion stream is switching poles at this phase. If there are two dominant poles this would indicate a beat period of 14 d; three poles suggest a beat period of 21 d, and so on. Though there are clear signs of the beat period at a multiple of 7 d, probably 14 d, there is significant variability of the light curve that does not seem to repeat at this period. Hence the process in which the matter threads the magnetic field and flows to the white dwarf surface may be somewhat chaotic.

Acknowledgments. Data for this work were contributed by Paula Szkody, Don Hoard, Mark Hammergren, Jeff Morgan, Eric Fierce, Knut Olsen, Paul Mason, Robert Rolleston, Eric Ruotsalainen, Elena Pavlenko, Nickolay Shakhovskoy, Sergei Shugarov, Ivan Andronov, Sergey Kolesnikov, Tim Naylor and Ed Schmidt. Additional contributions were made by Paul Mason, Elena Pavlenko, Ivan Andronov and Paula Szkody. Special thanks to George Wallerstein who arranged the MRO telescope schedule to make this project possible.

References

Mason, P.A., Chanmugam, G., 1992, in "The Vina Del Mar Workshop on Cataclysmic Variable Stars", ed. N. Vogt (San Franscisco: ASP conference series Vol. 29), p203
Remillard, R.A., Bradt, H.V., McClintock, J.E., et al., 1986, Ap. J., **302**, L11
Silber, A., Bradt, H.V., Ishida, M., et al., 1992, Ap. J., **389**, 704
Wynn, G.A., King, A.R., 1992, MNRAS, **255**, 83
Zucker, D.B., Raymond, J.C., Silber, A., 1995, Ap. J., **449**, 310

QPOS OF BY CAM IN LOW AND HIGH STATES

E. P. PAVLENKO[1], S. YU. SHUGAROV[2], S. V. ANTIPIN[2], D. A. SOKOLOV[2], G. BORISOV[2], I. BUDZINOVSKAYA[1]
1. Crimean Astrophysical Observatory, Crimea 334413, Ukraine. e-mail: pavlenko@crao.crimea.ua
2. Sternberg Astronomical Institute, Universitetsky avenue, 13, Moscow V-234, 119899, Russia. e-mail: shugarov@sai.msu.su

Abstract. We present the analysis of *BV* photometry of BY Cam based on observations carried out in Crimea. The long QPOs in brightness were detected in both high and low brightness states. Their typical periods, equal to 6, 12 and 24 m, are found. The amplitude of the QPOs depends on the phase of $P_{\rm beat}$.

1. Results

BY Cam is an AM Her type binary discovered by Remillard et al. (1986). It is the second, after the polar V1500 Cyg, to show asynchronism of the white dwarf's spin with respect to the binary and a fast synchronization of the components (Mason et al. 1995). BY Cam shows very complicated behaviour in optical light due to the existence of several periods: orbital, orbital side–band, spin and spin–orbit beat periods (Mason et al. 1995).

The observations of BY Cam were carried out at the Crimean Astrophysical Observatory and Sternberg Astronomical Institute with the 0.5, 0.6 and 1.25 m telescopes using the *BVR* system during 25 nights from 1989 up to 1994. The long–term *BVR* light curve is shown by Sokolov, Shugarov & Pavlenko (1996).

One of the most intriguing and unresolved problems in studying BY Cam is the evidence of the so-called long (tens of minutes) quasi-periodic brightness oscillations. They appear both in a low and high brightness state and have a varying amplitude, which sometimes increases up to 0.3 mag.

Adopting $P_{\rm beat} = 14.11$ d and $P_{\rm spin} = 0.1384238$ d (Mason et al. 1995), we have analysed the behaviour of the QPOs amplitude with the phase of $P_{\rm beat}$ (or with $O-C$ of $P_{\rm spin}$ for both minima and maxima). The elements:

A. Evans and J. H. Wood (eds.), Cataclysmic Variables and Related Objects, 217-218.

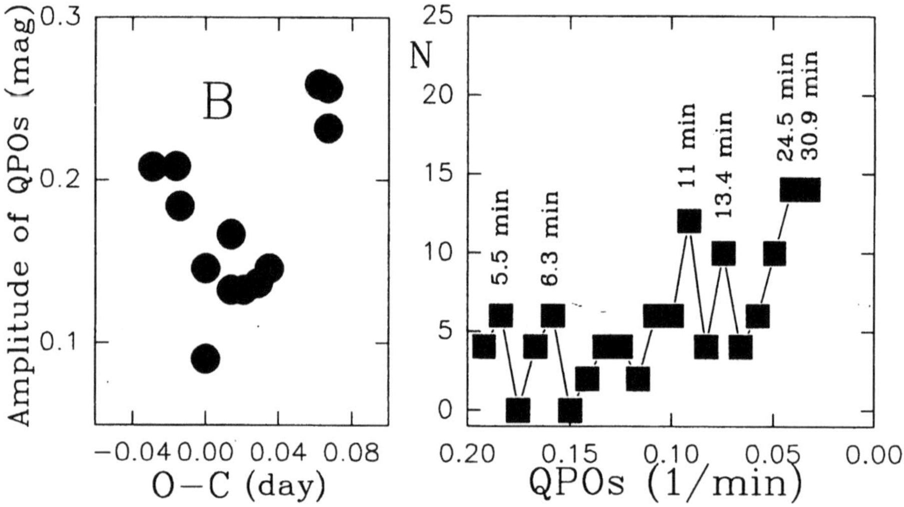

Figure 1. **Left**: Dependence of QPO amplitude on the $O - C$ of P_{spin}. **Right**: The histogram of the most significant QPOs detected for all observations in V.

$C_{max} = 2447540.288 + 0.1384238E$, $C_{min} = 2447591.443 + 0.1384238E$ were used. We have seen a strong dependence in B (Fig. 1 left) but not in V.

Over five years we have found several stable, (multiple?) QPOs (Fig. 1 right), which are close to 6, 12 and 24 m. Note that the asynchronous magnetic nova V1500 Cyg also exhibited QPOs with typical time-scales of 17...26 m (Pavlenko 1992). The analysis shows that these QPOs do not have a stable zero phase, producing a broad peak on a periodogram. But the intermediate brightness state exhibited a stable 5.8 m QPO over 100 d.

Acknowledgments. This research was supported by a grant from the Center for Cosmoparticle Physics 'Cosmion'. We also are grateful to the Keele organising committee for the possibility of attending this meeting.

References

Mason, P.A., Chanmugam, G., Andronov, I.L., et al., 1995, in "Cataclysmic Variables", eds A. Bianchini et al., Kluwer, p426

Pavlenko, E.P., 1992, Izv. Krym. Astroph. Obs., **86**, 55

Remillard, R.A., Bradt, H.V., McClintock, J.E., et al., 1986, Ap. J., **302**, L11

Sokolov, D.A., Shugarov S.Yu., Pavlenko E.P., 1996, these proceedings, p219

LONG-TERM BRIGHTNESS CHANGES OF TWO CVS

D. A. SOKOLOV[1], S. YU. SHUGAROV[1], E. P. PAVLENKO[2]

1. Sternberg Astronomical Institute, Universitetsky avenue, 13 119899, Moscow V-234, Russia, e-mail: shugarov@sai.msu.su
2. Crimean Astrophysical Observatory, Crimea 334413, Ukraine

Abstract. The long–term light curves of the AM Her type binary BY Cam and the nova-like variable PG 2133+115 are presented. The *BVR* observations were carried out at a 0.5 m telescope equipped with a high sensitive TV tube superisocon (Abramenko et al. 1978), the *UBV* photoelectric data were obtained at 0.6 and 1.25 m telescopes and the photographic estimates were made from the negatives obtained at the 0.4 m astrograph in Crimea.

Both stars show high and low brightness states. They show similar behaviour in some respects: the amplitude between high and low states was about two magnitudes in *B*, but the duration of a low state was approximately two months for BY Cam and 1...2 years for PG 2133+115.

1. PG 2133+115

Green, Ferguson & Liebert (1982) noted that this star has a UV excess and so is possibly a CV. We obtained *UBV* (1984–95) and photographic (1960–95) observations of this CV. On average, magnitudes and colours were as follows: $V = 14.7$ mag, $B - V = 0.0$ mag, $U - B = -0.9$ mag. Misselt & Shafter (1995) and the current authors did not detect the period of 2.9 h, found earlier from radial velocity variations. According to the photographic observations in 1969–70 the object was 2 mag fainter than the mean level and 1 mag light variations with a time-scale of 10...20 days were observed (see Fig. 1).

We suggest that PG 2133+115 is a CV which is principally in the 'on' brightness state and is observed at a comparatively small inclination angle.

A. Evans and J. H. Wood (eds.), Cataclysmic Variables and Related Objects, 219–220.

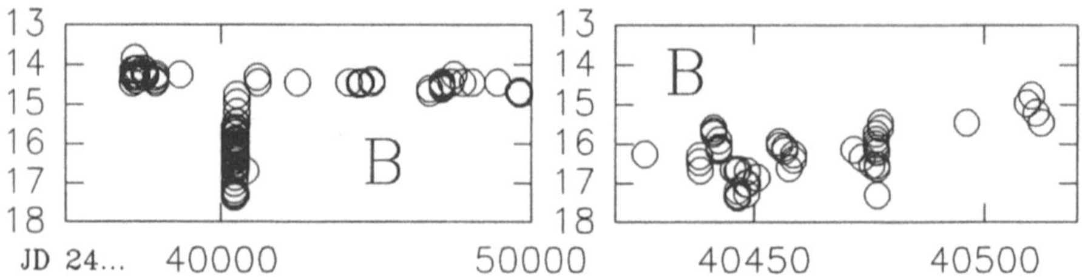

Figure 1. **Left:** PG 2133+115 in 1960-94. **Right:** The low state in 1969.

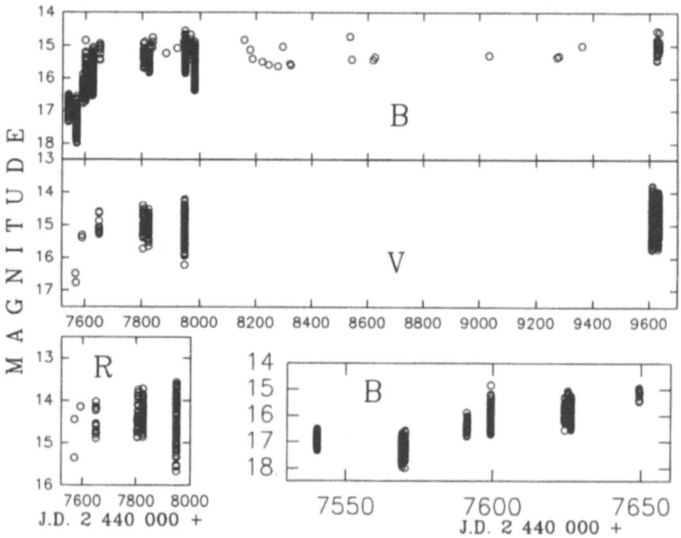

Figure 2. **Left:** BY Cam in 1989-94. **Right:** Low and intermediate states of BY Cam.

2. BY Cam

TV observations of BY Cam were obtained in 1989–90 and in 1994, photoelectric in 1994, and photographic estimates in 1953–93. The *BVR* light curve is shown in Fig. 2 (left), which demonstrates the low and high brightness states. In more detail the low and intermediate states are shown in Fig. 2 (right). The low state duration was a brief episode lasting two months only, returning to the high brightness level during three months. The maximal amplitude of the long-term variations was detected in *B* (2 mag).

References

Abramenko, A.M., Prokof'eva, V.V., Bondar, N.I. et al., 1988, Bull. of the Crimean Astroph. Obs., **78**, 189

Green, R.F., Ferguson, D., Liebert J., 1982, PASP, **94**, 560

Misselt, K.A., Shafter, A.W., 1995, AJ, **109**, 1757

OPTICAL FLICKERING AND SHOT NOISE IN AM HER SYSTEMS

S. LARSSON, B. LARSSON
Stockholm Observatory, S-13336 Saltsjöbaden, Sweden

1. Introduction

The emission from AM Her systems show a variety of time-variable phenomena, including flickering with a typical power spectrum slope of −1 to −2. The optical flickering is usually modeled as, or assumed to be, a shot noise process with many overlapping simultaneous shots. Panek (1980) found that, for AM Her, a consistent model could be constructed with randomly occurring 70...90 s rectangular shots. The power spectrum had a ν^{-2} shape above 0.02 Hz as expected, although it is not clear whether a break in the slope of the power spectrum was actually seen around 0.01 Hz. There are many additional reports in the literature of 'characteristic time-scales' on the order of tens of seconds. However we are not aware of any case where the reported time-scale has been supported by e.g. a break in the power density spectrum. It is easy to show that the usual method of estimating a 'characteristic time-scale' from the auto correlation function of de-trended data will suggest such a characteristic time-scale even if none is present in the original time series. The evidence for shot noise in AM Her light curves therefore needs a re-examination. In this paper we give a short summary of our analysis of optical flickering in V834 Cen, based on high speed photometry obtained at ESO in 1987. Details will be reported elsewhere.

2. Time domain analysis

All peaks in the light curves with some height above the noise level were identified and analyzed by directly fitting a few different pulse shapes. The distribution of pulse parameters were analyzed and compared to the corresponding results for simulated data.

A. Evans and J. H. Wood (eds.), Cataclysmic Variables and Related Objects, 221–222.

The following main results were obtained.

1. The distribution of pulse parameters indicates the presence of a distinct flickering component consisting of high amplitude, narrow pulses.
2. The distinct shots are best fitted by peaked pulses, such as triangular, or $\exp -|(t-t_0)/\tau|$. The mean asymmetry is close to zero.
3. The Full Width Half Maximum (FWHM) distribution for the distinct shots is strongly peaked around 6...7 s.
4. The amount of overlap for the distinct shots is small.

3. Frequency domain analysis

Power spectra were calculated from 1500 s data segments. These were then averaged for all the five observing nights to produce an overall mean power spectrum. The most important feature in the power spectrum is a break at about 0.03 Hz, which provides strong support for a shot noise model. Above this frequency the spectrum steepens to a power law of slope −2.5. This is significantly steeper than the 'canonical' value of −2 that is produced by, for example, sharp edged shots, such as one-sided exponentials.

4. Model

The pulse profiles implied by the data analysis correspond to very steep power spectra, ν^{-4} ! We have confirmed this by calculating power spectra for the individual observed 'flares'. We also calculated the power spectrum produced by double-sided exponentials with a distribution in pulse width taken from our time domain analysis. This was found to flatten the power spectrum slope at high frequencies from −4 to −2.4, which is very close to that of the data. The observed turn over around 0.03 Hz is also reproduced by this model. A power excess over the model at frequencies below 0.01 Hz can be explained by adding a ν^{-2} component, with the same strength as in our previous observations of V834 Cen in 1984 and 1985 when the distinct flickering component was absent. This two-component flickering model is consistent with both the time and frequency domain properties of the time series data. Although the model is not unique it represents a substantial improvement over earlier models that were based on much less direct information.

References

Panek, R., 1980, Ap. J., **241**, 1077

HIGH SPEED UV PHOTOMETRY OF THE AM HER SYSTEMS AM HER, VV PUP AND V834 CEN, WITH THE HUBBLE SPACE TELESCOPE

K. SCHAEFER[1], H. BOND[1], G. CHANMUGAM[2]
1. STScI, 3700 San Martin Dr., Baltimore, MD 21218, USA
2. Physics & Astronomy Dept., Louisiana State Univ., Baton Rouge, LA 70803, USA

Abstract. We have used the High Speed Photometer (HSP) aboard the *Hubble Space Telescope* to observe the magnetic cataclysmic variables VV Pup, AM Her, and V834 Cen in the UV (1400...3300 Å) with 0.01 s time resolution. We detected low frequency flickering in all three systems, and compare the time-scales with the predictions of King (1989). At higher frequencies we searched for shock oscillations from the accretion column(s) in these systems. The data were analyzed using the Gabor transform wavelet-like technique (Heil & Walnut 1989) to search for frequency evolution throughout each observation. Preliminary analysis suggests the detection of rapid UV quasi-periodic oscillations (QPOs) in VV Pup at 0.74 Hz, and at 4.4 Hz in V834 Cen. As in ground based observations, our observations failed to yield any rapid QPOs in AM Her itself.

1. Introduction

Theoretical studies of time dependent accretion onto highly magnetized white dwarfs predicted that the infalling gas forms a strong stand-off shock that is unstable for optically thin bremsstrahlung cooling (Langer, Chanmugam & Shaviv 1981, 1982). The shock front was predicted to oscillate up and down above the white dwarf surface, with period:

$$P \sim 1.1 \left(\frac{A}{10^{16}\mathrm{cm}^2}\right)\left(\frac{\dot{M}}{10^{16}\,\mathrm{g\,s^{-1}}}\right)^{-1} R_9^{-1}(M/\,\mathrm{M}_\odot)\ \ \mathrm{s},$$

where $\dot{M}$ is the accretion rate, A is the cross sectional area of the accretion column, M the white dwarf mass and R_9 the radius R in units of 10^9 cm.

A. Evans and J. H. Wood (eds.), Cataclysmic Variables and Related Objects, 223–224.

These and other studies led to the discovery of 1...2 s QPOs in VV Pup (Larsson 1989; Ramseyer et al. 1993) and V834 Cen (Larsson 1985; Middleditch et al. 1991). QPOs have not been detected from AM Her.

In addition to the rapid QPOs, large amplitude low frequency flickering has been seen in several AM Her binaries (see Cropper 1990). King (1989) has predicted that QPO (due to the motion and density of an ionization front near the L_1 point) in AM Her systems, with period:

$$P \sim 5.5\,10^{-2} P_{\rm orb}$$

may be seen. For AM Her, this corresponds to a period of ~ 600 s, and for VV Pup, ~ 300 s.

2. Analysis and results

Each ~ 30 m observation was analyzed for both periodic and quasi-periodic signals in the frequency range $0 < \nu$ (in Hz) < 50. Low amplitude QPO detections have been discovered at $\nu \sim 0.74$, 0.77 Hz for VV Pup (which was in a low state during the observations). VV Pup also exhibited long period flickering (~ 6 m), similar to the prediction by King's model.

AM Her has not exhibited any rapid QPO, but did display large amplitude flickering with a period of ~ 550 s. This is again close to the period predicted by King (1989).

V834 Cen also showed weak, long period flickering, where $3.2 < P$ (in m) < 8. Analysis of the UV data has not yielded any QPOs stronger than $\sim 3\,\sigma$, although analysis of the one V band observation (also obtained with the HSP) has yielded an interesting, but low amplitude 4.4 Hz detection. Further investigation of this frequency and other phenomena seen in these data is underway.

References

Cropper, M., 1990, Space Sci. Rev., **54**, 195
Heil, C.E., Walnut, D.F., 1989 SIAM Review, **31**, 4, 628
King, A.R., 1989, MNRAS, **241**, 365
Langer, S.H., Chanmugam, G., Shaviv, G., 1981, Ap. J., **311**, L23
Langer, S.H., Chanmugam, G., Shaviv, G., 1982, Ap. J., **258**, 289
Larsson, S., 1985, A&A, **14**, L1
Larsson, S., 1989, A&A, **21**, 146
Middleditch, J., Imamura, J.N., Wolff, M.T., Steiman-Cameron, T.Y., 1991, Ap. J., **382**, 315
Ramseyer, T.F., Robinson, E.L., Zhang, E., et al., 1993, MNRAS, **260**, 209

RECENT PROGRESS ON THE POLAR QS TEL: THE HST RESULTS

K. L. CLAYTON, S. R. ROSEN, J. P. OSBORNE
Leicester University, Leicester, LE1 7RH, UK

Abstract. We report on the Hubble Space Telescope (HST) observation of the polar, QS Tel, made during 1994 June. Orbital modulation is present in both the continuum and lines. A narrow dip is observed in the continuum folded light curve. The spectrum of the occulted source during this dip is broadly consistent in shape with the Rayleigh-Jeans tail of the EUV source.

1. Introduction

QS Tel (RE 1938−461), $P_{\rm orb} = 2.33$ h, was one of the brightest sources discovered in the EUV ROSAT WFC sky survey [1]. It has been seen to undergo a number of changes in its accretion geometry, having been seen accreting onto one pole in ROSAT observations [1][3] and onto two poles in an EUVE observation [3]. A deep, narrow dip has been found during one of the EUV/soft X-ray maxima, almost certainly caused by occultation of the accretion region by the gas stream from the secondary.

2. Observation and results

QS Tel was observed by the HST in 1994 June. Data were taken with the Faint Object Spectrograph (FOS) operated in rapid mode ($\Delta t = 20$ s). The G160L grating was employed to cover the 1154...2508 Å range at a resolution of 6.8 Å.

The mean spectrum shows a continuum rising slightly towards blue wavelengths, with strong emission lines of C IV, N V, Si IV and He II superposed. The continuum luminosity (1154...2508 Å) is $5.9 \pm 0.4\, 10^{31}$ erg s^{-1} (for a distance of 170 pc [4]), consistent with that observed by IUE in 1993 August [2].

A. Evans and J. H. Wood (eds.), Cataclysmic Variables and Related Objects, 225–226.

The orbital UV continuum light curve is simply modulated by a broad minimum at phase 0.7...0.8 with a depth of $\sim$ 18%. A 24% dip is observed covering phases 1.00...1.06, i.e., at the same phase as the dip observed in the ROSAT and EUVE data [3]. Modulation is present in the line fluxes, with a bright interval between phases $\sim$ 0.1...0.6. The line flux variation (peak to peak) is $\sim$ 45%. If a dip is present in the line flux, it is much broader and shallower than the continuum dip, spanning $\phi \sim 0.9\ldots1.2$.

The two brightest emission lines, C IV and He II, exhibit clear radial velocity motion, and can be characterised by a broad (velocity amplitude $\sim 400\,\mathrm{km\,s^{-1}}$) and a core (velocity amplitude $\sim 300\,\mathrm{km\,s^{-1}}$) component. We are unable to discern any difference in the velocity curves of the UV line components and their optical counterparts [3][4].

The radial velocity measurements of the two strongest emission lines find maximum red-shift (and thus inferior conjunction, assuming formation in the accretion flow close to the white dwarf) at phase 0.1 (both components) – this is also the phase of minimum line flux. Maximum line flux occurs at $\phi_{\rm orb}$=0.3...0.5, around the phase of first quadrature. This may reflect preferential photoionization of the stream on one side.

3. Discussion and conclusions

We do not know whether QS Tel was accreting onto one or two poles during this observation. However an IUE observation that took place only two weeks before the EUVE observation, which found QS Tel to be in a two pole accretion mode, had a similar luminosity to that of our data. If this implies a two pole accretion mode in our observation, the lack of a double peaked light curve argues against the polar regions contributing significantly to the flux.

The HST data shows a dip at phase 1.00...1.06. Subtracting the dip from the post-dip spectra reveals a strong far UV turn-up in the occulted source, which is consistent with the Rayleigh-Jeans tail of the EUVE black-body spectrum. The strength of the UV turn-up is only $\sim$ 20% of that expected on the basis of the EUVE observation; however QS Tel is known to vary by factors of $\lesssim 100$ in the EUV band [3]. The UV dip is probably caused by stream occultation of the white dwarf and polar accretion region.

References

1. Buckley, D.A.H., O'Donoghue, D., Hassall, B.J.M., et al. 1993, MNRAS, **262**, 93
2. de Martino, D., Buckley, D.A.H., Mouchet M., Mukai, K., 1995, A&A, **298** L5
3. Rosen, S.R., Mittaz, J.P.D., Buckley, D.A.H., et al., 1996, MNRAS, in press
4. Schwope, A.D., Thomas, H.-C., Burwitz, V., et al., 1995, A&A, **293**, 764

QQ VUL REVISITED

M. S. CATALÁN[1,2], S. D. DAVEY[2], R. C. SMITH[1],
D. H. P. JONES[3]
1. Dept. of Physics and Astronomy, University of St Andrews, Fife KY16 9SS, UK
2. Astronomy Centre, Division of Physics and Astronomy, University of Sussex, Brighton, BN1 9QH, UK
3. RGO, Madingley Rd., Cambridge, CB3 0EZ, UK

Abstract. With simultaneous blue ($\lambda\lambda$4200–5000 Å) and far–red ($\lambda\lambda$7950–8380 Å) phase-resolved, high resolution spectrophotometry, we re-examine the magnetic CV QQ Vul. A thorough radial velocity and flux analysis of all the observable features is carried out. We derive a complete set of system parameters and classify the donor star to be a spectral type M4.5 V. In addition to the Balmer, He I and He II emission lines in the blue range, arising from the accretion regions, we detect various emission features which solely follow the movement of the secondary star. The lines, which we interpret as reprocessed light from the illuminated hemisphere of the donor star, are Mg II at λ4481 Å, He II at λ8236 Å and C I at λ8335 Å. Intensity maps based on the line flux and radial velocity variations of the Na I absorption near λ8190 Å and Mg II emission feature were also made. These show that the line distribution on the surface of the donor is consistent with being symmetric. As a result of our analysis the position of the real zero phase was found to be at $\phi = 0.42$ of the linear polarisation ephemeris.

1. Introduction

QQ Vul is a long-period (222.5 m) AM Her type CV first detected with the HEAO-1 low-energy detectors (Nugent et al. 1983). Since its discovery, quite a number of studies have been published on this object in the optical and X-ray wavelengths (e.g. Nousek et al. 1984; Mukai & Charles 1987; Osborne et al. 1987). However, to date, its system parameters are still poorly known. We report recent spectroscopic observations of QQ Vul, paying particular attention to features that originate from the secondary, and derive its system parameters.

A. Evans and J. H. Wood (eds.), Cataclysmic Variables and Related Objects, 227–228.

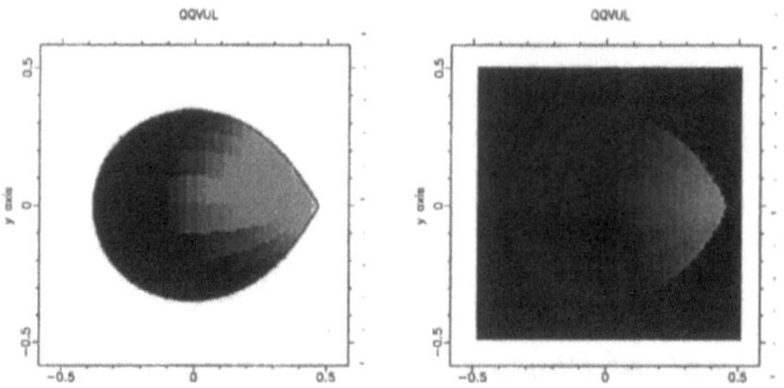

Figure 1. **Intensity distribution of the Na I (left) and Mg II (right) on the secondary star.**

2. Intensity maps and the spectral type of the secondary

Our spectral analysis has revealed that the emission spectrum seen is heavily peppered with lines that arise from the heated face of the donor star. In particular we found that the Mg II doublet at λ4481 Å, He II at λ8236 Å and C I at λ8335 Å have radial velocities and line flux variations consistent with an irradiation scenario. By adopting the same technique as in Davey & Smith (1992), we have obtained Na I and Mg II intensity maps of the donor star. Those derived from the line flux variations are shown in Fig. 1. Using the TiO band ratios method (Wade & Horne 1988), we have found the spectral type of the donor star in QQ Vul to be an M4.5 V.

3. Dynamical determination of the system parameters

The system parameters were found using the $v\ \sin(i)$ technique (see e.g. Martin et al. 1989). A preliminary estimate of the secondary star's true K-velocity was obtained from the average of the velocity semi-amplitudes derived from the Na I absorption feature, formed on the unirradiated regions of the star, and the Mg II emission originating from its illuminated side. We found K_2=174±12 km s^{-1}. Combined with our measurement of the rotational velocity $v_{\rm abs}\ \sin(i)$=112±10 km s^{-1} from the Na I lines, the mass ratio was found to be $q = 0.79 \pm 0.17$. This, in turn, yields a primary mass of M_1=0.42±0.18 M$_\odot$ and a secondary mass M_2=0.33±0.18 M$_\odot$.

References

Davey, S.C., Smith , R.S., 1992, MNRAS, **257**, 476
Martin J.S., Friend, M.T., Smith, R.C., et al.,1989, MNRAS, **240**, 519
Mukai, K., Charles, P.A., 1987, MNRAS, **226**, 209
Nousek, J.A., Takalo, L.O., Schmidt, G.D., et al., 1984, Ap. J., **277**, 682
Nugent, J.J., Jensen, K.A., Nousek, J.A., et al., 1983, Ap. J. Supp., **51**, 1
Osborne, J.P., Beuermann, K., Charles, P., et al., 1987, Ap. J., **315**, L123
Wade, R.A., Horne, K., 1988, Ap. J., **324**, 411

RX J1015.5+0904:
A NEW POLAR AT THE LOWER PERIOD LIMIT

V. BURWITZ[1], K. REINSCH[1], A. D. SCHWOPE[2],
K. BEUERMANN[1,3], S. MENGEL[2], P. NOTNI[2],
A. VAN TEESELING[1], H.-C. THOMAS[4]
1. Universitäts-Sternwarte, Geismarlandstr. 11, D-37083 Göttingen, Germany. (burwitz@usw051.dnet.gwdg.de)
2. Astrophysikalisches Institut Potsdam, Germany
3. MPI für Extraterrestrische Physik, Garching, Germany
4. MPI für Astrophysik, Garching, Germany

Abstract. We present X-ray and optical observations of the ROSAT discovered polar (AM Her binary) RX J1015.5+0904 and first results of our analysis of the X-ray and optical data. These results indicate a one-pole accretion geometry and an orbital period of $P = 79.88$ m for this $V \sim 17$ mag system.

1. Introduction

From our analysis of bright, soft, high-galactic-latitude ROSAT All-Sky Survey sources we extracted a set of AM Her candidates (Beuermann & Thomas 1993; Beuermann & Schwope 1994; Beuermann & Burwitz 1995). Follow-up optical and X-ray observations led to the discovery of 27 new polars, more than doubling the pre-ROSAT sample. Here we present a summary of our detailed analysis of our X-ray (ROSAT survey, pointed PSPC and HRI) and optical (photometric and spectroscopic) observations of the new polar RX J1015.5+0904 from our sample of ROSAT discovered AM Her binaries.

2. The ROSAT discovered polar RX J1015.5+0904

The orbital period of the $V \sim 17$ mag optical counterpart of RX J1015.5 +0904 derived from all the X-ray and optical data is $P = 79.88$ m, which is close to the lower limit for periods of cataclysmic variables. Our X-ray light curves indicate a one-pole accretion geometry with an absorption dip

A. Evans and J. H. Wood (eds.), Cataclysmic Variables and Related Objects, 229–230.

caused by the accretion stream crossing the line of sight (cf. Fig. 1). The optical light curve shows a double humped structure which is characteristic of cyclotron beaming. A complete analysis and more detailed discussion of our data will be presented in Burwitz et al. (1996).

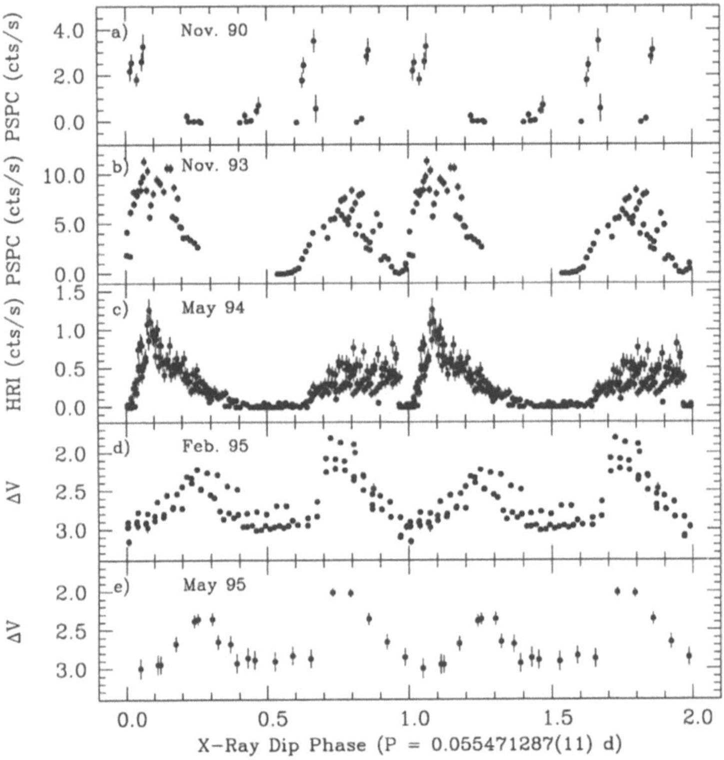

Figure 1. X-ray and optical light curves of RX J1015.5+0904. X-ray observations with ROSAT: (a) All-sky-survey (1990 November), (b) pointed PSPC observations (1993 November) and (c) pointed HRI observations (1994 May). Optical *V* CCD photometry obtained with: (d) the Dutch 90 cm telescope at La Silla in 1995 February, (e) the 70 cm telescope in Potsdam in 1995 May. Phase zero is defined by the center of the X-ray dip in the HRI light curve: T(dip) = HJD 2449488.35649(26) + 0.055471287(11) E.

References

Beuermann, K., Burwitz, V., 1995, in "Magnetic Cataclysmic Variables", eds D. Buckley, B. Warner, ASP Conf. Ser. **85**, 99

Beuermann, K., Schwope, A.D., 1994, in "Interacting Binary Stars", ed. A. Shafter, ASP Conf. Ser. **56**, 119

Beuermann, K., Thomas, H.-C., 1993, Adv. Space Res., **13(12)**, 115

Burwitz, V., Reinsch, K., Schwope, A.D., et al., 1996, A&A, in prep

SPACE OBSERVATIONS

FAR ULTRAVIOLET OBSERVATIONS OF DWARF NOVAE MADE WITH THE HOPKINS ULTRAVIOLET TELESCOPE

KNOX S. LONG
Space Telescope Science Institute
3700 San Martin Drive
Baltimore, MD 21218, USA

Abstract. Observations with the Hopkins Ultraviolet Telescope on the Astro-1 and Astro-2 space shuttle missions have provided the first set of moderate (3 Å) resolution far ultraviolet (FUV) spectra of dwarf novae to include the wavelength range between Lyα and the Lyman limit. Important lines which are detected in the HUT spectra of dwarf novae in this wavelength range include S VI $\lambda\lambda933, 945$, C III $\lambda977$, O VI $\lambda\lambda1032, 1038$, P V $\lambda\lambda1118, 1128$ and C III $\lambda1176$, as well as the higher order Lyman lines. The observations confirm earlier IUE observations that two dwarf novae – U Gem and VW Hyi – have quiescent FUV spectra dominated by the white dwarf, but suggest that the quiescent FUV emission from three other dwarf novae – SS Cyg, WX Hyi and YZ Cnc – are dominated by emission from a hot portion of the disk or a disk corona. The spectra obtained of the dwarf novae Z Cam and EM Cyg in outburst and also of the nova-like variable IX Vel can be modeled reasonably successfully in terms of steady state disks constructed by adding appropriately-weighted stellar model spectra.

1. Introduction

As a result of the observations made with *International Ultraviolet Explorer*, tremendous progress has been made in the last fifteen years in terms of characterizing the FUV spectra of disk-dominated cataclysmic variables at wavelengths down to and including Lyα. As a result, we now know that (i) the disk, a wind emerging from the inner portions of the system, the hot spot, and the white dwarf (WD) all contribute to the ultraviolet spectra of dwarf novae, (ii) the UV spectrum of a dwarf nova in outburst resembles that expected from an optically thick disk and (iii) the UV outburst is often delayed with respect to the optical outburst implying that, at least in

A. Evans and J. H. Wood (eds.), Cataclysmic Variables and Related Objects, 233–241.

some cases outbursts develop in the outer disk [see e.g. reviews by Córdova (1995); Córdova & Howarth (1987); Drew (1990)]. More recently, HST has been used to to obtain the first spatially resolved maps of accretion disks of CVs in the FUV and to obtain high signal-to-noise ratio, high resolution spectra of the strongest wind lines in several dwarf novae and nova-like systems. [See Horne et al. (1994); Robinson et al. (1995); Mason et al. (1995); Baptista et al. (1995), as well as numerous contributions to this conference.] However, like IUE, HST is limited to wavelengths longward of $\sim 1160\,$Å.

The Hopkins Ultraviolet Telescope consists of a 0.9 m f/2 primary feeding a prime-focus Rowland spectrograph with a photon-counting microchannel plate detector. Designed for use on the space shuttle, it was flown on Astro-1 in 1990 December and again in 1995 March. Optimized for the wavelength range between Lyα and the Lyman limit, it was used to obtain 830...1860 Å spectra with a resolution of $\sim 3\,$Å of a wide range of astrophysical objects, including 5 CVs on Astro–1 and 10 CVs on Astro-2. The only other spectra of CVs which include the wavelength range between Lyα and the Lyman limit were obtained with Voyager, which has a resolution of 18 Å and far less sensitivity (Polidan, Mauche & Wade 1990), and ORFEUS (Raymond et al. 1995), which has higher resolution, but has been used to observe only three CVs – AM Her, V3885 Sgr and Z Cam. A complete description of HUT and its performance on Astro-1 has been provided by Davidsen et al. (1992); improvements made to the instrument and its performance on Astro-2 are discussed by Kruk et al. (1995).

Here, I will summarize our analysis of the spectra of CVs obtained with HUT on Astro-1 and give an overview of the initial results of an analysis of the spectra obtained of dwarf novae on Astro-2.

2. Dwarf novae in quiescence

Five dwarf novae – SS Cyg, YZ Cnc, WX Hyi, U Gem, VW Hyi – were observed in quiescence on Astro-2. Three of these dwarf novae were actually observed several times during the mission, at intervals separated by several days. The time-averaged spectra of each of the dwarf novae are shown in Fig. 1. The spectra of the different dwarf novae differ considerably. U Gem and VW Hyi have spectra which can be characterized in terms of broad Lyman line absorption features and relatively narrow metal absorption lines, while SS Cyg, YZ Cnc, and WX Hyi have relatively flat continua and broad emission lines.

In the 'standard' theory of disks in dwarf novae, quiescent accretion disks are expected to have a temperature of $\lesssim 6000\,$K (Meyer & Meyer-Hofmeister 1982; Cannizzo, Ghosh & Wheeler 1982) and are expected to

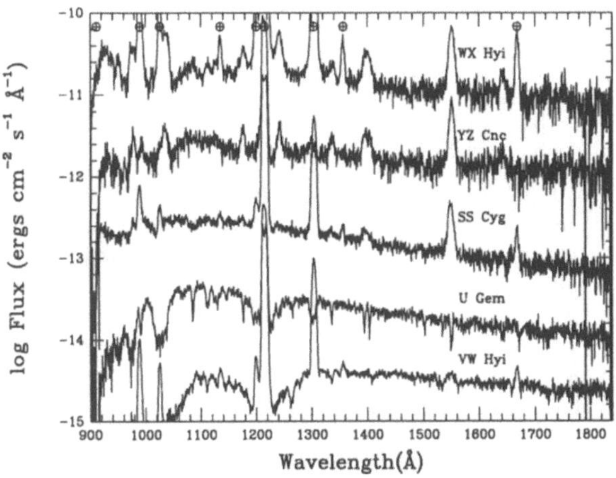

Figure 1. The spectra obtained of dwarf novae in quiescence with HUT on Astro-2. The spectra have been offset from one another for clarity. Lines which may be airglow are indicated by ⊕ on all figures. In the spectra of WX Hyi, YZ Cnc and SS Cyg, in which O VI λλ1032, 1038, N V λλ1239, 1243, and C IV λλ1548, 1551 are seen in emission, the Lyman lines are apparently absent and the continuum is bluer than in the spectra of U Gem and VW Hyi, in which these same lines are weak or absent.

be optically thin in the UV continuum. As a result, the WD is generally assumed to dominate the FUV light in dwarf novae. WDs have broad Lyman absorption lines as do HUT spectra of U Gem and VW Hyi. The UV flux from U Gem and VW Hyi is also about that expected from a WD at the estimated distances of U Gem and VW Hyi, given estimates of the temperature based on the widths of Lyα, as indeed IUE observations had indicated (Panek & Holm 1984, Mateo & Szkody 1984). Because the strong gravitational field causes metals to settle into the atmosphere on a time-scale of 1...3 yr (cf. Michaud 1987), absorption lines with equivalent widths of ~1 Å are not usually seen in the spectra of isolated WDs, unless the temperatures are very low (<12 000 K) or very high (>90 000 K). However, the time-scale to replenish the ($\sim 10^{-12}\,M_\odot$) atmosphere of a WD is short in a dwarf nova system and therefore prominent metal absorption lines are expected and indeed are observed in both U Gem and VW Hyi. Thus the HUT spectra provide strong support for the conclusions first reached with IUE that the WD dominates the FUV spectra of U Gem and VW Hyi.

U Gem was observed on both Astro-1 and Astro-2. The Astro-1 observation occurred just 10 d after the system had returned to optical quiescence following a normal outburst (Long et al. 1993). Model fits to the Astro-1 observations implied an average WD temperature of ~ 38 000 K and near solar abundances in the WD atmosphere. This was higher than the 30 000 K estimated from IUE spectra far from outburst (Panek & Holm 1984) and

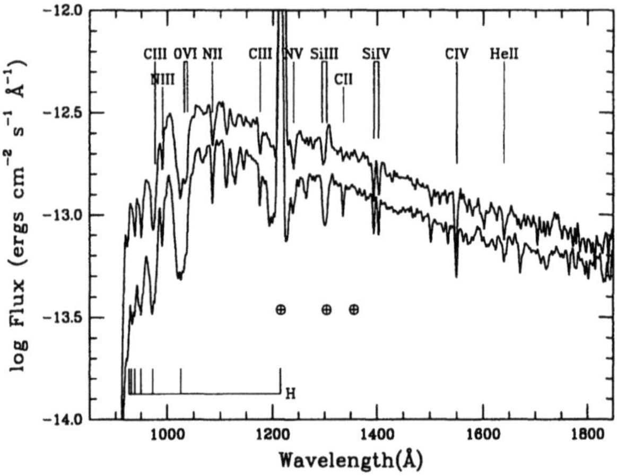

Figure 2. The spectra of U Gem obtained with HUT on Astro-1 (upper curve) and Astro-2 (lower curve). The flux at 1450 Å from U Gem as observed on Astro-2 185 d after the return to quiescence was about 30% less than on Astro-1.

suggested that the WD in U Gem is heated during outbursts and then cools between outbursts, consistent with evidence for a decline in the UV flux with time from outburst (Kiplinger, Sion & Szkody 1991). The Astro-1 data were also fit to two-temperature models in which 85% of the surface had a temperature of 30 000 K and 15% had a temperature of 57 000 K, which we suggested might arise from a viscously-heated differentially rotating atmosphere produced by the outburst. Other possible mechanisms for the cooling include compressional heating of outer portions of the WD by the weight of the matter accreted during outburst (Sion 1995) and the effects of a self-sustaining coronal flow onto the surface of the WD during (the early stages) of quiescence (Meyer & Meyer-Hofmeister 1994).

Our observation of U Gem on Astro-2 occurred at a time when U Gem had been at optical minimum for 185 d (Long, Blair & Raymond 1995). As shown in Fig. 2, the Astro-2 spectrum of U Gem resembles closely the spectrum obtained on Astro-1. However the flux observed on Astro-2 is less (70% at 1450 Å), the color temperature is lower, and the Lyman lines all have larger equivalent widths than on Astro-1. On Astro-2, the best fit single-temperature model was obtained with a temperature of 29 700 K and the flux observed corresponds to a WD radius of $5.4\,10^8$ (D/90 pc) cm, consistent with that expected from a 1.05 $M_\odot$ WD. These results are similar to those obtained with the Faint Object Spectrograph on *Hubble Space Telescope* ~ 13 and 70 d after outburst in which the flux at 1450 Å dropped 28% and in which single-temperature solar abundance model fits indicate a temperature drop from 39 400 K to 32 100 K (Long et al. 1994a).

The other quiescent dwarf nova in our sample that appears to be dominated by the WD, VW Hyi, was observed twice, 12 and 15 d after a normal outburst. As was true of U Gem, the spectrum contains broad absorption lines due to the Lyman series of hydrogen and numerous other absorption features which can be identified with transitions which are expected in a moderately-hot WD atmosphere with near solar abundances. Preliminary fits to the data from both spectra (restricted to the spectral range 1150...1840 Å) using WD model atmospheres suggest a temperature of about 18 000 K, similar to the value of 18 000±2000 K obtained by Mateo & Szkody (1984) from IUE observations and somewhat lower than the temperature of 22 000±1000 K inferred by Sion et al. (1995) from FOS observations. However, an 18 000 K model underpredicts the observed emission significantly shortward of 1050 Å, suggesting a departure from a simple WD atmosphere, which requires additional analysis.

In contrast to U Gem and VW Hyi, the spectra of SS Cyg, YZ Cnc, and WX Hyi are characterized by blue continua on which are superposed broad (typically 10 Å FWHM) emission features, which can be identified with resonance lines such as S VI $\lambda\lambda 933, 945$, O VI $\lambda\lambda 1032, 1038$, N V $\lambda\lambda 1239, 1243$, Si IV $\lambda\lambda 1394, 1403$ and C IV $\lambda\lambda 1548, 1551$. The signatures of the WD, namely broad Lyman and narrow metal absorption lines, appear to be absent, suggesting that some other component of the dwarf nova contributes significantly to these spectra. Longward of Lyα all of the HUT spectra of dwarf novae resemble spectra obtained with IUE [see e.g. la Dous (1990)]. Because of the fact that HUT covers the region shortward of Lyα, HUT should have been more sensitive to the WD than IUE. If we insist that the WDs dominate the continuum spectra, then they must be quite hot ($>$ 50 000 K for SS Cyg) so that the Lyman and metal lines will be weak enough to hide underneath the emission lines (and airglow features) in the HUT spectra.

The obvious alternative to the WD is a hot region of the disk which dominates the FUV emission of SS Cyg, YZ Cnc and WX Hyi. Unfortunately, there are few models for a hot component in a quiescent disk, the basic problem being of course that a hot portion of the disk should trigger the disk instability. One interesting possibility, recently investigated by Ko et al. (1996) as an extension of their studies of disks in X-ray binaries, is that X-ray and EUV radiation from the vicinity of the WD photoionizes and heats a surface layer on the disk to produce a hot corona. [As discussed by Meyer & Meyer-Hofmeister (1990), X-ray illumination of a disk does not induce, but tends to stabilize against disk instabilities.] Specifically Ko et al. have carried out self-consistent, non-LTE calculations of the vertical structure of the disk and used their models to predict the emergent line spectrum. Ko et al. compared their model predictions to ob-

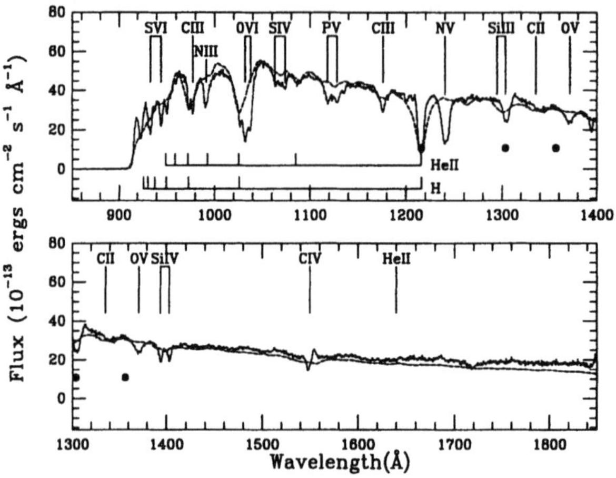

Figure 3. A comparison between the spectrum observed from Z Cam in 1995 March, near the peak of an optical outburst and a steady state accretion disk model with $\dot{M}$ of $4\,10^{17}$ g s^{-1} at a distance of 200 pc. The model reproduces the overall spectral shape and flux fairly well, although it is not expected to account for the absorption lines in the wind which produce broad absorption lines of high excitation ions such as O VI $\lambda\lambda$1032, 1038, N V $\lambda\lambda$1239, 1243, and C IV $\lambda\lambda$1548, 1551.

served IUE line fluxes for several dwarf novae, including SS Cyg and WX Hyi, and found that they were able to approximate the line ratios of N V $\lambda\lambda$1239, 1243, C II λ1335, Si IV $\lambda\lambda$1394, 1403, C IV $\lambda\lambda$1548, 1551, and He II λ1640 as long as the X-ray/EUV luminosity of the illuminating source was about 10^{32} erg s^{-1}. The same models appear to be able to account for C III λ977 and O VI $\lambda\lambda$1032, 1038 line fluxes in the HUT wavelength range as well (Ko 1995 private comm.). It is interesting that the predicted equivalent widths are also close to the equivalent widths we observe, indicating that the photoionized plasma may also contribute to the continuum we observe in these systems.

3. Observations of optically-thick disks

In the 'high' state, the FUV spectra of dwarf novae and other similar systems are dominated by the accretion disk as modified by a wind emerging from the inner disk. On Astro-1, HUT was used to observe one dwarf nova in outburst, Z Cam (Long et al. 1991), and two nova-like variables, IX Vel (Long et al. 1994b) and UX UMa. On Astro-2, Z Cam was observed in outburst again (in order to obtain simultaneous coverage with ASCA) and EM Cyg was observed three times (in attempt to follow a dwarf nova through an outburst).

The spectrum of Z Cam on Astro-2 is shown in Fig. 3. It is similar to

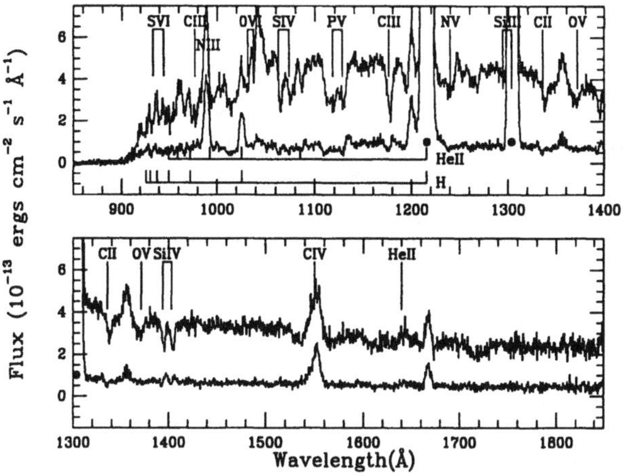

Figure 4. Spectra of EM Cyg obtained during the peak and declining phases of a relatively normal outburst in 1995 March.

the spectrum of Z Cam on Astro-1, and to the spectrum obtained of IX Vel. Prominent absorption features can be identified with the 1s-2p transitions of Li-like O VI, N V and C IV and of Na-like S VI, P V and Si IV. On Astro-2, the only line in the Z Cam spectrum showing evidence of P Cygni emission is C IV $\lambda\lambda 1548, 1551$. Presumably this reflects the fact that the C IV zone in the wind is more extended than that of the other prominent ions in the wind. The spectrum peaks near 1050 Å, although it extends to the Lyman limit.

We have attempted fits to the continua of the high state systems observed with HUT using steady-state accretion disk models constructed by summing stellar-model atmospheres and blackbodies. An example of such a fit for Z Cam on Astro-2 is shown in Fig. 3. In these preliminary fits to the data, we assumed a distance of 200 pc, an inclination angle of 57°, a WD mass of $0.7\,M_{\odot}$, $E(B-V) = 0.02$ and calculated models for a range of $\dot{M}$. The best-fitting model is shown as the dashed-line in Fig. 3 and provides a reasonable approximation to the continuum shape, including the turnover in the continuum below 1050 Å. Similar models constructed with blackbodies would not produce this turnover, which presumably is due to line-blanketing near the Lyman limit.

The spectra from the first and third observations of EM Cyg are shown in Fig. 4. The first spectrum was obtained near the peak of an optical outburst of the system, when the magnitude reported by the AAVSO was about 12.3; the second spectrum was obtained during the decline at magnitude 13.7. Longward of 1000 Å the continuum spectra are very similar aside from an overall decline in flux of about a factor of 5; shortward of 1000 Å the decline is greater indicating that the peak temperature in the disk has dropped. In the first spectrum, O VI λλ1032, 1038, N V λλ1239, 1243 and C IV λλ1548, 1551 appear to have the absorption and red-side emission features expected from a wind. In the second spectrum, O VI λλ1032, 1038 and N V λλ1239, 1243 have almost disappeared and C IV λλ1548, 1551 is purely in emission. Qualitatively, it is as if the effective size of the disk in the FUV has dropped by about a factor of two and the ionizing flux has declined at the energies necessary to produce O VI λλ1032, 1038 and N V λλ1239, 1243. Our initial attempts to model the disk spectra with steady state disks have been reasonably successful. For a distance of 300 pc, a model spectrum with an $\dot{M}$ of $3\,10^{16}$ g s^{-1} matches the overall shape of the first spectrum best, although the predicted flux is too small by a factor of 5. If one demands that the flux be correct, the implied $\dot{M}$ is larger, $2\,10^{17}$ g s^{-1}, but the model spectrum is bluer than is observed.

These problems are similar to those we encountered in attempting to fit the spectrum of the nova-like variable IX Vel from Astro-1, in which the 'best' fitting model (in the sense of approximating the spectral shape) produced had a flux a factor of 5 times lower than required, if the true distance to IX Vel is 95 pc. [Similar problems have have discussed by Wade (1988) based on the IUE color indices $\log(f_{\lambda 1460}/f_{\lambda 1800})$ and $\log(f_{\lambda 1800}/f_{\lambda 2800})$ of nova-like variables.] For IX Vel, as is likely to be true of EM Cyg, improved fits to the data, accounting for both the flux and the shape of the spectrum, could be obtained by modifying the accretion disk structure within 3 WD radii. However, it will be quite difficult to determine whether modifications of this sort represent real physical departures from a steady state disk until more progress is made in constructing models of disks which incorporate not only the vertical structure of the disk, but also radial momentum and energy flow, and perhaps most importantly the disk wind.

4. Conclusions

The spectra obtained with HUT on Astro-1 and Astro-2 are the first set of spectra of dwarf novae with sufficient spectral resolution to resolve individual absorption and emission lines in the wavelength range between Lyα and the Lyman limit. The analysis which has been done thus far demonstrates the importance of spectral coverage in this range for CVs, since models that fit the data in the IUE and HST wavelength ranges vary dramatically in the region shortward of Lyα. Much more work needs to be done in terms of computing realistic models for comparison with the data, especially to understand the non-WD contributions to the spectra of quiescent dwarf novae and to elucidate the physics of high state accretion disks.

Acknowledgments. The Astro observations of CVs would not have been possible without the dedicated support of Janet Mattei and the observers of the AAVSO, my colleagues on the HUT team, especially William Blair and John Raymond who are working with me to analyze the HUT CV data, and the NASA shuttle mission operations and support crew. This work is supported by NASA contract NAS 5-27000 to the Johns Hopkins University.

References

Baptista, R., Horne, K., Hilditch, R.W., et al., 1995, Ap. J., **448**, 395
Cannizzo, J.K., Ghosh, P., Wheeler, J.C., 1982, Ap. J., **260**, L83
Córdova, F.A., 1995, in "X-ray Binaries", eds. W. H. G. Lewin, J. van Paridijs, E.P.J. van den Heuvel, Cambridge: Cambridge Univ. Press, p331
Córdova, F.A., Howarth, I., 1987, in "Exploring the Universe with the IUE Satellite", eds. Y. Kondo, et al. D. Reidel, Dordrecht, p395
Davidsen, A.F., Long, K.S., Durrance, S.T., et al., 1992, Ap. J., **392**, 264
Drew, J E., 1990, in IAU Colloquium 122, "Physics of Classical Novae", eds. A Cassatella, R. Viotti, Springer-Verlag, Berlin, p228
Horne, K., Marsh, T.R., Cheng, F.H., et al., 1994, Ap. J., **426**, 294
Kiplinger, A.L., Sion, E.M., Szkody, P., 1991, Ap. J., **366**, 569
Ko, Y., Lee, Y.P., Schlegel, E.M., Kallman, T.R., 1996, Ap. J., **457**, 363
Kruk, J.W., Durrance, S.T., Kriss, G.A., et al., 1995, Ap. J., **454**, L1
la Dous, C., 1990, Sp. Sci. Rev., **52**, 203
Long, K.S., Blair, W.P., Davidsen, A.F., et al., 1991, Ap. J., **381**, L25
Long, K.S., Blair, W.P., Bowers, C.W., et al., 1993, Ap. J., **405**, 327
Long, K.S., Blair, W.P., Raymond, J.C., 1995, Ap. J., **454**, L39
Long, K.S., Sion, E.M., Huang, M., Szkody, P., 1994a, Ap. J., **424**, L49
Long, K.S., Wade, R.A., Blair, W.P., et al., 1994b, Ap. J., **426**, 704
Mason, K.O., Drew, J.E., Córdova, F.A., et al., 1995, MNRAS, **274**, 271
Mateo, M., Szkody, P., 1984 AJ, **89**, 863
Meyer, F., Meyer-Hofmeister, E., 1982, A&A, **106**, 34
Meyer, F., Meyer-Hofmeister, E., 1990, A&A, **239**, 214
Meyer, F., Meyer-Hofmeister, E., 1994, A&A, **288**, 175
Michaud, G., 1987, in IAU Colloqoquium 95, "Second Conference on Faint Blue Stars", eds A.G.D. Phillip, D.S. Hayes, J.W. Liebert (Schenectady: Davis), p249
Panek, R.J., Holm, A.V., 1984, Ap. J., **277**, 700
Polidan, R.S., Mauche, C.W., Wade R.W., 1990, Ap. J., **356**, 211
Raymond, J.C., Mauche, C.W., Bowyer, S., Hurwitz, M., 1995, Ap. J., **440**, 331
Robinson, E.L., Wood, J.H., Bless, R.C., et al., 1995, Ap. J., **443**, 295
Sion, E.M., 1995, Ap. J., **438**, 876
Sion, E.M., Szkody, P., Cheng, F., Huang, M., 1995, Ap. J., **444**, L97
Wade, R.A., 1988, Ap. J., **335**, 394

EUVE OBSERVATIONS OF VW HYI IN SUPEROUTBURST

CHRISTOPHER W. MAUCHE
Lawrence Livermore National Laboratory,
L-41, P.O. Box 808, Livermore, CA 94550, USA

Abstract. EUVE observed the SU UMa-type dwarf nova VW Hyi in superoutburst for an interval of nearly 2 d in 1994 June and produced EUV light curves and the first EUV spectrum of this important CV.

1. Observations

VW Hyi was observed in superoutburst with EUVE from 1994 June 2.13 to 4.10 UT (RJD 9 505.63 to 9 507.60; RJD = JD − 2 440 000). The observations took place $\approx$ 3.5 d after the rise of the optical flux on $\approx$ May 29.5 UT and $\approx$ 2 d after the peak of the optical flux at $V \approx 8.5$ mag on $\approx$ May 31.0 UT. The Deep Survey photometer (70...170 Å) collected data for less than one day before being shut off because of the high count rate ($\sim$ 20 counts s^{-1}) induced in the detector by source photons. Earth occultations and standard settings on the pipeline reduction software presently limits the useful amount of spectrometer data to 4280, 4177, and 1075 s for the SW (70...180 Å), MW (145...370 Å), and LW (290...760 Å) spectrometers, respectively.

As measured by the Deep Survey photometer and the spectrometers, the EUV flux of VW Hyi was highly variable during the observations. The upper panel of Fig. 1 shows the count rates recorded by the SW, MW, and LW spectrometers in the wavebands 80...170, 170...300, and 300...400 Å, respectively. Whereas the visual flux *fell* steadily from $V \approx 8.7$ mag to $V \approx 8.9$ mag during the observations, the EUV flux is seen to *increase*. On shorter time-scales, the EUV flux variations, in particular the flux diminutions, are accompanied by distinct spectral variations. The lower panel of Fig. 1 shows the ratio of the count rates in the 80...170 Å to 170...300 Å wavebands recorded by the SW and MW spectrometers as a function of time. Each diminution of the EUV flux is accompanied by a softening of

A. Evans and J. H. Wood (eds.), Cataclysmic Variables and Related Objects, 243–246.

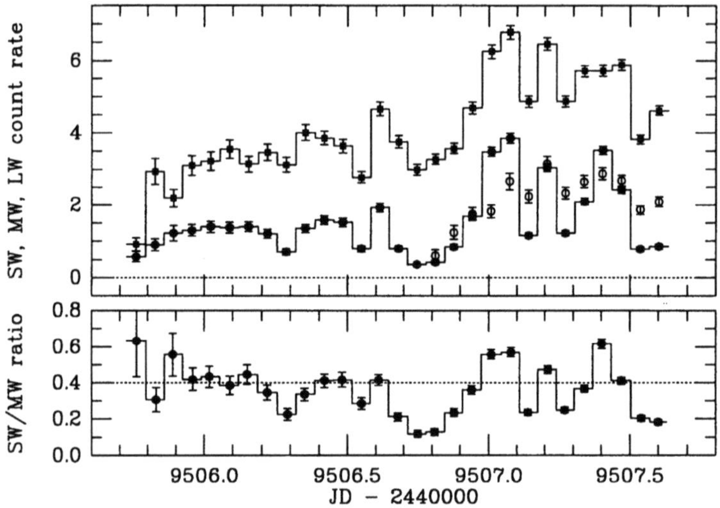

Figure 1. Upper panel: SW (100...170 Å: *filled circles*), MW (170...300 Å : *filled squares*), and LW (300...400 Å: *open circles*) count rate light curves of VW Hyi. Lower panel: SW over MW (hard over soft) count rate ratio.

the spectrum; just the reverse of the effect produced by variations in the column density.

Blithely ignoring these indications that the EUV spectrum of VW Hyi varied both in flux and shape during the observations, we constructed a net spectrum for the observation using data from each of the spectrometers. The result is shown in Fig. 2. The SW, MW and LW portions of the spectrum extend from 80...180, 160...360 and 300...420 Å, respectively, and are binned to 1, 2 and 4 Å, respectively; approximately twice the FWHM of the resolution of each spectrometer. There is residual signal in the LW spectrometer longward even of 420 Å, but this signal is most likely due to second-order flux; this effect is being investigated. The errors in the flux density due solely to counting statistics are everywhere less than $1\,10^{-12}$ erg cm^{-2} s^{-1} Å^{-1}, and are typically less than half this amount.

2. Discussion

After years of speculation about its nature (e.g. van der Woerd, Heise & Bateson 1986; Pringle et al. 1987; Mauche et al. 1991; van Teeseling, Verbunt & Heise 1993), it is gratifying finally to see the EUV spectrum of VW Hyi. Many aspects of the spectrum are consistent with previous measurements. The fact that the spectrum peaks longward of 170 Å is con-

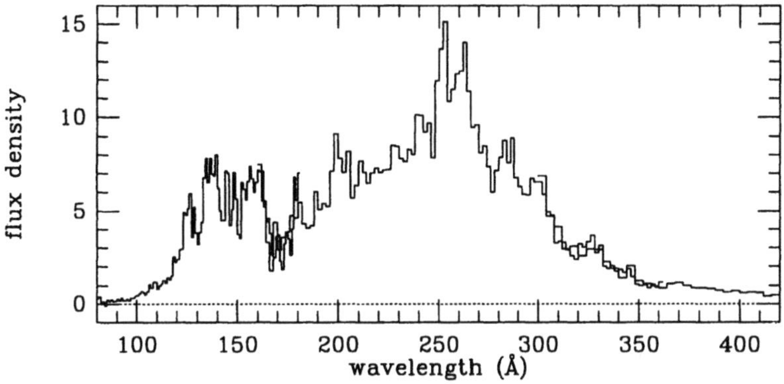

Figure 2. Net SW (80...180 Å), MW (160...360 Å) and LW (300...420 Å) spectra of VW Hyi. Units of flux density are 10^{-12} erg cm^{-2} s^{-1} Å^{-1}.

sistent with the fact that, in outburst, the EXOSAT Al-Par filter count rate is higher than the 3000 Lex or 4000 Lex count rates. That the spectrum extends to yet longer wavelengths is consistent with the low neutral hydrogen column density ($\approx 6\,10^{17}$ cm^{-2}) measured by Polidan, Mauche & Wade (1990). That the spectrum does not extend shortward of ≈ 100 Å (≈ 0.12 keV) is consistent with the non-detection in the 0.18...0.43 keV bandpass by HEAO-1 (Córdova et al. 1980) and with the *decrease* during outburst of the flux in the 0.1...2.4 keV bandpass measured by ROSAT (Wheatley et al. 1996). The brightness of the spectrum is inconsistent. The observed 80...420 Å flux is $1.5\,10^{-9}$ erg cm^{-2} s^{-1}, implying a luminosity of $7.7\,10^{32}\,(d/65\ \mathrm{pc})^2$ erg s^{-1}. For comparison, the luminosity of the accretion disk is ≈ 20 times higher (Mauche et al. 1991). The predicted EXOSAT Al-Par and 3000 Lex filter count rates are 28.3 and 15.5 counts s^{-1}, respectively. The highest measured EXOSAT Al-Par and 3000 Lex filter count rates appear to be 4.3 and 1.2 counts s^{-1}, respectively (van der Woerd & Heise 1987). Given the low net exposure on source, it is fortunate that VW Hyi was so bright during our observations.

It is fortunate also that the neutral hydrogen column density to VW Hyi is so low. Columns typical of other nearby CV systems ($N_{\rm H} \gtrsim$ few 10^{19} cm^{-2}; Mauche, Raymond & Córdova 1988), would render VW Hyi nearly unobservable: at $N_{\rm H} = 3\,10^{19}$ cm^{-2}, the optical depth of the ISM is 1.1 at 100 Å, 3.2 at 150 Å, 6.5 at 200 Å. SS Cyg (Mauche, Raymond & Mattei 1995) and U Gem (Long et al. 1995) both have higher columns, but are also intrinsically harder. If other, even nearby, CVs are as soft as VW Hyi, it will be hard to impossible to detect them in the EUV ($\lambda = 100\ldots912$ Å).

As is the case with SS Cyg and U Gem, the EUV spectrum of VW Hyi belies simple interpretation. Consideration of the following options is instructive of the problems facing us.

(i) A blackbody spectrum. To produce so little flux shortward of $\approx$ 100 Å, the temperature of a blackbody must be $T \lesssim 1\,10^5$ K ($kT \lesssim 10$ eV). To extinguish the flux longward of $\approx$ 350 Å, the neutral hydrogen column density then must be $N_{\rm H} \gtrsim 3\,10^{18}$ cm^{-2}, but such a model fails to reproduce the overall spectral distribution.

(ii) A modified blackbody spectrum. The ionization edges of C III–IV, N III–V, O III–IV, Ne II–IV, Mg III–IV, Si IV, S IV–VI and Fe IV–VI all lie in this bandpass, but are not apparent in the spectrum. This result probably does not constrain irradiation of the white dwarf by a hard continuum spectrum, since such models produce edges at wavelengths below 100 Å (van Teeseling, Heise & Paerels 1994).

(iii) The spectrum of an optically thin plasma. At $T \approx 1\ldots$few 10^5 K, the He II Lyman lines ($\lambda = 304, 256, 243, \ldots$ Å) and bound-free continuum ($\lambda < 228$ Å) are in emission, contrary to observations. At higher temperatures, particularly $T \gtrsim 5\,10^5$ K, the free-free continuum is weak, but strong emission lines begin to appear shortward of 100 Å.

Other processes likely affect the spectrum. The ISM and possibly the wind of VW Hyi will imprint a He II absorption edge at 228 Å onto the intrinsic spectrum. Measurement of, or upper limits on, such an edge will provide an estimate of the integrated column density of this ion. He I has autoionization resonances at $\lambda = 206, 195, 192, \ldots$ Å (Rumph, Bowyer & Vennes 1994). The first such resonance may be responsible for the emission/absorption feature at 206 Å.

Acknowledgements. The author is pleased to acknowledge the contributions to this research by the members, staff, and director, J. Mattei, of the AAVSO. This work was performed under the auspices of the U.S. Department of Energy by Lawrence Livermore Nat'l Laboratory under contract No. W-7405-Eng-48.

References

Córdova, F.A., Nugent, J.J., Klein, S.R., Garmire, G.P., 1980, MNRAS, **190**, 87
Long, K.S., Mauche, C.W., Szkody, P., Mattei, J.A., 1995, in "Cataclysmic Variables", eds A. Bianchini et al., Kluwer, p133
Mauche, C.W., Raymond, J.C., Córdova, F.A., 1988, Ap. J., **335**, 829
Mauche, C.W., Raymond, J.C., Mattei, J.A., 1995, Ap. J., **446**, 842
Mauche, C. W., 1991, Ap. J., **372**, 659
Polidan, R.S., Mauche, C.W., Wade, R.A., 1990, Ap. J., **356**, 211
Pringle, J.E., Bateson, F.M., Hassall, B.J.M., et al., 1987, MNRAS, **225**, 73
Rumph, T., Bowyer, S., Vennes, S., 1994, AJ, **107**, 2108
van der Woerd, H., Heise, J., 1987, MNRAS, **225**, 141
van der Woerd, H., Heise, J., Bateson, F., 1986, A&A, **156**, 252
van Teeseling, A., Heise, J., Paerels, F., 1994, A&A, **281**, 119
van Teeseling, A., Verbunt, F., Heise, J., 1993, A&A, **270**, 159
Wheatley, P.J., Verunt, F., Belloni, T., et al., 1996, A&A, in press.

THE POSSIBLE CONTRIBUTION FROM THE ACCRETION DISK DURING THE QUIESCENCE OF VW HYI

M. HUANG[1], E. M. SION[1], I. HUBENY[2], F. H. CHENG[3], P. SZKODY[4]
1. Villanova University, Villanova, PA 19085, USA
2. Laboratory of Astronomy & Solar Physics, Code 681, NASA/GSFC, Greenbelt, MD 20771, USA
3. Dept. of Astrophysics, Univ. of Maryland, Collge Park, MD 20742, USA
4. Dept. of Astronomy, FM-20, Univ. of Washington, Seattle, WA 98195, USA

Abstract. We have re-analyzed UV spectra of the dwarf nova VW Hyi during its quiescence, obtained with the *Faint Object Spectrograph* (FOS) on *Hubble Space Telescope* (HST), by considering the possible contribution from the accretion disk, which was not included in the previous analysis.

1. Introduction

Mateo & Szkody (1984) first pointed out the white dwarf origin of the IUE spectra of VW Hyi during its quiescence by identifying the sharp turnover near 1360 Å as a result of the Lyα absorption from a 18 000±2 000 K, $\log g = 8.0$ white dwarf atmosphere. Wade, Hubeny & Polidan (1994) confirmed the results of Mateo & Szkody (1984) by comparing the IUE data with synthetic spectra, but also pointed out that a composite model consisting of a white dwarf and an accretion disk in a steady state, with $\dot{M} = 10^{-10}\,M_\odot\,yr^{-1}$, describes the Ly$\alpha$ absorption equally well. Based on the assumption that the white dwarf is the dominant source of the UV flux during the quiescence of VW Hyi, Sion et al. (1995) analyzed two HST/FOS UV spectra obtained on 1993 November 13, approximately 10 days after its return to optical quiescence following a superoutburst. They used a grid of high surface gravity synthetic models constructed with TLUSTY and SYNSPEC, and concluded that the white dwarf has

A. Evans and J. H. Wood (eds.), Cataclysmic Variables and Related Objects, 247–248.

$T_{\rm eff}$ =22 000±1 000 K, log g = 8.0 ± 0.3, and chemical abundances as follows: $O/O_\odot$ = 0.3, $N/N_\odot$ = 5, $S/S_\odot$ = 0.2, and all other heavy elements 0.15 solar.

In the present work, we consider the possible contribution from the disk in addition to the white dwarf by re-examining the HST/FOS data analyzed by Sion et al. (1995).

2. Re-examination and analysis of the data

Comparing the best fit white dwarf model spectrum from Sion et al. (1995) with the observed spectra in the wavelength range $\lambda > 1400$ Å, we find that the continuum of the former is basically a straight line, while the latter shows two main features that the pure white dwarf model does not reproduce well, namely (i) the curved continuum between absorption features Si IVλ1400 and Si IIλ1530 and (ii) the possible Keplerian broadened emission feature centered at C IVλ1550. We compare the data with LTE, optically thick disk and white dwarf models. The *UV emitting area* of the disk in our best fit model has log g= 6.0, $T_{\rm eff}$=28 000 K, $r_{\rm in}$ = 1.0 $R_{\rm WD}$, $r_{\rm out}$ = 1.1 $R_{\rm WD}$, with $C/C_\odot$ = 20, $Si/Si_\odot$ = 15, and the abundance of the other elements being solar; the white dwarf has log g=8.0, $T_{\rm eff}$ = 19 000 K, with $Si/Si_\odot$ = 0.5 and other metals with their solar abundances. The high abundance of C and Si in the disk/belt may be a result of the enrichment on the surface of the secondary star due to possible past nova explosion event(s) (Stahl & Kolb 1994).

We are currently investigating other possible mechanisms, such as the quasi-molecular absorption, which may also contribute to the absorption feature near 1400 Å. The results of these investigations will be reported in due time.

References

Mateo, M., Szkody, P., 1984, AJ, **89**, 863

Sion, E.M., Szkody, P., Cheng, F.H., Huang, M., 1995, Ap. J., **444**, L97.

Stahl, R., Kolb, U., 1995, in Astronomy Poster Abstract XXII General Assembly of the IAU, ed. H. van Woerder, (Twin Press, Sliedrecht), p97

Wade, R.A., Hubeny. I., Polidan, R.S., 1994, in "Interacting Binary Stars", ed. A. Shafter, ASP Conf. Ser. **56**, 319

PRELIMINARY ANALYSIS OF A HUBBLE FOS SPECTRUM OF VW HYI IN QUIESCENCE: A DAZQ WHITE DWARF AND ACCRETION BELT/RING

E. M. SION[1], M. HUANG[1], F. H. CHENG[1], I. HUBENY[2]
P. SZKODY[3]
1. Villanova University, Villanova, PA 19085, USA
2. Laboratory of Astronomy & Solar Physics,
Code 681, NASA/GSFC, Greenbelt, MD 20771, USA
3. Dept. of Astronomy, FM-20,
Univ. of Washington, Seattle, WA 98195, USA

Abstract. We present a preliminary analysis of the UV spectrum of the dwarf nova VW Hyi during early quiescence, obtained with the Faint Object Spectrograph (FOS) on the Hubble Space Telescope (HST). The data were obtained approximately one day after the end of a normal outburst.

1. HST FOS spectrum and line measurements

We obtained a Hubble Space Telescope FOS spectrum of VW Hyi on 1995 January 28, 19:33:01 (UT time), approximately one day after the end of a normal outburst. The spectrum is dominated by a very broad Ly α profile and a rich metallic absorption line spectrum similar to our earlier FOS spectrum obtained ten days after superoutburst (see Sion et al. 1995; Huang et al. 1996). There is a moderately strong, broad C IV emission line associated with the Keplerian broadening of an accretion disk. A list of identified lines with equivalent widths (EW) and full widths at half-maximum (FWHM) are presented in Table 1.

2. Preliminary synthetic spectral analysis results

There are important differences between this new FOS spectrum and the one analyzed in Sion et al. (1995). Our new spectrum reveals that the Ly α absorption line is measurably narrower, the continuum hump be-

A. Evans and J. H. Wood (eds.), Cataclysmic Variables and Related Objects, 249–250.

TABLE 1. Absorption lines in the FOS spectrum of VW Hyi

Line	Wavelength (Å)	EW (Å)	FWHM(Å)
C III	1176.7	3.153	5.684
Si II	1261.2	1.685	3.161
	1265.2	2.083	3.929
C I	1278.55	3.448	9.406
O I + Si III + Si II	1303.42	9.079	18.78
Si III	1310.59	2.381	5.872
C II	1335.78	2.262	4.094
Si IV	1395.28	1.721	7.923
Si II	1527.52	1.712	5.463
	1533.77	2.929	9.868
C I	1562.14	1.151	4.267
C IV	1550.93	1.461	8.938
	1550.03*	5.541*	26.93*

* Emission line

tween 1400 Å and 1550 Å is not as pronounced, and the overall flux level is slightly higher. We interpret these differences as indicative of a higher overall surface temperature for the white dwarf one day after a normal outburst than its overall surface temperature ten days after superoutburst. Our best synthetic spectral fit to the data is a composite two-temperature model consisting of (i) a DAZQ white dwarf with $T_{\rm eff} = 22\,000$ K, $\log g = 8$, and abundances $C/C_{\odot} = 0.5$, $N/N_{\odot} = 5.0$, $O/O_{\odot} = 0.02$, $Si/Si_{\odot} = 0.2$, $Fe/Fe_{\odot} = 0.5$ with all other metals being 0.3 solar and (ii) a rapidly spinning accretion belt (or possibly inner disk ring) with $V_{\rm rot} = 3350\,{\rm km\,s^{-1}}$, $T_{\rm belt} = 28\,000$ K, $\log g = 6.0$, overabundance of C and Si by 20 times and 15 times solar, respectively, and a fractional belt area of approximately 6%. A detailed comparison of this second FOS spectrum, taken one day post-outburst, with the first FOS spectrum obtained ten days post superoutburst, is now in progress.

Acknowledgement. This work is supported by NASA grants GO-3836.01-91A, GO-6084.01-95A and NAGW-5726 to Villanova University.

References

Huang, M., Sion, E. M., Hubeny, I., et al., 1996, these proceedings, p247
Sion, E.M., Szkody, P., Cheng, F.H., Huang, M., 1995, Ap. J., **444**, L97

VW HYI: A RAPIDLY COOLING WHITE DWARF?

BORIS T. GÄNSICKE, KLAUS BEUERMANN
Universitäts-Sternwarte, Geismarlandstr. 11,
37083 Göttingen, FRG

Abstract. We estimate the post-outburst cooling time-scale of the white dwarf in VW Hyi, using the available quiescent IUE spectra. The determination of the white-dwarf temperature close to the outburst is hampered because disc emission still contributes to the SWP flux.

1. Introduction

Observations of the inter-outburst behavior of dwarf novae in the UV range are important in order to study the evolution of the disc/boundary layer as well as the possible cooling of the (accretion–heated) white dwarf. Alas, the white dwarf could be detected directly only in a small number of dwarf novae. The cooling of the white dwarf has been observed in U Gem (Long et al. 1994), WZ Sge (Sparks et al. 1993) and OY Car (Cheng et al. 1994). In the case of VW Hyi, Verbunt et al. 1987 detected a decline of the UV flux after an outburst, which has been interpreted by Meyer & Meyer-Hofmeister (1994) by the evaporation of the inner (hot, UV–emitting) accretion disk. On the other hand, Sion et al. (1995) clearly identify the white dwarf in HST spectra.

2. Analysis & results

We have retrieved all available IUE spectra of VW Hyi from the ULDA archive and selected 58 SWP/LWR/LWP spectra obtained in quiescence. The UV flux shows an exponential decline, decreasing markedly slower after a superoutburst (Fig. 1). The individual spectra can be fitted reasonably well with $\log g = 8$ white-dwarf model spectra by Ivan Hubeny, with the largest discrepancy in the LWP range. The radii derived for a distance of 65 pc (Fig. 1) are compatible with the assumed 0.6 $M_\odot$ white dwarf. Optical spectroscopy obtained in 1989 January shows the known orbital

A. Evans and J. H. Wood (eds.), Cataclysmic Variables and Related Objects, 251–252.

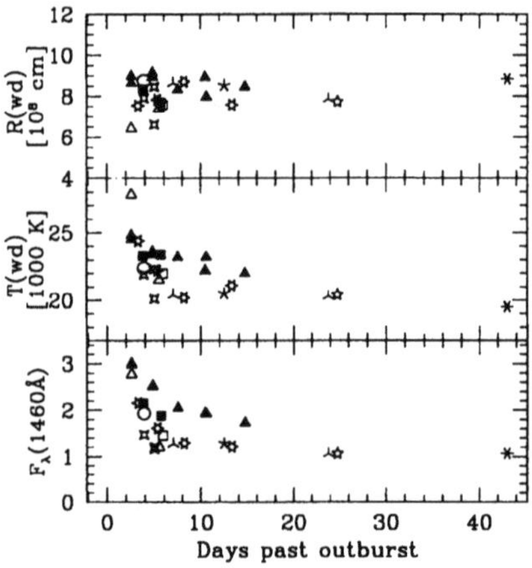

Figure 1: VW Hyi on the decline from the outburst. (a) UV flux in 10^{-13}ergs $cm^{-2}s^{-1}Å^{-1}$. Spectra taken after a normal outburst and after a superoutburst are indicated by open and filled symbols, respectively. (b) Effective temperatures derived from fitting white-dwarf models to the observed Lyα profile. (c) Radius of the emitting area at $d = 65$ pc.

modulation, presumably due to the varying aspect of the hot spot. Fitting the spot spectrum with Kurucz low-gravity models yields a temperature of $\sim 10\,000$ K, in agreement with the estimate of Mateo & Szkody (1984). The contribution of the hot-spot in the LWP range is $\sim$ 10%, consistent with the departure of the observed spectra from the white-dwarf models.

Taken at face value, the temperatures derived from fitting the observed Lyα profiles with white-dwarf models indicate that the white dwarf cools between two outbursts from $\sim 25\,000$ K to $\sim 20\,000$ K. The cooling timescale seems to be of the same order as the duration of the preceding outburst, i.e. $\sim$ 5 d after a normal outburst and $\sim$ 15 d after a superoutburst. However, our ongoing analysis of the IUE spectra confirms the results by Huang et al. (1996) who show that an HST spectrum taken 10 d after a normal outburst contains a non-negligible contribution of a disc remnant. For the two spectra closest to the previous outburst, we find that the disc still contributes $\sim$ 60% (normal , $\triangle$) and $\sim$ 30% (super, $\blacktriangle$) of the SWP flux, these spectra require a more detailed discussion. The spectra taken 24, 25 and 43 d after the outburst show no further evolution and are compatible with a white dwarf of $T_{\rm wd} \simeq 20\,000$ K without noticeable disc contribution.

References

Cheng, F.H., Marsh, T.R., Horne, K., Hubeny, I., 1994, AIP conf. proc. **308**, 197
Huang, M., Sion, E.M., Hubeny, I., et al., these proceedings, p247
Long K.S., Sion, E.M., Huang, M., Szkody, P.,1994, Ap. J., **424**, L49
Mateo, M., Szkody, P., 1984, AJ, **89**, 863
Meyer, F., Meyer-Hofmeister, E., 1994, A&A, **288**, 175
Sion, E.M., Szkody, P., Cheng, F.H., Huang, M., 1995, Ap. J., **444**, L97
Sparks, W.M., Sion, E.M., Starrfield, S.G., Austin, S., 1993, in "Cataclysmic Variables and Related Physics", eds O. Regev, G. Shaviv, Ann. Israel Phys. Soc., **10**, 96
Verbunt, F., Hassall, B.J.M., Pringle, J.E. et al., 1987, MNRAS, **225**, 113

ACCRETION DISK WINDS IN CATACLYSMIC VARIABLES

CHRISTIAN KNIGGE[1], JANET E. DREW[1],
KEITH O. MASON[2]
1. Nucl. & Astroph. Lab., Keble Road, Oxford, OX1 3RH, UK
2. Mullard Spa. Sci. Lab., Surrey, RH5 6NT, UK

Abstract. We present preliminary results of a modelling campaign based on HST eclipse observations of the wind-formed C IV resonance line in the eclipsing nova-like variable UX UMa. Within the framework of a simple kinematic accretion disk wind model, we are currently able to reproduce the shapes (but not the strengths) of the observed line profiles away from and during eclipse, as well as the behaviour of the integrated line flux light curve. The most important aspect of our modelling is that it suggests the presence of a vertically extended, relatively dense and slow moving transition region between the disk photosphere and the fast wind.

1. Introduction

Eclipse observations of the wind-formed UV resonance lines in high inclination CVs contain a wealth of information about the kinematics and geometry of the line forming region. Detailed modelling of such data – which can only be gathered with the *Hubble Space Telescope* (HST) – should produce a much improved physical picture of CV winds. Here we present preliminary results of the first modelling campaign of this kind, in which we are able to draw on HST eclipse observations of the C IV wind line in the nova-like variable UX UMa (Mason et al. 1995).

2. Model framework

The kinematic accretion disk wind model we use is that of Knigge, Woods & Drew (1995) and is illustrated in Fig. 1. In it, all streamlines converge at a fiducial point a distance, d, above/below the disk, which allows the geometric outflow collimation to be controlled by this single free parameter. A simple power law scaling with the local disk temperature is used to fix the local mass-loss rate per unit area from the disk: $\dot{m} \propto T_{\rm disk}(R)^{4\alpha}$.

A. Evans and J. H. Wood (eds.), Cataclysmic Variables and Related Objects, 253–256.

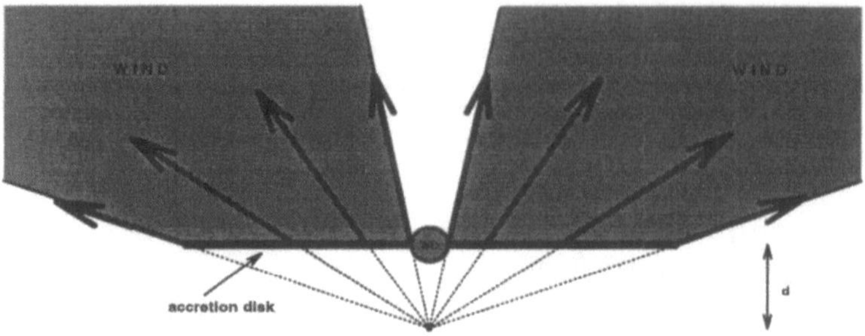

Figure 1. The geometry of our kinematic disk wind model.

The wind is assumed to rotate initially with the disk's Keplerian velocity and to conserve its specific angular momentum as it rises above the binary plane. For the non-rotational part of the wind velocity field we use a description which is similar in form to the β-type velocity law often used in models of hot star winds (Castor & Lamers 1979).

A three-dimensional Monte Carlo line transfer code is used to synthesize the line profile arising from our accretion disk wind models (Knigge et al. 1995). This ensures that multiple scattering in the outflow is treated self-consistently in these calculations.

The continuum light is assumed to be produced by a steady-state accretion disk with $\dot{M}_{\rm acc} \simeq 3\,10^{-8}\,{\rm M_\odot yr^{-1}}$ and a hot white dwarf (WD). The latter contributes about 20% to UX UMa's out-of-eclipse continuum flux, which corresponds to $T_{\rm WD} \simeq 8.5\,10^4\,{\rm K}$. The fit to the observed continuum light curve achieved by this simple model is shown in the left panel of Fig. 2. Here and throughout, Baptista et al.'s (1995) system parameters for UX UMa (most importantly $i = 71°$ and $q = 1$) have been used.

3. Preliminary results

Fig. 2 shows the light curves of UX UMa's UV continuum and integrated C IV line flux. Note that the shapes of the two light curves are radically different, with the net line flux exhibiting an increase above the out-of-eclipse level at phases close to mid-eclipse.

Some more insight can be gained by comparing the appearance of the C IV line around $\phi_{\rm orb} \simeq 0.9$ with that of the profile near $\phi_{\rm orb} \simeq 0$ (Fig. 3). The most striking aspect of this comparison is the presence of two absorption reversals close to the rest wavelengths of the resonance doublet in the pre-eclipse profile, which disappear during the eclipse to reveal a simpler, if asymmetric emission line.

The results of line profile syntheses for our current 'best-bet' model are also illustrated in Figs. 2 and 3. These show that the shapes of the

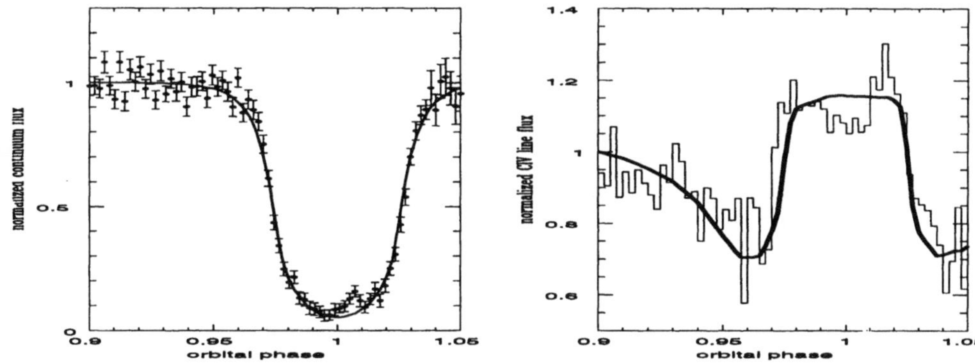

Figure 2. **Left panel:** UX UMa's normalised UV continuum light curve (points) and the fit achieved by our steady-state disk + hot WD model (solid line). **Right Panel:** The normalised C IV continuum-subtracted, integrated line flux light curve (histogram) and the fit achieved by our current best-bet accretion disk wind model (continuous line).

profiles and the line flux light curve are reproduced reasonably well, but that the model still underestimates the overall emission strength. A possible explanation for this failure is offered below.

Assuming that our modelling does reflect reality to some extent, what is the picture that emerges of the outflow in UX UMa?

First of all, its geometric collimation seems to be relatively well constrained by the data, with indicated inner and outer wind cone opening angles of about 5° and 65° ($d \simeq 15R_{\rm WD}$). The mass-loss power law index, α, on the other hand, has less influence on the emergent line profiles; in the model shown here it has been set equal to unity. Moreover, the wind acceleration beyond the sonic point (whose position in the outflow we will return to below) seems to be fairly gradual, with a velocity law index of $\beta = 4.5$ describing our current best model. The effective mass-loss rate inferred from our modelling (which is approximately equal to the true mass-loss rate if C IV is the dominant ionization stage in the outflow) is $\dot{M}_{\rm eff} \sim 0.01\dot{M}_{\rm acc}$

However, our most important finding thus far is that in order to reproduce the shape and eclipse behaviour of the absorption reversals in the observational line profiles, it is necessary to invoke the presence of a vertically extended, relatively dense and slow moving outflow region just above the accretion disk (which we model by moving the position of the sonic point of the wind away from the disk surface). This is not unreasonable, since such a region (i) can produce strong absorption effects by virtue of its high density, (ii) will absorb photons primarily at low velocities and thus close to line centre, (iii) lies close enough to the disk plane to be eclipsed by the secondary.

While alternative scenarios for the origin of the reversals – such as absorption in some sort of disk edge structure – cannot yet be ruled out,

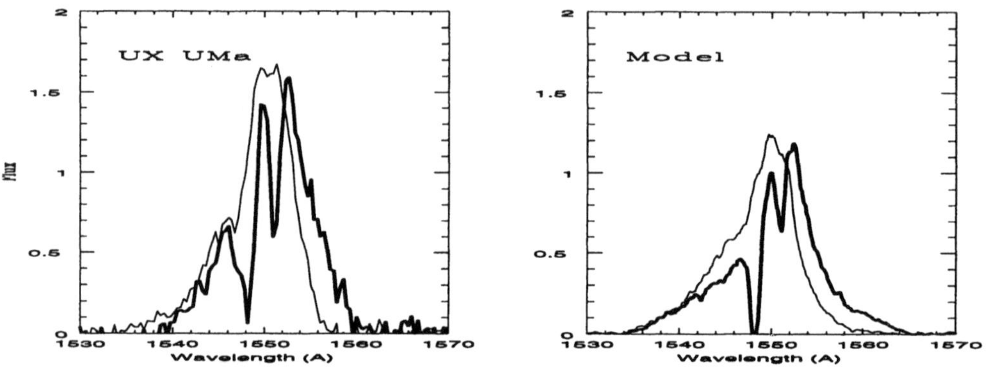

Figure 3. **Left panel:** UX UMa's average pre- (heavy line) and mid-eclipse (thin line) C IV line profiles. **Right Panel:** Pre- (heavy line) and mid-eclipse (thin line) C IV line profiles arising from our accretion disk wind model. All profiles have been continuum subtracted and normalised to a common scale.

our preferred interpretation is as evidence for the existence of a transition region between the disk photosphere and the fast moving outflow. This is in line with a recent analytical study of line-driven winds from accretion disks, in which the height of the sonic point above the disk plane was predicted to scale roughly as $z_s \sim R$ (Suleimanov 1995). [Our modelling of the position of the sonic point in the outflow is in fact loosely based on this scaling.] Our findings may also be related to the absorbing Fe-curtain which veils the WD in the dwarf nova OY Car during quiescence (Horne et al. 1994).

We finally note that, due to the high densities occuring in this transition region, line photons tend to scatter many times in our model before escaping. This means that the use of a Monte Carlo radiative transfer code is certainly warranted here. It also means, however, that the conservative scattering approximation we currently make in our calculations may not be sufficiently accurate. If thermal wind emission contributes significantly to the observed C IV line flux, this may help to explain the relative weakness of the model line profiles. This possibility is currently under examination.

References

Baptista, R., Horne, K., Hilditch, R.W., et al., 1995, Ap. J., **448**, 395
Castor, J.I., Lamers, H.J.G.L.M, 1979, Ap. J. Supp., **39**, 481
Horne, K., Marsh T.R., Cheng, F.H., et al., 1994, Ap. J., **426**, 294
Knigge, C., Woods, J.A., Drew, J.E., 1995 MNRAS, **273**, 225
Mason, K.O., Drew, J.E., Córdova, F.A., et al., 1995, MNRAS, **274**, 271
Suleimanov, V.F., 1995, Astr. Lett., **21**, 126

THE TRANSIENT EUV SKY AS OBSERVED BY ALEXIS

DIANE ROUSSEL-DUPRÉ, JEFFERY J. BLOCH

Astrophysics and Radiation Measurements Group, Los Alamos National Laboratory, P.O. Box 1663, MSD436, Los Alamos, NM 87545, USA

1. Introduction

Los Alamos National Laboratory's ALEXIS satellite, containing six wide field of view EUV telescopes was launched 1993 April 25. ALEXIS is a sky monitor/survey experiment, and with each 50 s rotation, the satellite's six telescopes (three energy bands 130, 178 and 186 Å) scan nearly half of the sky. Due to the damage sustained at launch by the satellite, the ALEXIS project team has had to spend over a year devising new methods to determine spacecraft attitude knowledge, essential for putting photons back on the sky correctly. These efforts have been successful. A recent review of the ALEXIS point source effort can be found in the Proceedings of IAU Colloquium 152, 1995. This paper reviews the transient systems that have been observed by ALEXIS which fall into two types: cataclysmic variables and unknown, short duration transients.

2. Transient search

Three methods are currently being used to detect EUV transients in the ALEXIS data: (i) archival search, (ii) daily automated sky map searches and (iii) manual sky map searches. Real time transient searches were initially done daily by hand. Starting 1995 January 1, these searches were automated, so now the 12, 24 and 48 h sky maps are searched for point sources. Point source locations found in the data by search algorithms are sent via e-mail to the science team for evaluation. Two superoutbursts from VW Hyi, one long outburst from U Gem, an EUV outburst from the AM Her system AR UMa, and fast transients ALEXIS J1139−685 and ALEXIS J1644−032 have been identified as significant transients in the ALEXIS data.

A. Evans and J. H. Wood (eds.), Cataclysmic Variables and Related Objects, 257–258.

3. Fast transients and cataclysmic variable outbursts

ALEXIS J1139−685 and ALEXIS J1644−032 both have similar time signatures and look to be potentially a new type of transient. Both systems were observed to be in outburst for of order 24...36 h. Target of opportunity observations were requested on EUVE for these systems and, in both cases, EUVE was able to slew to target and observe these systems starting less than 24...36 h after maximum light. However in both cases, 24 h of observations failed to detect the transients.

The ALEXIS VW Hyi observations are the first observations in the EUV with better than one day time resolution. The superoutbursts were observed primarily in the 176 Å telescope but also, weakly in the 130 Å telescope. A pre-outburst short duration enhancement was observed 2.75 d prior to the 1994 May/June superoutburst optical rise. The EUV superoutburst emission was delayed 0.75 d from the observed optical rise. Highly variable but persistent EUV emission was only observed for 7.5 d at the beginning of the outburst. The 1995 May superoutburst was quite different from the previous year. The primary differences are (i) the brightness of this outburst was a factor of two weaker and (ii) although the EUV rise started 1 d after the optical rise, the maximum flux was observed 4.8 d after the optical rise.

ALEXIS observed a EUV transient (ALEXIS J1114+43, 1ES 1113+432 or AR UMa) in 1994 late November. This outburst was probably already in progress when the telescope first scanned across it and lasted at least 16 d. The outburst was detected in all three energy channels (130, 178 and 186 Å). Ron Remillard obtained optical spectra of AR UMa at the end of the observed outburst confirming the identity as the EUV counterpart.

4. Summary

The ALEXIS satellite has been successful in identfying transients in real time. Two major types of transients have been observed: fast transients of duration 24...36 h and cataclysmic variables. Complete monitoring of transient outbursts, which has not been possible with previous pointed, staring satellites, is possible with ALEXIS. This unique capability has resulted in observations of a pre-superoutburst EUV flash in the CV system VW Hyi, as well as significant differences in EUV light curves from superoutburst to superoutburst. In addition, because ALEXIS is a monitoring satellite, it is able to detect CVs that have not been previously identified in the EUV (e.g. AR UMa) thus increasing our knowledge of these systems.

Acknowledgements. These results represent the dedicated efforts of C. Little, M. Kennison, K. Ramsey, S. Ryan, S. Stem, A. Hodapp, B. Edwards and T. Pfafman. This work was supported by the Department of Energy.

ASCA OBSERVATIONS OF CATACLYSMIC VARIABLES

MANABU ISHIDA[1], RYUICHI FUJIMOTO[1],
KEIICHI MATSUZAKI[2]
1. Institute of Space and Astronautical Science
3-1-1 Yoshinodai, Sagamihara, Kanagawa 229, Japan
2. Department of Physics, University of Tokyo
7-3-1 Hongo, Bunkyo-ku, Tokyo 113, Japan

1. Introduction

ASCA has observed ~20 cataclysmic variables since launch. In this review, we present the current status of ASCA observations of CVs and scientific results. In Sect. 2, we introduce the scientific instruments of ASCA, and show the list of CVs observed so far. Data analysis and results on magnetic and non-magnetic CVs are described in Sect. 3 and 4, respectively. In Sect. 5, we make a brief summary.

2. ASCA observations of CVs

ASCA was launched on 1993 February 20 on a M3S-II rocket from Kagoshima Space Center (Tanaka, Inoue & Holt 1994). ASCA is the fourth Japanese X-ray astronomy satellite. It is equipped with four equivalent X-ray telescopes, which have large effective area up to 10 keV. On the focal plane, two kinds of detector systems, each of which consists of two equivalent detectors, are placed. One is the SIS (Solid-state Imaging Spectrometer) which utilizes a X-ray CCD camera. The other is the GIS (Gas Imaging Spectrometer) which adopts the Gas Scintillation Proportional Counter. Although the effective area of ASCA is significantly smaller than that of Ginga, sensitivity for faint sources is improved by about 100 times because of its imaging capability. The SIS has higher energy resolution and larger effective area in the low energy band. The GIS, on the other hand, has a wider field of view, larger effective area in the high energy band, and higher time resolution. The two detector systems are thus complementary to each other. In Table 1, we show the list of CVs so far observed by ASCA. Up to 1995 June 30, a total of 28 CVs had been observed.

A. Evans and J. H. Wood (eds.), Cataclysmic Variables and Related Objects, 259–268.

TABLE 1. List of targets observed by ASCA to 1995 June

Type	Target	P.I.	Phase	Type	Target	P.I.	Phase
DQ	FO Aqr	Mukai	PV	NM	HT Cas	Mukai	AO-2
NM	SS Cyg	Nousek	PV	NM	V426 Oph	Watson	AO-2
DQ	EX Hya	Ishida	PV	NM	EC1931-5915	Schlegel	AO-2
AM	EF Eri	Osborne	PV	DQ	TX Col	Wheatley	AO-2
AM	AM Her	Gotthelf	PV	AM	BL Hyi	Fujimoto	AO-2
AM	VV Pup	Ishida	AO-1	AM	QQ Vul	Mukai	AO-2
NM	VW Hyi	Mauche	AO-1	AM	RX2107-05	Schlegel	AO-2
NM	TT Ari	Mauche	AO-1	NM	U Gem	Szkody	AO-2
AM	V834 Cen	Mukai	AO-1	DQ	RE0751+14	Fujimoto	AO-2
AM	BY Cam	Mauche	AO-1	DQ	GK Per	Mauche	AO-3
DQ	V1223 Sgr	Osborne	AO-1	AM	AM Her	Ishida	AO-3
DQ	AO Psc	Hellier	AO-2	NM	Z Cam	Mauche	AO-3
NM	RU Peg	Watson	AO-2	DQ	TV Col	Mauche	AO-3
NM	GP Com	Marsh	AO-2	NM	EI UMa	Watson	AO-3

3. Magnetic cataclysmic variables

ASCA has observed 17 MCVs (9 AM Her types and 8 DQ Her types) so far. In Fig. 1, we show the phase-average photon-counting SIS spectra of six sources. Owing to the high energy resolution of the SIS, $K\alpha$ emission lines from Mg, Si, S, Ar and Fe in hydrogenic and He-like ionization states are resolved. In the AM Her spectrum, the soft blackbody component is detected below 0.6 keV.

3.1. THE CONTINUUM EMISSION

The hard X-ray continuum emission of MCVs has been known to be very complex. Based on EXOSAT observations, Norton & Watson (1989) show that the low energy cutoff in the spectra of MCVs cannot be represented by an absorber with a single column density, and they have introduced the partial covering absorber model, which is composed of an absorber with two different hydrogen column densities covering the emission region. This model works well also for the MCV spectra obtained by Ginga in general (Ishida 1991).

However, it became gradually clear that the partial-covering model is sometimes too simplified a parameterization for the low energy cutoff of the MCV spectra. In Fig. 2a is shown the SIS spectrum of PQ Gem (=RE0751+14) together with a spectral model. The model consists of a single temperature (20 keV) thermal bremsstrahlung (Ishida 1991) covered

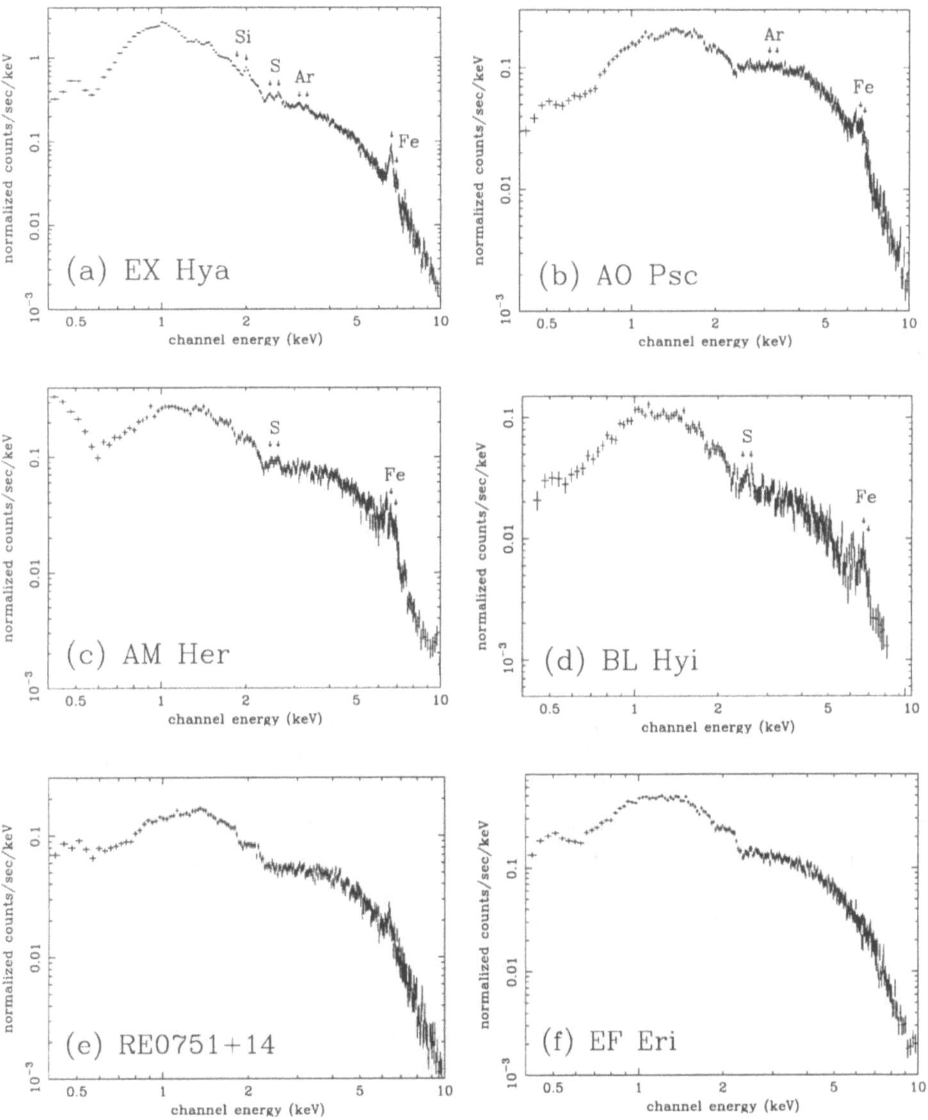

Figure 1. **Phase-average spectra of six MCVs taken by the SIS. Detector responses are not deconvolved.**

with a series of cold absorbing materials whose hydrogen column densities are $\sim 5\,10^{23}\,\mathrm{cm}^{-2}$, $\sim 3\,10^{22}\,\mathrm{cm}^{-2}$ and $< 5\,10^{20}\,\mathrm{cm}^{-2}$. The covering fractions are 62%, 17% and 21%, respectively. The fit is, however, still not acceptable, with $\chi^2_\nu = 1.40$. This situation is quite similar to FO Aqr (Mukai,

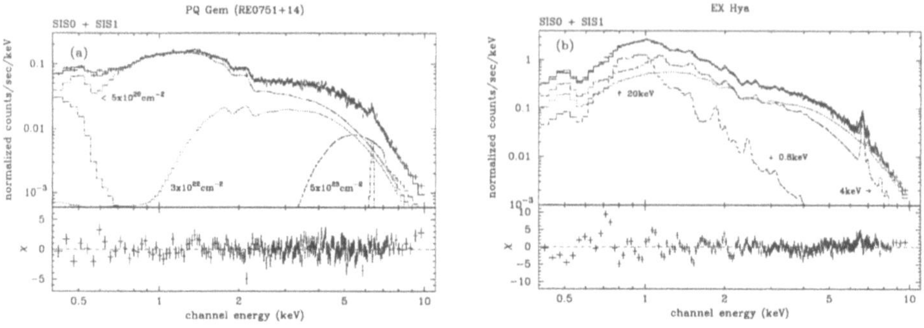

Figure 2. The SIS spectra of (a) PQ Gem and (b) EX Hya. The hydrogen column densities are indicated in (a), and the temperatures of thin thermal plasma are indicated in (b) for each spectral component.

Ishida & Osborne 1994), and possibly to AM Her (Matsuzaki et al. 1996). Moreover, it becomes clear that the reflection of the hard X-ray emission by the white dwarf surface probably makes a significant contribution to the observed spectrum of AM Her based on the Ginga observation (Beardmore et al. 1995).

In Fig. 2b, we show the spectrum of EX Hya, together with a model. Since the spectrum undergoes very small amount of absorption, ($N_H < 7\,10^{20}\,\mathrm{cm}^{-2}$), EX Hya is the best target to investigate the temperature distribution of the post-shock hot plasma. The spectra obtained by the Einstein SSS (Singh & Swank 1993) and Ginga (Ishida et al. 1994) are both explained well by two optically thin thermal plasma emission components with temperatures of $\sim$ 1 keV and $\sim$ 10keV. However, this model does not work for the ASCA SIS spectrum. In Fig. 2b, we have introduced a series of the optically thin thermal plasma emission components with the temperatures of ~0.8 keV, ~4.0 keV and 20 keV, but the fit is not acceptable ($\chi^2_\nu = 3.04$). This result indicates that we observe a continuous distribution of temperature caused by radiative cooling in the post-shock region.

The hard X-ray continuum of MCV is thus composed of multiple emission components with the photoelectric absorption represented by a continuous distribution of hydrogen column densities and a reflection component. It is therefore very difficult to make a quantitative discussion of the continuum emission. One possible way to resolve this difficulty is to construct an emission model from the post-shock plasma by taking into account the profiles of the temperature and the density (Aizu 1973). This will hopefully enable us to discuss the temperature distribution and the absorption/reflection separately.

3.2. DETERMINATION OF THE MASS OF THE WHITE DWARF

Since the accretion flow becomes highly supersonic close to the white dwarf surface, the temperature just below the shock front is calculated according to the strong shock condition. With the aid of the mass-radius relation, the shock temperature, T_S, becomes

$$kT_S = \frac{3}{8}\frac{GM}{R}\mu m_H = 22\left(\frac{M}{0.6\,M_\odot}\right)^{4/3} \text{keV}, \tag{1}$$

where μ is the mean molecular weight (0.62 for material with solar abundance) and m_H is the mass of a hydrogen atom. So if T_S is obtained, we can determine the mass of the white dwarf. We have developed a method to do this with $K\alpha$ emission lines from heavy elements in the post-shock plasma (Ishida & Fujimoto 1995; Fujimoto & Ishida 1995).

The observed line emissions are a superposition of the emission from the entire post-shock plasma. Since the temperature and the density profiles of the post-shock region are solved (e.g. Aizu 1973), the intensity of each emission line is given solely by the boundary conditions. The intensity of the $K\alpha$ line from the element with atomic number Z and the ionization state z is given by

$$\begin{aligned} I_{Z,z} &= \frac{1}{4\pi D^2}\int_0^H A_Z\, \varepsilon_{Z,z}\,[T(x)]n(x)^2\, S dx \\ &= \frac{A_Z S}{4\pi D^2}\int_{T_B}^{T_S} \varepsilon_{Z,z}(T)\, n(x(T))^2 \frac{dx}{dT}\, dT \quad \text{cm}^{-2}\text{s}^{-1}; \end{aligned} \tag{2}$$

here T_B is the temperature at the base of the accretion column, S is the cross section of the accretion column, D the distance, H the height of the shock front measured from the white dwarf surface, A_Z is the abundance of the element with atomic number Z relative to the solar value, $\varepsilon_{Z,z}$ is the line emissivity in the case of solar abundance and n is the electron number density. Since $\varepsilon_{Z,z}(T)$ is tabulated in Mewe, Gronenschild & van den Oord (1985), and the temperature and the density profiles are given by Aizu (1973), the expected intensity is calculated once T_S, n_S, T_B, D, A_Z and S are given. Observationally, both hydrogenic and He-like $K\alpha$ lines from the same element are usually detected. Therefore, by taking their intensity ratio, we can eliminate n_S, D, A_Z and S from our scheme:

$$R_Z(T_S, T_B) = \frac{I_{Z,Z-1}}{I_{Z,Z-2}}. \tag{3}$$

Now that R_Z depends just on the temperatures at the shock front and at the base of the accretion column, we can constrain them if we can measure

TABLE 2. T_S and mass of the white dwarf.

Star	T_S (keV)	Mass ($M_\odot$)	Elements Used
EX Hya	15±2	0.47±0.04	Si, S, Ar, Fe
AO Psc	12±4	0.4±0.1	Ar, Fe

TABLE 3. Detected emission lines

	Mg	Si	S	Ar	Fe
EX Hya	*	*	*	*	*
V1223 Sgr	*	*	*	*	*
AM Her	–	*	*	–	*
BL Hyi	–	*	*	–	*
AO Psc	–	–	–	*	*
FO Aqr	–	–	–	–	*
EF Eri	–	–	–	–	*
PQ Gem	–	–	–	–	–

* ··· detection (>90%)

the line intensity ratio of at least two elements. This method is currently successful for EX Hya and AO Psc, for which the results are summarized in Table 2.

Note that if the magnetic field of the white dwarf is as strong as 10 MG, cyclotron emission makes a significant contribution to the cooling of the post-shock plasma (Wu, Chanmugam & Shaviv 1994). To take this effect correctly into account, we have to know the absolute value of the density at the shock, for which we have to await ASTRO-E. Our method is thus applicable now only for intermediate polars. We are going to apply our method to other intermediate polars. Note also that the $K\alpha$ emission lines are not detected from all the MCVs. The reason for this is discussed in the next section.

3.3. PHOTOIONIZATION EFFECT

As we have shown in Fig. 1, $K\alpha$ emission lines from heavy elements are detected from several sources. In Table 3, we show a list of detected emission lines. From this table, it is understood that the emission lines from light elements are more difficult to be detected. One possible explanation is that the plasma is still too hot for completely ionized light elements to start to recombine with electrons, even at the base of the accretion column. Another possiblity is to assume abundance anomalies; heavier elements are more abundant than lighter elements in general, which is rather ad hoc.

Both these ideas are, however, faced with difficulty in explaining the spectrum of PQ Gem (Fig. 1e). In the PQ Gem spectrum, only $K\alpha$ emission line is detected at 6.4 keV with an equivalent width (EW) of $\sim$ 100 eV, which originates from the pre-shock cold accretion column and the surface of the white dwarf via fluorescence. The temperature of the post-shock plasma in PQ Gem is measured to be 20 keV by the Ginga observation (Ishida 1991). At this temperature, the $K\alpha$ emission line from hydrogenic iron should be detected with an EW of 200 eV. So the plasma is cool enough to radiate the emission line, but the line is absent due to other reasons. The abundance anomaly is also rejected, because the EW of the fluorescent iron $K\alpha$ line indicates that the abundance of the accreting matter is close to the solar value (Beardmore et al. 1995).

The last possibility is the photoionization effect in the post-shock hot plasma. In the low density plasma ($n_e < 10^{10}\,\mathrm{cm}^{-3}$), the ionization equilibrium of iron is determined by a balance between electron collisional ionization and radiative recombination, and photoionization can be neglected. The condition for this is that

$$S_{Z,z}n_e \gg \int_{E_0}^{\infty} f(E)\sigma_{Z,z}(E)\,dE, \tag{4}$$

where $S_{Z,z}$ is the electron-collisional ionization rate coefficient (in $\mathrm{cm^3 s^{-1}}$; Mewe & Gronenschild 1981 and references therein), E_0 is the edge energy, f is the photon flux, and $\sigma_{Z,z}(E)$ is the photoionization cross section (Band et al. 1990). It is understood from equation (4) that the photoinization becomes dominant for high density, because $f \propto n^2$.

In Fig. 3, we show a distribution map of the collisional ionization-dominant and photoionization-dominant regions on the density versus temperature plane for hydrogenic and He-like iron. Here we consider a cylinder with height and diameter of 1% of the white dwarf radius. Evaluation is made at the center of this cylinder filled with uniform plasma with the density and the temperature indicated in the abscissa and the ordinate, respectively. The track of the density and the temperature indicated by the Aizu model in the case of $kT_S = 20\,\mathrm{keV}$ and $n_S = 10^{16}\,\mathrm{cm}^{-3}$ is shown. For hydrogenic and He-like iron the plasma, first in the collision-dominant region, gradually moves into the photoinization-dominant region as it descends the accretion column.

If photoionization becomes important, the ionization distribution shifts to higher ionization degree than what is expected from the collisional ionization only, and the He-like and hydrogenic elements will eventually be

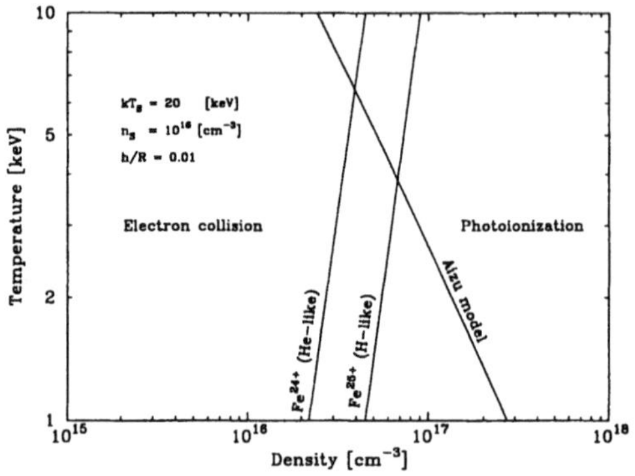

Figure 3. The collisional ionization-dominant and the photoionization-dominant region in the density and the temperature plane. Note that the result is dependent on the geometry (see text).

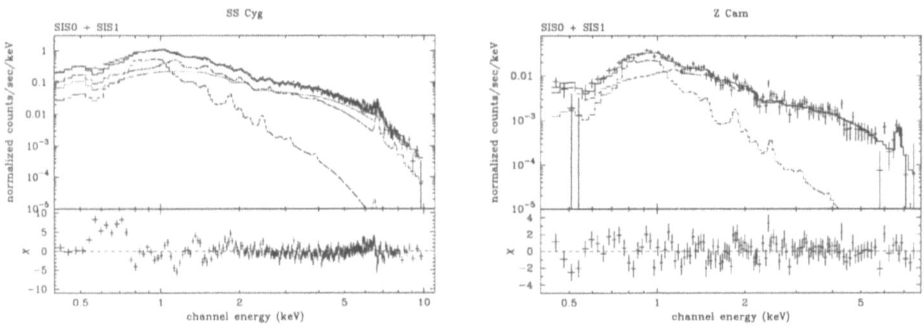

Figure 4. The SIS spectra of SS Cyg and Z Cam

completely ionized, leading to the absence of the emission lines. Note that this effect is more significant for light elements, because, although both $S_{Z,z}$ and $\sigma_{Z,z}$ are both proportional to Z^{-2}, the photon flux is always greater at lower energies. Also, $K\alpha$ lines of lighter elements are emitted from lower portions of the accretion column, where the density is higher. Depression of emission lines from light elements can be attributed to the photoionization effect in the post-shock plasma.

4. Non-magnetic cataclysmic variables

4.1. SS CYG

In Fig. 4, we show the SIS spectrum of SS Cyg. At the time of the ASCA observation (1993 May 27), SS Cyg was nearly at the end of the peak of a so-called anomalous outburst (Nousek et al. 1994). It is well known from EXOSAT observations (Watson, King & Heise 1985) that the hard X-ray flux drops when the source enters into outburst. In the ASCA observation, the flux is measured to be $2.7\,10^{-11}\,\mathrm{erg\,s^{-1}\,cm^{-2}}$ in the 2...10 keV range, which is smaller than that measured by Ginga ($8.3\,10^{-11}\,\mathrm{erg\,s^{-1}\,cm^{-2}}$ in quiescence) by a factor of 3 (Yoshida, Inoue & Osaki 1992).

Yoshida et al. (1992) have introduced a two temperature (~ 7 keV and ~45 keV) plasma emission model based on Ginga observations. Because of the high energy resolution of the SIS, the spectrum can no longer be fit with this model. In Fig. 4, we have introduced a three temperature (~0.8 keV, ~3 keV and ~20 keV) plasma model, but there still remain systematic residuals below 2 keV, which correspond to the $K\alpha$ emission lines from Ne, Mg and Si. The situation is quite similar to EX Hya (Sect. 3.1). It is remarkable that the EW of the iron $K\alpha$ emission line is 735 ± 90 eV, which is twice as large as that in quiescence ($355{\pm}70$ eV). This means the average temperature of the plasma in outburst is significantly lower than that in quiescence, because the EW of the iron $K\alpha$ emission line is a monotonically decreasing function of the plasma temperature above 1 keV. Assuming solar abundance, the average plasma temperature is 10 keV and 18 keV during the ASCA and the Ginga observation, respectively. The hard X-ray flux is proportional to $\sqrt{T} n_e^2 V$, where V is the volume of the optically thin boundary layer. Considering drops of the temperature and the flux by a factor of 2 and 3, respectively, during the outburst, the emission measure $n_e^2 V$ decreases by a factor of 2. However, the density is considered to be higher than in the quiescence. This means that the volume of the boundary layer should significantly shrink when the source enters into outburst.

4.2. Z CAM

Z Cam was observed in 1995 March 8...12, also nearly at the end of the peak of the outburst. The SIS spectrum (Fig. 4 right) is complex like that of SS Cyg, and we have introduced a two temperature (~0.8 keV and ~7 keV) plasma emission model. The fit is, however, not acceptable ($\chi^2_\nu = 1.34$). The observed flux is $6.5\,10^{-13}\,\mathrm{erg\,s^{-1}\,cm^{-2}}$ in the 2...10 keV energy band.

5. Summary

With the emission line information obtained by ASCA, we have discussed the mass of the white dwarf and the possibility of the photoionization effect for MCVs.

ASCA observations have revealed that the hard X-rays from the boundary layer in dwarf novae are the emission from a multi-temperature plasma. The ASCA observations can potentially improve the understanding of the structure of the boundary layer.

References

1. Aizu, K., 1973, Prog. Theoret. Phys., **49**, 1184
2. Band, I.M., Trzhaskovskaya, M.B., Verner, D.A., Yakovlev, D.G., 1990, A&A, **237**, 267
3. Beardmore, A.P., Done, C., Osborne, J.P., Ishida, M., 1995, MNRAS, **272**, 749
4. Fujimoto, R., Ishida, M., 1995, in "Magnetic Cataclysmic Variables", eds D. Buckley, B. Warner, ASP Conf. Ser. **85**, p136
5. Ishida, M., 1991, Ph.D. thesis, University of Tokyo/ ISAS RN 505
6. Ishida, M., Makishima, K., Mukai, K., Masai, K., 1994, MNRAS, **266**, 367
7. Ishida, M., Fujimoto, R., 1995, in "Magnetic Cataclysmic Variables", eds D. Buckley, B. Warner, ASP Conf. Ser. **85**, p132
8. Matsuzaki, K., Ishida, M., Fujimoto, R., Kitamura, H., 1996, in "Proceedings of The 11th Colloquium on UV and X-ray Spectroscopy of Astrophysical and Laboratory Plasma", Universal Academy Press, Tokyo, in press
9. Mewe, R., Gronenschild, E.H.B.M., 1981, A&A Suppl., **45**, 11
10. Mewe, R, Gronenschild, E.H.B.M., van den Oord, G.H.J., 1985, A&A Suppl., **62**, 197
11. Mukai, K., Ishida, M., Osborne, J.P., 1994, PASJ, **46**, L87
12. Norton, A.J., Watson, M.G., 1989, MNRAS, **237**, 853
13. Nousek, J.A., Baluta, C.J., Corbet, R.H.D., et al., 1994, ApJ, **436**, L19
14. Singh, J., Swank, J., 1993, MNRAS, **262**, 1000
15. Tanaka, Y., Inoue, H., Holt, S.S., 1994, PASJ, **46**, L37
16. Watson, M.G., King, A.R., Heise, J., 1985, Space Sci. Rev., **40**, 127
17. Wu, K.W., Chanmugam, G., Shaviv, G., 1994, ApJ, **426**, 664
18. Yoshida, K., Inoue, H., Osaki, Y., 1992, PASJ, **44**, 537

AN ASCA OBSERVATION OF THE ECLIPSING DWARF NOVA HT CAS

K. MUKAI[1], E. M. SCHLEGEL[1,2], J. H. SWANK[1], T. NAYLOR[3], JANET H. WOOD[3]
1. Laboratory for High Energy Astrophysics, NASA Goddard Space Flight Center, Greenbelt, MD 20771, USA
2. Also Universities Space Research Association
3. Department of Physics, University of Keele, Keele, Staffordshire, ST5 5BG, UK

Abstract. We report on a 1-day ASCA observation of the eclipsing dwarf nova HT Cas. We confirm the presence of an X-ray eclipse, which is narrow and deep. The data are consistent with the X-rays originating entirely from the immediate neighborhood of the white dwarf. We draw some preliminary conclusions on the boundary layer and other relevant issues.

1. Introduction

Hard X-ray emission in quiescent dwarf novae (e.g. van Teeseling 1996) is most often interpreted as arising from the optically thin boundary layer between the accretion disk and the white dwarf. However, there has been little direct observational evidence for this interpretation. Moreover, the disk instability theory of dwarf nova outbursts predicts a mass accretion rate onto the white dwarf that is much lower than is required by X-ray observations. Instead, coronal siphon flow models suggest that the boundary layer in the classical sense may not exist (Meyer & Meyer-Hofmeister 1994). There is thus a strong motivation for further observational studies of the hard X-rays from quiescent dwarf novae, particularly eclipsing systems which provide a clean experiment.

We have chosen HT Cas as our target, because it is probably the brightest system in hard X-rays among well-studied, deeply eclipsing dwarf novae (i.e. systems in which the white dwarf is eclipsed, not just parts of the accretion disk). In a previous study, Wood et al. (1995) observed HT Cas with ROSAT and revealed the presence of an X-ray eclipse; however, HT Cas was in an unusual low state at the time. This led to lower-than-expected

A. Evans and J. H. Wood (eds.), Cataclysmic Variables and Related Objects, 269–272.

data quality; also a doubt remained about the relevance of the data to the normal state. To follow up, we utilized the ASCA satellite (see e.g. Ishida 1996) which has imaging (i.e. low background) capabilities up to ~10 keV, with roughly 4 times higher effective area than ROSAT.

2. Observation

HT Cas was observed with ASCA between 1994 September 6 17:25 UT and 1994 September 7 17:10 UT. The observation was planned to maximize eclipse coverage: 6 eclipses were covered completely as well as parts of 2 more. Standard screening criteria have been applied to produce 31 ks of good SIS data and 34 ks of good GIS data. Light curves and spectra were then extracted from circular regions with ~3.5 and ~6.0 arcmin radii respectively, for SIS and GIS. Background subtraction is a major issue for fainter sources: blank sky and local background subtraction methods have been tried to give an indication of systematics.

In addition, optical CCD photometry was performed several times during the August... October period at Keele Observatory. HT Cas was found to be in normal quiescence throughout this period (V~16.5 mag out of eclipse). For phasing, we apply a correction of ~30 s to the ephemeris of Horne, Wood & Stiening (1991), as derived from recent optical photometry (Wood et al. 1995).

3. Results

The folded ASCA light curve of HT Cas is shown in Fig. 1. We discuss the out-of-eclipse behavior first, then discuss the eclipse.

3.1. THE OUT-OF-ECLIPSE BEHAVIOR

The orbital phase interval 0.2...0.4 was covered rather poorly, hence the light curve is noisier during this time. Taking this into consideration, we have found no evidence for a systematic variability outside the eclipse in folded and unfolded light curves. We have also examined light curves for soft ($E < 2$ keV) and hard ($E > 2$ keV) bands, and found no systematic variability in hardness ratio.

We have therefore extracted the average out-of-eclipse spectra from the 4 instruments, and fitted a variety of standard models. The best fit temperature with the Raymond-Smith model is $10.5^{+1.5}_{-1.4}$ keV with local background subtraction, or $8.7^{+1.1}_{-1.0}$ keV with blank sky background, i.e. HT Cas had a hard spectrum for a non-magnetic CV at the time of the ASCA observation. At such temperatures, plasma models predict a spectrum dominated by the bremsstrahlung continuum, with 6.7 keV (He-like) and 6.97 keV

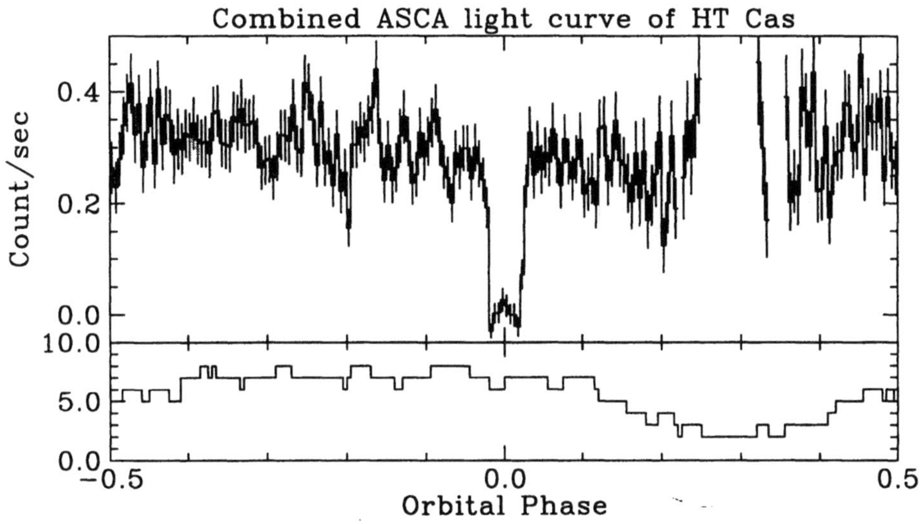

Figure 1. Folded X-ray light curve of HT Cas: data from all 4 ASCA instruments have been combined, folded on the ephemeris of Horne et al. with 30 s correction derived from recent optical photometry. The lower panel shows the number of different cycles that were combined to derive the value for that bin. There are 200 phase bins of ~32 s per bin.

(H-like) lines of Fe K, which is consistent with the bremsstrahlung fit we obtained; the lines were approximated by a single, broad Gaussian as the statistical quality of the data was not good enough for a multi-component fit. We therefore conclude that the X-ray emission from HT Cas was dominated by ~10 keV plasma. Our attempt to add a second, lower-temperature component did not result in an improved fit.

The source is absorbed with $N_{\rm H} = 3.3 \pm 0.3\,10^{21}\,{\rm cm}^{-2}$, independent of whether bremsstrahlung or the Raymond-Smith model was used. The measured flux (average of out-of-eclipse data) is $3.5\,10^{-12}\,{\rm ergs\,cm^{-2}s^{-1}}$ in the 2–10 keV range, corresponding to an inferred bolometric flux of $1.07\,10^{-11}$ ${\rm ergs\,cm^{-2}s^{-1}}$. Adopting a distance of 165 pc, the hard X-ray luminosity of HT Cas was $\sim 3.54\,10^{31}\,{\rm ergs\,s^{-1}}$; if this was half the accretion luminosity onto a $0.64\,{\rm M}_\odot$ (Horne et al. 1991) white dwarf, then an accretion rate was $7.1\,10^{14}\,{\rm g\,s^{-1}}$ or $1.1\,10^{-11}\,{\rm M}_\odot\,{\rm yr}^{-1}$.

3.2. THE ECLIPSE

The eclipse appears total and narrow (Fig. 1). Here we report preliminary results from our ongoing effort to quantify this statement.

First, how total is the eclipse? The errors are dominated by the uncertainties in the true background level. We have therefore measured the mid-eclipse flux in two types of instruments independently, using two background subtraction methods, and two energy bands. With the SIS, we obtain a more tightly constrained result: $-2.2\pm5.3\%$ of the out-of-eclipse average in the 0.7...5 keV band (which was selected by eye to maximize signal-to-noise) using local or blank sky background. The values for GIS were $3.8\pm10.1\%$ and $1.8\pm10.3\%$, respectively, using local and blank sky backgrounds. Thus the eclipse is consistent with being total, although a very low residual flux (up to $\sim$3%) cannot be excluded.

A piece-wise linear fit to the light curve indicates the eclipse duration (mid-ingress to mid-egress) of 0.045 cycle, as compared with 0.0489 cycle for the optical eclipse of the white dwarf photosphere (Horne et al. 1991). As a first approximation, a spherical emission region with a radius $R_{\rm wd}$ (as determined in the optical), using the system geometry derived in the optical, can reproduce the parameters of the X-ray eclipse. Thus, we conclude that the X-ray emission is localized to the white dwarf and has a size comparable to it. However, there are indications that the X-ray eclipse is perhaps too narrow: our continued investigation will hopefully resolve this mystery.

4. Discussion

The ASCA spectrum of HT Cas indicates an absorption much greater than is seen in the ROSAT spectrum. It indicates that, at the time of the ASCA observation, there was an absorbing cloud within the binary system. Given the high inclination of HT Cas, it is reasonable to suppose that an increased accretion rate is accompanied by a thickening of the disk or the boundary layer, resulting in the observed increase in absorption.

ROSAT and ASCA observations have established that much of the hard X-ray flux in HT Cas in quiescence originates very near the white dwarf, in normal quiescence and in a low state. The ASCA measurement of the flux indicates that the accretion rate onto the white dwarf must be relatively high. The boundary layer model of hard X-ray emission is a natural interpretation for the X-ray data, although maintaining this level of accretion may cause problems for the disk instability model of dwarf nova outbursts. The coronal siphon model will be consistent with the data only if the extended accretion disk corona that is an essential ingredient of the model is shown not to emit a significant fraction of the X-rays.

References

Horne, K., Wood, J.H., Steining, R.F., 1991, Ap. J., **378**, 271
Ishida, M., 1996, these proceedings, p259
Meyer, F., Meyer-Hofmeister, E., 1994, A&A, **288**, 175
van Teeseling, A., 1996, these proceedings, p273
Wood, J.H., Naylor, T., Hassall, B.J.M., Ramseyer, T.F., 1995, MNRAS, **273**, 772

ROSAT OBSERVATIONS OF NON-MAGNETIC CVS

A. VAN TEESELING[1], F. VERBUNT[2], K. BEUERMANN[1]
1. Universitäts-Sternwarte Göttingen, Germany
2. Astronomical Institute, Utrecht, The Netherlands

1. Introduction

In non-magnetic cataclysmic variables the accreted matter forms an accretion disk around the white dwarf. In the boundary layer between the white dwarf and the accretion disk the accreted matter decelerates from Keplerian velocities to the rotation velocity of the white dwarf. If the accretion rate is high the boundary layer would be optically thick and cool ($T \sim 10^5$ K), and if the accretion rate is low the boundary layer would be optically thin and hot ($T \sim 10^8$ K) (Pringle & Savonije 1979).

There are several observational problems with this simple picture: a soft X-ray component could only be detected so far in 5 dwarf novae in outburst and not in any nova-like variable. Also in high-accretion-rate systems there is a hot optically thin X-ray source, which has, however, an X-ray luminosity which is much less than the UV luminosity of the system (van Teeseling & Verbunt 1994). Finally, there is evidence for orbital modulation in the X-rays from some systems (e.g. van Teeseling et al. 1995).

We have analysed ROSAT PSPC observations of more than 40 non-magnetic cataclysmic variables (van Teeseling, Beuermann & Verbunt 1996). With this large sample we could investigate how these observations constrain theoretical models for the X-ray source in non-magnetic cataclysmic variables. Here we summarize some results.

2. Results

Fig. 1 shows that the X-ray count ratios of non-magnetic cataclysmic variables form a narrow band. The count ratios are consistent with absorbed optically thin spectra. Bremsstrahlung temperatures, inferred from fits to the ROSAT spectra, have been plotted against the UV+optical luminosity in Fig. 2. There appears to be no correlation between the temperature and the mass accretion rate in the system. This is not consistent with simple boundary-layer theories (Pringle & Savonije 1979; Tylenda 1981).

There is a significant anti-correlation between the inferred emission mea-

A. Evans and J. H. Wood (eds.), Cataclysmic Variables and Related Objects, 273–276.

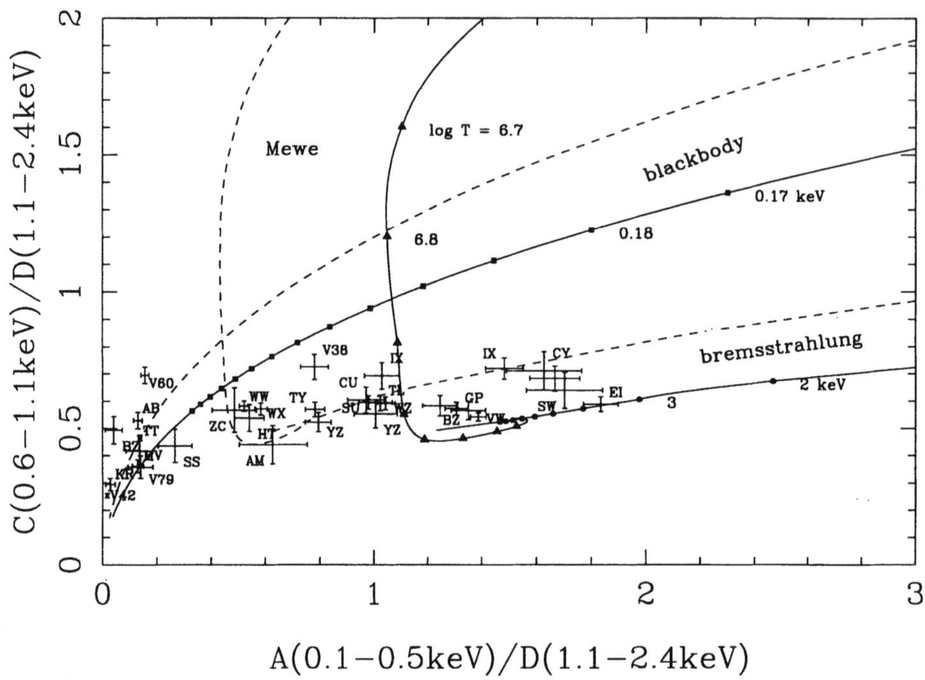

Figure 1. ROSAT PSPC count ratios of non-magnetic cataclysmic variables, with 1-σ error bars. Systems are identified with the first letters of their name. Also plotted are the predicted count ratios for Kaastra & Mewe (1993) spectra, blackbody spectra and thermal bremsstrahlung spectra. The theoretical spectra form bands with the solid boundary corresponding to zero interstellar absorption and the dashed boundary corresponding to a hydrogen column density of $n_{\rm H} = 10^{20}\,{\rm cm}^{-2}$. Absorption moves the predicted spectra almost horizontally to the left. The difference (in keV or $\log T$) between two subsequent marked points is constant. The highest blackbody temperature is 10 keV and the highest bremsstrahlung temperature is 90 keV.

sure (or count rate×distance^2) (EM) and the orbital inclination. This is illustrated in Fig. 3, which contains 2 UX UMa systems, 2 Z Cam systems, 2 VY Scl systems, one non-SU UMa system, and 9 SU UMa systems. Fig. 3 also contains the candidate DQ Her systems TT Ari, SW UMa, WZ Sge, and V426 Oph. The anti-correlation in Fig. 3 implies that the X-ray sources in all these systems are of similar nature. We can exclude all models in which the X-rays are emitted in a relatively large, optically thin volume. An optically thin boundary layer which is partly obscured by an optically thick inner accretion disk would have EM $\propto 1/\tan i$. Just as an illustration we have plotted the relation ${\rm EM} = 2\,10^{54}/\tan i\,{\rm cm}^{-3}$.

For high-$\dot{M}$ systems the 'bolometric' X-ray flux is much lower than the UV+optical flux (Fig. 2). The X-ray luminosity is independent of the accretion rate $\dot{M}$, while the UV luminosity increases with increasing $\dot{M}$.

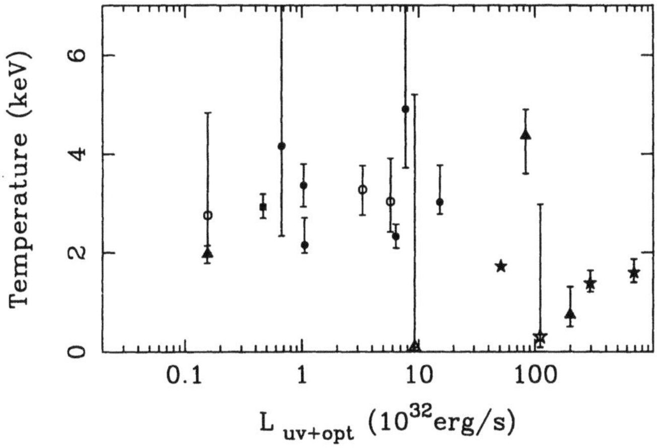

Figure 2. Bremsstrahlung temperature as a function of the ultraviolet+optical luminosity. Different symbols are used for subclasses: ⋆ = UX UMa, • = SU UMa, and ▲ = other. For the open symbols the luminosity has been derived from the optical magnitude only.

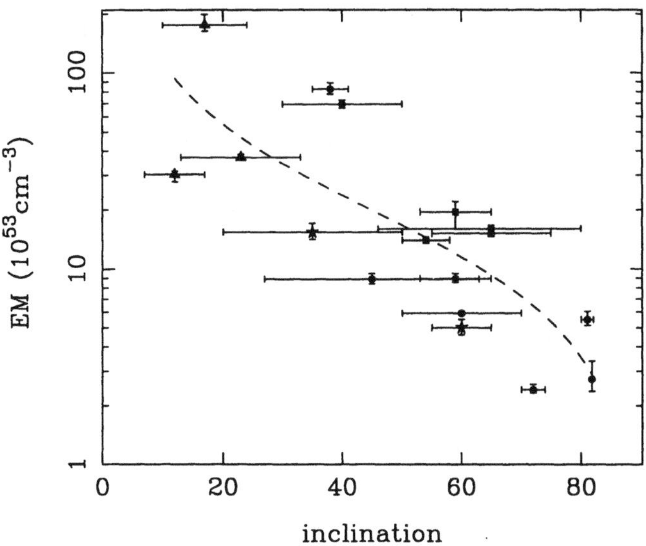

Figure 3. Observable emission measure as a function of the orbital inclination. The errors in the emission measures do not include the uncertainty in the distances. Different symbols are used for subclasses: ⋆ = UX UMa, • = SU UMa, ■ = Z Cam, and ▲ = other. The dashed line is $EM = 2\,10^{54} / \tan i \, cm^{-3}$.

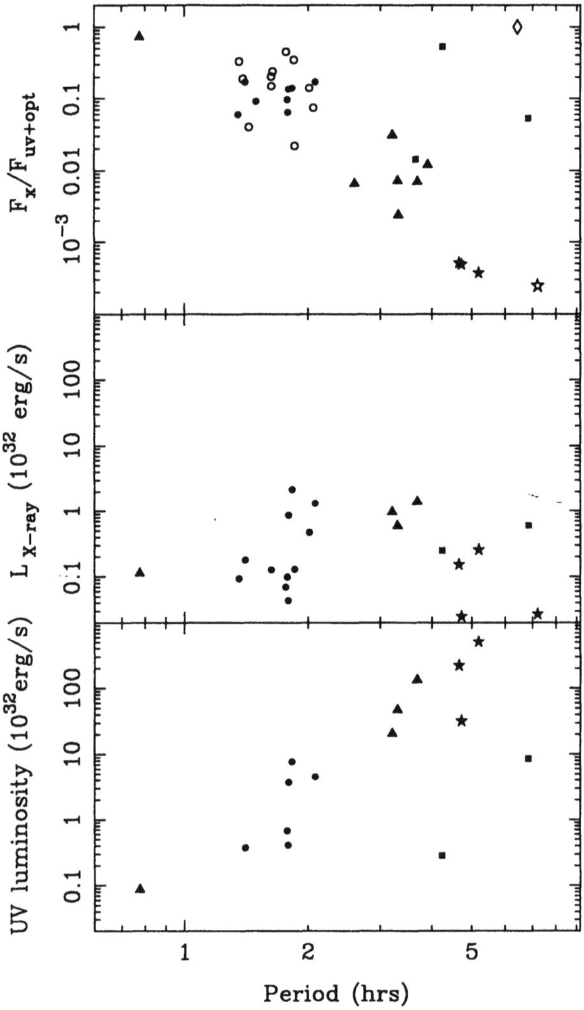

Figure 4. The top panel shows the ratio of X-ray flux to ultraviolet+optical flux. For the open symbols the ratio has been derived from the optical magnitude only. The middle and bottom panels show the bolometric X-ray luminosity and the (912-3000Å) UV luminosity. The UV flux is given by Verbunt (1987) and the distances by Warner (1987). Different symbols are used for subclasses: ⋆ = UX UMa, • = SU UMa, ◇ = U Gem, ▪ = Z Cam, and ▲ = other.

References

Kaastra, J.S., Mewe, R., 1993, Legacy, **3**, 16
Pringle, J.E., Savonije, G.J., 1979, MNRAS, **187**, 777
Tylenda, R., 1981, Acta. Astr., **31**, 267
van Teeseling, A., Verbunt, F., 1994, A&A, **292**, 519
van Teeseling, A., Drake, J.J., Drew, J.E., et al., 1995, A&A, **300**, 808
van Teeseling, A., Beuermann, K., Verbunt, F., 1996, A&A, submitted
Verbunt, F., 1987, A&AS, **71**, 339
Warner, B., 1987, MNRAS, **227**, 23

COMPTEL OBSERVATIONS OF X-RAY BINARIES

A. F. IYUDIN
MPI für extraterrestrische Physik, Garching, FRG

The COMPTEL experiment on the Compton Gamma-Ray Observatory measures $\gamma-$radiation in the energy range from 0.75 to 30 MeV [10]. X-ray binaries form one category of potential γ-ray sources. Up to now only two X-ray binaries, Cyg X-1 and Nova Per, have been seen by COMPTEL.

Cyg X-1. Preliminary results were reported by [6] and indicate that (i) there was no evidence for any hardening of the spectrum near 1 MeV, (ii) the plasma temperature suggested by a Wien spectral model was much higher than that implied by hard X-ray observations. This latter conclusion seems to require a revision in the standard spectral model for Cyg X-1.

The $\gamma-$ray flux from Cyg X-1 is constant up to several MeV. The time history of the flux in the 0.75...2.0 MeV energy range shows that there are no significant variations near 1 MeV. The average photon spectrum derived from all observations up to 1994 July shows evidence for significant emission extending well above 2 MeV, with a data point in the 2...5 MeV range at $4.3\,\sigma$ significance. However, the observed flux near 1 MeV is significantly below that reported by HEAO-3 and several other balloon observations [6]. The COMPTEL spectral data show a broad line-like feature around $\sim$ 4 MeV with the best fit of a single power law plus Gaussian. If we assume that the Gaussian represents a single broadened nuclear line of $^{12}C^*$, then the width of 1.38 MeV demands a temperature of $\sim 2\,10^{12}$ K [7].

Nova Per 1992. The X-ray transient GRO J0422+32 is also known as Nova Per 1992 [8]. COMPTEL observed this object twice during ToO viewing periods 1992 August 12...20 and 1992 September 1...17, and then later in 1993...94. First results on the COMPTEL detection of $\gamma-$ray emission from this source were reported by [4], [5]. During the first observation, which started when the X-ray flux was at maximum, Nova Per was detected with flux in the 1...2 MeV energy range that was higher than expected, based on extrapolation of the standard Comptonisation model [11] fits to the contemporaneous OSSE [2] and SIGMA [9] data. Nova Per was seen up to $\sim$ 2 MeV during a primary outburst in 1992 Aug...Sep by COMPTEL. There are indications that this binary had secondary outbursts in visual

A. Evans and J. H. Wood (eds.), Cataclysmic Variables and Related Objects, 277–278.

light, with period $\sim$ 125 d. The spectrum also shows large deviations from the standard Comptonisation model in the MeV energy range.

Preliminary results of the continued analysis of all available COMPTEL data for the period 1992...94 can be summarised as follows: γ–ray emission at 1...2 MeV ($\sim 3\,\sigma$) was detected during the main outburst [4], [5] and later during viewing period (VP) 325 (1994 April 22 ... May 5) with $\sim 4.3\,\sigma$ significance in the 1.5...2.1 MeV band. VP325 presumably corresponds to the time of the nova recurrent optical activity (time of VP325 – time of main outburst = 5 periods of $\sim$ 125 d, as derived by [1] and [3]). For all VPs for which no activity of Nova Per 1992 was observed only upper limits on the γ–ray flux, which are significantly ($\gtrsim 3\,\sigma$) lower than the flux in the same energy intervals during active periods, were derived.

A spectral fit of the available data was performed with the standard Comptonisation model of [11], and with the more generalised Comptonisation model of [12], and resulted in electron temperatures $\sim$ 100 keV [5]. These results are in accordance with the higher temperatures ($\sim$ 200 keV) found by COMPTEL for Cyg X-1 [6]. The quality of the fits with the use of the Wien type spectrum were generally not very good and show significant deviations from the Comptonisation models at the higher energy intervals (300...700 keV) of SIGMA ($5.8\,\sigma$ deviation) [9] and at the COMPTEL energies ($2.8\,\sigma$) [5].

Conclusion: Both objects (i) belong to the systems with a black hole companion and (ii) show a large deviation in the spectrum from the Compton model fit. Nova Per shows excess emission above the continuum in relatively narrow energy bands. The presence of nonthermal emission processes in both objects seems unavoidable.

Acknowledgement. This work was presented on behalf of the COMPTEL collaboration.

References

1. Callanan, P.J., Garcia, M.R., McClintock, J.E., et al., 1995, Ap. J., **441**, 786
2. Cameron, R.A., et al., 1992, IAUC 5587
3. Chevalier, C., Ilovaisky, S.A., 1994, preprint No. 83, Observatoire de Haute-Provence
4. van Dijk, R., et al., 1994, in AIP 304 Conf. Proceed., "The Second Compton Symposium", eds. C. E. Fichtel, N. Gehrels, J. P. Norris, (New York: AIP), p197
5. van Dijk, R., Bennett, K., Collmar, W., et al., 1995, A&A, **296**, L33
6. McConnell, M.L., Forrest, D., Ryan, J., et al., 1994, Ap. J., **424**, 933
7. McConnell, M.L., et al., 1995, presented at the Third Compton Symposium, München
8. Paciesas, W.C., et al., 1993, in AIP 304 Conf. Proceed., "The Second Compton Symposium", eds. C. E. Fichtel, N. Gehrels, J. P. Norris, (New York: AIP), p365
9. Roques, J.P., Bouchet, L., Jourdain, E., et al., 1994, Ap. J. Supp., **92**, 451
10. Schönfelder, V., Aarts, H., Bennett, K., et al., 1993, Ap. J. Supp., **86**, 657
11. Sunyaev, R.A., Titarchuk, L.G., 1980, A&A, **86**, 121
12. Titarchuk, L., 1994, Ap. J., **434**, 570

NOVAE & SYMBIOTIC STARS

CLASSICAL NOVA EVOLUTION: CLUES FROM SOFT X-RAY EMISSION

JAMES MACDONALD
University of Delaware
Department of Physics and Astronomy, Newark, DE 19716, USA

Abstract. Multi-wavelength observations have shown that, after optical decline, the stellar remnants of classical nova outbursts evolve at constant, near-Eddington, bolometric luminosity to high effective temperature ($> 2\,10^5$ K), before turning off. Here we briefly review the observations of classical novae in this phase of evolution, and discuss what the soft X-ray observations tell us about the mass of the underlying white dwarf and the rate of mass change due to stellar winds and accretion from the stellar companion.

1. Introduction

A classical nova outburst occurs when a white dwarf in a close binary system accretes sufficient hydrogen-rich material from a non-degenerate companion to trigger a thermonuclear runaway (Starrfield et al. 1974). The ensuing rapid release of nuclear energy causes the photospheric luminosity to increase to greater than the Eddington limit, $L_{\rm Edd}$, and the white dwarf envelope, of mass $M_{\rm env}$, to expand to red giant dimensions on a dynamical time-scale, ejecting a mass, $M_{\rm ej}$, of between 10^{-6} and 10^{-4} $M_{\odot}$. Early hydrodynamic simulations found that only a fraction of the envelope is ejected in the initial explosion. The remaining envelope quickly returns to hydrostatic equilibrium and enters a steady nuclear burning phase with a constant bolometric luminosity close to $L_{\rm Edd}$.

Observationally the recognition that a constant bolometric luminosity phase is a common characteristic of classical novae followed upon a combination of ultraviolet and infrared observations of Nova FH Ser 1970 (Hyland & Neugebauer 1970; Geisel, Kleinmann & Low 1970; Gallagher & Code

A. Evans and J. H. Wood (eds.), Cataclysmic Variables and Related Objects, 281–287.

1974; Gallagher & Starrfield 1976). For this moderately fast nova, the visual luminosity fell quite rapidly, with the dominance of the UV luminosity after the first week reflecting the rapidly increasing photospheric temperature. The fall in the UV after approximately 60...80 d and the simultaneous emergence of the infrared is due to absorption and re-emission by dust grains in the ejecta. During the first 100 d, the sum of UV, visual and IR luminosities were roughly constant (Gallagher & Starrfield 1978).

Although estimates of the time to turn-off of the nuclear burning were uncertain, it was immediately clear that evolution of the stellar remnant on a purely nuclear time-scale was too slow to be consistent with the observations. Turn-off requires some mechanism of mass loss (Starrfield 1979; MacDonald, Fujimoto & Truran 1985, hereafter MFT). In the next section, we briefly review the theory of the turn-off of classical novae. In particular, the stellar remnants are expected to evolve to high effective temperatures ($T_{\rm eff} > 2\,10^5$ K) and be copious emitters of EUV/soft X-ray photons. In theory, measurement of the maximum $T_{\rm eff}$, attained at turn-off, can be used to determine the mass of the white dwarf, and the time to reach maximum $T_{\rm eff}$ constrains the rate of mass loss from the stellar remnant and also the rate of any accretion from the companion star. We apply these ideas to two well observed novae, GQ Mus and V1974 Cyg.

2. The evolution to turn-off

In discussing the evolution of the stellar remnant, it is useful to divide the nova outburst into three phases: (A) envelope expansion in response to the thermonuclear runaway, (B) common envelope evolution and (C) steady nuclear burning at constant bolometric luminosity, with possible wind mass loss or accretion from an irradiated companion.

Of crucial importance to the lifetime of phase B is the amount of envelope material remaining on the stellar remnant after phase A. Recent hydrodynamic calculations of phase A differ significantly in the fraction of envelope material that is ejected. For example, Starrfield et al. (1992), in simulations of Nova Her 1991, find $M_{\rm ej}/M_{\rm env}$ in the range 0.09...0.67, whereas Prialnik & Shara (1995) find $M_{\rm ej}/M_{\rm env}$ between 0.92 and 1.00. We suspect that this significant disagreement is due in part to Prialnik & Shara (1995) completely suppressing convective energy transport in hydrodynamic phases of evolution and to their method of removal of mass zones that are escaping from the star (Prialnik & Kovetz 1995). Both these effects work in the direction of increasing the amount of ejected material.

If, as the calculations of Starrfield and collaborators suggest, a significant fraction of the envelope remains after phase A, then the stellar remnant is larger than the orbital separation and phase B is entered. As shown by

MacDonald (1980) and MFT, dynamical friction quickly ejects almost all of the material lying outside the orbit at about the escape speed in the vicinity of the orbit. For this phase to occur, M_A, the envelope mass at the end of phase A, must satisfy (MFT)

$$\frac{M_A}{M_\odot} \gtrsim 2\,10^{-5} \left(\frac{M_{WD}}{M_\odot}\right)^{-4.5} \left(\frac{S}{10^{11}\,\mathrm{cm}}\right)^{0.3}$$

where M_{WD} is the white dwarf mass and S is the orbital separation. For a typical S of 10^{11} cm, this is satisfied for all the simulations of Starrfield et al. (1992). Indirect observational support for the existence of phase B in classical novae comes from comparison with very slow novae such as RR Tel and RT Ser, in which the companion is a giant star (Feast & Glass 1974) and the orbital separation too large for a significant common envelope phase. The very slow novae have long turn-off times and, more importantly, very low expansion velocities of order $100\,\mathrm{km\,s^{-1}}$, whereas classical novae have expansion velocities of order the orbital velocity of the companion star or greater (MacDonald 1980). Paresce et al. (1995) find that the kinematics and morphology of the shell ejected by V1974 Cyg, determined from their HST FOC observations, are consistent with the shaping expected from a common envelope phase.

As shown by MFT, the lifetimes of phases A and B are much shorter than that of phase C and hence it is the lifetime of phase C that determines t_{rem}, the time for turnoff of classical nova remnant. In the absence of wind mass loss, the only process to consume envelope material is nuclear burning. On the basis of polytropic models for the envelope, MFT then estimate, for a composition representative of novae ejecta, that

$$t_{rem} \simeq 20 \left(\frac{M_{WD}}{M_\odot}\right)^{-6.3} \mathrm{yr},$$

which has a very strong dependence on white dwarf mass. However it is unrealistic to assume that the hot remnant with a luminosity close to the Eddington limit will not have a wind, whether optically thick (Bath 1978, Kato 1983, Kato & Hachisu 1994) or line-driven (MacDonald 1984, MFT). Furthermore for systems with small orbital periods, irradiation of the companion by the hot remnant may induce accretion onto the white dwarf, prolonging phase C (Ögelman et al. 1993). If an independent estimate of M_{WD} can be obtained then comparison of t_{rem} with the nuclear burning time-scale in principle gives the rate of mass loss or mass gain, which can be compared to the theoretical prediction for e.g. optically thick winds. Determinations of M_{WD} by standard spectroscopic techniques have large errors, primarily because the radial velocity of the white dwarf cannot be directly measured and must instead be inferred from accretion disk emission lines (see Warner 1990 for a review). Attempts to obtain $M_{\rm WD}$ from a mass-luminosity relation similar to that for AGB stars are hampered by uncertainties in the distance. However, MFT argue that $T_{\rm max}$, the maximum effective temperature reached by the remnant before turning-off mainly depends on $M_{\rm WD}$ and their polytropic model gives a dependence

$$T_{\rm max} \approx 6.6\,10^5 \left(\frac{M_{\rm WD}}{\rm M_\odot}\right)^{1.6} \ {\rm K},$$

which agrees with the detailed stellar models of Fujimoto & Iben (1992) within 10%. Hence determination of $T_{\rm max}$ from fitting stellar atmosphere models to the results of a campaign of monitoring classical novae at soft X-ray energies has the potential of determining $M_{\rm WD}$.

3. The soft X-ray observations

The detection of soft X-rays from three classical novae, Nova GQ Mus 1983, Nova PW Vul 1984 and Nova QU Vul 1984, at times ranging from 100 to 1000 d after optical maximum with the EXOSAT satellite (Ögelman, Beuermann & Krautter 1984; Ögelman, Krautter & Beuermann 1987, hereafter OKB), has provided observational verification that the constant bolometric luminosity phase does extend to high effective temperatures ($> 2\,10^5$ K), as originally suggested by MFT.

Since the majority of the EXOSAT observations were made with only the thin Lexan filter on the low energy telescope, there is, in general, insufficient data in the X-ray measurements alone to discriminate between emission from a hot photosphere and emission from shocked circumstellar gas. OKB also find that the single observation of Nova Vul 1984 #2 with the Boron filter, 307 d after optical maximum, is not significant enough to exclude or support a thermal bremsstrahlung spectrum. However, they favor emission from a hot photosphere on the basis of consideration of plasma cooling times, and the absence of coronal [Fe XIV] λ5303 lines that should be emitted from shocked gas (Kurtz, Vanden Bout & Angel 1972). Krautter & Williams (1989) rule out shocked circumstellar gas by considering the degree and temporal evolution of the ionization of the ejecta of GQ Mus and argue that it is consistent with photo-ionization from a hot radiation source. EXOSAT observations made two months after the 1985 January outburst of the recurrent nova RS Oph (Mason et al. 1987) have been interpreted as due to X-rays from a shock front moving into a circumstellar envelope formed by the pre-outburst wind of the red giant companion (O'Brien, Kahn & Bode 1987). Even so, the X-ray emission 200 d after outburst is consistent with an origin on the surface of a white dwarf (Mason et al. 1987).

The greater energy resolution of the ROSAT PSPC permits a clearer distinction between a hot photosphere and shocked gas. The X-ray emission from GQ Mus in 1992 February, 9 yr after optical maximum, has a very soft spectrum and is consistent with a photospheric origin, at near L_{Edd} (Ögelman et al. 1993). ROSAT spectra of Nova Her 1991 (Lloyd et al. 1992), and Nova V1974 Cyg 1992 (Krautter, Ögelman & Starrfield 1992), 5 and 60 d after optical maximum respectively, are hard and the implied luminosities are below 10^{-2} L_{Edd}, perhaps indicating an origin in shocked gas. GQ Mus was observed twice again with the ROSAT PSPC in 1993 (Shanley et al. 1995). The count rates fell sharply from 1992 February to 1993 January and GQ Mus was not detected in 1993 August/September, indicating that GQ Mus had turned off 10 yr after optical maximum. Nova V1974 Cyg was monitored with ROSAT throughout its active phase in X-rays and was found to turn off 18 months after optical maximum (Starrfield et al. 1995).

In anticipation of these ROSAT detections of classical novae, MacDonald & Vennes (1991, hereafter MV) performed model atmosphere calculations of the X-ray emission from the photospheres of hot, high gravity stars, using representative values for the white dwarf mass and luminosity of $M_{WD} = 1.2\,M_{\odot}$ and $L_{WD} = 4.2\,10^4\,L_{\odot}$. Stellar atmospheres in local thermodynamic equilibrium were calculated for a range of effective temperatures from 10^5 to 10^6 K (the corresponding range in $\log g$ is from 4.85 to 8.85) and two compositions representative of the ejecta of CNO novae and ONe novae. An important result of this study is that blackbody fits to the X-ray spectra can grossly underestimate, by as much as a factor of 2, the true effective temperature. Indeed for GQ Mus, Ögelman et al. (1993) found the best blackbody fit temperature of $2.5\,10^5$ K gave a bolometric flux that was about a factor 100 too high to be consistent with the Eddington limit. However, within the 2 and $3\,\sigma$ error limits, they could find a blackbody fit with higher temperature ($\simeq 3.5\,10^5$ K) and lower bolometric flux. We estimate that use of MV model atmospheres (see Fig. 3 of MV) would give $T_{eff} = 4\,10^5$ K. Starrfield et al. (1995) also find that the UV colours of GQ Mus from 1987 to 1994 are inconsistent with blackbodies. Furthermore, Krautter & Starrfield (1995) find that blackbody fits to the V1974 Cyg ROSAT data give super-Eddington fluxes, and that this problem is cured by use of the MV fluxes.

V1974 Cyg was also detected 280 d after visual maximum by EUVE in the all-sky survey. The count rates were $0.108 \pm 0.010\,s^{-1}$ and $0.091 \pm 0.010\,s^{-1}$ in the Lexan/Boron and Al/C/Ti filters respectively (Bowyer et al. 1994). A preliminary analysis of these data was given by MacDonald & Vennes (1994), who found T_{eff} to be between $3\,10^5$ and $3.5\,10^5$ K and the column density about $1.8\,10^{21}\,cm^{-2}$. Stringfellow & Bowyer (1994), using better calibrated effective area curves for the two filters, find $T_{\rm eff} = 4.0 \pm 0.5\,10^5$ K and $N_{\rm H} \simeq 3\,10^{21}$ cm^{-2}. We note that the temperatures found for these two objects and also for QU Vul from analysis of the EXOSAT data by MV, are remarkably similar. Pequignot (1996) has also determined $T_{\rm eff}$ as a function of time for GQ Mus from photoionization models for the emission lines, assuming a blackbody source of ionizing photons. Although his results depend somewhat on the evolution of the nebular density, he finds that the stellar remnant reaches high temperatures relatively quickly ($\simeq 3\,10^5$ K after one year) and then remains constant or increases slowly over the next 2000 d. The initial relatively rapid temperature increase may indicate that mass loss at a high rate continued even after the stellar remnant was inside its Roche lobe. The long duration of essentially constant temperature is perhaps explained by an episode of accretion as suggested by Ögelman et al. (1993).

Using the above values for $T_{\rm max}$, our estimate for $M_{\rm WD}$ is $0.75\,{\rm M}_\odot$ for both stars. Furthermore, to get the short turnoff times requires mass loss at average rates of $8\,10^{-6}$ ${\rm M}_\odot\,{\rm yr}^{-1}$ for GQ Mus and $5\,10^{-5}$ ${\rm M}_\odot\,{\rm yr}^{-1}$ for V1974 Cyg. The estimated mass is much lower than thermonuclear runaway theory requires for classical nova outbursts (MacDonald 1983). A possible resolution of this discrepancy is that there is an optically thick outflow from the surface which results in a more extended envelope and lower $T_{\rm eff}$ than for a hydrostatic envelope. A second possibility is that accretion of near solar composition material before or during the soft X-ray phase covers up the metal-enriched envelope so that the use of the model atmospheres calculated by MV for metal-enriched compositions gives misleading temperature estimates.

References

Bath, G.T., 1978, MNRAS, **182**, 35
Bowyer, S., Lieu, R., Lampton, M., et al., 1994, Ap. J. Supp., **93**, 569
Feast, M.W., Glass, I.S., 1974, MNRAS, **167**, 81
Fujimoto, M.Y., Iben, I., 1992, AJ, **399**, 646
Gallagher, J.S., Code, A.D., 1974, Ap. J., **189**, 303
Gallagher, J.S., Starrfield, S., 1976, MNRAS, **176**, 3
Gallagher, J.S., Starrfield, S., 1978, ARAA, **16**, 171
Geisel, S.L., Kleinmann, D.E., Low, F.J., 1970, Ap. J., **161**, L101
Hyland, A.R., Neugebauer, G., 1970, Ap. J., **160**, L177
Kato, M., 1983, PASJ, **35**, 507
Kato, M., Hachisu, I., 1994, Ap. J., **437**, 802
Krautter, J., Ögelman, H., Starrfield, S., 1992, IAUC 5550
Krautter, J., Starrfield, S., 1995, private communication
Krautter, J., Williams, R.E., 1989, Ap. J., **341**, 968
Kurtz, D.W., Vanden Bout, P.A., Angel, J.R.P., 1972, Ap. J., **178**, 701
Lloyd, H.M., O'Brien, T.J., Bode, M.F., et al., 1992, Nature, **356**, 222
MacDonald, J., 1980, MNRAS, **191**, 933

MacDonald, J., 1983, Ap. J., **267**, 732
MacDonald, J., 1984, Ap. J., **283**, 241
MacDonald, J., Fujimoto, M.Y., Truran, J.W., 1985, Ap. J., **294**, 263
MacDonald, J., Vennes, S., 1991, Ap. J., **373**, L51
MacDonald, J., Vennes, S., 1994, AAS meeting Minneapolis, MN, May 29 - June 2
Mason, K.O., Cordova, F.A., Bode, M.F., Barr, P., 1987, in "RS Ophiuchi (1985) and the Recurrent Nova Phenomenon", ed. M.F. Bode, VNU Science Press, Utrecht, p167
O'Brien, T., Kahn, F.D., Bode, M.F., 1987, in "RS Ophiuchi (1985) and the Recurrent Nova Phenomenon", ed. M.F. Bode, VNU Science Press, Utrecht, p177
Ögelman, H., Beuermann, K., Krautter, J., 1984, Ap. J., **287**, L31
Ögelman, H., Krautter, J., Beuermann, K., 1987, A&A, **177**, 110
Ögelman, H., Orio, M., Krautter, J., Starrfield, S., 1993, Nature, **361**, 331
Paresce, F., Livio, M., Hack, W., Korista, K., 1995, A&A, **299**, 823
Pequignot, D., 1996, these proceedings, p305
Prialnik, D., Shara, M., 1995, AJ, **109**, 1735
Prialnik, D., Kovetz, A., 1995, Ap. J., **445**, 789
Shanley, L., Ögelman, H., Gallagher, J.S., et al., 1995, Ap. J., **438**, L95
Starrfield, S., 1979, in "White Dwarfs and Variable Degenerate Stars", ed. H.M. Van Horn and V. Weidemann, p274
Starrfield, S., Truran, J.W., Sparks, W. M., Kutter, G.S., 1974, Ap. J., **176**, 169
Starrfield, S., Shore, S.N., Sparks, W.M., et al., 1992, Ap. J., **391**, L71
Starrfield, S., et al., 1996, in IAU Colloquium 152, "Astrophysics in the Extreme Ultraviolet", Cambridge University Press, Cambridge, in press
Stringfellow, S.G., Bowyer, S., 1994, in "Interacting Binary Stars", ed. A. Shafter, ASP Conf. Ser. **56**, 315
Warner, B., 1990 in IAU Colloquium 122, "Physics of Classical Novae", eds A. Cassatella, R. Viotti, Springer-Verlag, Berlin, p24

THE ROSAT OBSERVATIONS OF CLASSICAL NOVAE

M. ORIO[1,2], H. ÖGELMAN[1], S. BALMAN[1]
1. Department of Physics, U Wisconsin, Madison, USA
2. Osservatorio Astronomico di Torino, Italy

We observed a number of classical and recurrent novae in the Galaxy and the LMC with the ROSAT X-ray telescope and searched the archival data for other serendipitous observations. Preliminary results show that only 9 out of 37 observed objects were bright enough in X-rays to be detected with ROSAT, either in outburst or at quiescence.

Three basic mechanisms can cause X-ray emission from classical or recurrent novae. The first is hot hydrogen burning in a thin shell of the remnant envelope left on the white dwarf after the nova explosion. Hydrogen burning post–novae should be blackbody like emitters at nearly Eddington luminosity (as per the 'supersoft' X-ray sources). In our sample, only GQ Mus (Nova Mus 1983, see Ögelman et al. 1993; Shanley et al. 1995) and V1974 Cyg 1992 (Krautter et al. 1996) had these characteristics. Remarkably, among 10 LMC novae that had an outburst in the last 47 yr *none* was detected as a 'supersoft' X-ray source. The $3\,\sigma$ upper limits for the blackbody temperatures of the post-nova white dwarfs are mainly in the range $20\ldots30\,\mathrm{eV}$. A post–nova can also emit X-rays because of shocks occurring in the ejected shell (e.g. O'Brien et al. 1994). Three out of four classical novae that were observed in outburst displayed a hard X-ray component in the ROSAT energy band, which might be due to a shocked shell. Finally, X-ray emission is expected from quiescent nearby novae because of accretion. Only four nearby accreting sources were detected; the ROSAT upper limits for the non-detected quiescent novae are $L_{\mathrm{x}} < 10^{31}\ldots10^{32}\,\mathrm{erg\,s^{-1}}$, assuming a thermal plasma at kT = a few keV.

References

Krautter, J., Ögelman, H., Starrfield, S., et al., 1996, Ap. J., **456**, 788
O' Brien, T. J., Lloyd, H. M., Bode, M. F., 1994, MNRAS, **271**, 155
Ögelman, H., Orio, M., Krautter, J., Starrfield, S., 1993, Nature, **361**, 331
Shanley, L. Ögelman, H., Gallagher, J., et al., 1995, Ap. J., **438**, L95

A. Evans and J. H. Wood (eds.), Cataclysmic Variables and Related Objects, 289.

OBSERVATIONS OF NOVAE IN M51, M87, AND M101: A PRELIMINARY REPORT

A. W. SHAFTER[1], R. CIARDULLO[2], C. J. PRITCHET[3]
1. Dept. of Astronomy, San Diego State Univ., San Diego, CA 92182, USA
2. Dept. of Astronomy & Astrophysics, Penn State Univ., University Park, PA, 16802, USA
3. Dept. of Physics, Univ. of Victoria, Victoria, BC, Canada

Abstract. We report preliminary results from a new program of observations of novae in nearby galaxies. The principal goal of the program is to explore the population of nova progenitors through observations of galaxies having a wide range of Hubble types. Here we present preliminary results from observations of the spiral systems M51 and M101, and the giant elliptical, M87. Based on available data, we find no evidence that the luminosity specific nova rate varies across the Hubble sequence.

1. Introduction

Establishing the stellar population of close binaries is important to our understanding of how binary star formation and evolution may depend on the underlying stellar population. The study of novae is well suited for this purpose because their eruptions enable them to be readily identified, not only in the Galaxy, but, more importantly, in external galaxies as well. Despite its importance, the stellar population of novae is poorly understood. Galactic data appear to be consistent with novae belonging to an old disk population (Patterson 1984). However, it is clear that the Galactic data are of limited usefulness in establishing the population of novae because these data are heavily biased by selection effects, principally interstellar absorption. It seems clear that our best opportunity to explore the population of novae is to study their spatial distribution in nearby galaxies.

Novae have been observed in M31 since the pioneering work of Hubble in the early years of this century. In recent years novae have been studied in several additional galaxies. Among these galaxies are late-type systems

A. Evans and J. H. Wood (eds.), Cataclysmic Variables and Related Objects, 291–294.

such as the Magellanic Clouds (Graham 1979) and M33 (Della Valle et al. 1994). A testament to the current state of confusion concerning the population of novae is evidenced by comparing the summaries of nova rates in extragalactic systems recently published by Ciardullo et al. (1990a), and by Della Valle et al. (1994). These two groups come to very different conclusions regarding the population of novae despite reviewing the same data in some cases. Ciardullo et al. conclude that the luminosity specific nova rate is essentially independent of the underlying stellar population, while Della Valle et al. argue that disk dominated stellar systems such as M33 and the LMC are more prolific nova producers. In an attempt to resolve the confusion, we have decided to undertake a study of novae in several additional galaxies, including the massive disk-dominated systems M51 and M101. These galaxies have the advantage that their absolute nova rates should be significantly higher than their low mass counterparts M33 and the LMC, making an estimation of their luminosity specific nova rates more accurate and reliable. In this paper, we report preliminary results of observations of M51, M87, and M101.

2. Observations and results

Observations were obtained in 1994 May and 1995 April using the Kitt Peak National Observatory 4 m Mayall telescope. A 2048×2048 CCD detector was used at prime focus, yielding a field-of-view of $\sim 16 \times 16$ arcmin^2. The field-of-view was sufficient to cover essentially all of M51 and M87, and most of M101. Following Ciardullo et al. (1987), we imaged the galaxies in Hα and in the continuum (broad-band R). Total integration times were approximately 1 h for M51 and M101, and approximately 3 h for M87, with the precise integration times being set by the seeing and our desire to reach a limiting absolute magnitude in Hα of -7.5 (the Hα magnitude scale is defined in Ciardullo et al. (1987)). Observations of spectrophotometric standard stars revealed limiting Hα apparent magnitudes of 23.1, 23.5, and 23.0 for M51, M87, and M101. Adopting μ_o=31.03 and μ_o=28.65 for M87 and M101 (de Vaucouleurs 1993), and estimating μ_o=29.1 for M51, we reached limiting absolute magnitudes of -6.0, -7.5, and -5.7, for M51, M87, and M101, respectively.

For the spiral galaxies M51 and M101, novae were identified by blinking Hα images from 1994 and 1995. In the case of M87, we took advantage of the fact that novae can be identified by single epoch observations, as they are the only bright Hα sources thought to be present with the possible exception of low excitation planetary nebulae. The planetary nebulae were identified by their strong OIII λ5007 emission as part of another project (Ciardullo & Jacoby 1996), and were excluded from our sample. During the first two seasons of observation, a total of 6 novae were found in M51, with another 6 identified in M101. In the case of M87, 14 novae were discovered during a single epoch of observation in 1995 April.

To estimate the global nova rate of the galaxy, the observed nova rate must first be corrected for incompleteness caused by a bright and spatially varying background, for the fraction of the total galactic luminosity sampled, and extinction internal to the galaxy. We were insensitive to novae within a radius of 23 arcsec from the nucleus of M87 because of the high surface brightness. Based on the surface photometry of Cohen (1986), we estimate that our observations cover 78% of M87's luminosity. In the case of M51 and M101, the completeness is harder to quantify. Here we assume no correction for incompleteness due the background luminosity, but, in the case of M101 we estimate that our 16×16 arcmin^2 field excludes 10% of the galactic light. Extinction along the line of sight to these three galaxies is small, and has been neglected. Furthermore, we assume no significant absorption internal to M87, however, in the case of M51 and M101, the internal extinction is patchy and can be significant. Here we assume that the extinction internal to M51 and M101 is sufficient to obscure novae on the back sides of these galaxies. Thus, we apply correction factors of 50% to both our M51 and M101 observations. We acknowledge that the correction factors for the spiral systems are large and uncertain, and thus our initial estimates for the global nova rates for these galaxies will be more uncertain than that for M87.

Finally, to estimate the nova rate from multi-epoch observations, it is necessary to determine the effective survey time. As shown by Ciardullo et al. (1990b) the effective survey time can be estimated from the limiting absolute magnitude of the survey. From a fit of the time (τ_c) M31 novae remain brighter than a specified absolute magnitude in Hα (M_c), they found: $\log\ \tau_c = 0.48M_c + 0.56$. Here we depart slightly from Ciardullo et al. and adopt a non-linear fit to the same M31 data: $\log\ \tau_c = -0.0958 - 0.907M_c - 0.0843M_c^2$. The latter relationship gives a better estimate of the limited nova lifetime at faint magnitudes. For the single epoch of observation of M87, we find a global control time of 90 d. For two epochs covering M51 and M101, we find control times of 410 d and 436 d, respectively. These effective survey times in conjunction with the correction factors determined above lead to global nova rates of approximately 11 ± 4 novae per year for both the M51 and M101 systems, and 73 ± 19 for the giant elliptical galaxy, M87. We stress that the error estimates reflect only the Poission statistics of the number of novae detected, and do not reflect the uncertainties in the correction factors for incompleteness. These factors are particularly large for the spiral systems.

To compare nova rates from different galaxies it is necessary to normalize the observed nova rate by the luminosity of the galaxy. Since nova systems are composed of low-mass, low-luminosity stars, it seems most appropriate to normalize by the infrared luminosity of the galaxy. The luminosity at optical wavelengths is dominated by radiation from a relatively small number of young, massive, and very luminous Population I stars. For consistency with the previous studies of Ciardullo et al. (1990a) and Della Valle et al. (1994) we compute the luminosity specific nova rate, ρ_k, defined as the number of novae per year per 10^{10} $L_\odot$ in the infrared K band. The K magnitudes were computed in a consistent manner using the integrated B-band luminosities and the $B-V$ colors from the *RC3* (de Vaucouleurs et al. 1991), and the $V-K$ colors from the Hubble means of Aaronson (1978). Using the distances given earlier, we find $\rho_k = 1.20 \pm 0.55$, $\rho_k = 1.55 \pm 0.48$, and $\rho_k = 1.63 \pm 0.73$ for M51, M87, and M101, respectively.

3. Conclusions

In our initial analysis, we find no significant difference between the luminosity specific nova rates for the three galaxies studied. Furthermore, the nova rates appear to be consistent with those determined for other galaxies with measured nova rates. We conclude, therefore, in agreement with the results of Ciardullo et al. (1990a), that the present evidence does not support the contention that luminosity specific nova rates vary significantly across the Hubble sequence. The only caveat is that our preliminary nova rates for the spiral systems M51 and M101 are relatively uncertain and could be somewhat higher if we have underestimated the effects of internal extinction in these galaxies. Further work is in progress to improve our estimate the nova rates in these late-type galaxies.

Acknowledgements. We thank K. A. Misselt for his help with data analysis. This work has been supported by NSF grant AST-9315280 (A.W.S.).

References

Aaronson, M., 1978, Ap. J., **221**, L103
Ciardullo, R., Ford, H.C., Neill, J.D., et al., 1987, Ap. J., **318**, 520
Ciardullo, R., Ford, H.C., Williams, R.E, et al., 1990a, AJ, **99**, 1079
Ciardullo, R., Shafter, A.W., Ford, H.C., et al., 1990b, Ap. J., **356**, 472
Ciardullo, R., Jacoby, G.H., 1996, in preparation
Cohen, J., 1986, AJ, **92**, 1039
Della Valle, M., Rosino, L., Bianchini, A., Livio, M., 1994, A&A, **287**, 403
de Vaucouleurs, G., de Vaucouleurs, A., Corwin, H.G., et al., 1991 "Third Reference Catalogue of Bright Galaxies" (Springer-Verlag: New York).
de Vaucouleurs, G., 1993, Ap. J., **415**, 10
Graham, J.A., 1979, in "Changing Trends in Variable Star Research", IAU Colloq. 46, p96
Patterson, J., 1984, Ap. J. Supp., **54**, 443

ARE NOVAE STANDARD CANDLES?

SHLOMI PISTINNER[1], GIORA SHAVIV[2],
PETER H. HAUSCHILDT[3], SUMNER STARRFIELD[3]
1. Department of Physics, Ben-Gurion University, Beer Sheba 84105, Israel
2. Department of Physics, Technion, Haifa 32000, Israel
3. Department of Physics & Astronomy, Arizona State University, Tempe, AZ 85287-1504, USA

1. Introduction

Novae in outburst can be used as distance indicators. The distance measurement process is considerably simplified if novae could be regarded as standard candles. The term standard candles in the context of this paper means that the absolute magnitude at maximum for different novae is, to a good approximation, constant. The $t_3 - M_B$ diagram (Schmidet 1957) implies that this is the case for M31 and LMC novae (Della Valle & Livio 1995). However, this conclusion does not necessarily apply to non Local Group novae (Livio 1992; Della Valle & Livio 1995); for example the Virgo novae are not similar to Local Group novae (Della Valle & Livio 1995). Here we focus on a question of principle, namely: *Does the fact that novae have similar spectra imply that they have similar luminosities?*

Our discussion is limited to the very early phase in which a pseudo-photosphere forms within the ejecta, the so-called photospheric phase. Classical methods of stellar atmosphere analysis cannot be applied directly to nova atmospheres in this stage, because they are expanding and because of their large spatial extent. Hauschildt et al. (1992, 1994, 1995) have devised methods to simulate the expanding atmospheres of novae in this phase and thereby, obtained elemental abundances, effective temperatures, and atmospheric structures.

The detailed physics of a nova model atmospheres has been recently discussed by Hauschildt et al. (1995). The synthetic spectra they have calculated indicate that, over *certain wavelength ranges*, in which continuum processes dominate line processes, the spectra are relatively *insensitive* to

A. Evans and J. H. Wood (eds.), Cataclysmic Variables and Related Objects, 295–298.

the input parameters. In a recent work Pistinner et al. (1995) explored the sensitivity of early nova spectra to the input parameters. They identify spectral regions in which the continuum spectra are insensitive to the input parameters and certain wavebands in which they are sensitive to the input parameters. Pistinner et al. (1995) have identified the dilution factor as a small number, using it in a small parameter expansion they were able to find an analytical solution to the radiative transfer equation and provide predictions. These predictions account for the insensitivity of the spectra to the input parameters.

Nova atmospheres differ from normal stellar atmospheres by their motion. However, another important fact is frequently overlooked. Namely, nova atmospheres differ by being dominated by scattering and not by absorption in the continuum. The purpose of this paper is to study the effects resulting from this fact.

2. Continuum similarities of nova atmospheres

For the sake of brevity the element of motion will not be included in this presentation. The element of motion can be rigorously accounted for (Pistinner et al. 1995). Consider the static configuration shown in Fig. 1. Three distinct zones are marked: (a) the 'photosphere' (marked in Fig. 1 as ZA) where $\tau_{\text{ext}} = \tau_{\text{scat}} + \tau_{\text{abs}} \leq 1$ and $\tau_{\text{scat}} \gg \tau_{\text{abs}}$, where $\tau_{\text{ext}}, \tau_{\text{scat}}, \tau_{\text{abs}}$ are the optical depth for extinction, scattering and absorption in the continuum respectively; (b) a region (marked ZB) where $\tau_{\text{scat}} \geq \tau_{\text{abs}}$; and (c) a region (marked ZC) where absorption dominates. This particular configuration leads to the 'continuum similarity' (CS) of synthetic novae spectra over large range of wavelengths. By CS we mean that the continuum spectra are insensitive to the input parameters of the model atmospheres used to calculate the spectra. Specifically, significant changes to the input parameters do not result in significant changes to the emergent spectra. Another way to define CS is by noting that there are circumstances where different combinations of input parameters yield practically identical spectra.

An example of similar models is given in Table 1. In particular the spectra are insensitive to the luminosity (in the above case the luminosities change by a factor of 10!), implying that high similarities between the *observed* spectra of different novae *cannot* be used to argue that all nova have similar luminosities.

Therefore, any argument about novae being standard candles based on high spectral similarities is patently invalid. This does not imply that novae in the Local Group are not approximately standard candles. Other techniques, which involve the temporal changes in the brightness of the nova, may indeed show that these novae have comparable luminosities. However,

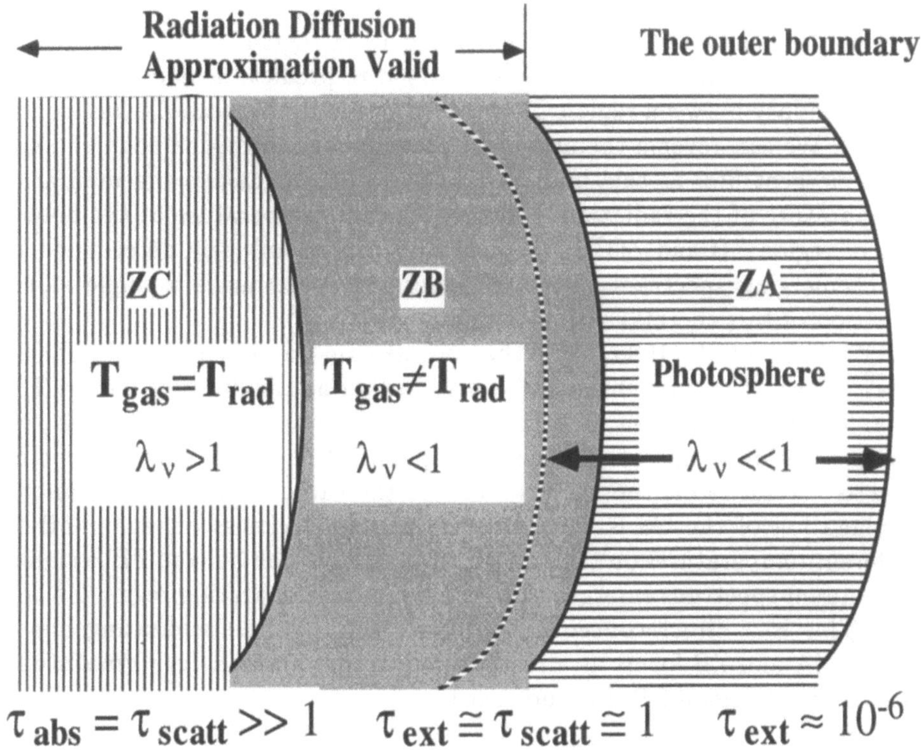

Figure 1. Different zones in a nova atmosphere.

TABLE 1. Models

T_{eff}	R_{phot}
15 000 K	$4.6\,10^{12}$ cm
18 000 K	$1.0\,10^{12}$ cm

the extrapolation of these relations to other galaxies has been shown to yield larger scatter than for the Local Group (Della Valle & Livio 1995).

To understand the underlying physical reason for the similarity effect consider first region ZA. In this region both the lines and the continua are nearly purely scattering due to the low density of the envelope. This is the well known Schuster (1905) problem. It can be shown that this region has rather small imprints of the continuum and the lines.

The continuum formation region is region ZB, in this region the memory of the blackbody radiation emerging from region ZC is processed. A synthetic spectrum calculation is performed to predict the emergent flux at the boundary between region ZB and ZA (since in this case ZA has a small effect). We can reverse the argument now and ask, provided that we know the emergent flux at this boundary, can we determine the thermalization temperature of the radiation – namely, the gas temperature at the boundary of region ZB and ZC? If this quantity can be uniquely determined for each emerging spectrum then each emerging spectrum is a single valued function of this thermalization temperature.

Pistinner et al. (1995) show that in this case the only physical quantity that governs the problem is a quantity known as the dilution factor. This quantity can be written as a *combination* of input parameters. Thus, the thermalization temperature is not a unique function of the input parameters and as a result so is the spectra. This is the cause of the CS. A heuristic explanation of the CS is that, by varying the thickness of region ZB for different temperatures on the surface separating region ZC and ZB, a similar continuum spectrum emerges from the boundary separation region ZA and ZB. This is a direct consequence of the fact that the separating line of region ZA and ZB forms as a result of scattering process. In normal stellar atmospheres region ZB does not exist!

3. Conclusion

Similar spectroscopic appearance of *observed* novae can not be used to argue that they have similar luminosities and thus are standard candles. The element abundance and the luminosity can be determined only to within a given transformation. The UV spectral range is not susceptible to spectral similarities and thus, high quality UV observations might aid in removing the above discussed ambiguity.

References

Della Valle, M., Livio, M., 1995, Ap. J., **452**, 704
Hauschildt, P.H., Wehrse, R., Starrfield, S., Shaviv, G., 1992, Ap. J., **393**, 307
Hauschildt, P.H., Starrfield, S., Austin, S., et al., 1994, Ap. J., **422**, 831
Hauschildt, P.H., Starrfield, S., Shore, M., et al., 1995, Ap. J., **447**, 829
Livio, M., 1992, Ap. J., **393**, 516
Pistinner, S., Shaviv, G., Hauschildt, P.H., Starrfield, S., 1995, Ap. J., **451**, 724
Schmidet, Th., 1957, Z. Astrophys., **41**, 182
Schuster, A., 1905, Ap. J., **21**, 1

TNR ON TOP OF ACCRETING WDS: 2-D SIMULATIONS

S. A. GLASNER[1], E. LIVNE[1], J. W. TRURAN[2]
1. Racah Institute of Physics, The Hebrew University, Jerusalem, Israel
2. University of Chicago, Lab. for Astrophysics and Space Research, Enrico Fermi Institute, Chicago, IL 60637, USA

Spherically symmetric models for classical novae as thermonuclear runaways (TNR) of hydrogen-rich, accreted, envelopes on top of cold white dwarfs (WD) are in very good agreement with observations (Starrfield, Sparks & Truran 1974, 1985; Prialnik, Shara & Shaviv 1978; Prialnik 1986; MacDonald 1980). Multi-dimensional investigations are needed since the burning rate is very sensitive to the temperature and small fluctuations in temperature can lead to large differences between the flash times of different points on the burning shell; these differences can be much larger than the rise time of the nova light curve. Convection is assumed to be the mechanism that couples the burning process at different points in the shell so that the overall phenomena is spherically symmetric.

We present here 2-D simulations of the accreted envelope for which the only perturbations are those induced by the turbulent nature of the model. Livne (1993) developed an implicit hydro code (VULCAN) which enables such simulation of long term subsonic flows. The initial model consists of a solar mass C–O WD at the stage where its luminosity is $L = 0.01\,L_{\odot}$.

Using a 1-D code, we evolve the accreting model in time up to a stage where the temperature at the base of the envelope is already $1\,10^8$ K. We now map the 1-D hydrogen shell on a 2-D grid with identical radial zones. A few underlying C-O zones from the 1-D model are also included in the 2-D simulation in order to permit overshoot mixing of C-O rich matter by the convective flow into the hydrogen rich accreted envelope. Within a few seconds several convective cells are built spontaneously from the numerical noise. During the first 70 s nothing violent happens, burning proceeds smoothly while the temperature at the base of the hydrogen shell rises from 10^8 K to $1.1\,10^8$ K. Due to the interaction between downward currents and burning, temperature fluctuations of 10–20 % develop within each burning shell. After 70 s the flow changes its manner and becomes violent. In addition to the intense burning all around the base of the envelope, lo-

A. Evans and J. H. Wood (eds.), Cataclysmic Variables and Related Objects, 299–300.

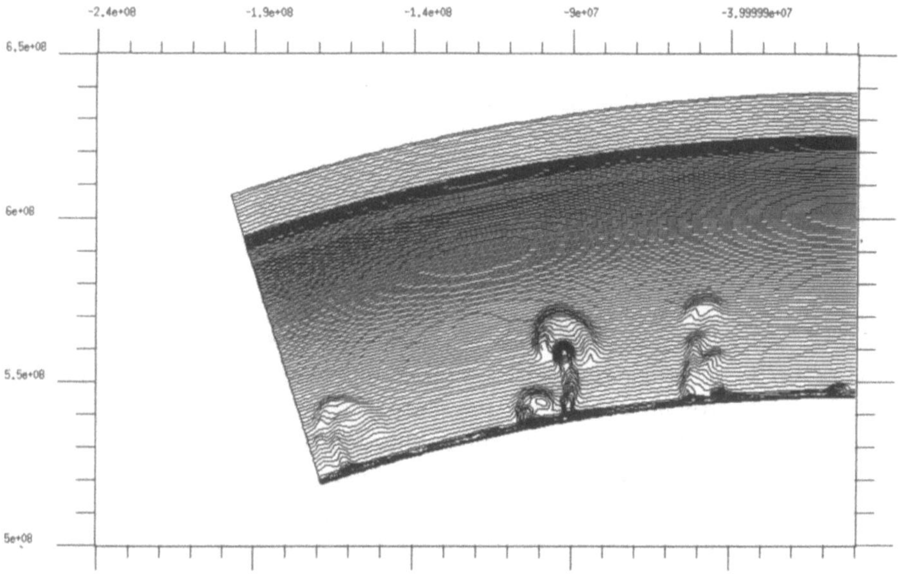

Figure 1. Contours of temperature at time 70 s; axes are cylindrical coordinates R and Z

cal eruptions of very fast burning appear at different points. Convective cells and convective velocities are larger than those predicted by 1-D Mixing Length Theory, convection seems to maintain the burning, rather than regulate it. Since the speed of sound is faster than the matter velocities, pressure waves travel faster than any flow pattern and the pressure quickly equalizes at each spherical shell. Therefore, although the burning process is not spherically symmetric, the overall expansion is almost spherical.

References

Livne, E., 1993, Ap. J., **412**, 634
MacDonald, J., 1980, MNRAS, **191**, 933
Prialnik, D., Shara, M.M., Shaviv, G., 1978, A&A, **62**, 339
Prialnik, D., 1986, Ap. J., **310**, 222
Starrfield, S., Sparks, W.M., Truran, J.W., 1974, Ap. J. Supp., **28**, 247
Starrfield, S., Sparks, W.M., Truran, J.W., 1985, Ap. J., **291**, 136

NOVA EJECTA ABUNDANCES RESULTING FROM MULTI–CYCLE EVOLUTIONARY CALCULATIONS

DINA PRIALNIK[1], ATTAY KOVETZ[2]
1. Department of Geophysics and Planetary Sciences
2. School of Physics and Astronomy
Tel Aviv University, Ramat Aviv 69978, Israel

1. Introduction

The theory of nova outbursts invokes accretion of hydrogen-rich material by a white dwarf (WD) in a close binary system, due to Roche lobe overflow of a red dwarf companion, leading to periodic outbursts — powered by thermonuclear runaways — that result in mass ejection. Observed characteristics of novae are reproduced by varying the values of three basic and independent parameters: the accreting white dwarf's mass, its core temperature, and the mass transfer rate. In a recent study (Prialnik & Kovetz 1995) we have carried out a systematic investigation of nova outbursts on WD progenitors composed of C and O in equal mass fractions, concluding that the entire range of *observed* nova characteristics could be accounted for. A subset of 34 out of 64 evolutionary sequences (obtained with different parameter combinations) reproduced classical novae, and here we focus on the break-up of the heavy element abundances in the ejecta of these models, in comparison with observations, based on a recent analysis of observed nova abundances by Livio & Truran (1994).

The source of heavy elements in nova ejecta is material dredged up from the WD core (by the diffusion-convection mechanism in our case). Most calculations to date have assumed C and O in equal mass fractions, as indicated by Iben & Tutukov (1985)'s extensive study of close binary evolution, which shows that a WD formed in such a system should have, essentially, a C to O ratio very close to 1. An interesting question is, to what extent is the WD composition reflected in the ejecta abundances? And, does the WD composition affect other nova characteristics? A different question is, do novae constitute a significant source of the relatively rare nucleides ^{13}C, ^{15}N and ^{17}O, as concluded by previous calculations and observations?

A. Evans and J. H. Wood (eds.), Cataclysmic Variables and Related Objects, 301–302.

2. Results of evolutionary calculations

2.1. HALF-C, HALF-O PROGENITORS

1. The observed range of variation of C, N, and O abundances is reproduced by the models, leaving out, however, the so-called 'ONeMg novae', which are believed to have ONeMg WD progenitors. Our calculations confirm that their abundances cannot be obtained with CO WD progenitors.
2. The general trend (exhibited by calculations) is for C and a fraction of the O to be converted into N. However, there are many cases of severely depleted O both in calculations and in observations.
3. In spite of the fact that the nuclear reactions network reaches up to ^{31}P, and of the high temperatures attained at outburst, the only abundant species in the ejecta are the isotopes of C, N and O. Neon, the next most important element observed in novae, is entirely insignificant. Hence, it appears that the only way to obtain *any* Ne in the ejecta would be to have it present in the WD composition. Indeed, according to Iben & Tutukov, CO WDs formed in close binary systems do contain a few percent of Ne.
4. There is no marked correlation between the calculated abundances and the basic nova parameters. Nevertheless, extremely high C should be associated with massive WDs and extremely high O with low accretion rates.
5. The isotopes ^{13}C and ^{17}O are in all cases significantly overabundant; ^{15}N, however, is sometimes greatly enhanced and sometimes underabundant.

2.2. PURE-C AND PURE-O PROGENITORS

1. The WD composition is *not* reflected in the abundances of the ejecta: the ejecta of pure-C WDs can be almost completely devoid of C; and similarly for pure-O WDs. The ejecta *cannot* be used to infer the WD composition.
2. There are two observed novae (HR Del and QU Vul) with exceptionally high O/N ratios and no detected C. Such abundances can only be obtained from a WD progenitor with a high O abundance in its outer layers.
3. The Ne abundance is still very low, even for the pure-O progenitors.
4. As to the effect of the progenitor's composition on other nova characteristics, the most important change consists of higher ejected (and accreted) masses — in some cases by a factor of 4 — obtained for pure-O WDs. These higher masses may be a step towards resolving the discrepancy between calculated and observationally derived masses of nova shells.

References

Iben, I. Jr., Tutukov, A.V., 1985, Ap. J. Supp., **58**, 661
Livio, M., Truran, J.W., 1994, Ap. J., **425**, 797
Prialnik, D., Kovetz, A., 1995, Ap. J., **445**, 789

THE GAMMA–RAY SPECTRUM OF CLASSICAL NOVAE

M. HERNANZ[1], J. JOSÉ[2], J. GÓMEZ[1]
1. Centre d'Estudis Avançats de Blanes (CSIC), Camí de Sta. Bàrbara s/n, 17300 Blanes (Girona), Spain
2. Dept. Física i Eng. Nuclear, EUPVG, UPC, Avda. Víctor Balaguer s/n, 08800 Vilanova i la G. (Barcelona), Spain

Abstract. We report on preliminary hydrodynamic simulations of thermonuclear runaways in the hydrogen-rich envelopes of accreting white dwarfs, leading to nova outbursts. A one-dimensional, Lagrangian, implicit, hydrodynamic code (Kutter & Sparks 1972) has been used to reproduce the *gross* features of a classical nova, from the onset of accretion, through the progress of the outburst up to the expansion stage, where significant mass loss takes place. Special interest is focused on the synthesis of ^{22}Na and ^{26}Al, for their interest in γ–ray line astrophysics. The γ–ray spectrum is built by means of a Monte Carlo simulation.

1. A nova outburst: from the onset of accretion to mass ejection

The evolution of a $1.25\,M_\odot$ carbon–oxygen white dwarf star, accreting matter at a rate of $\dot{M}_{\rm acc} = 1.6\,10^{-8}\,M_\odot\,{\rm yr}^{-1}$ has been computed. The accreted matter is assumed to be enhanced in CNO nuclei ($X = 0.46, Y = 0.13, Z = 0.41$) in order to power a fast nova. Detailed nucleosynthesis is obtained by following the evolution of 60 nuclei, ranging from ^{1}H to ^{28}Si, and linked through an updated network that includes more than 260 nuclear reactions. An envelope of $M_{\rm env} \simeq 1\,10^{-5}\,M_\odot$ is accreted on top of the white dwarf star through a hydrostatic phase. As soon as the temperature at the bottom of the accreted envelope reaches $T_{\rm b} \simeq 4\,10^7\,$K, convection settles and extends gradually towards the star's surface. A peak energy generation of $1.4\,10^{15}\,{\rm erg\,g^{-1}\,s^{-1}}$, followed by a peak temperature of $T_{\rm max} \simeq 1.9\,10^8\,$K are obtained, 529 yr from the beginning of the computations. At that time, convection reaches the surface, allowing the transport of short-lived β^+–unstable nuclei (^{13}N, ^{14}O, ^{15}O, ^{17}F) to the cool,

A. Evans and J. H. Wood (eds.), Cataclysmic Variables and Related Objects, 303–304.

outer layers. The ejection phase is powered by the delayed energy emission from the β^+–unstable nuclei. A mass $7.4\,10^{-6}\,M_\odot$ is ejected (72% of the envelope's mass) with velocities ranging from 958...4010 km s^{-1}. The maximum visual and bolometric magnitudes attained during the outburst are $M_{\rm vis} = -6.4$ and $M_{\rm bol} = -9.4$. The abundances, by mass, of ^{22}Na and ^{26}Al synthesized are $6.3\,10^{-4}$ and $9.8\,10^{-3}$, respectively.

2. Gamma-ray spectrum

During the nova outburst, several unstable nuclei are synthesized with abundances high enough to produce a strong γ–ray emission in their decay (Livio et al. 1992, Starrfield et al. 1992). These nuclei are divided into (i) short lived positron emitters which take part in the CNO cycle (^{13}N, ^{14}O, ^{15}O and ^{17}F) and (ii) ^{26}Al and ^{22}Na (Weiss & Truran 1990). Only the emission produced by the decay of ^{22}Na might be detectable by COMPTEL or INTEGRAL. The ^{22}Na decays following the chain $^{22}\mathrm{Na} \rightarrow {}^{22}\mathrm{Ne}$, producing a well defined γ–ray line at 1270 keV, and positrons which annihilate emitting 511 keV photon pairs. By means of a Monte Carlo code (Ambwani & Sutherland 1988), the spectra resulting from the interaction of these photons with the nova envelope have been calculated. The following physics is included in the code: (i) e^-–e^+ pair production, (ii) photoelectric absorption, (iii) non-conservative Compton scattering, (iv) e^-–e^+ annihilation, and (v) positronium formation. The calculations have been carried out at different times starting at 2300 s after the onset of the outburst. In all cases, the spectra are obtained by using the channels defined by the expected resolution of INTEGRAL and also, for comparison purposes, the spectra have been convolved with the COMPTEL (CGRO) spectral response. At 250 pc the continuum would not be detectable by COMPTEL, while INTEGRAL would detect part of it from several hours to 1 or 2 d after outburst. Concerning the 1270 keV line, an intensity of $2\,10^{-5}$ counts cm^{-2} s^{-1} would be obtained for our model for a nova at a distance of 1 kpc.

References

Ambwani, K., Sutherland P., 1988, Ap. J., **325**, 820
Kutter, G. S., Sparks, W. M., 1972, Ap. J., **175**, 407
Livio, M., Mastichiadis, A., Ogelman, H., Truran, J. W., 1992, Ap. J., **394**, 217
Starrfield, S., Shore, S. N., Sparks, W. M., et al., 1992, Ap. J., **391**, L71
Weiss, A., Truran, J. W., 1990, A&A, **238**, 178

THE NATURE OF GQ MUS AS DEPICTED FROM PHOTOIONIZATION MODELS OF THE SHELL

D. PÉQUIGNOT, C. MORISSET
Observatoire de Paris-Meudon,
F-92195 Meudon Principal Cedex, France

Consequences of a detailed photoionization model study of the shell ejected by Nova Mus 1983 (GQ Mus) are presented for the period from early 1983 to 1990 (Morisset & Péquignot 1996).

The drastic time variation of the emission line spectrum, including the transition from a nebular to a coronal stage in 1986, can be quantitatively understood in terms of a smooth evolution of both the expanding shell and the hot thermal source. This transition is due to a decrease of the shell density with time, and not to an increase of the source effective temperature.

The model shell comprises two components so that not all directions become optically thin to ionizing radiation simultaneously. The range of density and geometrical thickness of the *emitting* material is of the order of a factor 2.

If the *emitting* mass of the shell were conserved after 1984, the gas density would have to decrease much faster than t^{-3}, where t is the time since the explosion, implying a continuous *differential acceleration* of the material across a cloud and, according to models, a *steady* t^{-1} *decrease* of L_*, the luminosity of the central source. Alternatively, it is advocated that the rapid decline of Hβ with time after 1984 is due to an *evaporation* of the clouds in an intercloud medium heated by a fast wind from the source. Consistent with the existence of this wind, L_* is then found from photoionization models to remain nearly constant for at least five years.

Assuming blackbody emission, the central-source power $L_* = 1.75 \pm 0.2\,10^{38}\,\mathrm{erg\,s^{-1}}$ and the effective temperature $T_* = 2.7 \pm 0.2\,10^5\,\mathrm{K}$ are determined in 1984 without reference to either the distance to GQ Mus or the precise reddening correction. T_* is slightly smaller using stellar atmosphere models.

By the end of the radiation-bounded period (day 500), the mass of the shell is

$$\frac{M_{\rm shell}}{\rm M_\odot} = (7.3 \pm 1)\,10^{-5}\left(\frac{D}{3\,\rm kpc}\right)^2 10^{3.1(E_{\rm B-V}-0.5)},$$

A. Evans and J. H. Wood (eds.), Cataclysmic Variables and Related Objects, 305–306.

D being the distance to the nova and $E_{\rm B-V}$ the interstellar differential extinction. The shell is composed of $\sim 2\,10^4$ clouds filling altogether $\sim 1\%$ of the volume, the size of an individual cloud being comparable to the sound-crossing length at the temperature of the photoionized material. The expected fraction of clouds shielded from the source by other clouds is consistent with this fraction being identified with the thicker component of the model.

The post-nova shrank down to Roche lobe dimensions in less than two months. Three months after the explosion, T_* already exceeds $1.7\,10^5$ K. In a second outburst spanning days 110...170, the remnant swelled back to common-envelope dimensions. Following this event, the evolution of the source slowed down considerably.

The slow rise of T_* from the 6th month after outburst to turn off 10 yr later, is marginally consistent with hydrogen burning and no accretion for a white-dwarf mass $M_{\rm WD} \sim 0.8\,{\rm M}_\odot$. However both the luminosity derived from photoionization models of the 1984 data and the initial behaviour of the source indicate $M_{\rm WD}$ in excess of $1.10\,{\rm M}_\odot$.

A large $M_{\rm WD}$ may in principle be reconciled with the evolution of T_* and H-burning if a continuous replenishment of the white-dwarf envelope from a hydrogen-rich companion is postulated. However the H-burning scenario would seem to require a pathologically large and fine-tuned accretion rate to account for observation. Accretion power itself is not able to sustain the luminosity.

Alternatively, it is speculated that the exceptionally long duration of the luminous phase may be due to a rare event, namely *post-outburst burning of helium*, perhaps through the reaction $^{12}{\rm C}(\alpha, \gamma)^{16}{\rm O}$ in a He-C-O layer built up as a result of elemental diffusion during many previous cycles of the nova. In this scenario, the second outburst spanning days 110...170 after explosion might be the consequence of a *helium flash*, developing soon after the first ebbing of the common envelope and soon before the normal end of H-burning.

References

Morisset C., Péquignot D., 1996, A&A, in press.

NOVA OUTBURSTS ON ROTATING OBLATE WHITE DWARFS

A. D. SCOTT
Department of Physics, Keele University, Keele, ST5 5BG, UK

1. Introduction

Observations of spatially resolved nova shells and reconstructions from emission line profiles show that a ring/blob asymmetry in the ejecta of novae is very common (Martin 1989). The polar blobs expand more rapidly than the equatorial rings, giving the ejecta an overall prolate geometry.

Observations of the formation of carbonaceous and silicate dusts in the same nova provide unambiguous evidence that abundance gradients exist in the ejecta. Carbonaceous dusts form where carbon is more abundant than oxygen and vice versa for silicates (Evans 1990; Gehrz et al. 1992; Smith, Aitken & Roche 1994; Snijders et al. 1987).

2. Proposition — shaping of ejecta

1. Since the polar radius of an oblate white dwarf is less than the equatorial radius, the effective surface gravity and the pressure at the core–envelope interface at the equator is less than at the poles. The electron degeneracy at the base of the accreted envelope and hence the peak temperature of the thermonuclear runaway (TNR) will therefore be stronger at the poles, all other things being equal (Livio 1992). Where the peak temperature is higher, the violence of the TNR outburst is stronger. Hence material will be ejected at higher velocities in the polar regions than at the equatorial regions, producing prolate ejecta.
2. The larger effective gravity and smaller white dwarf radius at the poles implies that the **critical point** in stellar wind theory is located at a smaller radius at the poles than at the equator. Hence the terminal wind velocities will be greater at the poles than at the equator, which again produces prolate ejecta (Kato & Hachisu 1994, for example).

A. Evans and J. H. Wood (eds.), Cataclysmic Variables and Related Objects, 307–308.

3. The nova wind is expanding while rotating. The polar region is able to expand freely while the equatorial flow feels a progressively stronger centrifugal barrier (Shore 1992). This follows from the de Laval nozzle analogy used in stellar wind theory. Centrifugal forces act to partially counter gravitation toward the equator; the equatorial wind will therefore be slower than the polar wind.

3. Proposition — abundance gradients

Evolutionary calculations of the TNR off ONeMg white dwarfs of different masses (Starrfield et al. 1992) show that the C/O ratio increases with the degree of electron degeneracy at the base of the envelope at the start of the TNR. It is therefore proposed that the C/O abundance ratio is greater in the polar blobs than in the equatorial rings.

Thus, for the first time, the shape of the nova ejecta *and* the abundance gradients can be explained *simultaneously* as a consequence of the same phenomenon, i.e. the rotation and distortion of the white dwarf.

Observational investigations, whether of the underlying binary system or the resolved ejected shell of the nova, are now needed to test this hypothesis. Theoretical investigations, combining hydrodynamical simulations of the nova outburst with models of nucleosynthesis and energy production as a function of latitude on the white dwarf need to be constructed.

Acknowledgements. I thank Professor A. Evans, my thesis supervisor, for his support and encouragement, and Dr. T. Shahbaz for getting me interested in rotating white dwarfs. I also thank Drs. W. Welsh and H. W. Duerbeck for useful discussions. The STARLINK project and the Department of Physics at the University of Keele are thanked for their provision of facilities.

References

Evans, A., 1990, in "Physics of Classical Novae" Proc. IAU Coll. 122, eds Cassatella, A., Viotti, R., Springer–Verlag, p252

Gehrz, R.D., Jones, T.J., Woodward, C.E., et al., 1992, Ap. J., **400**, 671

Kato, M., Hachisu, I., 1994, Ap. J., **437**, 802

Livio, M., 1992, Ap. J., **393**, 516

Martin, P.G., 1989, in "Classical Novae", eds Bode, M. F., Evans A., Chichester: Wiley & Sons, p73

Shore, S.N., 1992, An Introduction to Astrophysical Hydrodynamics, San Diego: Academic Press

Smith, C.H., Aitken, D.K., Roche, P.F., 1994, MNRAS, **267**, 225

Snijders, M.A.J., Batt, T.J., Roche, P.F., et al. 1987, MNRAS, **228**, 329

Starrfield, S., Shore, S.N., Sparks, W.M., et al., 1992, Ap. J., **391**, L71

OPTICAL IMAGING OF OLD NOVA SHELLS

T. J. O'BRIEN, A. J. SLAVIN
Astrophysics Group, Liverpool John Moores University, Byrom St, Liverpool L3 3AF, UK

Abstract. A deep, predominantly Hα, imaging survey of 37 classical nova remnants has just been completed using the William Herschel and Anglo-Australian Telescopes. This has resulted in the detection of new nebular remnants and the resolution of detailed structure in previously known remnants. These data indicate a correlation between nova speed class and the ellipticity of the resulting remnants – those of faster novae tend to comprise randomly distributed clumps of ejecta superposed on spherically symmetric diffuse material, whilst slower novae produce more structured ellipsoidal remnants with at least one, and sometimes several, rings of enhanced emission.

1. Motivation

During outburst classical novae eject $\sim 10^{-5} \ldots 10^{-4}$ $M_\odot$ of material at speeds of up to thousands of km s^{-1}; after outburst they will therefore be surrounded by an expanding shell of ejecta. The study of these nebular remnants is important for several reasons, for example distance determination via expansion parallax, their contribution to the interstellar medium, elemental abundances and the nature of the outburst mechanism, and the study of physical processes such as dust formation. In addition their structure appears to reflect processes which take place during outburst on the scale of the binary system, the study of which may have wider applicability to, for example, planetary nebulae and some supernovae.

Despite their importance nebular remnants have been detected around only a small fraction of classical novae (see Fig. 1), whilst images of those remnants which have been detected are dispersed through the literature and of variable quality. The first results of a new deep optical imaging survey of novae in the northern and southern sky are presented in this paper and in Slavin, O'Brien & Dunlop (1995).

A. Evans and J. H. Wood (eds.), Cataclysmic Variables and Related Objects, 309–312.

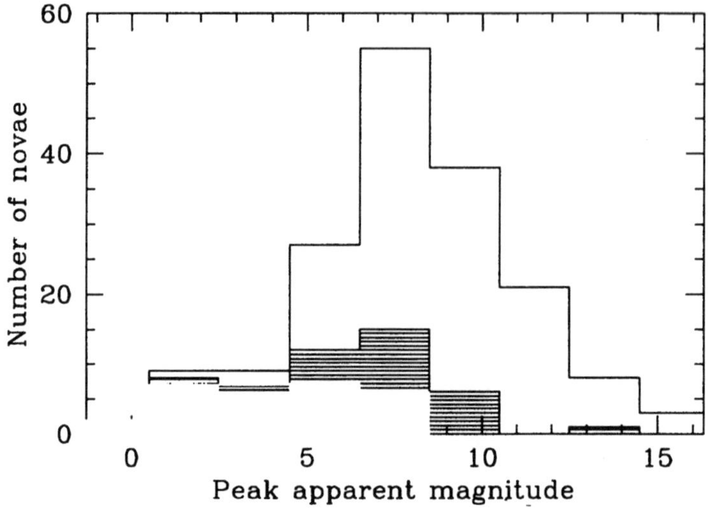

Figure 1. The frequency distribution of classical novae as a function of their peak apparent magnitude. The underlying histogram contains 172 classical novae; data from Bode & Evans (1989) and Duerbeck (1987). The shaded histogram shows the subset containing 51 novae which have been searched for shells, and the solid histogram shows 28 of these which have actually had shells detected. See Sect. 2, Slavin et al. (1995), Wade (1989).

2. The observations

Observations were made using CCD cameras on the 4.2 m William Herschel Telescope (1993 September 9–11) and the 3.9 m Anglo-Australian Telescope (1995 February 23–25). A total of 37 novae were imaged using Hα narrow-band filters. New features were resolved in 12 previously known remnants whilst 6 new nebular remnants (NQ Vul, V450 Cyg, HS Pup, DY Pup, RR Cha, V842 Cen) were discovered. There were 16 non-detections. Images of extended objects were taken in several narrow bands around Hα, giving some crude velocity resolution, as well as in other emission lines, including [N II] and [O III].

Fig. 2 shows the Hα images of the shells of DQ Her and RR Pic. It also shows the result of subtracting the elliptically-symmetric component of each shell (see Slavin et al. 1995 for details). These images reveal that in each nova the faint halo surrounding the bright shell contains long tails originating in clumps within the shell.

The new remnants were discovered by comparing the novae with stars of similar peak brightness on the same CCD frame. For example, Fig. 3 shows contour plots of DY Pup and V359 Cen; it is obvious that DY Pup is extended and that V359 Cen is indistinguishable from a typical point source.

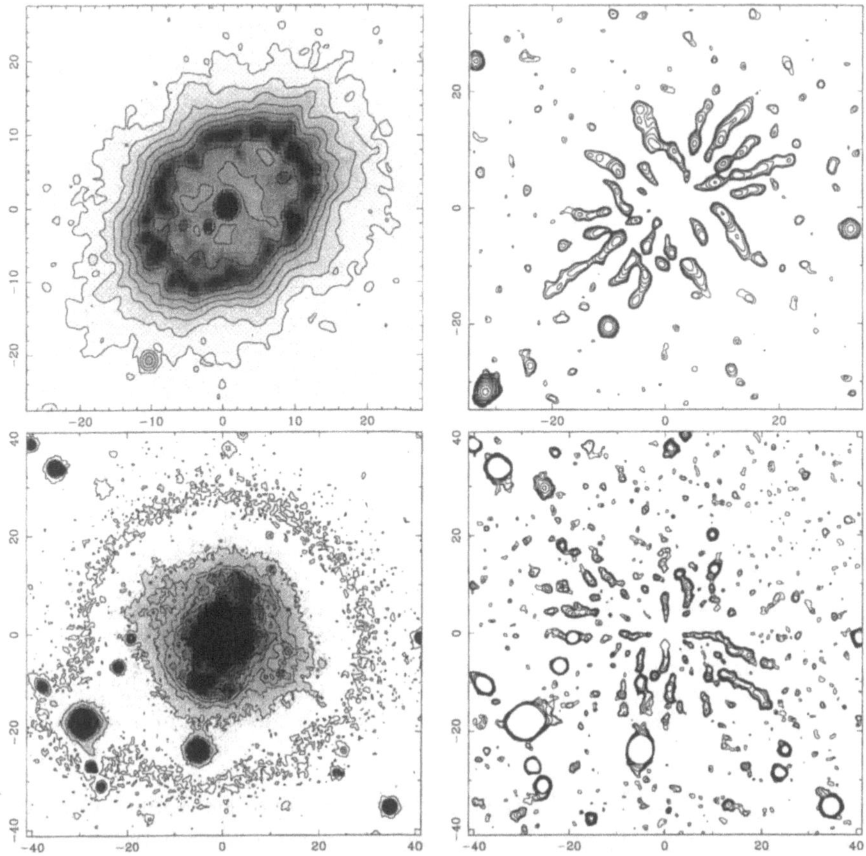

Figure 2. **Hα images of DQ Her and RR Pic showing the ellipsoidal clumpy shells, halos and ring structures. Each is accompanied by a contour plot of the 'tails' revealed by removing the elliptically-symmetric component of the remnant. Axes are marked in arcsec, north is up, east to the left.**

3. Discussion

These data show that the fastest novae are surrounded by almost spherically-distributed diffuse material containing typically four to six brighter clumps (each clump presumably composed of much smaller blobs). In contrast, the remnants of slower novae have sculpted axisymmetric structures comprising of ellipsoidal shells and equatorial rings; tropical rings are also seen in several objects. Assuming the equatorial rings are circular and lie in the orbital plane of the system, orbital inclinations can be derived, e.g. $81° \pm 4°$ for DQ Her, $58° \pm 14°$ for FH Ser, $70° \pm 4°$ for RR Pic.

This work has posed several interesting problems which remain to be resolved by further observation and modelling. For example, the exact mech-

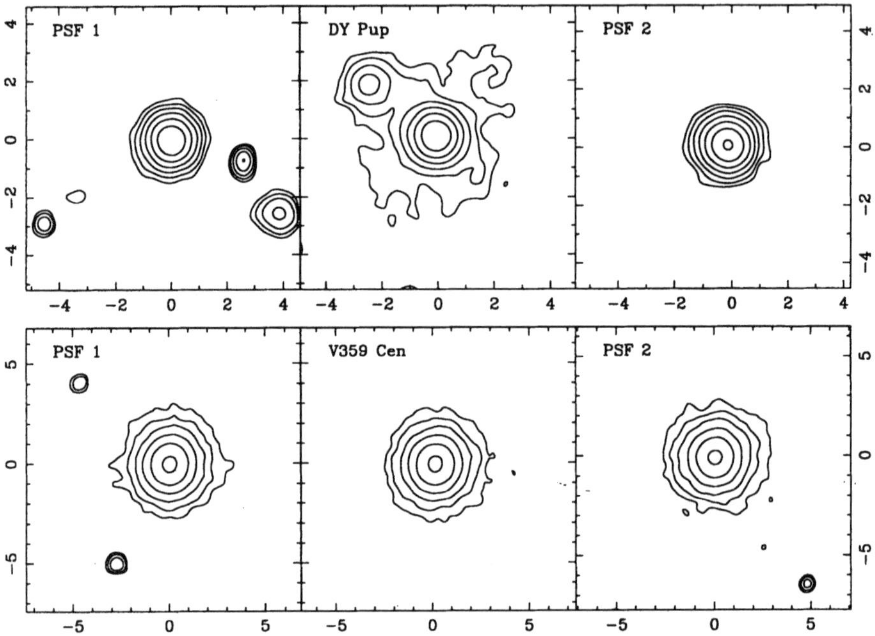

Figure 3. Contour plots of DY Pup and V359 Cen. In each case the nova is compared with two stars taken off the same frame; contour levels are set at fixed fractions of the peak brightness of each object. Axes are marked in arcsec, north is up, east to the left.

anism which shapes nova shells is unclear (although see Lloyd, O'Brien & Bode 1996). As a study of old nova shells is essentially archaeological, photometry and spectroscopy are required throughout the outburst in order to study ejection and shaping as it takes place; robotic telescopes may have an important role to play here. Further high-resolution imaging, using the HST for example, is necessary to better resolve structures such as clumps, rings and tails and to complement radio imaging of the early phases of the outburst. In addition, spectroscopy of old shells will provide kinematical data, aiding distance determination and calibration of the maximum magnitude-rate of decline relation. It will also help determine whether tails represent clumps being ablated by a wind or regions in the halo shadowed from the central ionising source.

References

Bode, M.F., Evans, A., 1989, Classical Novae, Wiley & Sons Ltd., Chichester

Duerbeck, H.W., 1987, Space Science Rev., **45**, Nos. 1–2

Lloyd, H.M., O'Brien, T.J., Bode, M.F., 1996, these proceedings, p313

Wade, R.A., 1989, in "Physics of Classical Novae", Proc. IAU Coll. 122, eds. Cassatella A. Viotti R., p. 179, Springer-Verlag, Berlin

Slavin, A.J., O'Brien, T.J., Dunlop, J.S., 1995, MNRAS, **276**, 353

SHAPING OF NOVA SHELLS BY BINARY MOTION

H. M. LLOYD, T. J. O'BRIEN, M. F. BODE
Astrophysics Group, Liverpool John Moores University, Byrom Street, Liverpool L3 3AF, UK

Abstract. Several years after outburst, extended optical emission can be seen around many classical novae. Common features include shells, polar 'blobs', equatorial rings and, at least in the case of DQ Her (1934), tropical rings above and below the 'equator'. We present hydrodynamic calculations of the dynamical effects of the underlying binary system on the material ejected during outburst, and show that many of the features observed in the optical remnants of novae can be reproduced. Polar blobs, banded shells and spherical shells can all be obtained in the model, depending on the speed class of the nova.

There is strong evidence to support the fact that the velocity of ejection in novae increases with time (see e.g. Friedjung 1994) which will naturally lead to a situation in which fast moving ejecta will catch up with slower moving material ejected earlier in the outburst, sweeping it up into a shell and shock-heating it to X-ray temperatures. A spherically symmetric calculation of this 'interacting winds' model was performed by O'Brien, Lloyd & Bode (1994) in order to model the X-ray emission from V838 Her. Here we extend the model to axisymmetry to allow for the effects of the binary system, which will deposit energy and angular momentum into the flow via a gravitational drag force. The hydrodynamic code uses a spherical polar grid, and the ϕ velocity is allowed to be non-zero (the '2.5-D' approximation). Previous 2-D calculations of these effects have only been performed for slow novae (Livio et al. 1990).

The mass-loss is assumed to take place in a wind, with a velocity which increases with time. Orbital energy and angular momentum are added into the flow at the orbital radius of the secondary star. Four runs of the code were performed with velocities appropriate for slow (Run 1), moderately fast (Runs 2 and 3) and fast (Run 4) novae. Fig. 1 shows model images obtained from the four runs (at times after the acceleration of the wind

A. Evans and J. H. Wood (eds.), Cataclysmic Variables and Related Objects, 313–314.

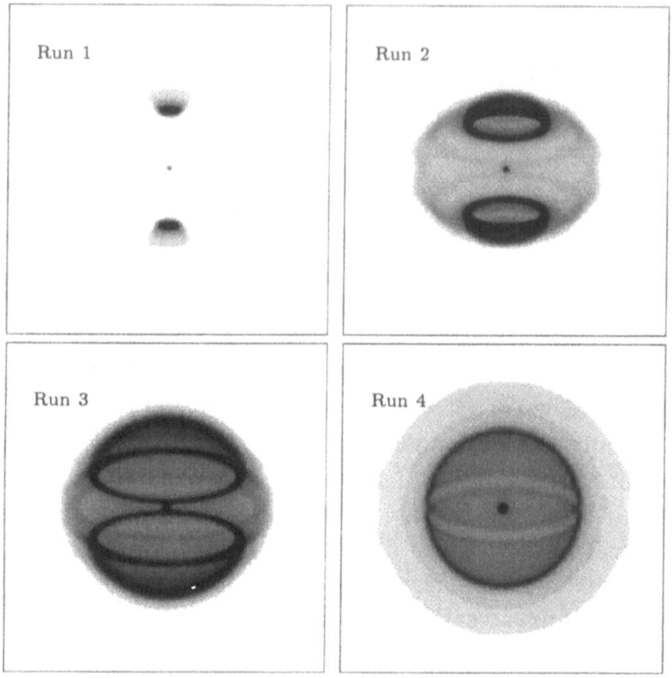

Figure 1. Synthetic images obtained from the hydrodynamic models.

is complete) by taking the emission coefficient to be proportional to the square of the gas density. Run 3 shows a banded shell similar to that seen in DQ Her (Slavin, O'Brien & Dunlop 1995) whereas Run 2 shows polar rings. Both these structures are formed via the mechanism described by Lloyd, O'Brien & Kahn (1995) in their model for the nebula of SN1987a. The slow wind is shaped into a double cone structure which is then swept up by the fast wind to give a pair of rings. The effect of the drag force is greater for lower velocities and, in the case of Run 1, the cone is forced to collapse onto the symmetry axis and a bipolar structure is obtained (such as that seen in HR Del, Slavin, O'Brien & Dunlop 1994). The mass loss in the fast nova model (Run 4) is relatively unaffected by the drag force and results in an almost spherical shell.

References

Friedjung, M., 1994, Mem. S. A. It., **65**, 1
Livio, M., Shankar, A., Burkert, A., Truran, J. W., 1990, Ap. J., **356**, 250
Lloyd, H. M., O'Brien, T. J., Kahn, F. D., 1995, MNRAS, **273**, L19
O'Brien, T. J., Lloyd, H. M., Bode, M. F., 1994, MNRAS, **271**, 155
Slavin, A. J., O'Brien, T. J., Dunlop, J. S., 1994, MNRAS, **266**, L55
Slavin, A. J., O'Brien, T. J., Dunlop, J. S., 1995, MNRAS, **276**, 353

ULTRAVIOLET SPECTROPOLARIMETRY OF CLASSICAL NOVAE IN OUTBURST: EVIDENCE FOR ASPHERICAL EJECTA

J. J. JOHNSON[1], C. M. ANDERSON[1], K. S. BJORKMAN[1], A. D. CODE[1], G. K. FOX[1], A. J. WEITENBECK[1], K. WOOD[1], R. J. EDGAR[1], W. T. SANDERS[1], B. BABLER[1], M. MEADE[1], N. E. B. ZELLNER[1], G. C. CLAYTON[2], R. SCHULTE-LADBECK[3], O. L. LUPIE[4]
1. UW-Madison, Space Astronomy Lab
2. Univeristy of Colorado,CASA
3. University of Pittsburgh
4. Space Telescope Science Institute

Abstract. We used the Wisconsin Ultraviolet Photo-Polarimeter Experiment during the Astro-2 mission aboard the Space Shuttle Endeavour, to obtain ultraviolet spectropolarimetry of three classical novae that had recently gone into outburst. All three novae appear to have intrinsic polarization, with polarization changes across emission lines. This result indicates that, geometrically, the ejecta were quite aspherical.

1. Observations

On 1995 March 2, the Astro-2 mission was launched aboard the Space Shuttle Endeavour. Part of the Astro-2 package was the Wisconsin Ultraviolet Photo-Polarimeter Experiment (WUPPE). WUPPE is a 0.5 m Cassegrain telescope with a spectropolarimeter that covers the wavelength range 1450...3300 Å with spectral resolution 5 Å. One of the many science programs that the WUPPE team had was to observe cataclysmic variables. We were very fortunate to have three bright classical novae go into outburst within a month and a half of launch. These were Nova Cir 1995, Nova Aql 1995, and Nova Cen 1995. They were observed, respectively, 44, 26, and 15 d after discovery.

After we subtracted the polarization due to the interstellar medium, there was still polarization in the continuum as well as variations in the polarization and position angle through the emission lines. In particular we

A. Evans and J. H. Wood (eds.), Cataclysmic Variables and Related Objects, 315–316.

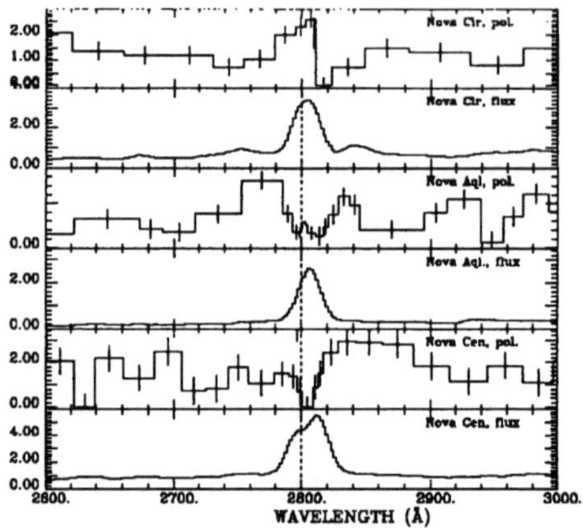

Figure 1. Polarization and flux through the Mg II line

saw that the three novae were all different in their behavior through the Mg II line at 2800 Å. Fig. 1 shows the variation in line shape and polarization behavior through this line. In particular, note how the polarization rises through the emission line in Nova Cir. Since this line is formed via resonance scattering, the implication is that the geometry of the scattering region differs markedly from the geometry of the region producing the continuum polarization and also differs from the geometry of the region producing the other emission lines.

2. Conclusions

Conclusions at this stage are tentative as the analysis of our data has just begun. However, some features are already evident:
(i) all three novae show evidence of intrinsic polarization which means that there is asymmetry in the ejecta;
(ii) the lines are formed in a region separate from the source of the polarized light *except* in the case of the Mg II line for Nova Cir, which appears to be formed in yet another aspherical region.
The combination of spectral and polarization analysis will allow us to build up a 3-D 'snapshot' picture of the nova eruption.

A UNIFIED MODEL FOR THE RADIO AND OPTICAL IMAGES AND SPECTRA OF NOVA V1974 CYG 1992

R. M. HJELLMING
National Radio Astronomy Observatory†
Socorro, NM 87801-0387, USA

Abstract. A model for nova shells which is a combination of wind model and spherically symmetric, Hubble flow shell model, fits the light curve, image, and nebular line shape data for Nova V1974 Cyg 1992. The results show that after outburst the shell is ejected by a wind-like outflow with a linear velocity gradient, and that after 180 d this outflow stops. The shape of the elliptical inner boundary at this time determines the subsequent emission measure structure seen in images, light curves, and nebular line profiles.

1. Introduction

The first classical novae to be observed during a major portion of their radio light curves were HR Del 1967, FH Ser 1970, and V1500 Cyg 1975 (Hjellming et al. 1988). Two models that have been used to fit the data for these and other novae: a variable wind model (Kwok 1983) where the ejecta are continuous with a constant velocity and a mass loss rate which is proportional to the inverse of the time since outburst; and a Hubble flow shell model (Hjellming et al. 1988) where a finite shell is ejected with a velocity gradient proportional to the radius. Both models assumed spherical symmetry. In this paper we discuss the modifications that are needed to fit the images and spectral line shapes of Nova V1974 Cyg 1992.

Until now models used to interpret radio light curves have assumed spherical symmetry. For the variable wind model (Kwok 1983) a constant velocity ($v_{\rm wind}$) wind with a mass loss rate inversely proportional to the time since outburst ($t - t_0$) gives a density profile in the nova shell described by $\rho(t) = \dot{M}(t_0)/[(t/t_0 - 1)v_{\rm wind}\, 4\pi\mu R^2)]$, where R is the radius, μ is the mean

†The National Radio Astronomy Observatory is operated by Associated Universities, Inc. under a cooperative agreement with the National Science Foundation.

A. Evans and J. H. Wood (eds.), Cataclysmic Variables and Related Objects, 317–320.

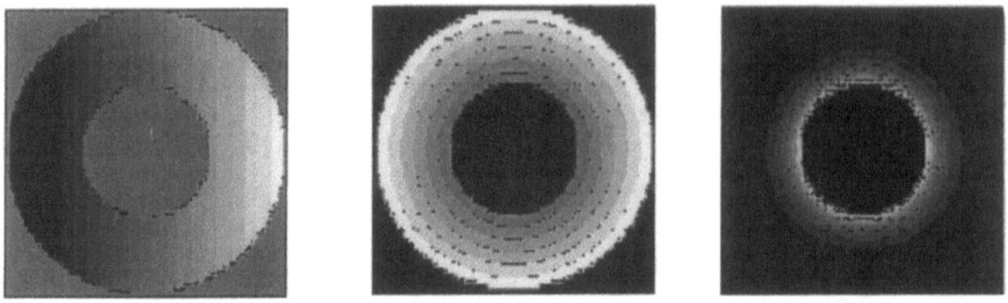

Figure 1. Images of the internal structure of a shell with a linear velocity gradient showing: (left) radial velocities as seen by an observer to the left of the page; (centre) a cross-section of the velocity struction in the shell, where the velocities are proportional to radius; and (right) the emission measure image of the shell as seen from any direction.

molecular weight, and $\dot{M}(t_0)$ is the initial mass loss rate. For the Hubble flow shell model (Hjellming et al. 1988) a finite shell is ejected with inner and outer radii $R_1 = v_1(t - t_0) + R_{10}$ and $R_2 = v_2(t - t_0) + R_{20}$, where $v_2 \geq v_1$, and inside the shell $v \propto R$, so $\rho(t) = M_{\rm shell}/[4\pi\mu R^2(R_2(t) - R_1(t))]$. Both models give density profiles that are inversely proportional to both $(t - t_0)$ and R^2, so they mostly have the same properties. However, for the pure wind model a finite inner radius never develops and there is no velocity gradient across the shell; the lack of finite inner radius at later times implies that a completely optically thin, detached shell does not develop, and the late decay curves do not match the observations (Hjellming 1990, 1995).

For a spherically symmetric nova shell with a linear velocity gradient and a finite inner radius, the radial velocity, velocity (or radius), and emission measure structure of the shell is shown in Fig. 1. These figures indicate why the observable properties of optically thin shells are largely determined by the mass near the inner shell boundary. If one combines the emission measure and radial velocity structure to predict a line profile, there is only one nebular line shape predicted by these models: $\phi(v) = \{[v_1(t)v_2]/\{4[v_1(t) - v_2]\}\}[v^{-2} - v_2^{-2}]$ for $v_1(t) \leq |v| \leq v_2$, and $\phi(v) = \phi[v_1(t)]$ for $|v| \leq v_1$. While v_2 is a constant, $v_1(t)$ can vary until shell ejection stops and the inner boundary is formed.

Another important point relates to the generally higher masses derived from radio light curves or images (Hjellming 1995). Part of the reason for this is obvious from Fig. 1, because while the shell has equal mass in equal increments of radius, the four inverse powers of radius determining the contributions to the emission measure mean that any observations of the optically thin shell will be dominated by the innermost part of the shell.

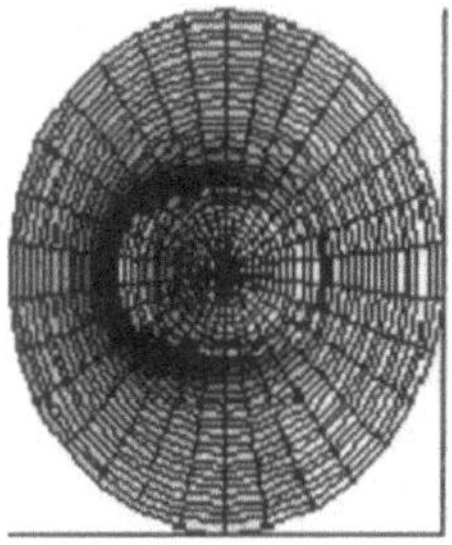

Figure 2. A ruled surface display for a model of the optically depth structure of Nova V1974 Cyg 1992 at 313 d after outburst, showing the different elliptical inner and outer boundaries; it should be rotated counterclockwise by 25° to match the position angle of the nova shell.

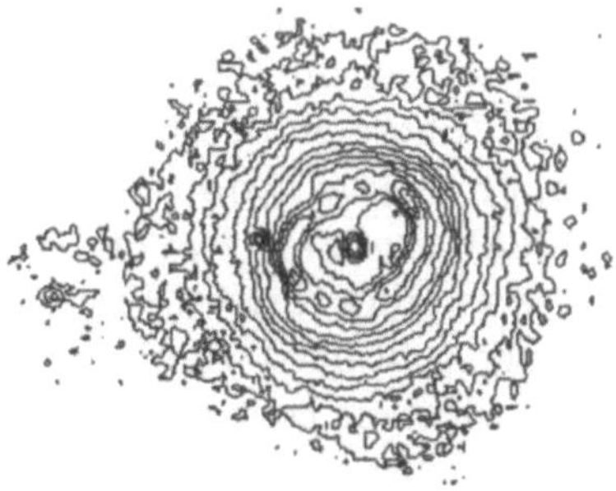

Figure 3. A contour display of the 1994 January 9 HST image of Nova V1974 Cyg 1992, where adjacent contour levels differ by a factor of 1.58. The major axis of the brightest part of the inner ellipse is 0.38″.

2. Observations and models for Nova V1974 Cyg 1992

As discussed by Hjellming (1995) the extensive Merlin (Pavelin et al. 1993), VLA, and HST images of V1974 Cyg 1992 make it necessary to adopt models which deviate from spherical symmetry, and it was proposed that a modification of the spherical shell model, where the inner and outer boundaries are allowed to be different ellipses, would solve the problem of the early shape appearing elliptical with a major axis at position angle 25°, while the late VLA and HST images show a roughly elliptical shape with a position angle of 115°. Computations of the radiative transfer through shells with linear velocity gradients and different ellipsoidal inner and outer boundaries have been carried out. Fig. 2 shows a ruled surface display of the optical depth 313 d after outburst for a model that fits the radio light curves and produces roughly the right shapes for the optically thick and optically thin states. In this model a spherically symmetric shell with inner and outer apparent angular velocities of 0.08 and $0.2''\,\mathrm{yr}^{-1}$ is replaced by ellipsoids with triaxial angular velocities of 0.07, 0.06, and $0.09''\,\mathrm{yr}^{-1}$ for the inner boundary and 0.18, 0.18, and $0.23'' \mathrm{yr}^{-1}$ for the outer boundary. where the middle number is the line of sight component. The remaining parameters of the model are $M_{\mathrm{shell}} = 3.1\,10^{-4}\,\mathrm{M}_{\odot}$ for a mean molecular weight of 1.273, a distance of 2 kpc, and temperature that increases linearly with time from about 5000 K at 40 d after outburst to about 26 000 K at 300 d after outburst, as discussed by Hjellming (1995). However, one additional factor must be included in the modeling to fit all the infrared, millimetre and centimetre light curves, in addition to the changes in nebular line profiles. One must assume that the inner shell regions remain near the surface of the white dwarf until about 180 d after outburst, and at that time shell ejection effectively ceases and the above-mentioned inner shell parameters become valid. This same elliptical boundary model then correctly predicts nebular line profiles that change from the early, flat topped (except for fluctuations due to clumps) line shapes seen in 1992 September to the narrower line shapes seen in 1993 April (Shore et al. 1993).

While Fig. 2 is an approximate fit to the needed radio shapes of V1974 Cyg 1992 (Hjellming 1995), there are two other indications that the ellipsoidal boundary model is roughly correct. Fig. 3 is a contour display of the 1994 January 9 HST image (obtained from the STScI/HST archive) of this nova, in the form where adjacent contour levels differ by a factor of 1.58. The roughly even spacing of the contours indicates that the brightness distribution is a power law, and is more extended than the bright inner part of the shell. Most of the outer contours are elliptical with a position angle consistent with 25°, and there is a gradual transition from to roughly spherially symmetric, and then to the innermost, brightest portions which are ellipses with a position angle consistent with 115°.

References

Hjellming, R.M., 1990, in "Physics of Classical Novae", eds Cassatella, A., Viotti, R., Springer-Verlag, New York, p169

Hjellming, R.M., 1995, in "Cataclysmic Variables", eds A. Bianchini et al., Kluwer, p139

Hjellming, R.M., Wade, C.M., Vandenberg, N.R., Newell, R.T., 1988, AJ, **84**, 1619

Kwok, S, 1983, MNRAS, **202**, 1149

Pavelin, S., Davis, R.J., Morrision, L.V., et al., 1993, Nature, **363**, 424

Shore, S.N., Sonneborn, G., Starrfield, S., et al., 1993, AJ, **106**, 2408

V1974 CYG (NOVA CYG 1992) - A POSSIBLE LINK TO SU UMA STARS

A. RETTER, E. M. LEIBOWITZ, E. O. OFEK
Tel-Aviv University, Israel

V1974 Cyg erupted on 1992 February 19, and was the brightest nova since V1500 Cyg. It soon became an object of massive international multi-wavelength observational study.

Some major photometric features of the nova were investigated by De-Young & Schmidt (1994), and Semeniuk et al. (1994, 1995). They found two periodicities in the nova light curve. While the shorter 0.081263 d period was present during the 1993 and 1994 observations, the longer 0.0850 d period was only observed from 1994 July. Semeniuk et al. (1995) interpreted the shorter period as the orbital period of the binary system, and the longer one as caused by the rotation of the white dwarf due to a magnetic field.

We observed V1974 Cyg during 8 nights at the end of 1994. The photometry was carried out with the 1 m telescope at the Wise Observatory, using the Tektronix 1K×1K CCD camera. We observed the star with the standard *BVRI* filters, concentrating mainly on the *I* band.

Our results also confirm undoubtedly the presence of two independent periods: 0.08127 d and 0.08505 d. The persistence of the shorter period over more than one year, and the possible eclipse shaped structure of its folded light curve at the end of 1994, suggest that it is the orbital period of the binary system. The second period is about 4.7 percent longer than the assumed orbital period. The period range and the difference between them are similar to those found in SU UMa stars.

The well known relation between the two periods of SU UMa stars is shown in Fig. 1 (Stolz & Schoembs 1981, 1984), where the point representing the two periods of V1974 Cyg (asterisk) is also marked. The correspondence is remarkable.

This fact suggests a possible linkage between the two periods of the classical nova V1974 Cyg and the superhump phenomenon in SU UMa stars. Note that the orbital period of the nova falls at the middle of the range of the binary periods of the SU UMa stars. Further indication of the link comes from the fact that the light curve shape of the longer period of

A. Evans and J. H. Wood (eds.), Cataclysmic Variables and Related Objects, 321–322.

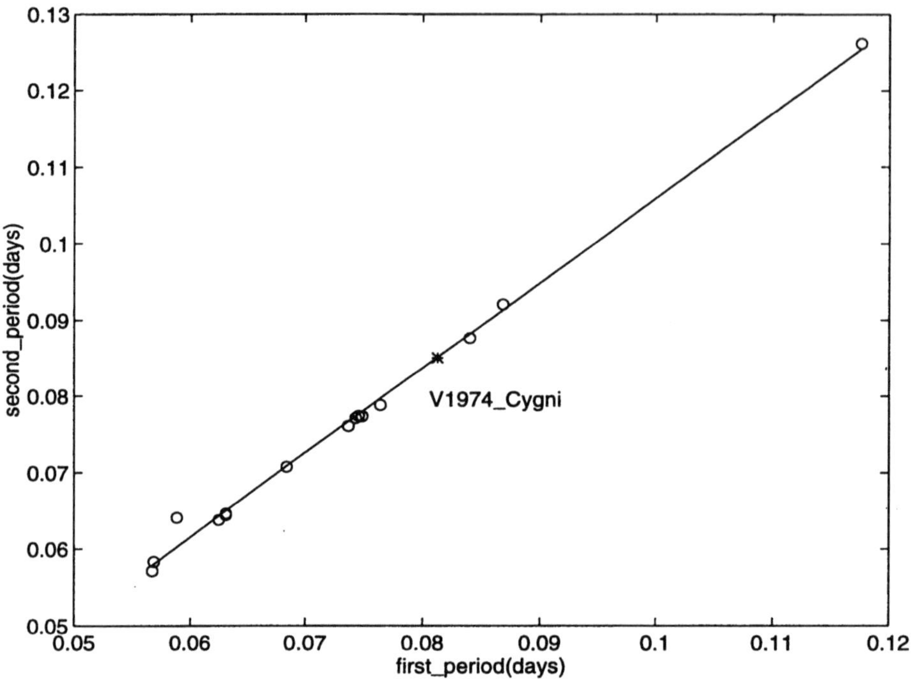

Figure 1. The relation between the orbital and superhump periods in SU UMa stars. The asterisk corresponds to V1974 Cyg's periodicities.

the nova is very much like the shape of the superhump periodicity in a few SU UMa stars (see e.g. la Dous 1993).

The appearance of a new periodicity in the nova power spectrum in the summer of 1994 suggests that an additional light source became dominant. The observational facts compatible with the SU UMa phenomenon suggest, that it is a precessing accretion disk. According to this picture, the nova might experience a long SU UMA like superoutburst a few months after the eruption of the nova.

References

DeYoung, J.A., Schmidt, R.E., 1994, Ap. J., **431**, L47

la Dous, C., 1993, in "Cataclysmic Variables & Related Objects", ed. Hack M., la Dous, C., Centre National de la Recherce Scientifique, Paris, France, p49

Semeniuk, I., Pych, W., Olech, A., Ruszkowski, M., 1994, Acta Astron., **44**, 277

Semeniuk, I., DeYoung, J.A., Pych, W., et al., 1995, Acta Astron., **45**, 365

Stolz, B., Schoembs, R., 1981, IBVS 2029

Stolz, B., Schoembs, R., 1984, A&A, **132**, 187

UBV PHOTOMETRY OF V705 CAS (NOVA CAS 1993) DURING THE FIRST YEAR OF THE OUTBURST

K. PETRÍK[1], L. HRIC[2], Z. URBAN[2], D. HANŽL[3], D. CHOCHOL[2]

1.Hlohovec Observatory, Sládkovičova 41, 920 01 Hlohovec, The Slovak Republic. e-mail: astropet@ta3.sk

2. Astronomical Institute of the Slovak Academy of Sciences, 059 60 Tatranská Lomnica, The Slovak Republic. e-mail: hric@ta3.sk, urban@ta3.sk, chochol@ta3.sk

3. N. Copernicus Observatory and Planetarium, 616 00 Brno, The Czech Republic. e-mail: hanzld@aci.cvut.cz

Nova Cas 1993 (=V705 Cas) was discovered ten days before it reached its maximum brightness in the V colour ($V_{\rm max} = 5.53$ mag and $B_{\rm max} = 6.14$ mag) on 1993 December 17.29 (Munari et al. 1994b). The pre-nova varied around 18...18.5 mag and 17...17.5 mag in B and V, respectively. No more prominent light fluctuations were found (Munari et al. 1994a). The nova started to fade rapidly, entering the deep minimum typical for the transition period of the DQ Her-subclass of novae in 1994 mid-February. The time development of the UV luminosity of the nova was discussed by Hauschildt et al. (1994). In this paper, we analyze our original *UBV* photometric data and all the available data obtained by the other authors. We derive the basic parameters of the nova and present the results of our search for short-term periodicities.

Our observational material was obtained at Hlohovec, Skalnaté Pleso, Stará Lesná and Brno N. Copernicus Observatories in the interval between 1993 December 17, and 1994 December 1.

The courses of our B and V light curves, as well as that of the U light curve, clearly confirm the membership of V705 Cas in the rather special and not so frequent subclass of novae designated as DQ Her-type. After the brightness maximum, the nova entered the early decline stage characterized by moderately fast light decrease with short-term variations (days to tens of days) superimposed on it. Approximately fifty days after the maximum, a sudden brightness decrease in all three colours is observed. The break in the light curve is well illustrated by rapid change in the course of the $U - B$ index. The nova reached the bottom of this deep minimum (16.1 mag in

A. Evans and J. H. Wood (eds.), Cataclysmic Variables and Related Objects, 323–324.

the visual) 103 d after the maximum. The brightness of the nova then rose again up to about 12.2 mag. The nova further oscillated irregularly by a few tenths of magnitude around this brightness up to the present.

We have used our light curves for the derivation of the following basic parameters of the nova: $t_{2,\mathrm{B}}$ =40 d, $t_{2,\mathrm{V}}$ =33 d, $t_{3,\mathrm{B}}$ =63 d, $t_{3,\mathrm{V}}$ =61 d. Using the formulae derived by Cohen (1985), Capaccioli et al. (1989), Schmidt (1957) and van den Bergh & Younger (1987), we determined the following value for the mean absolute V magnitude of the nova at the maximum: $M_\mathrm{V} = -7.05 \pm 0.22$ mag. On the basis of formulae by Livio (1992) and Pfau (1976), we obtained the following mean value for the mean absolute B magnitude at maximum: $M_\mathrm{B} = -7.03 \pm 0.20$ mag. Using the empirical formula developed by Livio (1992), we derived the mass of the white dwarf component of the nova: $M_\mathrm{WD} = 0.75 \pm 0.04\,\mathrm{M}_\odot$. Having derived the above quantity, we were able to compute the relevant Eddington luminosity, which is $M_\mathrm{bol}^\mathrm{Edd} = -6.44 \pm 0.06$ mag. Adopting the bolometric correction B.C.$=-0.25$, the corresponding Eddington luminosity for V is $M_\mathrm{V}^\mathrm{Edd} = -6.19$ mag. This means that the outburst of V705 Cas was a super-Eddington one. We obtained the mean colour excess value: $E_\mathrm{B-V} = 0.45 \pm 0.09$. The corresponding value of extinction is 1.40 ± 0.28, which further leads to the distance determination of $r = 1.73 \pm 0.2$ kpc.

We have probed the light curves of the nova during the early decline for the presence of periodicities in the period interval 0.01...100 d. The application of the minimization of the phase dispersion method (Stellingwerf 1978), as well as of Fourier analysis, resulted in the detection of the statistically significant period of 0.14 d. The period could reflect either the orbital motion of the underlying binary or the rotation of the white dwarf, or else oscillations within the expanding nova envelope. Its value fits well into the observed orbital period distribution of novae.

References

Capaccioli, M., Della Valle, M., D'Onofrio, M., Rosino, L., 1989, AJ, **97**, 1622
Cohen, J.G., 1985, Ap. J., **292**, 90
Hauschildt, P., Starrfield, S., Shore, S., et al., 1994, AJ, **108**, 1108
Livio, M., 1992, Ap. J., **393**, 516
Munari, V., Hric, L., Tomov, T., Hazucha, P., 1994a, IBVS 3977
Munari, U., Tomov, T., Antonov, A., et al., 1994b, IBVS 4005
Pfau, W., 1976, A&A, **50**, 113
Schmidt, Th., 1957, Z. Astrophys., **41**, 182
Stellingwerf, R.F., 1978, Ap. J., **224**, 953
van den Bergh, S., Younger, P.F., 1987, A&A, **70**, 125

INFRARED SPECTROSCOPY OF NOVA V705 CAS

A. EVANS[1], T. R. GEBALLE[2], J. M. C. RAWLINGS[3],
A. D. SCOTT[1]
1. Department of Physics, Keele University, UK
2. Joint Astronomy Centre, Hilo, Hawaii, USA
3. Department of Physics & Astronomy, University College, London, UK

1. Introduction

Nova V705 Cas was discovered on 1993 December 7, several days before maximum light. The light curve indicates that $t_2 \simeq 40$ d (see e.g. [4]), characteristic of dust-forming novae. The light curve went into decline on day 58, indicating the formation of an optically thick dust shell. IR spectra have been obtained regularly since outburst using the cooled grating spectrometer CGS4 on the United Kingdom InfraRed Telescope (UKIRT). We present here the results of ongoing IR spectroscopic monitoring of this nova.

2. Early detection of CO

Emission by CO in the $K-$band was detected around maximum light (see Fig. 1a), the earliest detection of a molecule in the ejecta of an erupting nova. The presence of CO points to the presence of a cool, neutral region at the outer edge of the ejecta, in which chemical reactions occur. Assuming that the CO is optically thin, we find a CO mass of $2.8\,10^{-10}\,M_\odot$, temperature 4300 K, and that $\sim 10^{-4}$ (by number) of the C in the ejecta was in the form of CO [1]. The time-scale for forming CO in nova ejecta, as per the chemical models of [5, 6], is a few hours. The presence of CO, and its characteristics, provide strong support for these models.

3. UIR emission

The so-called Unidentified InfraRed (UIR) features were first seen on day 252 (see Fig. 1b); they were not present on day 158. They are different to those appearing in other Galactic UIR sources, such as the Orion Bar.

A. Evans and J. H. Wood (eds.), Cataclysmic Variables and Related Objects, 325–326.

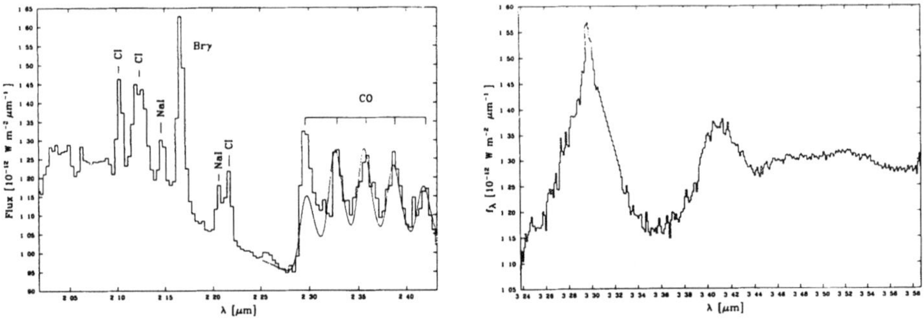

Figure 1. (a – left): CO emission at day –1; curve is model fit to optically thin CO emission. (b – right): UIR emission in the 3 μm band at day 252.

The 3.4 μm feature is anomalously strong by comparison with the 3.28 μm feature; the 11.3 μm feature is also present, and is accompanied by a 8 μm feature, not normally associated with UIR emission. Previous work [2] has argued that, in novae, these features arise in hydrogenated carbon grains rather than free-flying PAH molecules and that, in view of the abundance anomalies in novae, nova UIR features would differ substantially from those normally seen in Galactic objects such as the Orion Bar.

4. Line emission

The early spectrum is dominated by low excitation emission lines, particularly C I. Later spectra reveal the presence of [Si VI] λ1.96, 2.22 and [Ne II] λ12.8. However there is no evidence that this nova displayed a coronal phase similar to that displayed by other recent novae [3].

Spectroscopic monitoring, at near-IR (*JHKL*) and at 10 μm, is continuing, and will be supplemented by ISO observations in due course.

Acknowledgements. We thank UKIRT observers for taking the early spectra of Nova Cas. UKIRT is operated by the U.K. PPARC.

References

1. Evans, A., Geballe, T.R., Rawlings, J.M.C., Scott, A. D., 1996, MNRAS, in press
2. Evans, A., Rawlings, J.M.C., 1994, MNRAS, **269**, 427
3. Greenhouse, M.A., Grasdalen, G.L., Woodward, C.E. et al., 1990, Ap. J., **352**, 307
4. Petrík, K., Hric, L., Urban, Z., et al., 1996, these proceedings, p323
5. Rawlings, J.M.C., 1988, MNRAS, **232**, 507
6. Rawlings, J.M.C., Williams, D.A., 1989, MNRAS, **240**, 729

DETECTION OF THE IRRADIATED RED DWARF IN WY SGE (NOVA 1783)

M. W. SOMERS[1], K. MUKAI[2], T. NAYLOR[1], F. A. RINGWALD[1]
1. Department of Physics, Keele University, ST5 5BG, UK
2. Office for Guest Investigator Programs, Code 668, NASA/ Goddard Space Flight Center, Greenbelt, MD 20771, USA

Abstract. We present infrared (IR) photometry and optical spectroscopy of the eclipsing old nova WY Sge. According to hibernation theorists the disc in a system a few centuries after outburst should be significantly fainter than in other, more recent, old novae but similar to the discs of dwarf novae in quiescence. Despite the apparent faintness of the late type star we have sufficient information to infer that the face closest to the white dwarf (WD) is irradiated.

1. Introduction

WY Sge is an object of considerable interest because it is the oldest securely recovered nova, and thus may help our understanding of the evolution of classical nova systems between outbursts. Shara et al. (1986) suggest that a few hundred years after outburst old novae should fade considerably. Naylor et al. (1992) have argued against such a model. The IR photometry (see also Somers et al. 1996), and optical spectroscopy presented here, show that WY Sge is not in hibernation.

2. Results

2.1. INFRARED PHOTOMETRY

The flux remaining in eclipse provides an upper limit to the contribution of the late type star. It implies that the late type star in WY Sge provides less than 50% of the flux in the K−band. Comparison with dwarf novae (Szkody & Mateo 1986) shows that all have the secondary star contributing around 63% or more of the flux at K (assuming all the disc light is eclipsed). If we limit ourselves to systems with a period greater than 3 h (and thus a

A. Evans and J. H. Wood (eds.), Cataclysmic Variables and Related Objects, 327–328.

similarly sized secondary to WY Sge) the secondary star provides $\gtrsim 76\%$ of the light in all cases, with $\gtrsim 91\%$ for the two eclipsing systems (U Gem, IP Peg). The fraction of light from the late type star in the non-eclipsing nova-like variable IX Vel (Haug 1988) in the K−band is about 20...40%, in agreement with WY Sge.

At K maximum light occurs at phase 0.32 ± 0.01 which implies that the out of eclipse modulation is not purely ellipsoidal. We have modelled this in terms of a gravity and limb darkened Roche lobe being irradiated by a source at the position of the WD and shaded by the disc.

2.2. SPECTROSCOPY

The spectra taken in the red arm of ISIS on the WHT were shifted and added on the K_2 velocity derived from the narrow line emission. Several absorption features are apparent, most noticeably the TiO bands. We use the flux density deficit method in the TiO bands (Wade & Horne 1988) to calculate the fraction of light originating from the late type star and its spectral type. The secondary is of spectral type M1–M5, with a best estimate of M4. The ISIS blue arm spectra covered the region around He II and Hβ. In the individual spectra it is apparent that there is a sharp component of the Hβ profile that moves with phase. The narrow component is phased correctly to be tracing the velocity of the late type star; it is anti-phased with the inner disc, which roughly coincides with the velocity of the WD. Narrow line emission is indicative of irradiation of the inner face of the late type star.

3. Conclusions

We show that, contrary to the predictions of hibernation theory, WY Sge appears to have a brighter disc, at both IR and optical wavelengths, than other similar period and inclination dwarf novae. The disc is of comparable brightness as that in the nova-like variable IX Vel.

We also find that the secondary is irradiated on the face closest to the WD. The origin of the irradiation is unclear; it is either the remnant heating from the cooling WD after the nova explosion or accretion irradiation.

References

Haug, K., 1988, MNRAS, **235**, 1385
Naylor, T., Charles, P. A., Mukai, K., Evans, A., 1992, MNRAS, **258**, 449
Shara, M. M., Livio, M., Moffat A. F. J., Orio, M., 1986, Ap. J., **311**, 163
Somers, M. W., Mukai, K., Naylor, T., 1996, MNRAS, in preparation
Szkody, P., Mateo, M., 1986, AJ, **92**, 483
Wade, R. A., Horne, K., 1988, Ap. J., **324**, 411

THE SPECTRUM OF THE OLD NOVA V603 AQL OBSERVED WITH THE HST

M. FRIEDJUNG[1], P. L. SELVELLI[2], A. CASSATELLA[1]
1. Institut d'Astrophysique (CNRS), 98 bis Boulevard Arago, 75014 Paris, France
2. Osservatorio Astronomico, Via G,B. Tiepolo 11, 34131 Trieste, Italy
3. Instituto Astrofisica Spaziale (CNR), Via E. Fermi 21, 00044 Frascati, Italy

Abstract. Spectra of the old nova V603 Aql taken with the HST–GHRS did not reveal resonance doublet absorption components arising in material ejected in the 1918 outburst. However absorption components, probably produced by a disk wind, were observed, though most line emission probably comes from regions associated with an accretion disk. Most strikingly, dramatic line profile variations on time-scales of not much more than 10 m were seen.

1. Introduction

V603 Aql is a bright old nova, whose eruption was observed in 1918. Due to its brightness it appears a favorable target to search for lines produced by absorption in the ejected envelope as well as due to the disk wind. In the present work the regions of the C IV, N V and Si IV resonance doublets were observed in 1994 Febuary with the Goddard High Resolution Spectrograph of the HST at a resolution of $2\,10^4$. Each exposure was programmed to be split into 4 sub-exposures of 11 m (C IV), 10.5 m (N V) and 8 m (Si IV), in order to reduce fixed pattern noise of the diode array.

2. The search for absorption lines due to the ejected envelope

Absorption components might be expected at a blue-shift near −1900 km s^{-1}, corresponding to the blue-shift of the envelope ejected in 1918. None were seen on the summed spectra. If spherical symmetry of the enve-

A. Evans and J. H. Wood (eds.), Cataclysmic Variables and Related Objects, 329–330.

lope is assumed, with C and N abundances and ionization fractions of recent novae following the consistent analysis of Andreä, Drechsel & Starrfield (1994), we find a maximum envelope mass $\sim 2\,10^{-5}\,M_{\odot}$. The mass could be considerably larger if the envelope had broken up into dense clumps.

3. Line profiles due to the central remnant

Broad spectral features were however detected. These were blue-shifted absorption for the N V doublet, a P Cygni absorption and emission profile for C IV and mainly emission for Si IV. The summed emission profiles are not far from symmetric and, in the case of C IV, very strong compared with the blue-shifted absorption. Comparing them with what are perhaps the best calculated disk wind profiles of Knigge, Woods & Drew (1995) for a probable orbital inclination of 17° (Ritter 1990), the emission is far too strong. It is hard to explain such emission unless most came from regions near the disk in Keplerian orbits. The emission line widths are sufficiently large to indicate emission near the white dwarf, suggesting that at least most of the disk is not broken up by a magnetic field. On the other hand the absorption components and the disk emission are explainable as due to a disk wind.

The most remarkable result however is the variations in the line profiles between the sub-exposures. In particular the blue-shifted absorption components of N V and C IV even disappeared at times; C IV showed the strongest absorption on a sub-exposure following one with no absorption. The time-scale, not much more than $\sim 10\,$m, is short compared with the orbital period of 199 m. The simplest explanation is that of rapid variations in the state of ionization.

References

Andreä, J., Drechsel, H., Starrfield, S., 1994, A&A, **291**, 869
Knigge, C., Woods, J.A., Drew, J.E., 1995, MNRAS, **273**, 225
Ritter, H., 1990, A&AS, **85**, 1179

A NEW TYPE OF ASYMMETRIC OUTFLOW IN THE RECURRENT NOVA RS OPH IN QUIESCENCE

T. IIJIMA
Astronomical Observatory of Padova, Asiago Section
Osservatorio Astrofisico, I-36012 Asiago (Vi), Italy

Abstract. H I emission lines in the spectrum of the recurrent nova and symbiotic star RS Oph in quiescence are sometimes accompanied by blue- or red-shifted broad emission components. The displacements of the broad components from the narrow lines change from -1300 to $+1200\,\mathrm{km\,s^{-1}}$, but possibly not continuously. The broad components seem to appear in some restricted regions on the spectra. An asymmetry between the blue- and red-shifts is noticed.

RS Oph is a binary system consisting of a white dwarf and a mass-losing M type giant (e.g. Bruch 1986; Garcia 1986; Dobrzycka & Kenyon 1994). Recently, broad emission components of H I lines were found in some quiescent spectra, and their peculiar properties have been determined, e.g. they appear on the blue side of the narrow lines or on the red side, but not on both sides simultaneously; furthermore, their wavelengths, profiles, and intensities change with a time-scale of days (Iijima et al. 1994; hereafter Paper I). Here results of new observations in 1993 are reported. The criteria of the observations are the same as in Paper I. The broad components usually show double-peaked profiles. The last five columns of Table 1 give displacements of individual peaks of the broad components of Hα and Hβ relative to the narrow lines and of the center wavelengths between the two peaks v_{cent}. As in Paper I, the more distant peak is called the primary peak. The observational error is about $\pm 30\,\mathrm{km\,s^{-1}}$. A histogram of v_{cent} is shown in Fig. 1, where the results presented in Paper I are also included. Since there is a strong narrow emission line at the center, displacements less than $\pm 800\,\mathrm{km\,s^{-1}}$ are not detectable. As seen in the figure, there are four prominent peaks in the histogram located at, from left to right, -1230 ± 25, -1140 ± 15, $+930\pm20$, and $+1020\pm20\,\mathrm{km\,s^{-1}}$. It seems that the broad components do not move continuously, but appear in these restricted

A. Evans and J. H. Wood (eds.), Cataclysmic Variables and Related Objects, 331–332.

TABLE 1. Spectroscopic observations of RS Oph in 1993

Date	Exp. (m)	JD − 2 440 000	Spectral range (nm)	$H\alpha^p$	$H\alpha^s$	$H\beta^p$ (km s^{-1})	$H\beta^s$	v_{cent}
July 28	20	9197.45	398 – 510			−1405	−733	−1069
July 28	5	9197.48	571 – 683	−1496	−1014			−1255
Aug. 13	20	9213.41	392 – 504			−1540	−769	−1155
Aug. 13	5	9213.45	580 – 692	−1514	−993			−1254
Aug. 14	20	9214.41	396 – 508			−1458	−843	−1150
Aug. 14	5	9214.45	574 – 686	−1508	−976			−1242

Exp.: Exposure time; p: primary peak, s: secondary peak, v_{cent}: mean of p and s

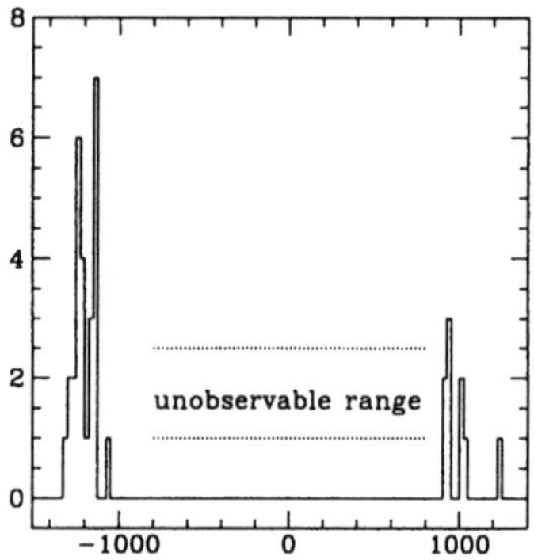

Figure 1. Histogram of radial velocities of the center of broad emission components. Horizontal scale in km s^{-1}.

regions. The lack of blue-shifts around $-1000\,\mathrm{km\,s^{-1}}$ is noticeable. This result suggests that there is an asymmetry between the blue- and red-shifts, or the center of the symmetry is blue-shifted by about $100\,\mathrm{km\,s^{-1}}$. It seems difficult to explain these asymmetric and discrete displacements with known models of bipolar gas outflow. There might be a new type of gas outflow in this system. Among numerous symbiotic stars only this object and CH Cyg (Paper I; Kuczawska, Mikolajewski & Kirejczyk 1992; Leedjärv 1993) have similar broad emissions.

References

Bruch A., 1986, A&A, **167**, 91
Dobrzycka D., Kenyon S.J., 1994, AJ, **108**, 2259
Garcia M.R., 1986, AJ, **91**, 1400
Iijima T., Strafella F., Sabbadin F., Bianchini A., 1994, A&A, **283**, 919
Kuczawska E., Mikolajewski M., Kirejczyk K., 1992, IBVS 3806
Leedjärv L., 1993, IBVS 3951

V1016 CYG: PROPER MOTION OF RADIO EMISSION

S. P. S. EYRES,[1] S. M. DOUGHERTY,[2] R. J. DAVIS,[1]
M. F. BODE,[2] H. M. LLOYD,[2] H. T. KENNY[3]
1. University of Manchester, NRAL, Jodrell Bank, Macclesfield, Cheshire, SK11 9DL, UK
2. Astrophysics Group, Liverpool John Moores University, Liverpool, L3 3AF, UK
3. Royal Military College, Kingston, Canada

Abstract. We present MERLIN maps of V1016 Cyg at 6 cm, observed on 1992 July 21 and 1995 March 23. Previously discovered bipolar structure is resolved into three distinct peaks. Evolution and proper motion of these peaks has been measured.

1. Discussion

The symbiotic nova V1016 Cyg is one of the strongest radio emitters in that class of stars, HM Sge being the only comparable radio source of this type in the northern hemisphere (see Eyres et al. 1995a). Following optical outburst in 1965 (FitzGerald et al. 1966), V1016 Cyg has been well observed in the radio and Harris & Scott (1976) first resolved a shell, believing it to have a bipolar appearance. Here we present MERLIN observations at 6 cm made on 1992 July 21 and 1995 March 23. The bipolar features are resolved into three peaks, and the evolution of these peaks can be followed.

Fig. 1 shows the maps resulting from each observation. We identify three components in the earlier map which can be traced to the later one. The northern ridge is labelled N, and contracts between the two observations. The southern features are labelled SE and SW. The SE feature appears to enlarge, and brightens overall, while the SW feature moves significantly between the two maps. Along with data at 18 cm, V1016 Cyg has many similarities in detailed radio structure to HM Sge (Eyres et al. 1995b).

Over the 975 d between observations, we have been able to measure proper motions of three identifiable features in the data. These are given in Table 1. Given the errors, only SW can be said to have moved with any

A. Evans and J. H. Wood (eds.), Cataclysmic Variables and Related Objects, 333–334.

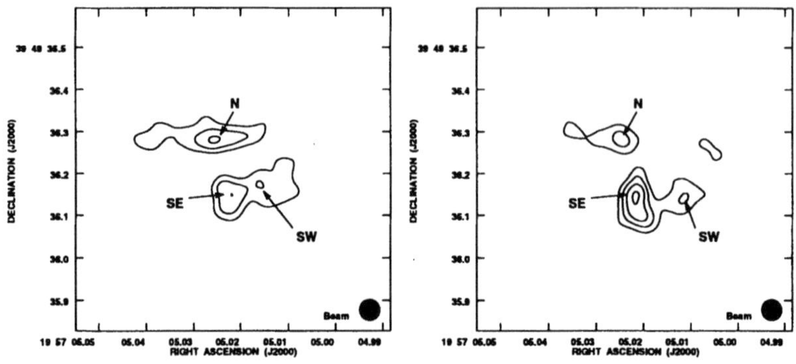

Figure 1. Radio maps of V1016 Cyg made using MERLIN at 6 cm on 1992 July 21 (left) and 1995 March 23 (right). Contour levels are 4, 6, 8 and 10 times the rms noise of 135 μJy beam^{-1}. The three features are labelled as N, SE, and SW, for reference to Table 1.

TABLE 1. Proper motion in milli-arcsec (mas), position angle measured from north through east. Errors were $\sim$ 10 mas in the positions from which the motions have been derived.

Feature	ΔRA (mas)	Δdec (mas)	Total Motion (mas)	Position Angle
N	−15	0	15	270°
SE	0	0	0	–
SW	−60	−45	75	233°

certainty. The association of the various radio components with the stellar components or outflow will require a more detailed analysis.

Acknowledgements. MERLIN is a national facility operated by the University of Manchester on behalf of PPARC. HML and SMD are supported by a Research Assistantships, and SPSE is supported by a Research Studentship Award from PPARC.

References

Eyres, S.P.S., Kenny, H.T., Cohen, R.J., et al., 1995a, MNRAS, **274**, 317

Eyres, S.P.S., Davis, R.J., Kenny, H.T., et al., 1995b, in "Radio Emission from the Sun and Stars", eds. Taylor A.R., Paredes J.M., A. S. P., San Francisco, in press

FitzGerald, M.P., Houk, N., McCuckey, S.W., Hoffleit, D., 1966, Ap. J., **144**, 1135

Harris, S., Scott, P.F., 1976, MNRAS, **175**, 371

EVOLUTION OF AN ACCRETION DISK IN THE SYMBIOTIC BINARY CI CYG

JOANNA MIKOŁAJEWSKA
Copernicus Astronomical Center
Bartycka 18, PL-00716 Warsaw, Poland

Abstract. We have combined IUE spectra with optical spectroscopic and photometric data collected over a few orbits of the symbiotic binary CI Cyg to follow the evolution of a disk and boundary layer during a full activity cycle. Our results indicate an extended optically thin disk during quiescent periods, which evolves into an optically thick state in an eruption; a classical boundary layer at the inner edge of the disk ionizes a surrounding H II region in quiescence, and this emission fades during the rise to visual maximum in outburst. This evolution of $\dot{M}$ rivals that observed in classical CVs and pre-main sequence stars, so symbiotic systems like CI Cyg represent another opportunity to study the physics of disk accretion.

1. Accretion models for symbiotic stars

Symbiotic stars are commonly accepted as long period ($P_{\rm orb} \sim 200...1000\,$d) interacting binaries similar to cataclysmic variable stars. They consist of an evolved red giant star, a hot companion, and an ionized nebula. The hot components in many systems show occasional 2–3 mag eruptions, characterized by dramatic changes of the optical spectrum. In particular, an A–F supergiant star appears to replace the red giant at optical maximum and most high-ionization emission lines disappear. The cause of these eruptions remains controversial, as both accretion-powered events and thermonuclear runaways (TNR) in material accreted by a white dwarf can account for many features observed at a typical symbiotic outburst.

In accretion-powered systems, two physical mechanisms can change the accretion rate ($\dot{M}$) through an accretion disk, and as a result the hot component's luminosity; mass-transfer instabilities and disk instabilities (see e.g. Mikołajewska & Kenyon 1992a for details). The 2–3 mag optical out-

A. Evans and J. H. Wood (eds.), Cataclysmic Variables and Related Objects, 335–338.

bursts require main sequence (MS) accretors (Bath & Pringle 1982; Kenyon & Webbink 1984) unless $\dot{M}$ exceeds a value appropriate to the Eddington limit. Accretion-driven eruptions can be distinguished observationally from TNR events by an increase in the emitted flux at all wavelengths, rather than flux redistribution that characterizes the rise to optical maximum and early optical decline in the TNR case. MS accretors develop A–F supergiant optical spectra at maximum as do degenerate TNR events or, if $\dot{M}$ increases, steadily burning white dwarfs but the UV spectra more closely resemble late B or early A supergiants. This behavior is due to the radial temperature profile in the disk: $T_{\rm eff}$ decreases from $\sim 20\,000 \ldots 30\,000$ K in the inner disk to $\sim 5\,000 \ldots 10\,000$ K in the outer disk. Finally, most disk models predict that a boundary layer (BL) between the disk and slowly rotating central star radiates 50 % of the accretion energy. Since the BL temperature exceeds $\sim 10^5$ K for MS accretors, it emits mostly in the EUV range. This radiation photoionizes gas surrounding the disk, and possibly also the disk material, to produce optical and UV emission lines. The emission line fluxes provide a reasonably good estimate of the EUV luminosity, so the evolution of the BL can be determined with optical and UV spectra. An important prediction of the model is that the EUV luminosity from the BL should be comparable to the optical + UV luminosity from the disk.

2. CI Cyg

A good example of a symbiotic system with an accretion disk is CI Cyg, which underwent a major outburst in 1975 and recently returned to quiescence. The system contains an M5 II giant, $M_{\rm g} \sim 1.5\ {\rm M}_\odot$, transferring material into a large accretion disk around a low mass, $M_{\rm h} \sim 0.5\ {\rm M}_\odot$, MS companion (Kenyon et al. 1991). The *UBV* photometry (Belyakina 1991 and references therein) demonstrated that CI Cyg is an eclipsing system with $P_{\rm orb} = 855$ d. The characteristic double bump visible in the quiescent *VRI* light curves (Belyakina 1992) indicates that the giant is tidally distorted, and that mass transfer occurs *via* Roche lobe outflow.

The early phases of the last eruption were observed only at optical and IR wavelengths, but the latter stages of the decline and the present quiescent phase were also well covered with the IUE. These data represent a unique opportunity to investigate the evolution of an accreting symbiotic system over a full activity cycle.

The 1975 outburst was characterized by the appearance of a bright optical continuum, increase of Balmer emission line fluxes, and the absence of high-ionization lines. The outburst also caused a significant change in the eclipses: they became narrow with well defined phases of contact, contrary to the broad minima observed at quiescence (Belyakina 1991; Kenyon et al.

1991). The hot component's luminosity increased from $L_{\rm h} \sim$ a few $\times 10^2$ $L_\odot$ at quiescence to $\sim 10^4$ $L_\odot$ at maximum.

The optical colors at maximum were inconsistent with a single temperature stellar photosphere; the apparent spectral type varied from B7 ($U-B$) through A5–A6 ($B-V$) to F0 ($V-R$ and $R-I$). During the late decline (1980), optical spectrophotometry indicated the presence of a totally eclipsed red continuum source ($T_{\rm c} \sim 4000$ K, $R \sim 50$ $R_\odot$; Mikołajewska & Mikołajewski 1983), while the totally eclipsed short wavelength IUE continuum was consistent with a late B – early A spectral type. The eclipses allow us to estimate the outburst radius of the hot component in the orbital plane. This radius exceeds the spherically symmetric radius inferred from the total luminosity by a factor of $\sim 2 \ldots 3$. If the hot component is an ellipsoid, then the radius of the major axis exceeds the radius of the minor axis by a factor of $\sim 3-5$ (Mikołajewska & Kenyon 1992a). The large values of $H/R \sim 0.2-0.3$ for the disk in CI Cyg agrees with Bath & Pringle's (1982) predictions for high–$\dot{M}$ disks surrounding MS stars.

As the system declines from the visual maximum, strong high-ionization lines return. These emission lines require a source with a temperature of $\sim 10^5$ K, and EUV luminosity comparable to the optical + UV luminosity. We associate this source with a boundary layer. Kenyon & Webbink (1984; see also Plavec & Hubeny 1994) fit the UV continuum at that time with optically thick steady-disk models, and these models suggest $\dot{M}$ declines with time from $\dot{M}$ $\sim$ a few $\times 10^{-4}$ to a few $\times 10^{-5} M_\odot\,{\rm yr}^{-1}$. The EUV luminosity estimated from strong H I, He I and He II emission lines follows the disk luminosity, with $L_{\rm EUV} \approx L_{\rm disk}$ (Kenyon et al. 1991).

The eclipses in H I, He I and He II emission lines are broad at quiescence and narrow during outburst, which indicates that the size of the H II and He III region in the orbital plane shrinks in outburst (Mikołajewska 1985; Kenyon et al. 1991). Such behavior again points to the presence of an optically (and geometrically) thick accretion disk that prevents ionization in the orbital plane.

In 1983 CI Cyg entered the quiet phase. The color temperature of the short wavelength IUE continuum increased, while the long wavelength IUE and near UV continuum was dominated by strong *bound-free + free-free* emission. We interpret this as a transition from an optically thick to an optically (and geometrically) thin – at least in the optical range – disk. The UV continuum then became dominated by hot radiation from the boundary layer and the innermost accretion disk. The transition was also accompanied by development of strong high-ionization forbidden emission lines – [Ne V], [Fe VII] and [Mg V]. These forbidden lines cannot be produced in the same regions as the permitted emission lines, because they are not affected by the eclipses. In fact, the forbidden line ratios require lower densities,

$n_e \sim 10^8\,\mathrm{cm}^{-3}$, and higher temperatures, $T_e \sim 10^5\,\mathrm{K}$, than the values, $n_e \sim 10^{10}\,\mathrm{cm}^{-3}$, $T_e \sim 15\,000\,\mathrm{K}$, derived from the eclipsed permitted and intercombination lines (Kenyon et al. 1991). These results suggest a model in which the permitted and intercombination lines form in a photoionized region surrounding the hot component, and forbidden lines form in a larger shock-heated region produced by interacting winds or an accretion disk 'corona'.

3. Concluding remarks

The outburst behavior of CI Cyg – an increase in the emitted flux at all wavelengths, an ellipsoidal hot component, the variation of spectral type with wavelength – points to accretion-powered outburst with an MS accretor. The IUE and optical observations suggest that the disk and boundary layer each emit $\sim 50\%$ of the total luminosity during the decline, which also agrees with simple disk models. There are however two deviations from the behavior predicted by theory: EUV emission from the BL vanishes near visual maximum, and a large mechanically heated region develops during the late decline from maximum. Similar behavior has occured in outbursts of AX Per, a 'sister' system of CI Cyg (Mikołajewska & Kenyon 1992b). The disappearance of BL emission is probably related to changes in the physical conditions at the disk's inner edge as $\dot{M}$ increases in outburst. In fact, the BL emission initially tracks the rise in $\dot{M}$, but declines as $\dot{M}$ exceeds $\sim 10^{-4}\,\mathrm{M}_\odot\,\mathrm{yr}^{-1}$. Such behavior is also observed in pre-MS accretion disks: low $\dot{M}$ objects (T Tau stars) appear to have a standard BL, while in high $\dot{M}$ objects (FU Ori stars) the BL emission is missing (Kenyon 1994).

Acknowledgement. This study was supported by KBN Research Grant No. 2 P304 007 06.

References

Bath, G. T., Pringle, J. E., 1982, MNRAS, **201**, 345
Belyakina, T. S., 1991, Izv. Krym. Astrofiz. Obs., **83**, 118
Belyakina, T. S., 1992, Izv. Krym. Astrofiz. Obs., **85**, 62
Kenyon, S. J., 1994, in "Interacting Binary Stars", ed. A. Shafter, ASP Conf. Ser. **56**, 63
Kenyon, S. J, Webbink, R. E., 1984, Ap. J., **279**, 252
Kenyon, S. J., Oliversen, N. A., Mikołajewska, J., et al., 1991, AJ, **101**, 637
Mikołajewska, J., 1985, Acta Astr., **35**, 65
Mikołajewska, J., Kenyon, S. J., 1992a, MNRAS, **256**, 177
Mikołajewska, J., Kenyon, S. J., 1992b, AJ, **103**, 579
Mikołajewska, J., Mikołajewski, M., 1983, Acta Astr., **33**, 403
Plavec, M. J., Hubeny, I., 1994, in "Interacting Binary Stars", ed. A. Shafter, ASP Conf. Ser. **56**, 87

THE PECULIAR SYMBIOTIC STAR CH CYG – A BRIGHT COUNTERPART OF CATACLYSMIC VARIABLES?

LAURITS LEEDJÄRV
Tartu Observatory
EE 2444 Tõravere, Estonia

1. Introduction

CH Cyg has begun to show symbiotic features only since 1963. Up to now four outbursts of CH Cyg have been recorded, the fifth going on at the present time. CH Cyg reveals a lot of peculiarities as compared to 'normal' symbiotic stars (extended inactive periods between the 'symbiotic' periods; the most extended active period in 1977–1986 including the ejection of bipolar radio jets; a stage of increased activity since 1992, with peculiar profiles of the Balmer emission lines etc.); see Mikołajewski, Mikołajewska & Khudyakova (1990) for further references.

Classical models for explaining the outbursts of symbiotic stars (e.g. Mikołajewska & Kenyon 1992) are unable to account for the behaviour of CH Cyg. The magnetic rotator model has been proposed and developed by Mikołajewski & Mikołajewska (1988, 1996). The model includes a wind accreting, magnetic ($B \sim 10^7$ G), rapidly rotating ($P \sim 8$ m) white dwarf (WD) as the hot component of the symbiotic system. A strong magnetic field prevents matter from the red giant's wind being accreted onto the WD surface in the usual way. Instead, three stages of accretion could be considered: inactive, propeller and accretor state.

2. Recent manifestations of the propeller action

The latest active period of CH Cyg observed since early 1992 is characterized by a new type of peculiarity: high-velocity (up to $\gtrsim 1000\,\mathrm{km\,s^{-1}}$) blue- and red-shifted broad emission components of Balmer lines. Examples of the spectra of CH Cyg obtained during 1992–1994 with the 1.5 m telescope at the Tartu Observatory have been published by Leedjärv & Mikołajewski (1995).

A. Evans and J. H. Wood (eds.), Cataclysmic Variables and Related Objects, 339–340.

The additional emission (occasionally also absorption) components are variable on a time-scale of days and, as shown by Tomov et al. (1995), even hours and minutes. Most of the time only blue-shifted features are visible. The magnetic rotator model is as follows. Matter accreted by the WD is expelled back by the rapidly rotating magnetosphere of the WD. This outward flow of matter is not quite continuous, instead we can see individual blobs of matter, rejected by the propeller. Most of the rejected blobs should be expelled towards us, and the blue-shifted emission and absorption features are visible. Possible flip-flop instabilities of the accretion wake (e.g. Livio et al. 1991) on a time-scale of weeks or months can produce blobs which could reach the magnetosphere on the receeding hemisphere of the propeller, causing red-shifted emission components (without absorption) occasionally.

During active periods brightness variations of CH Cyg on a time-scale of minutes can be observed, especially at shorter wavelengths. This kind of flickering is common for many types of cataclysmic variables, including the recurrent novae T CrB and RS Oph. During the present propeller state for example, Tomov et al. (1996) have announced flickering with amplitude ~ 0.7 mag in B light in 1994 June. We observed CH Cyg during several nights in 1995 May with the 60 cm telescope and the double-beam, sky-chopping photometer at the Tuorla Observatory, Finland. In spite of the low brightness level of CH Cyg ($V \approx 9$ mag), flickering is still present, reaching amplitude ~ 0.6 mag in U and ~ 0.4 mag in V.

Emission line profiles of CH Cyg observed since 1992 are quite unique among variable stars. Only the recurrent nova RS Oph has demonstrated similar features (Iijima et al. 1994). On the other hand, the propeller phenomenon considered mostly in the case of neutron stars, could be more widespread among magnetic white dwarfs. CH Cyg as a relatively bright star ($V \approx 6 - 9$ mag) could serve as a testing site of interactions between matter and magnetic field, relevant to many types of much fainter cataclysmic variables.

References

Iijima, T., Strafella, F., Sabaddin, F., Bianchini, A., 1994, A&A, **283**, 919
Leedjärv, L., Mikołajewski, M., 1995, A&A, **300**, 189
Livio, M., Soker, N., Matsuda, T., Anzer, U., 1991, MNRAS, **253**, 633
Mikołajewska, J., Kenyon, S.J., 1992, MNRAS, **256**, 177
Mikołajewski, M., Mikołajewska, J., 1988, in "The Symbiotic Phenomenon", Mikołajewska J. et al. (eds.), Kluwer, p233
Mikołajewski, M., Mikołajewska, J., 1996, MNRAS, submitted
Mikołajewski, M., Mikołajewska, J., Khudyakova, T.N., 1990, A&A, **235**, 219
Tomov, T., Kolev, D., Munari, U., Antov, A., 1996, MNRAS, submitted

POSSIBLE PERIODIC COMPONENTS IN THE FLICKERING OF CH CYG AND MWC 560

M. MIKOŁAJEWSKI[1], T. TOMOV[2], A. DAPERGOLAS[3],
Y. BELLAS-VELIDIS[3]
1. Nicolaus Copernicus University, Torun, Poland
2. National Astronomical Observatory Rozhen, Bulgaria
3. National Observatory of Athens, Greece

1. Discussion

CH Cyg and MWC 560 are very peculiar symbiotic binaries consisting of an M giant and a white dwarf companion. The systems have many features in common. In particular, both show occasional eruptions with sub-Eddington luminosity, followed by the appearance of high-velocity jets. Recently, Mikołajewski, Mikołajewska & Tomov (1995) have proposed a model of activity for these systems with a rapidly rotating magnetic white dwarf accreting from the M giant's wind. Two states of activity are possible: *propeller* and *accretor*. Although, both stars spend most time in the propeller state, occasional *propeller-accretor* (and vice versa) transitions may give rise to bipolar outflows and rapid jet formations. Mikołajewski et al. (1995) also proposed to distinguish a new subclass of interacting binaries – propeller stars – with CH Cyg and MWC 560 being the prototypes.

One of the most important common features of both stars is the presence of rapid variability (flickering) during active phases with time-scales from minutes to hours, and amplitudes up to 50% of the blue continuum flux (Mikołajewski et al. 1990; Tomov et al. 1996). However, some periodicities with an amplitude not exceeding 5% appear relatively frequently.

Our observations were carried out using the 1.2 m Kryonerian telescope of the Athens National Astronomical Observatory. The U observations of CH Cyg were performed during five nights between 1989 August 1 and 6. MWC 560 was observed in U and V filters during four nights between 1993 January 18 and 23. The total duration of observations was 14.9 and 8.7 h for MWC 560 and CH Cyg, respectively.

Power spectra of both stars' observations show several peaks corresponding to periods of several minutes for CH Cyg and of tens of min-

A. Evans and J. H. Wood (eds.), Cataclysmic Variables and Related Objects, 341–342.

utes for MWC 560. Most of them represent time-scales of variability rather than real periods. They correspond to nearly regularly spaced flares that appear on several occasions and become incoherent after 2–4 cycles. The only peak present on all periodograms appears around 1.9 mHz for CH Cyg, and around 0.24 mHz for MWC 560, respectively. To determine precisely the periods corresponding to them, we have analysed power spectra of all merged data for both stars. We have found possible coherent oscillations with $P = 529.24$ s for CH Cyg and $P = 4225.3$ s for MWC 560, respectively.

The period found for CH Cyg – of about 530 s – is close to previously reported values between 498 and 512 s (Mikołajewski et al. 1990); but it is about five times less than the period recently reported by Hoard (1993). We believe that the 2600 s period found by Hoard (1993) represents the typical time-scale of a low frequency flickering component rather than a real coherent periodicity. The 70 m period of MWC 560 belongs to the sharp maximum of possible quasi-periods found in almost fifty different runs of flickering by Tomov et al. (1996). Michalitsianos, Perez & Shore (1993) have also found a ~1 h periodicity in their data, very close to our coherent oscillation.

The flickering phenomenon observed in binary systems with orbital periods of several years (Mikołajewski et al. 1995 and refs. therein) can be understood only in terms of instabilities of accretion onto a rapid magnetic rotator. The time-scales and amplitudes of these rapid fluctuations should reflect the interaction between magnetosphere and accreted matter. Since Keplerian rotation near the magnetosphere is close to the period of the rotator one can expect quasi-variability on time-scales close to that period.

Both MWC 560 and CH Cyg probably contain white dwarfs with magnetic fields of about $10^7 \ldots 10^8$ G. During our observations the stars were at relatively low phases of activity – in the propeller state, when the accretion is stopped by the centrifugal barrier. Simultaneously, a relatively small amount of matter around the rotator makes visible the regions close to the magnetosphere. That is why we can interprete the detected coherent periods as the rotational periods of an accreting white dwarf – or regions very close to it – in both systems.

Acknowledgement. This study was supported by KBN Research Grant No. 2 P304 007 06.

References

Hoard, D.W., 1993, PASP, **105**, 1232
Michalitsianos, A.G., Perez, M., Shore, N., 1993, Ap. J., **259**, 244
Mikołajewski, M., Mikołajewska, J., Tomov, T., et al., 1990, Acta Astron., **40**, 129
Mikołajewski, M., Mikołajewska, J., Tomov, T., 1995, in "Compact Stars in Binaries", Proc. IAU Symp. 165, J. van Paradijs et al. (eds.), in press
Tomov, T., Kolev, D., Ivanov, M., et al., 1996, A&A, in press

A CLOSE LOOK AT THE OUTBURST OF THE VERY SLOW SYMBIOTIC NOVA PU VUL – A TRIPLE SYSTEM

D. CHOCHOL[1], S. TAMURA[2], O. KANAMITSU[3],
T. PRIBULLA[1], A. TAJITSU[2]

1. Astronomical Institute of the Slovak Academy of Sciences, 059 60 Tatranská Lomnica, Slovakia
2. Astronomical Institute, Faculty of Science, Tohoku University, Aoba-ku, Sendai, Miyagi 980, Japan
3. Fukuoka University of Education, 729 Akama, Munakata-shi, Fukuoka 811-41, Japan

The very slow nova PU Vul was discovered by Honda and Kuwano in 1979 after the rise from $B \sim 15\ldots16$ mag to a maximum $B \sim 9$ mag. The object has been studied by many authors (for references see Belyakina et al. 1989; Vogel & Nussbaumer 1992) but a clear picture of the system was not given. Two minima of different duration were observed in the system in the active stage, in 1980–81 and 1993–94 (Kolotilov et al. 1995). Nussbaumer & Vogel (1996) ascribed the minima to eclipses of the hot component by the cool M giant in a binary with orbital period of 4900±100 d. Our period analysis of the preoutburst B data (Liller & Liller 1979) by Stellingwerf's (1978) method revealed a period of 4858 d. We found the third minimum in 1926–28 (lasting for 500...630 d) in these data. The centers of the three minima were used to improve the ephemeris:

$$\mathrm{JD\ (min.)} = 2425032(\pm 30) + 4882(\pm 10)E\ .$$

The different lengths of the eclipses, their asymmetry and their longer duration than expected from the standard characteristics of the components in a binary, can all be explained by supposing that PU Vul is a hierarchical triple system. The observed minima are caused by the eclipses of the white dwarf in the inner orbit by the M giant in the outer orbit. The parameters of the inner orbit found from the radial velocity curve of the white dwarf, mimicking an F supergiant from 1979 to 1987 (for references see Kanamitsu et al. 1991), are given in Fig. 1. The spectroscopic orbital period of 761 d is close to the most significant period found in B (768 d) and V (782 d) photometric data obtained in the years 1982–87. Analysis of the U data revealed the most significant period of 328 d, which corresponds to the mass transfer bursts from the inner M giant onto the hot component.

A. Evans and J. H. Wood (eds.), Cataclysmic Variables and Related Objects, 343–344.

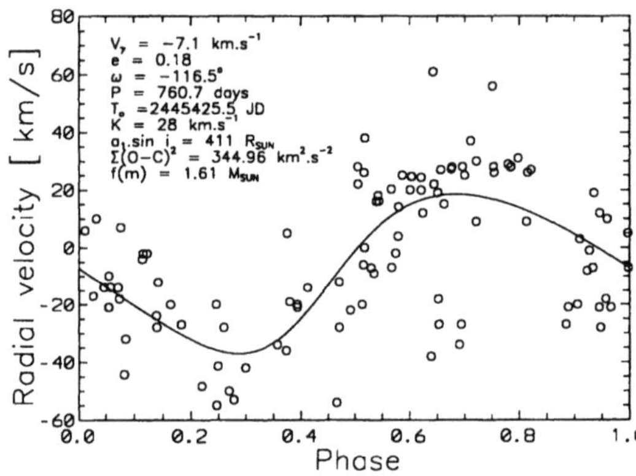

Figure 1. Spectroscopic orbit of the slow symbiotic nova PU Vul

An expanding atmosphere of the hot star formed a common envelope surrounding the inner binary system. After the flare and onset of the fast wind in 1987 July, the envelope was ejected. From that time PU Vul behaves as a typical very slow nova, with $[t_2]_V = 1016\,d$ and $[t_2]_B = 1123\,d$. Using the $M_B - t_2$ relation, we have calculated $M_B = -5.83$. The corresponding mass of the white dwarf, calculated using the relation of Livio (1992), is $M_{wd} = 0.57\,M_\odot$. Emission line profiles during the nebular stage of the nova (after 1991) show an expanding envelope in the form of an equatorial ring and polar blobs. The radial velocities of these substructures were used to determine the inclination angle of the inner orbit as $i = 78°$. Using the values of i, M_{wd} and the mass function $f(M_{wd}, M_{giant}, \sin i) = 1.6\,M_\odot$, we have determined the mass of the inner M giant as $M = 2.6\,M_\odot$.

Analysis of photometric and spectroscopic data have revealed that the very slow, eclipsing symbiotic nova PU Vul is a hierarchical triple system consisting of a white dwarf ($M = 0.57\,M_\odot$) and a M giant ($M = 2.6\,M_\odot$) in an inner orbit ($P = 761\,d$), with the second M giant in an outer orbit ($P = 4882\,d$). The mass ratio of the outer to inner orbit is $X = 6.4$. Among the group of hierarchical triple systems that includes λ Tau ($X = 8.3$) and CH Cyg ($X = 7.0$), PU Vul has the most extreme mass ratio.

References

Belyakina, T. S., Bondar, N. I., Chochol, D., et al., 1989, A&A, **223**, 119
Kanamitsu, O., Yamashita, Y., Norimoto, Y., et al., 1991, PASJ, **43**, 523
Kolotilov, E. A., Munari U., Yudin B. F., 1995, MNRAS, **275**, 185
Liller, M. H., Liller, W., 1979, AJ, **84**, 1357
Livio, M., 1992, Ap. J., **393**, 516
Nussbaumer, H., Vogel, M., 1996, A&A, submitted
Stellingwerf, R. F., 1978, Ap. J., **224**, 923
Vogel, M., Nussbaumer, H., 1992, A&A, **259**, 525

A UNIQUE SYMBIOTIC-LIKE/CATACLYSMIC TRIPLE STAR 4 DRA (=CQ DRA): SIX YEARS OF UBV MONITORING

Z. URBAN[1], L. HRIC[1], L. LEEDJÄRV[2]
1. Astronomical Institute of the Slovak Academy of Sciences, 059 60 Tatranská Lomnica, Slovakia
e-mail: urban@ta3.sk, hric@ta3.sk
2. Tartu Astrophysical Observatory, 2444 Tõravere, Estonia
e-mail: leed@jupiter.aai.ee

The bright ($m_v \approx 5$) M3III star 4 Dra (HD108907, HR4765, CQ Dra) is a unique triple system combining in itself the basic structural properties of a symbiotic-like binary and of a cataclysmic one. The red giant 4 Dra A forms a wide binary ($P_{orb} \approx 1703$ d) with a hot companion 4 Dra B which seems to be a compact cataclysmic binary system (4 Dra BC, probable P_{orb} of about 4 h).

We have monitored the 4 Dra system in *UBV* at the Skalnaté Pleso and Stará Lesná Observatories and at the Tartu Astrophysical Observatory during a total of 205 nights between 1989 March 12 and 1995 June 20; a few additional data points were obtained at the Bucharest Observatory.

Our long-term *UBV* light curves (LCs) show:

- The modulation of the *U* brightness with the 1703 d period.
- Two sudden *U* brightenings, centered at phases 0.1 and 0.0, respectively, with similar amplitudes of about 0.3 mag but of apparently different duration, their peak-to-peak separation being about 1583 d.
- Pulsation-like cycles well-correlated in all three colours, superimposed in the *U* on the long-term orbital modulation, with mean period of 340 ± 10 d (almost exactly one fifth of the 1703-d period).
- Irregular variations in all three colours, with an amplitude of 0.1–0.2 mag, often simultaneous or quasi-simultaneous, of similar brightness-change profiles and typical time-scale of tens of days.
- Two specific intervals (so-called plateaux) in the *B* and *V* LCs where the level of the irregular variability appears to be supressed. They display a mild steady brightness increase. Both are located near the periastron and thus correlated with the sudden brightenings in *U*.
- The globally minimal values in *U*, *B* and *V* occur near phase 0.5.

A. Evans and J. H. Wood (eds.), Cataclysmic Variables and Related Objects, 345–346.

The phenomenological model of the wide 4 Dra A+BC system we use is based on the published spectroscopic orbital solution and our derivation of the corresponding Roche geometry. Although the red giant star seems to be well inside its Roche lobe, given the extended and unstable atmospheres of red giants and the observed intense mass loss via stellar winds from these stars, almost all of its critical Roche volume could be filled by gas of various density and temperature. This could occur especially during the periastron passage time when some kind of intense tidal interaction could still occur, despite the rather large separation of 4 Dra A and 4 Dra BC even at that time. A stimulated mass loss from the red giant's extended amosphere towards the hot component, or some collimation of the outflowing stellar wind at least, seems to be quite plausible then. We thus conclude that, as well as being an, as yet, unique testing site for the evaluation of the possible parameters of the precursor systems to cataclysmic variables, 4 Dra A+BC provides us also with an equally unique environment for studying the modulated wind accretion by a cataclysmic *binary*.

We attribute the long-term modulation of the *U* LC of the 4 Dra A+BC system to some variety of reflection effect. Near-periastron brightenings in the *U* could be ascribed either to the Rayleigh scattering of the light coming from 4 Dra BC in the red giant's amosphere, or to the irradiation of a cloud enveloping 4 Dra BC as a result of elevated mass influx during the near-periastron stage by a cataclysmic binary inside it, or perhaps even to a direct response of this cataclysmic binary to an increased wind accretion. The interval of about 1583 d between the two events, different from the radial-velocity-determined orbital period of 1703 d, suggests rather a physical dynamical interpretation than some kind of an 'orbital clock'. The pulsation-like cycles are red giant pulsations. The irregular light variations reflect some processes within the extended atmosphere of the red giant component. Their conspicuous supression near the periastron suggests invisibility of the relevant source region or perhaps a kind of switch-off or masking of the parent mechanisms by the dynamical processes occurring at this stage. The global *UBV* minimum at about phase 0.5 corresponds to a partial eclipse of the reflection effect region by 4 Dra BC and its circumsystem material.

We refer the interested reader to the full-length version of this paper (to be submitted to Astron. and Astrophys.) for all the details and references.

Acknowledgements. We thank D. Reimers for communicating his results to us prior to publication, and R.F. Griffin for a consultation.

X-RAY BINARIES

OBSERVATIONS OF LOW MASS X-RAY BINARIES

ALAN P. SMALE
USRA Research Scientist
Laboratory for High Energy Astrophysics, Code 668
NASA/GSFC, Greenbelt, MD 20771, USA
alan@osiris.gsfc.nasa.gov

Abstract. In this paper I review the properties and behavior of low-mass X-ray binary systems (LMXBs) that contain neutron stars (NS), concentrating on the Galactic bulge sources and bursters. I describe the observed characteristics of LMXBs including their light curves, spectra, eclipses, dips, bursts, flares, pulsations, QPO, long-term periodicities and orbital period changes, and explain how fast timing results and the distinction between 'Z' and 'atoll'-type sources provide the key to a unified model of LMXB behavior.

1. Introduction

A LMXB consists of a neutron star (NS) primary, an accretion disk surrounding the primary in the plane of the orbit, and a late type (F, G, K or M) dwarf (or degenerate dwarf) secondary. The primary can be identified as a NS by direct observations of bursts, pulsations, or derived masses or mass ratios, or by analogy with other sources with similar characteristics. The late-type secondaries can occasionally (i.e. for transients) be observed directly through optical/IR photometry and spectroscopy, but are more often inferred through their derived masses. The duration and properties of X-ray eclipse transitions may also supply supporting evidence for their nature. Late-type secondaries do not have natural winds sufficient to power the X-ray source, and significant mass transfer occurs only when the companion fills its Roche lobe. This mass is passed to the NS through an accretion disk in a way familiar to CV observers. X-ray heating of the accretion disk (and sometimes the companion star) dominates the optical emission. In addition, X-ray heating can evaporate material from the disk, forming an accretion disk corona (ADC) above its central regions.

A. Evans and J. H. Wood (eds.), Cataclysmic Variables and Related Objects, 349–358.

In total, there are ~200 X-ray binaries known, ~125 of which are LMXBs. Of these, 48 have known optical (or IR) counterparts, 32 have known orbital periods, 40 are known Type I burst sources, 40 are transients (12 of which burst when active), 17 are found in globular clusters, and 6 are known Z-sources. LMXBs are preferentially located in the Galactic plane, towards the Galactic center, or in clusters, away from regions of star formation. Many LMXBs lie in regions of severe interstellar absorption and have no known counterparts.

Historically there have been many classifications of LMXBs, based on flux level, spectral behavior, etc. Most such classifications can be considered obsolete. Observationally, sources divide into classes based largely on inclination effects and intrinsic luminosity. Physically, recent analysis of fast timing and spectral data show that there is only one really useful classification of LMXBs, based on accretion rate and magnetic field strength:

(i) Bright Galactic bulge sources have luminosities of $L_x \sim 10^{38}$ erg s^{-1}, show X-ray flares lasting from minutes to hours, and often show a Z-type behavior on a color-color diagram (see later). The known orbital periods of these systems so far are 19 h (Sco X–1), 9.8 d (Cyg X–2), and 14 d (Sco X–2), indicating that the secondaries in these systems are evolved. The evolution of the companion drives the mass transfer in these systems.

(ii) Lower-luminosity burst sources have $L_x \sim 10^{36} \ldots 10^{37}$ erg s^{-1} (and are thus only 'lower-luminosity' compared to bulge sources, not compared to CVs!). These sources often emit sharp, bright bursts lasting ~seconds, and display atoll-shaped behavior in their color-color diagrams. Typical orbital periods in these systems range from 3...15 h, indicating 'normal' dwarf companions (although three have $P_{orb} < 1$ h and degenerate secondaries). In these systems it is magnetic braking and/or gravitational radiation that shrinks the orbit and drives the mass transfer. Dippers and transients are generally burst sources. All globular cluster LMXBs are burst sources.

My goal in this review is to supply a swift overview of various types of LMXB behavior for the non-expert, and in the process I will sometimes generalize without pointing out exceptions to the 'rules'.

2. Orbital modulations

At X-ray and optical energies, the appearance of the light curves from LMXBs and the existence of observable modulations depends strongly upon the source inclination. One empirical conclusion that we can draw very strongly from the data is that the accretion disks are *not* thin.

$i > 80°$. At high inclinations the central source is obscured by the disk edge, and the observed X-rays are scattered through an accretion disk corona, producing a light curve with a smooth sinusoidal modulation. In addition to this modulation, we see a partial eclipse of the ADC by the companion. Modeling of the data from X1822−371, the canonical ADC system, shows that the corona must be extended such that it covers roughly the central third of the accretion disk. The optical light curve can be reproduced using a model with four components: the disk itself, its inner and outer rims, and the heated face of the secondary.

$75° < i < 80°$. At these inclinations we *do* see the central source, although our view of it is frequently blocked by the companion star, producing sharp regular eclipses, as well as by azimuthal structure on the accretion disk, producing irregularly-shaped deep dips which lead the eclipses by $\sim$0.2 in phase. The eclipses provide information about the size of the companion and the system geometry, and provide fiducial markers for timing studies of the orbital ephemeris and possible orbital period changes. The optical minima correspond to the times of X-ray eclipse. X0748−676 is a prime example of a source in this inclination range.

$70° < i < 75°$. Here, X-ray dips are observed but no eclipses. Optical light curves become roughly sinusoidal, with minima occurring 0.2 in phase after the dips. Examples: X1254−690, X1916−053.

$i < 70°$. At inclinations lower than 70° we cease to see such interesting effects. At X-ray energies no eclipses or dips are observed, and an orbital modulation is rarely seen. In the optical, the light curves are generally sinusoidal (e.g. X1735−444, X1636−536), with the exception of those sources in which ellipsoidal variations are observed from giant companions.

From these observations, an empirical model of a LMXB can be built up with all its important regions fixed in the binary's rest frame. If $\phi = 0$ is the phase at which the NS is 'behind' the secondary (i.e. furthest from the observer), then eclipses of the central source (or partial eclipses by of the ADC) occur at this phase, and the accretion stream from the companion strikes the disk at $\phi = 0.8 \ldots 0.85$, producing dips. Irregularly-shaped azimuthal structure associated with the impact point rises $\sim$15° above the disk edge. Most of the DC optical emission comes from reprocessing in the disk, with an AC component from the heated face.

This will be a familiar picture for CV experts, as CVs also have strongly inclination-dependent modulations, sometimes show eclipses and dips, and often have activity associated with the stream impact point.

X-ray dips provide a wealth of information not limited to their role in defining the geometry of the system. From their irregular, ragged appearance and cycle-to-cycle differences in width and depth we know that the disk structure is dynamic. Also, spectral fitting results show that dips are associated with (but cannot be completely described by) increases in absorption of up to $N_H \sim 10^{23}$ cm^{-2}. (The dips at similar phases seen in a number of IPs are generally associated with much smaller N_H values $\lesssim 10^{21}$ cm^{-2}.)

More complex and satisfying models show that spectra from dippers can be fit with a two-component model where one (the 'absorbed' component) is observed directly from the central source, obscured by the intervening material, and the other ('unabsorbed' component) is Thomson-scattered into the line of sight. The decrease in the normalization of the absorbed component relative to its value in quiescence (i.e. the non-dipping, non-bursting state) provides an estimate of the electron density of the absorbing medium, which can be combined with N_H to give a measure of the metal abundance of the material, assuming no photoionization occurs; such estimates show a metal underabundance of 1.6...2.5 in X1916−053, 2...7 in X0748−676, and up to 600 in X1755−338.

While the empirical 'model' of azimuthal structure seems secure, it poses difficulties for more theoretical models, which show that it is difficult to sustain the necessary vertical structure in a physically realistic way.

3. X-ray bursts

Type I X-ray bursts are caused by the thermonuclear detonation of material accreted onto the surface layers of the neutron star. In some accretion regimes, the nuclear burning in this material is unstable, and the catastrophic ignition and combustion of this material shows up observationally as a powerful burst from the system with a rise time of ~1 s and a decay time of ~10 s. Bursts can be spectrally modeled using variations on a theme of blackbody emission. There are many differences in the details of bursts from different sources, and even from the same source: some bursts show photospheric radius expansion, which allows us (in principle) to measure both the mass and radius of the NS, directly examining the NS equation of state; double-peaked and triple-peaked bursts have been observed, some an apparent effect due to radius expansion, others a real effect due to variations in energy generation; and in bursts from some systems (X1608−522, X1636−536, X1747−214) a transient absorption feature is observed in some bursts, perhaps due to a gravitationally red-shifted resonance line from the NS atmosphere.

Type II bursts are only seen in the Rapid Burster, and are caused by an as-yet poorly understood accretion instability mechanism. An excellent review of both types of X-ray burst behavior can be found in [4].

4. X-ray spectroscopy: continuum and line emission

Loosely speaking, the X-ray (1...20 keV) continuum spectrum from LMXBs can be defined as 'soft', with a characteristic thermal bremsstrahlung temperature of $kT \sim 5\ldots10\,\mathrm{keV}$, as compared with the 'hard' spectrum of the MXRB pulsars, which have energy indices of $\alpha \sim 0\ldots1$.

In practice, higher-quality spectra show that for the brighter bulge/Z sources a more complex two-component model is required, although there is some controversy about the spectral characterization and the physics underlying it [12]. The so-called 'Western model' consists of a Comptonized component from the scattering of low-energy photons on high-energy electrons, plus a $0.5\ldots2\,\mathrm{keV}$ blackbody which provides between 2 and 40% of the observed flux. Flares in the brighter sources can be modeled as an increase in this blackbody component, with the Comptonized component staying roughly constant. The 'Eastern model' to the same or similar data consists of a disk blackbody component, where the spectrum is built up by summing the blackbody contributions from an accretion disk in which the temperature varies as a function of radius ($T \propto r^{-3/4}$), plus a $2\ldots3\,\mathrm{keV}$ blackbody. In both cases the blackbody component is associated with the boundary layer between the disk and the neutron star.

For the lower-luminosity systems the continuum spectra can generally be fit with a power-law with energy index $\alpha = 1$ and a high energy cutoff in the range $kT = 5\ldots20\,\mathrm{keV}$, although there are some exceptions; X1820$-$303 and X1608$-$522 have episodes in which a blackbody is required and the spectral behavior more closely resembles the bright bulge sources.

Iron $K$$-$line emission at 6.7 keV is a feature of many LMXBs, including Sco X-1, Cyg X-2, and GX 5-1 as well as several fainter sources. In general this emission is broad, with a FWHM $\sim 0.5\ldots1\,\mathrm{keV}$, but weak, with equivalent widths (EW) $\sim 50\ldots100\,\mathrm{eV}$. These lines have been resolved using EXOSAT GSPC, BBXRT, and now ASCA. At lower energies, line emission from Fe L, O and N K, and other species between 0.5 and 1.5 keV have been detected with the Einstein OGS and SSS, and ASCA [9]. A spectacular new result from ASCA is the discovery of strong Neon Lyman α, β emission at $\sim$1.0 keV from the compact degenerate LMXB X1626$-$67 [1], along with O Lyα and Heα at 0.65, 0.58 keV. The data from the ASCA and Astro-E eras place severe demands upon the ongoing effort at many institutions to produce detailed theoretical plasma codes.

5. Optical spectroscopy

As seen from the NS, the accretion disk of the average LMXB subtends an angle of $\pm10\ldots15°$. Thus, the disk 'sees' about 25% of the source flux, or $\sim10^{37}$ erg s^{-1}, as compared to the $\sim10^{33}$ erg s^{-1} produced by the M-dwarf companion, and so the optical continuum emission observed from LMXBs is dominated by reprocessing in the disk; the spectrum shows a bluish continuum, with an effective temperature of $\sim3\,10^4$ K. The absolute magnitudes of LMXBs depends on the scale of the system; evolved systems such as Cyg X–2 have $M_v = -2.0$ mag, while other Z-sources have $M_v = 0$ mag, bursters $M_v = 1 \ldots 2$ mag, and the very short period systems (X1916–053, X1626–67, X1820–303, $P_{orb} < 1$ h) $M_v = 4 \ldots 5$ mag [11]. Optical line emission from LMXBs is rather paltry compared to the spectacular riches that CV observers are used to. He II λ4686 is generally the strongest line observed, with a mean strength of EW = 5.5 ± 3.8 Å (the error represents the scatter across the class of LMXBs). The Bowen C III/N III blend at λλ4640 is also observed, with EW = 2.8 ± 2.1 Å, and sometimes emission from the Balmer series (EW(Hβ) = 2.6±3.7 Å). The lines are formed variously in the heated face of the secondary, in the accretion disk, and/or in the accretion stream, depending upon the source.

Considering the energetics of the systems, the search for spectral features that may be associated with the secondary is a tricky proposition. The most detailed work to date has been done by Shahbaz et al. [7], who obtained spectra of a sample of 14 LMXBs between 4000...9000 Å, and fit the spectra with combinations of standard accretion disk models plus companion star templates. Stellar features were detected only from systems with evolved companions (e.g. GX 1+4, Cyg X–2) and sources with small accretion rates (e.g. transients in quiescence Cen X–4, Aql X–1).

6. Timing analysis

6.1. NEUTRON STAR SPIN PERIODS

Pulsars are common in MXRB, and the Ritter catalogue [6] contains 19 CVs for which the white dwarf spin period is known, with P_{spin} ranging from 28...7500 s. However, in LMXBs the pulsars appear to be a collection of exceptions to the rule rather than a separate class of objects in their own right.

Her X–1 (P_{orb}= 40.8 h, P_{spin}= 1.24 s) shows a 35-day high/low cycle and optical brightness variations that march through the light curve every 35-day period, characteristics much more typical of an MXRB. X1627–67 (P_{orb}= 0.70 h, P_{spin}= 7.7 s) has a degenerate companion in an ultra-compact binary. GX 1+4 (P_{spin}= 114 s) is a symbiotic system with an M6 III giant counterpart. The observed magnetic fields in Her X–1 and GX 1+4 are 3 orders of magnitude higher than the 'normal' range for LMXBs. Each of these systems, then, probably has an evolutionary history that deviates from the norm.

For more 'standard' LMXBs the neutron star spin period is unknown. X-ray burst theory constrains fields to $B \lesssim 10^{10}$ G, while horizontal branch QPO (see below) require $B \sim 10^9$ G. The expected NS spin period depends upon the accretion torques over the lifetime of the system, and physical models and observational results from radio pulsars and QPO suggest that NS spin periods should be in the range 1...10 ms. Sophisticated searches for these periods have been performed on a sample of the brightest LMXBs, including all the Z-sources, to 512 Hz, and have placed upper limits on the pulse fractions of 0.3...5%.

6.2. Z/ATOLL SOURCES, QPO, AND NOISE

Timing analysis has nevertheless proved crucial to interpreting LMXB behavior, through the study of quasi-periodic oscillations (QPO) and related noise phenomena. QPO are defined as peaked noise in a frequency-power plot, where the width of the peak is less than half of the peak frequency. (36 CVs also show QPO, but their time-scale and interpretation are rather different.)

A principal diagnostic tool is the color-color diagram, where the higher-energy hardness ratio is plotted against a lower-energy hardness ratio to produce a diagram similar in concept to the Hertzprung-Russell diagram. Six of the brighter LMXBs (GX 5–1, Cyg X–2, Sco X–1, GX 17+2, GX 340 +0, GX 349–2=Sco X–2) describe a Z-shaped track on such a diagram [2]. Sources move along the Z continuously, typically tracing out a complete Z in ∼days, rather than jumping tracks from one state to another. The Z-shape persists from year to year, although some evidence for a secular shift in the position of the entire Z is seen in Cyg X–2 [3].

Fig. 1 shows an example of such a Z-shaped curve, with the Horizontal, Normal, and Flaring branches (HB, NB, FB) labeled. Mass transfer increases in the direction shown by the arrow. In the HB, QPO are observed with a peak frequency $F_{\mathbf{peak}}$ of 13...55 Hz, with $F_{\mathbf{peak}}$ increasing with $\dot{M}$. NB/FB QPO appear at a lower frequency range of 6...20 Hz; HB and NB QPO are sometimes observed simultaneously, indicating that they have different physical origins.

As the lower left portion of the figure shows, substantial noise components are also seen from the Z sources: high frequency noise (HFN), with a cutoff energy of 50...100 Hz; low frequency noise (LFN) at 2...20 Hz (seen only on the HB), and very low frequency noise (VLFN) that appears most strongly in the flaring branch, with a power index of $\alpha = 1.5\ldots2.0$.

Other sources, including some bright bulge sources as well as all the bursters, show a different pattern in the color-color diagram, resembling an atoll divided into three areas: Upper Banana, Lower Banana, and Island (UB, LB, IS). Atoll sources do not display QPO, although they do show noise characteristics; HFN is seen extending from 1...100 Hz, occasionally with a broad band-limited peak that mimics (but is different from) QPO behavior. VLFN is also seen in the atoll sources.

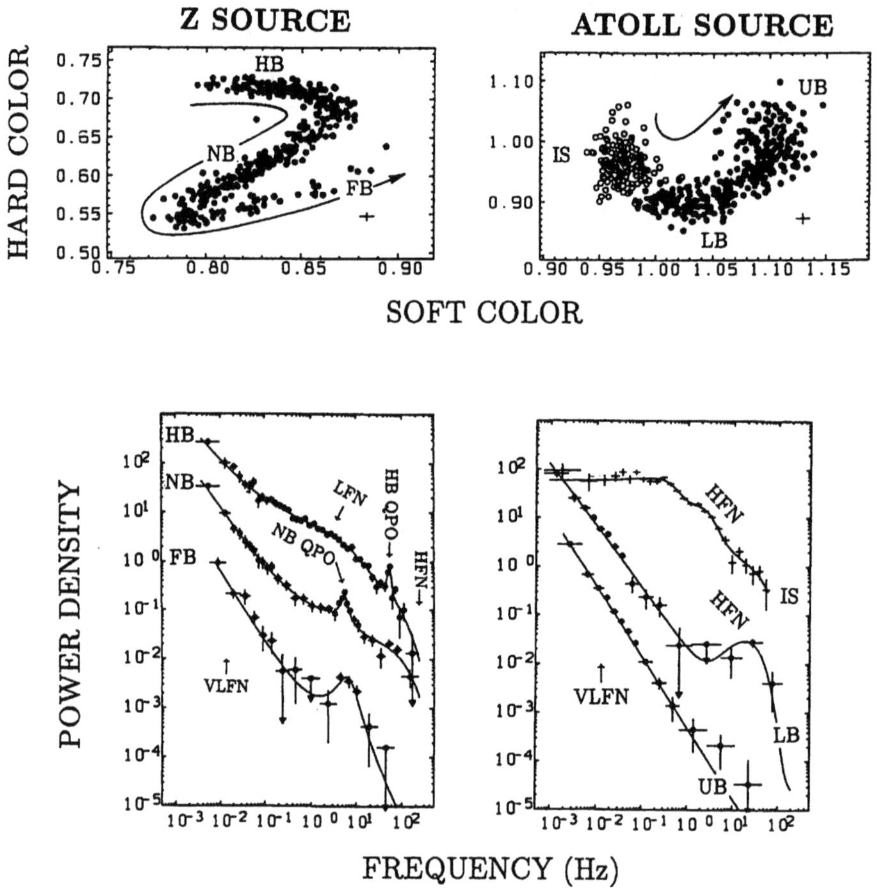

Figure 1. The behavior of Z and atoll sources in the color-color diagram, and in frequency space [10]. (Figure provided courtesy of Michiel van der Klis.)

HB QPO are normally interpreted in terms of the beat frequency model. Plasma at the inner edge of the accretion disk interacts with the magnetosphere of the rapidly-spinning neutron star at the beat frequency between the Keplerian velocity of the inner disk material and the NS spin frequency. This model gives an expected $P_{\rm spin}$ in the range 3...20 ms, which is in line with the predictions given in the previous section.

Models suggest that the HB covers accretion rates between 0.5...0.9 $\dot{M}_{\rm Edd}$. As the accretion rate rises above this, radiation pressure becomes critical in the inner disk and the system approaches spherical accretion. The NB QPO seen in this regime may be caused by oscillations in the accretion flow, perhaps due to feedback between X-ray production and accretion. As the source goes over $\dot{M}_{\mathrm{Edd}}$ and into the FB, even this radial inflow becomes unstable and non-radial oscillation modes may become important. VLFN may be due to either the time-scale on which the source moves along the Z, or the time-scale of convective burning fronts on the NS surface.

Based on the ideas above a simple, consistent physical model has been proposed [5]. In this model, the Z sources contain NS with magnetic fields $B \sim 10^9$ G, which generate soft photons by high-harmonic cyclotron emission. These photons are Comptonized by electrons in a hot central corona (HCC) around the 30 km radius magnetosphere, producing a flat spectrum with a cutoff energy equivalent to the electron temperature (~10 keV). From this beginning, numerical models reproduce the spectral changes observed as the mass accretion rate increases, including the Z-shaped tracks on the color-color diagram. The same model can also generate the behavior seen from atoll sources, if the magnetic field is an order of magnitude weaker and the NS accretes at a much lower rate of 0.01 ... 0.1 $\dot{M}_{\mathrm{Edd}}$.

The emerging picture of LMXB behavior has some pleasing consistencies. Pulsars show noise spectra similar to the band-limited HFN at 10^2 Hz seen in Z-sources, supporting the idea that this noise originates near the magnetosphere. Atoll sources share noise characteristics at 1...10 Hz with both Z-sources and black hole candidates, supporting the origin of this noise in the inner (radiation-dominated) disk. Thus, a basic unified picture of timing and spectral behavior in all the X-ray binaries requires just two structures: the magnetosphere, and the inner disk region.

6.3. LONGER-TERM TIMING PHENOMENA: CYCLES AND ORBITAL PERIOD CHANGES

In MXRBs firm evidence exists for long-term cycles from 30...300 d in LMC X–4, SS 433, LMC X–3, Cyg X–1, and Her X–1, probably due to the changing aspect of a tilted, twisted accretion disk, sometimes in retrograde precession. In LMXBs long-term cycles are rare and have been confirmed in only two systems, X1820–303 ($P_{\mathrm{long}} = 175$ d) and X1916–053 ($P_{\mathrm{long}} = 199$ d). In these systems disk precession is unlikely, and the generally-accepted model involves intrinsic variations in mass transfer. One possibility is that the mass transfer events are analogous to SU UMa-type superoutbursts, for which the bright spot luminosity also implies enhanced mass transfer with recurrence rates of 130...400 d.

Three LMXBs (X1820–303, X0748–676, X1822–371) show orbital period changes either greatly in excess of what would be expected through evolutionary models, or with the wrong sign. Some possible physical causes are: X-ray heating of the companion, causing mass loss, radius changes, and/or simulated wind accretion; stellar magnetic activity; hierarchical triple systems, modulating the orbital period derivations by Keplerian motion; and solar-type cycles in the companion which alter its radius and therefore the orbital period. The last of these explanations has been invoked in CVs, ~20 of which show some evidence for $P_{\rm long} = 1 \ldots 20$ yr. A review of precession and other long-term cyclic behavior in X-ray binaries can be found in [8].

7. Conclusions

A few key parameters define the observed phenomenology of both LMXBs and CVs. The differences in behavior between subclasses of each source type largely result from differences in: (i) accretion rate; (ii) mass ratio of components; (iii) magnetic field strengths, and the existence and location of the magnetosphere; (iv) source inclination, and the characteristics, extent and edge structure of disks; and (v) X-ray heating.

The phenomenology of LMXBs now seems to be well understood, at least of time-scales of ~days down to ~ a few ms. The Z-sources display QPO and various noise characteristics, have higher accretion rates (0.6 ... 1.1 $\dot{M}_{\rm Edd}$) and higher magnetic field strengths ($B \sim 10^9 \ldots 10^{10}$ G), and show flaring behavior and two-component spectra. The atoll sources have lower accretion rates (0.01 ... 0.1 $\dot{M}_{\rm Edd}$) and lower magnetic field strengths ($B \sim 10^8 \ldots 10^9$ G), and frequently (but not always) show Type I bursts and single-component continuum spectra. Iron line emission at 6.7 keV and other line features at lower energies are observed in both classes. Dips and bursts are predominantly seen in the atoll sources.

References

1. Angelini, L., White, N.E., Nagase, F., et al., 1995, Ap. J., **449**, L41
2. Hasinger, G., van der Klis, M., 1989,, A&A, **225**, 79
3. Kuulkers, E., van der Klis, M., Vaughan, B., 1996, A&A, in press
4. Lewin, W.H.G., van Paradijs, J., Taam, R.E., 1993, Spa. Sci. Rev., **62**, 223
5. Psaltis, D., Lamb, F.K., Miller, G.S., 1995, Ap. J. Lett., **454**, L137
6. Ritter, H., Kolb, U., 1995, in "X-ray Binaries", eds. Lewin, W.H.G., van Paradijs, J. and van den Heuvel, E.P.J., CUP, p578
7. Shahbaz, T., et al., 1995, in preparation; Shahbaz, T., Ph.D. thesis, U. Keele.
8. Smale, A.P., 1994, in "The Evolution of X-ray Binaries", eds. S.S. Holt and C.S. Day, AIP, New York, p525
9. Smale, A.P., White, N.E., Angelini, L., et al., 1996, in preparation.
10. van der Klis, M., 1994, Ap. J. Supp., **92**, 511
11. van Paradijs, J., McClintock, J.E., 1995, in "X-ray Binaries", eds. Lewin, W.H.G., van Paradijs, J., van den Heuvel, E.P.J., CUP, p58
12. White, N.E., Stella, L., Parmar, A.N., 1988, Ap. J., **324**, 363

Due to space constraints I have only supplied references to other reviews and to papers that are pivotal, very recent, or unusually obscure.

GENERATION OF X-RAY FLUCTUATIONS IN X-RAY BINARIES

S. MINESHIGE[1], M. TAKEUCHI[1], H. NEGORO[2]
1. Department of Astronomy, Faculty of Science, Kyoto University, Sakyo-ku, Kyoto 606-01, Japan
2. Institute of Space and Astronautical Science, Yoshinodai, Sagamihara 229, Japan

Abstract. Short-term aperiodic fluctuations with $1/f$-like power spectral density (PSD), where f is the frequency, are characteristics of X-ray radiation from X-ray binaries. To investigate a mechanism creating fluctuations, we construct a cellular-automaton model for accretion disks based on the concept of self organized criticality (SOC). In this model, mass accretion takes place either by an avalanche triggered when the local mass density exceeds some critical value, or by a gradual diffusion occurring regardless of the critical condition. With this model, we can reproduce the observational PSD, the distributions of peak intensities of X-ray shots, and non-random temporal distribution of the shots. Possible accretion disk models producing $1/f$-like X-ray fluctuations are discussed. We conclude that such disks are likely to be advection dominated.

1. Introduction

A long time has passed since the discovery of short-term X-ray variations from stellar black-hole candidates (see the review by Oda 1977). Through power spectrum analysis it has been claimed that such X-ray time variations are similar to $1/f$ noise. A similar $1/f$-like PSD was found also in neutron-star X-ray binaries in the low-luminosity state (Makishima 1988; van der Klis 1994). Their origin remains as one of the big mysteries of X-ray astronomy.

We have proposed a simple cellular-automaton model based on the notion of self-organized criticality (Bak, Tang & Wiesenfeld 1988). In this model, we conjecture that the inner portion of the disk is composed of numerous reservoirs with finite capacities. By imposing the cellular-automaton

A. Evans and J. H. Wood (eds.), Cataclysmic Variables and Related Objects, 359–362.

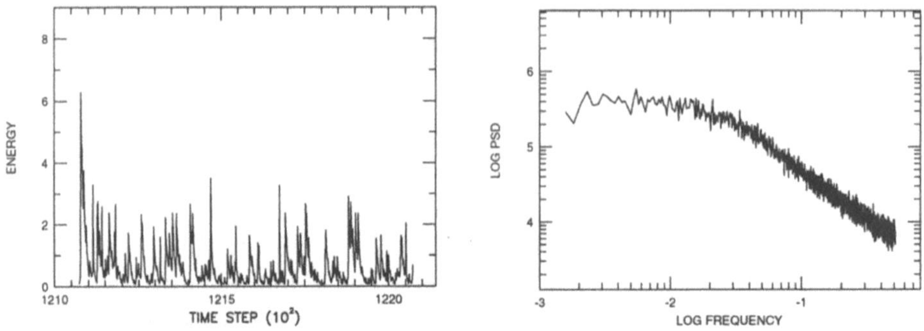

Figure 1. Typical light curves (left) and their PSD (right) produced by the Monte-Carlo simulations based on the SOC model. We set $m' = m/10$.

rule that mass accretion occurs when a mass density exceeds some critical value, we have succeeded in producing the basic features of X-ray fluctuations. In the following, we briefly outline the essence of our model (cf. Mineshige, Ouchi & Nishimori 1994; Mineshige, Takeuchi & Nishimori 1994).

2. Cellular-automaton procedures

(i) Divide the inner portions of an accretion disk into numerous small segments (or cells);
(ii) input a mass of m into an outermost cell;
(iii) when a mass density exceeds some critical value, let three mass particles ($3m$) fall from that cell into the inner three adjacent cells (avalanche);
(iv) in addition to procedure (iii), let a mass of m' ($\ll m$) fall into the adjacent inner cell at each ring (gradual diffusion);
(v) calculate X-ray luminosities, assuming that a certain fraction of the released potential energy is dissipated via radiation;
(vi) repeat (i)–(v) more than 10^5 times, plot light curves, calculate PSDs.

3. SOC model and the X-ray observations

The system then evolves to, and stays at, a self-organized critical state, exhibiting the following characteristics:
(i) aperiodic variations appear in the X-ray light curves (Fig. 1; left). Their PSDs show a $1/f$-like decline at large frequencies (Fig. 1; right);
(ii) the peak intensities (p) of individual shots are smoothly distributed over a wide range (Fig. 2). Roughly, the distribution is $D(p) \propto p^{-8}$ both in our SOC model with $m' = m/10$ and the observations.

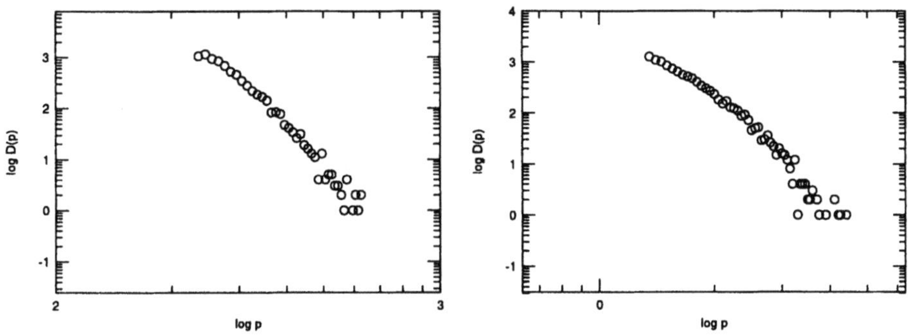

Figure 2. The observational (left) and theoretical (right) peak-intensity distribution of shots (from Negoro et al. 1995).

(iii) large shots tend to occur after an interval and this interval (waiting time) increases as p increases. This is because once a large shot occurs, a long waiting time is needed to refill an accretion disk with gas until a next large event takes place.

These three characters appear both in our SOC model (Mineshige et al. 1994b) and in the Ginga observations of Cyg X–1 (Negoro et al. 1995).

4. Discussion

Through the present analysis, it has been established that X-ray shots of Cyg X–1 occur according to a certain rule, although it looks as if they would randomly appear. What can we learn from this analysis about the structure of accretion flow onto compact objects?

First, the origin of the seed fluctuations is most likely to be magnetic activity, since magnetic fields tend to dissipate energy sporadically via reconnection, even if energy input is continuous. In fact, Lu & Hamilton (1991) claimed to reproduce power-law occurrence rates of solar flares by applying the concept of SOC to the solar magnetic fields. In the conventional picture of solar-type accretion disks in which magnetic activities are only enhanced in the chromosphere-corona, however, large-amplitude fluctuations are not expected. We thus conjecture that the disk is of low-β, (where β is the ratio of gas pressure to magnetic pressure) i.e., magnetic pressure dominated (cf. Shibata, Tajima & Matsumoto 1990).

Second, there exist the so-called advection-dominated disks characterized by hot temperatures (and thus hard spectra), low radiation emissivity, and large infall velocity – sometimes comparable to the azimuthal velocity (Narayan & Yi 1994; Abramowicz et al. 1995). Such disks are favored

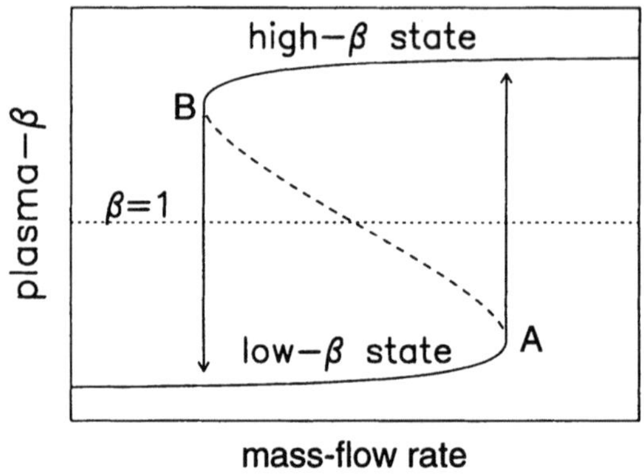

Figure 3. Hysteresis relation between the low-β, advection-dominated branch with enhanced variability (lower) and the high-β, standard branch with little variability (upper).

to produce fluctuations according to our model (Mineshige, Kusunose & Matsumoto 1995), partly because then magnetic activity can be enhanced, and partly because mass flow then occurs both in the azimuthal and radial directions (a basic assumption of our model; see Sect. 2). The main energy release thus occurs by sporadic reconnection events, leading to magnetic eruptions like solar flares.

Finally, black-hole accretion disks seem to alternate between the optically thin, advection dominated state with enhanced variability and the optically thick, standard state with little variability (Fig. 3). This will explain hard spectra of transient black-hole candidates during the rise and late decay phase, and soft spectra in the early decay phase (Mineshige et al. 1995).

References

Abramowicz, M.A., Chen, X., Kato, S., et al., 1995, Ap. J., **438**, L37
Bak, P., Tang, C., Wiesenfeld, K., 1988, Phys. Rev., **A38**, 364
Lu, E.T., Hamilton, R.J., 1991, Ap. J., **380**, L89
Makishima, K., 1988, in "Physics of Neutron Stars and Black Holes" ed. Y. Tanaka, Universal Academy Press, Dordrecht, p175
Mineshige, S., Kusunose, M., Matsumoto, R., 1995, Ap. J., **445**, L43
Mineshige, S., Ouchi, N.B., Nishimori, H., 1994a, PASJ, **46**, 97
Mineshige, S., Takeuchi, M., Nishimori, H., 1994b, Ap. J., **435**, L125
Narayan, R., Yi, I., 1994, Ap. J., **428**, L13
Negoro, H., Kitamoto, S., Takeuchi, M., Mineshige, S., 1995, Ap. J., **452**, L49
Oda, M., 1977, Space Sci. Rev., **20**, 757
Shibata, K., Tajima, T., Matsumoto, R., 1990, Ap. J., **350**, 295
van der Klis, M., 1994, Ap. J. Supp., **92**, 511

CYG X-2: ITS RECURRENT X-RAY BEHAVIOUR

E. KUULKERS[1,2], M. VAN DER KLIS[1]
1. Astronomical Institute "Anton Pannekoek", University of Amsterdam & Center for High-Energy Astrophysics, Kruislaan 403, 1098 SJ Amsterdam, The Netherlands
2. ESA/ESTEC, Astrophysics Division (SA), P.O. Box 299, 2200 AG Noordwijk, The Netherlands

Abstract. We find evidence for the occurrence of three different types of behaviour which as their basic distinction have different X-ray brightness levels. They are found to be recurrent.

1. Introduction

Z-sources are the brightest known low-mass X-ray binaries (Hasinger & van der Klis 1989): Sco X-1, GX 5–1, GX 17+2, GX 349+2, GX 340+0 and Cyg X-2. Most of these sources trace out a 'Z' in the so-called X-ray colour-colour diagram [CD; analogous to the $(U - B)$ versus $(B - V)$ diagram]. The three limbs of the Z are called horizontal branch (HB), normal branch (NB) and flaring branch (FB), from top to bottom, respectively.

2. Results

We analyzed the X-ray spectral variations of Cyg X-2 using all EXOSAT ME (~2...20 keV) data of this source between 1983 and 1985. We describe these variations in terms of the shifts and shape changes of the 'Z' pattern in X-ray colour-colour and X-ray hardness-intensity diagrams [HID; analogous to, e.g. $(U - B)$ versus V). For a full report we refer to Kuulkers, van der Klis & Vaughan (1996).

Colour, or hardness, is defined as the ratio of the dead-time corrected count rates in two X-ray energy bands, while intensity is defined as the total dead-time, background and collimator-response corrected count rate in the X-ray energy band of interest.

A CD and HID of all the observations are displayed in Fig. 1. A close look at the data reveals that Cyg X-2 varies between three intensity levels:

A. Evans and J. H. Wood (eds.), Cataclysmic Variables and Related Objects, 363–364.

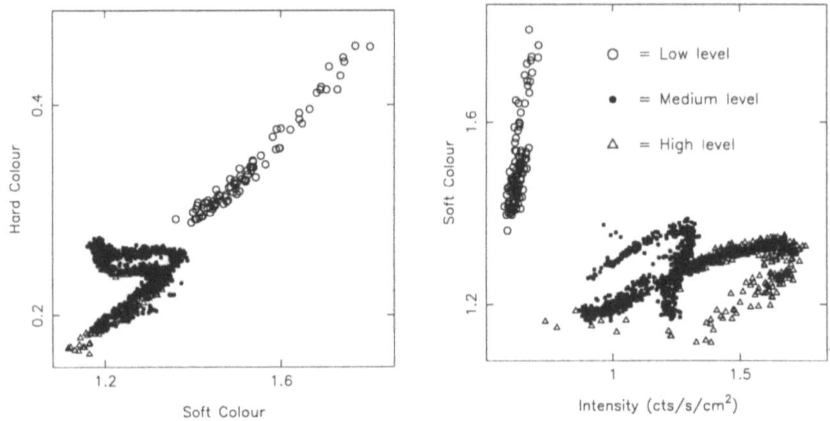

Figure 1. **Colour-colour (left) and hardness-intensity (right) diagram of all EXOSAT ME observations of Cyg X-2. [Soft colour: (3.1...6.4 keV)/(0.9...3.1 keV); hard colour: (6.4...19.7 keV)/(3.1–6.4 keV); intensity: 0.9...19.7 keV]**

'high', 'medium' and 'low'. During a *high* level episode the Z pattern is shifted to higher overall intensity ($\gtrsim 1.3\,\mathrm{counts\,s^{-1}\,cm^{-2}}$) with respect to the pattern of a *medium* level episode (between 0.7 and $1.3\,\mathrm{counts\,s^{-1}\,cm^{-2}}$). There are also differences in the *shape* of the Z pattern between the *high* and *medium* level. During the *medium* level both the CD and HID show a FB. In the upper part of the FB of this episode the intensity drops while the colours increase. During one of the *high* levels a branch connected to the lower part of the NB is seen in the HID, but *not* in the CD. We interpret this as FB behaviour. In this branch the intensity drops while the colours remain the same. During a *low* level episode, when the source showed the lowest observed overall intensity ($\lesssim 0.7\,\mathrm{counts\,s^{-1}\,cm^{-2}}$), no Z shape was seen but only one large, curved branch.

The Ginga LAC observations of Hasinger et al. (1990) show episodes reminiscent of the *medium* and *high* levels. The Einstein MPC data of Vrtilek et al. (1986) also shows evidence for the occurrence of the three different modes of behaviour, i.e. *low*, *medium* and *high* episodes.

We investigated whether the occurrence of the three levels is periodic as would be expected from e.g. a precessing disk. We found, however, no periodic recurrence time for the different episodes using the EXOSAT, Ginga and Einstein MPC data.

References

Hasinger, G., van der Klis, M., 1989, A&A, **225**, 79
Hasinger, G., van der Klis, M., Ebisawa, K., et al., 1990, A&A, **235**, 131
Kuulkers, E., van der Klis, M., Vaughan, B.A., 1996, A&AS, in press
Vrtilek, S.D., Kahn, S.M., Grindlay, J.E., et al., 1986, Ap. J., **307**, 698

THE ORBITAL PERIOD OF THE GALACTIC Z SOURCE SCO X-2

K. A. SOUTHWELL, J. CASARES, P. A. CHARLES
Department of Astrophysics, Nuclear Physics Building,
Keble Road, Oxford, OX1 3RH, UK

Abstract. Sco X-2 is a bright Galactic bulge Z source, whose optical counterpart was discovered in 1990 [?], and observed to exhibit strong Hα emission. We have therefore obtained spectroscopic data in order to determine the orbital period and investigate the nature of the companion star. From an analysis of the Hα emission line velocities, we find a most likely orbital period of 13.94 d. This implies an evolved companion, and is consistent with the currently known orbital periods in other Z sources.

1. Results and discussion

Spectroscopy of Sco X-2 was obtained in 1991 July/August and 1992 July with the WHT on La Palma. Narrow Hα emission is seen, but no structure is resolvable. A period search of the line velocities was undertaken by folding the data on a series of trial periods, and determining the χ^2 of a sinusoidal fit. In Fig. 1, we show the results of this search. The minimum χ^2 of 1.25 is attained for $P_{\rm orb} = 13.94 \pm 0.10$ d. We note there is little to

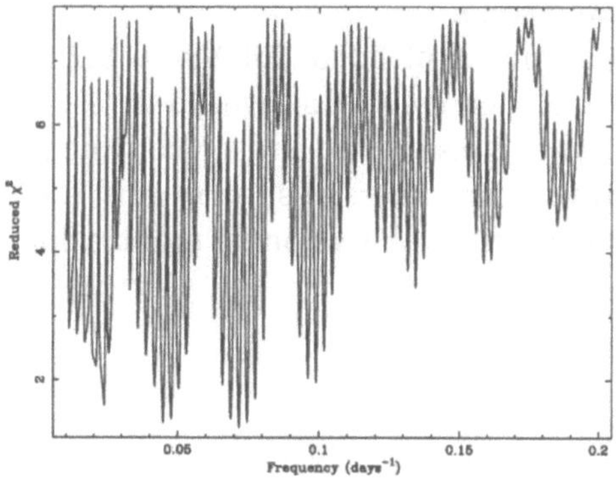

Figure 1. Minimised χ^2 versus trial frequency of sinusoidal fit to radial velocity data.

A. Evans and J. H. Wood (eds.), Cataclysmic Variables and Related Objects, 365–366.

choose between this and the $\chi^2 = 1.33$ value ($P = 22.25 \pm 0.16$ d). However, [?] derived a likely upper limit of $P \sim 19$ d for Sco X-2; furthermore, our radial velocities are reasonably fitted by a sinusoid when folded on 13.94 d, but show no obvious modulation when folded on 22.25 d. We therefore consider 13.94 d to be the more likely orbital period, although we regard this result with caution, given our poor sampling. In Fig. 2, we show the radial velocity curve produced by folding the data on this period. We find $\gamma = -227 \pm 6\,\mathrm{km\,s^{-1}}$ and $K = 70 \pm 8\,\mathrm{km\,s^{-1}}$ for the systemic velocity and semi-amplitude respectively. Adopting the relation $P^2(\mathrm{h}) \approx 110\,\overline{\rho}$, where

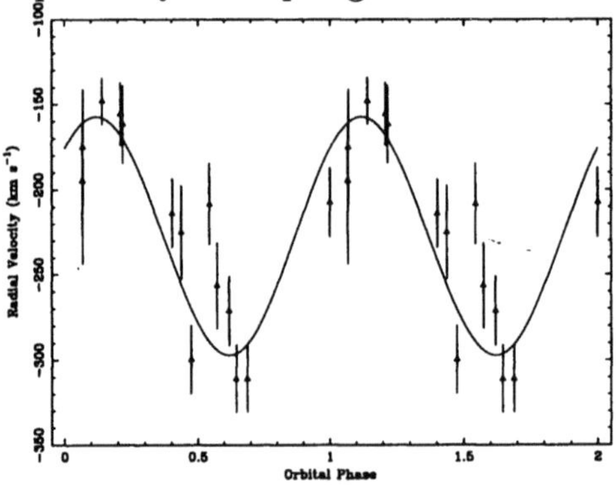

Figure 2. Radial velocity curve for Sco X-2, produced by folding the data on a period of 13.94 d. A sinusoidal fit is shown.

$\overline{\rho}$ is the mean density, and using the tabulation of [?] yields a spectral type of K0 III. We derive a value for the mass function, $f(M) = 0.5\,\mathrm{M_\odot}$. The H$\alpha$ emission could arise in the accretion disk (in which case the narrowness of the line suggests a low inclination), or in the X-ray heated face of the secondary. However, the mass function does not allow us to favour one of these sites over the other, particularly since K may be biased (either by a bright spot in the former case, or by the heating of the inner face of the companion star in the latter).

We find equivalent widths of 3.8 ± 0.4 Å and 9.2 ± 5.8 Å for our 1991 and 1992 observations respectively. However, our results are somewhat erratic due to poor signal-to-noise ratio, and no convincing periodicities were found. The Hα FWHM in our average spectrum is 7.3 Å corresponding to $327\,\mathrm{km\,s^{-1}}$. Doppler correcting the velocity modulation results in a value of 5.6 Å ($281\,\mathrm{km\,s^{-1}}$).

References

1. Allen, C.W., 1973, Astrophysical Quantities, The Athlone Press, University of London.
2. Penninx, W., Augusteijn, T., 1991, A&A, **246**, L81

OPTICAL COUNTERPART OF THE LMXB GX 13+1

P. J. GROOT[1], M. VAN DER KLIS[1], J. VAN PARADIJS[1,2],
T. AUGUSTEIJN[3], M. BERGER[1]
1. Astronomical Institute 'Anton Pannekoek'/CHEAF, University of Amsterdam, Amsterdam, The Netherlands
2. Physics Department, University of Alabama, Alabama, USA
3. ESO, La Silla, Chile

We have observed the IR counterpart of the Atoll type LMXB GX 13+1, (Naylor, Charles & Longmore 1991), from 1993 July 11...30, with the Dutch 90 cm telescope at ESO using a Gunn z (0.9–1.0 μm) filter. Over this period 110 usable images (integration time 12 m) were obtained. The data were reduced using the ESO–MIDAS/DAPHOT reduction package. A Lomb-Scargle period search was performed with the PERIOD package.

Fig. 1 shows our Gunn z band image with stars labeled as in Naylor et al. (1991). They propose their star 101 as the IR counterpart. Fig. 2 shows the light curve we obtained for this star. The star is clearly variable on a time-scale of $\sim$ 10 d. To estimate any period more accurately we have made a Lomb-Scargle statistics periodigram (see Fig. 3); maximum power is at 12.6 d. We estimate the error to be $\sim$ 1 day.

The time interval covered by our observations (18 d) is too short to establish the periodicity of the variation. If the orbital period is long as is suggested by our results it may be difficult to understand the difference between Atoll and Z sources on the basis of simple evolutionary arguments (see e.g. Lewin, van Paradijs & van der Klis 1988).

Acknowledgements. We thank Jeroen de Jong, Vik Dhillon and Gene Magnier for the use of their programs.

References

Naylor, T., Charles, P.A., Longmore, A.J., 1991, MNRAS, **252**, 203
Lewin, H.G., van Paradijs, J., van der Klis, M., 1988, Sp. Sc. Rev., **46**, 273

A. Evans and J. H. Wood (eds.), Cataclysmic Variables and Related Objects, 367–368.

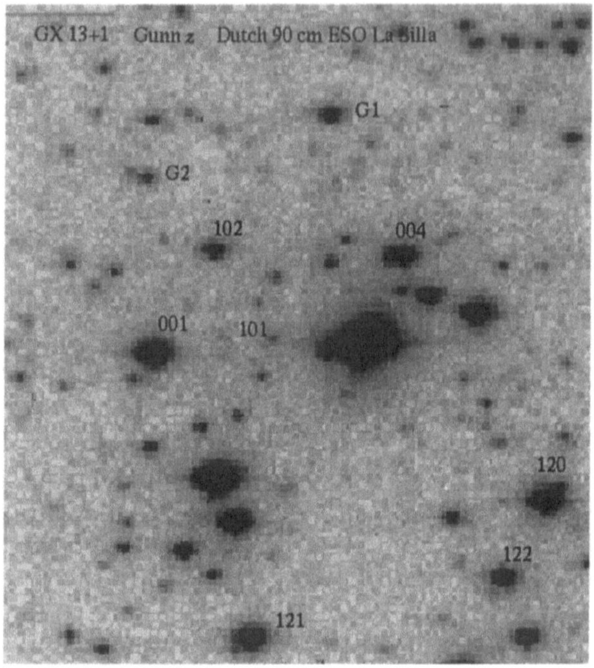

Figure 1. Gunn z band image of GX 13+1. Star 101 is the IR counterpart.

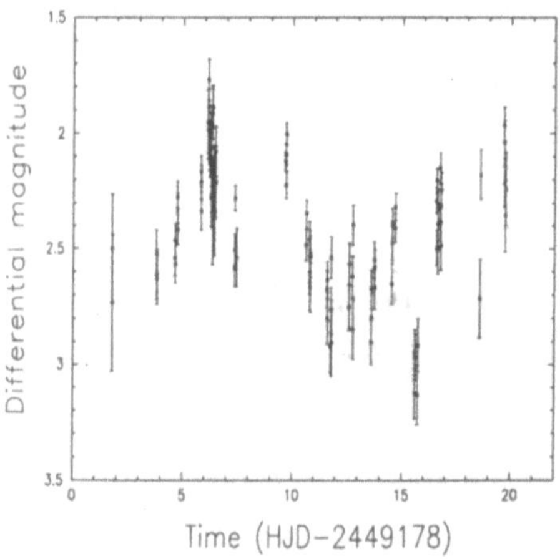

Figure 2. Light curve of star 101 in Fig. 1. Magnitudes in units of average magnitude of comparison stars.

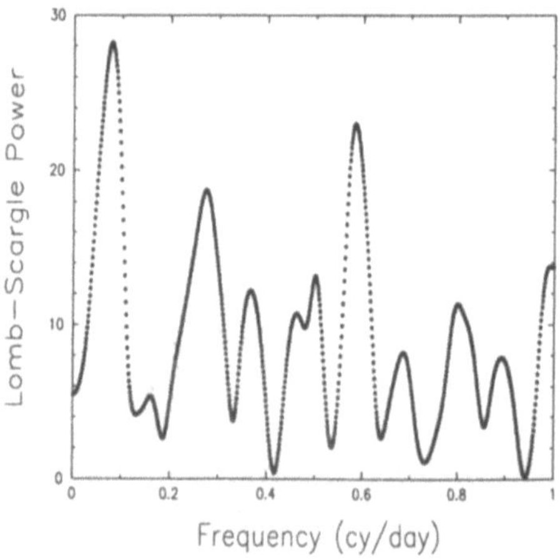

Figure 3. Lomb-Scargle statistics power spectrum of the light curve in Fig. 2.

DETECTION OF THE SECONDARY STAR IN X1822−371

E. T. HARLAFTIS[1], P. A. CHARLES[2], K. HORNE[1]
1. School of Physics & Astronomy, University of St. Andrews, St. Andrews KY16 9SS, UK
2. Astrophysics, Department of Physics, University of Oxford, Keble Rd, Oxford OX1 3RH, UK

Abstract. We have detected He I absorption from the companion star to X1822−371 and find a lower limit to its K−velocity of $230 \pm 50\,\mathrm{km\,s^{-1}}$. We interpret the He I as arising on the X–ray heated inner face of the companion star.

1. Discussion

The prototype accretion disc corona (ADC) source is X1822−371 whose orbital light curve on its 5.57 h period indicates significant structure in the outer disc at orbital phases 0.8 and 0.3 (Hellier & Mason 1989). We observed X1822−371 with the AAT in 1993 June (5 orbital cycles for a total of 110 spectra) with the aim of investigating the disc structure and the elusive nature of the companion star. A full account of these observations will be given in Harlaftis et al. (1996, in preparation).

An image of the variation of the Hα profile on 1993 June 26 is given in Fig. 1 as trailed spectra (intensity scale −0.1 to 0.5 mJy). The Hα profile of X1822−371 is complex, with a double-peaked emission profile. The absorption or emission wings of the profiles do not show any binary motion. The peaks of the profiles appear to become stronger at phases 0.4 and 0.8. In the core of the emission profile, there is absorption which shows maximum intensity at phases 0.6 and 0.2 and could perhaps be part of a weak, complex, sinusoidal-like wave. During eclipse, the velocity separation of the Hα peaks is only $460 \pm 20\,\mathrm{km\,s^{-1}}$. Note that the strength of the eclipse profile appears stronger because there is no absorption below the subtracted continuum.

We find the weak He I triplet 5875.6 Å in absorption moving with the binary cycle. The trailed spectra (23 bins resulted from all spectra) are shown

A. Evans and J. H. Wood (eds.), Cataclysmic Variables and Related Objects, 369–370.

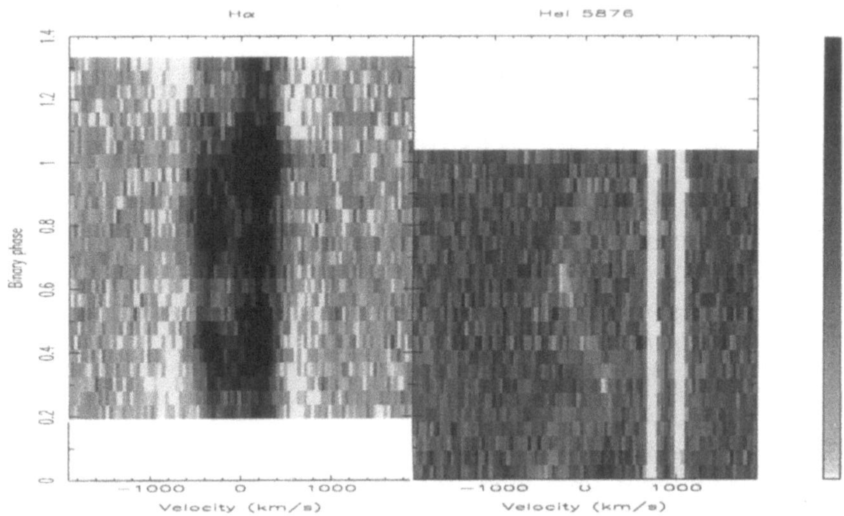

Figure 1. The trailed spectra of Hα (left) and He I λ5876 and the interstellar Na D lines (right). The He I absorption line moves from red to blue at phase 0.5, indicating its association with the companion star.

in Fig. 1 together with the NaD interstellar lines for comparison (intensity scale −0.1 to 0.1 mJy). The sinusoidal wave crosses zero velocity from red-to-blue at phase 0.5 and is much narrower between phases 0.3…0.7 which indicates the companion star as its source. This is consistent in phase with Cowley et al. (1982) who claimed that the Hδ absorption core was moving in anti-phase to the He II emission (the latter arising in the accretion disc). The presence of Balmer and weak He I lines in absorption make the companion star spectrum resemble that of an early B-star (~20 000 K). However, during eclipse there is no trace of He I in absorption (e.g. a B9 star) which indicates that the back of the companion star is cooler than the inner face by ~10 000 K. The above observations suggest that the hard X-rays from the vicinity of the compact object irradiate the photosphere of the companion star severely enough to change its spectral characteristics from those possibly of an M-star to those of a B-star. Extraction of the radial velocity curve between phases 0.3…0.8 using Gaussian fits gives a semi-amplitude of ~230±50 km s^{-1}, which with modelling of the irradiation of the inner face of the companion star will constrain the mass function of the compact object.

References

Cowley, A., et al. 1982, Ap. J., **255**, 596
Hellier, C., Mason, K.O., 1989, MNRAS, **239**, 715

RADIO EMITTING X-RAY BINARY STARS

R. E. SPENCER
University of Manchester
Jodrell Bank
Macclesfield
SK11 9DL, UK

1. Introduction

Some of the most astrophysically interesting objects are found among the radio-emitting X-ray binary stars (REXRB). The class includes the well-studied objects such as SS 433, Cyg X–3 and Sco X–1. The recent discoveries of relativistic ejection of radio knots in the X-ray transients 1915+105 (Mirabel & Rodriguez 1994) and 1655–40 (Hjellming & Rupen 1995) well illustrate the extreme nature of some of these objects.

X-ray binaries are semi-detached binary stars in which matter is transfered from a more or less normal star onto a neutron star or black hole. X-ray satellites have detected large numbers of these objects (193 in a recent catalogue by van Paradijs 1995). However only a small fraction of these are known to have radio emission (e.g. Hjellming 1988).

2. Radio properties

The radio emission mechanism is almost certainly non-thermal and due to synchrotron radiation, since the sources have high brightness temperatures and in some cases show polarized emission. Energetic electrons and magnetic fields are therefore required. The relationship between the radio and X-ray emission remains unclear, though some sources with Z–shaped X-ray colour-colour diagrams have stronger radio emission in the radio when in the horizontal branch (e.g. GX 5+1, Tan et al. 1992).

The REXRBs found to date contain a wide range of different types of source, although all obtain their energy as a result of accretion onto a compact object. There is as yet no clear demarcation in the radio behaviour

A. Evans and J. H. Wood (eds.), Cataclysmic Variables and Related Objects, 371–374.

between high mass XRBs (Table 1), low mass XRBs (Table 2) and black hole candidates. Flux densities vary from a barely detectable 0.1 mJy to 20 Jy, giving radio luminosities which are $\sim 10^6$ times smaller than those at X-ray wavelengths, so at first sight the radio emission is insignificant.

TABLE 1. The High Mass REXRB

Object		Orbital Period	Radio Flux mJy	Dist. kpc	Radio Lum. erg s^{-1}	X-ray Lum. erg s^{-1}
γ Cass.	0053+604	2.3 h	0.3	0.22	10^{27}	10^{33}
LSI+61°303	0236+610	26.5 d	25..240	2.3	$10^{31}..10^{32}$	10^{33}
V725 Tau	0535+262	111 d	0.2	1.8	10^{28}	$10^{34}..10^{37}$
SS 433	1909+048	13.1 d	200..2000	4.9	$10^{32}..10^{33}$	10^{35}
GRS	1915+105	?	<0.2..700	12.5	$10^{31}..10^{34}$	10^{38}
Cyg X–1	1956+350	5.6 d	7..20	2.5	$10^{30}..10^{31}$	10^{37}
Cyg X–3	2030+407	4.8 h	50..20000	10	$10^{32}..10^{35}$	10^{37}

However this ignores the contribution to the energetics made by the relativistic electrons and magnetic fields necessary for the production of the radio emission. Estimates for the minimum energy in these components can be made (see e.g Pacholczyk 1970) using $E_{\rm min} = (B^2/8\pi)(7/3)$Volume, where B is the magnetic field calculated assuming equipartition. The ratio of proton to electron energy, and the filling factor, have both been assumed to be unity in the calculations. Thus the minimum energy, and hence the power required, can be found if the time-scale for the generation of an extended component is known. The time-scale can be easily determined from the radio flux variability when components are seen to be formed and to evolve.

Only a very few REXRBs turn out to have known radio structures, due to most being unresolved by the VLA and too weak to be studied by VLBI. Most objects which have known structures are listed in Table 3, the only exception being Cir X–1. It is striking that all these objects have collimated flows in their radio structure and that three (J1655–40, SS 433 and 1915+105) have well developed jets.

Using data given by the papers in Table 3, we can work out the equipartition magnetic fields (mostly around 30 mG) and minimum energies of the electrons and magnetic fields required to give the observed radio emission. Using the rise times of the radio flares we can then find the minimum power required as shown in Table 3.

TABLE 2. The Low Mass REXRB

Object		Orbital Period	Radio Flux mJy	Dist. kpc	Radio Lum. erg s^{-1}	X-ray Lum. erg s^{-1}
J0422+32	0422+327	5.1 h	4..8	1..6	$\sim 10^{30}$	$10^{36}..10^{37}$
V1055 Ori	0614+091	5 d	3..40	2.5	$10^{30}..10^{31}$	10^{36}
Nova Mon	0620–003	7.8 h	5..300	0.6	$10^{28}..10^{30}$	$10^{32}..10^{38}$
Nova Mus	1124–684	10.4 h	2..170			
Cen X–4	1455–314	15.1 h	1..8			
Sco X–1	1617–155	0.79 d	1..10	2	$10^{29}..10^{30}$	$10^{37}..10^{38}$
GRO	J1655–40	2.62 d	2..7000	3.2	$10^{30}..10^{33}$	10^{37}
GX9+9	1728–169	4.2 h	0.4	10	10^{30}	10^{38}
GX9+1	1758–205		0.4			10^{38}
GX13+1	1811–171		0.1..1	7	$10^{29}..10^{30}$	10^{38}
GX17+2	1813–140		0.5..13	8.5	$10^{29}..10^{31}$	10^{38}
Ser X–1	1837+049		1	10	10^{30}	10^{38}
Aql X–1	1908+005	19 h	0.1..0.4			10^{36}
QZ Vul	2000+251	8.3 h	0.4..8	2	$10^{29}..10^{30}$	10^{38}
V404 Cyg	2023+338	6.5 d	0.3..1640	3.	$10^{29}..10^{33}$	10^{38}
Cyg X–2	2142+380	9.8 d	0.6..12	8	$10^{31}..10^{32}$	10^{38}
GT2318+620	2318+620		6..40	3..6	$10^{31}..10^{32}$	$10^{34}..10^{35}$

TABLE 3. Energetics of REXRB

Object	Type	Refs. for radio	Radio Rise-time	X-ray Lum. erg s^{-1}	Min. Radio Power erg s^{-1}
LSI+61°303	ns + Be	1	4 d	10^{33}	10^{34}
GRO J1655-40	BH?	2	< 10 d	10^{37}	10^{37}
SS 433	ns + WN	3	1 d	10^{35}	10^{37}
GRS 1915+105	not known	4	< 2 d	10^{38}	10^{39}
Cyg X–3	ns+WR	5	$\sim$ 6 h	10^{37}	10^{39}

1 - Massi et al. 1993; 2 - Tingay et al. 1995; 3 - Jowett 1995; 4 - Mirabel & Rodriguez 1995; 5 - Spencer et al. 1986.

3. Conclusion

We can see that the minimum powers required for the radio emission are comparable or even in excess of the X-ray luminosity, and much greater than the observed radio luminosity. Therefore the radio emission, contrary to common belief, is energetically important and may even dominate the system. Note that SS 433 also has a high kinetic energy in thermal material responsible for the moving optical lines of $\sim 10^{41}$ erg s^{-1} and suggests that even higher powers are required. The extraction of such large luminosity from the binary system should therefore be considered when accretion models are being discussed.

References

Hjellming, R.M., 1988, in "Galactic and Extragalactic Radio Astronomy", eds G. Verschuur, K. Kellermann, Springer-Verlag, New York, p381

Hjellming, R.M., Rupen, M.F., 1995, Nature, **375**, 464

Jowett, 1995, PhD thesis, University of Manchester, and private communication

Massi, M., Paredes, J.M., Estalella, R., Felli, M., 1993, A&A, **269**, 249

Mirabel I.F., Rodriguez, L.F., 1994, Nature, **371**, 46

Pacholcyzk, A.G., 1970, Radio Astrophysics, Freeman, San Fransisco

Spencer, R.E., Swinney, R.W., Johnston, K.J., Hjellming, R.M., 1986, Ap. J., **309**, 707

Tan, J., Lewin, W., Hjellming, R., et al., 1992, Ap. J., **385**, 314

Tingay, S.J., Jauncey, D.L., Preston, R.A., et al., 1995, Nature, **374**, 141

van Paradijs, J., 1995, in "X-ray Binaries", eds W. Lewin, J. van Paradijs, E. van den Heuvel, Cambridge University Press, p536

MERLIN OBSERVATIONS OF EXTENDED EMISSION AROUND GRS 1915+105.

S. J. NEWELL, R. E. SPENCER, F. H. JOWETT
NRAL, Jodrell Bank, Cheshire, SK11 9DL, UK

Abstract. We present MERLIN observations of the exciting X-ray transient GRS 1915+105. Previous radio observations of GRS 1915 have shown that it is a source of superluminal radio jets. The emission of anti-parallel radio knots has led to this being described as a 'microquasar'. MERLIN observations presented here show further details of the structure. The two main features observed are a bright extension in the direction of the superluminal jet and also, an extended region of low brightness which is elongated almost at right angles to the jet. This latter feature could either be an associated nebulosity or the remains of a jet which has undergone a radical change in direction.

GRS 1915+105 was originally detected as an X-ray transient by the WATCH all-sky monitor on board the GRANAT satellite (Castro-Tirado, Brandt & Lund 1992). Its importance was not fully realised, however, until VLA observations were taken immediately after a flaring event in 1994 March (Mirabel & Rodriguez 1994; hereafter MR94). These observations showed a pair of anti–parallel radio knots (in the NW–SE direction) which were moving ballistically out from the central core of the system. The radio knots were emitted inclined to the line of sight and were undergoing relativistic beaming. The apparent proper motions of the two knots gave apparent velocities of the knots on the plane of the sky of $1.25 \pm 0.15\,c$ and $0.65 \pm 0.08\,c$ (MR94). From this an intrinsic velocity of $0.92\,c$ at a jet angle 70° to the line of sight was deduced (MR94). This velocity is obviously far greater than the values seen in the other X-ray binaries and so required further study.

We observed GRS 1915 using the full MERLIN array at 408 MHz on 1994 November 28, a few days after a small outburst. The results of mapping the data using natural weighting are very interesting (Fig. 1). The resolution

A. Evans and J. H. Wood (eds.), Cataclysmic Variables and Related Objects, 375–376.

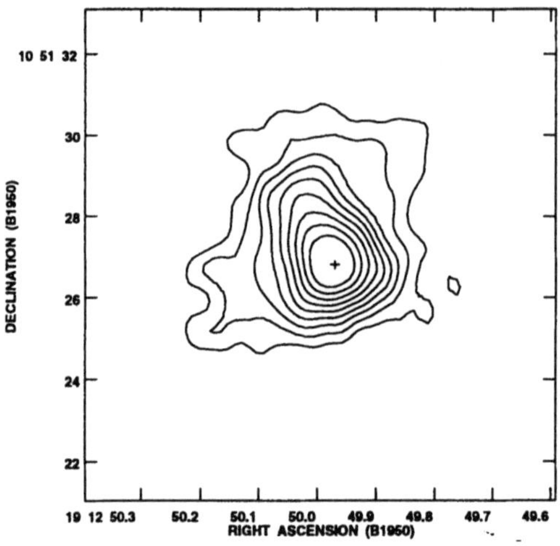

Figure 1. MERLIN map of GRS 1915. The NW–SE orientation which was observed in the original maps (MR94) can be seen. The new extension at right angles to the jet is also quite clear. The peak flux in this map is 15.3 mJy beam^{-1}. The contours are at −3.0, 4.0, 5.0, 6.0, 7.0, 8.0, 9.0, 10.0, 11.0 and 12.0 times 1.1169 mJy beam^{-1}.

of this map picks up a region of lower brightness which surrounds the source. This could be caused by:

1. Precession of the jet as seen in SS433. It is not clear, however, how we link the closely aligned 'jet' from the high resolution map with the idea that this jet has undergone precession. It is possible that we may be seeing the jet from an earlier period prior to the current alignment but this is not at all certain and requires more investigation
2. A low brightness nebulosity surrounding the source. This could be a gaseous halo around the source, or something more akin to a supernova remnant. Possibly this is material dumped into the ISM by leakage from the jets or released by the interaction of the jets with a supernova remnant.

We have verified the presence of a jet in this source which appears to agree with the observations of Mirabel & Rodriguez (1994). We have also added to the questions to be answered with our discovery of the low brightness nebulosity surrounding the source.

References

Castro-Tirado, A.J., Brandt, S, Lund, N, 1992, IAUC 5590
Mirabel, I.F., Rodriguez, L.F., 1994, Nature, **371**, 46 (MR94)

ACCRETION DISK DYNAMICS OF HER X–1

S. D. VRTILEK[1,2], F. H. CHENG[1]
1. University of Maryland, College Park, MD 20742, USA
2. CfA, 60 Garden St., Cambridge, MA 02138, USA

1. Introduction

The Her X–1/HZ Her system consists of a neutron star with a rotation period of 1.24 s orbiting a 'normal' stellar companion every 1.7 d. Variation in optical and ultraviolet magnitude is attributed to X-ray heating on the side facing the active neutron star with some contribution from X-ray heating of the disk. The system also has a 35-day period in which the X-ray emission displays so-called main and short 'on' states separated by relative X-ray quiescence. The 1.7-day optical flux variations continue throughout the 35-day period. No clear consensus exists as to the origin of the 35-day period; a popular interpretation is that of occultation of the compact X-ray source by a tilted, precessing, accretion disk.

Each temporal state has its own distinctive spectral behavior. During the main on state the pulse-averaged spectrum is a flat power law with an excess strong thermal component at low energies (a characteristic of many binary X-ray pulsars). Spectral features are detected at 6.4 keV attributed to Fe K–shell, and between 35...58 keV attributed to electron-cyclotron resonance implying a surface magnetic field of $(2\ldots5)\,10^{12}$ G. The pulse profile changes as a function of energy: the pulse peak shifts in phase by 180° between low and high energies and shows less structure in the low energy band. The spectrum also changes as a function of pulse phase: the soft thermal component and the Fe line flux are strongest during the off-peak of the pulse. This implies that the soft X-rays are likely to be reprocessed hard X-rays and the Fe line emission originates at the reprocessing site.

2. The 1993 multiwavelength campaign

The launch of EUVE opened up a new region of the spectrum that is particularly useful for study of the strong soft component. The unusu-

A. Evans and J. H. Wood (eds.), Cataclysmic Variables and Related Objects, 377–380.

ally high galactic latitude and correspondingly low interstellar absorption ($N_H \sim 10^{20}$ cm^{-2}), towards Her X–1/HZ Her make it the *only* X-ray binary system EUVE is able to study. Observations at other wavelengths are also important since the soft component is likely to be reprocessed radiation; our campaign employed IUE, ROSAT, ASCA, GRO/BATSE, and several ground-based observatories in addition to EUVE.

The campaign was scheduled during the high on state of the 35-day cycle but the X-ray flux observed was a factor of 50 below the expected level (Vrtilek et al. 1994). For the second time in Her X–1's recorded history the system had been caught in a so-called 'anomalous low' state. A substantial, unexpected drop in flux at X-ray energies, with no substantial change in absorbing column density; little or no change in UV and optical fluxes; an increase in pulse period; and no pulsed emission above 1.0 keV are some characteristics of the anomalous low state. With the exception of the increase in pulse period and the degree of X-ray flux reduction these characteristics are similar to those observed during 'normal' 35-day low states. Our observations are particularly exciting because the simultaneous multiwavelength coverage enables us to understand the 'lost' X-ray flux.

The ROSAT pulse profile and energy dependence show the normal main on state behavior during the highest X-ray flux observed (a factor of 2 below the expected): a shift of pulse phase by about 180° at 0.9 keV is clearly detected. The pulse period measured by ROSAT, 1.237749(1)s, shows a sharp increase (7 μs) from the previous measurement contrary to the usual spin-up. During the lowest X-ray flux observed neither ROSAT nor ASCA detected pulsed emission above 0.9 keV.

Long term observations of Her X–1 by BATSE suggest that the hard X-ray flux of Her X–1 undergoes episodes of reduction that are correlated with spin-down. The correlation between times of low hard X-ray luminosity and episodes of spin-down are predicted by models such as those of Ghosh & Lamb (1979); an *increase* in mass accretion rate ($\dot{M}$) causes the Alfven radius at which material joins the magnetosphere to move inward, thereby *reducing* the drag-producing torques of the field lines threading the disk outside the corotation radius, leading to spin-up.

3. Modelling of the UV and optical continuum

The model used to fit the IUE data is described in detail by Cheng, Vrtilek & Raymond (1995; 1996). We adopt the system geometry of Howarth & Wilson (1983) where changes in X-ray, UV and optical flux during the 35-day cycle are due to a tilted, precessing, accretion disk. The X-ray heating code is adapted from one used to model the UV flux of low-mass X-ray binaries (LMXBs; Vrtilek et al. 1990). A major difference is that, while the

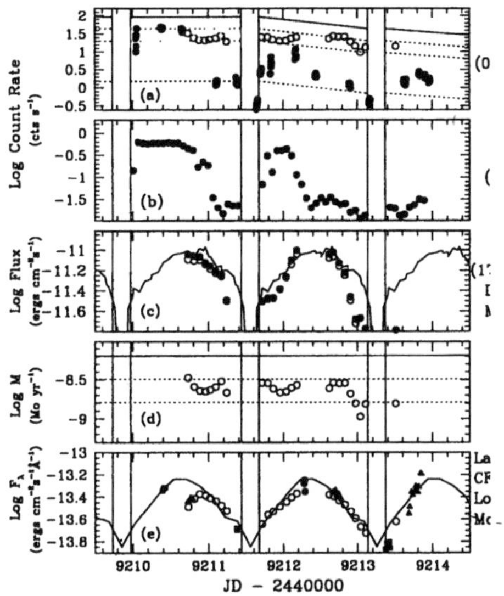

Figure 1. Model with changes in mass accretion rate alone

contribution from the star is negligible in LMXBs, it constitutes 70% of the total flux in UV and optical bands for Her X–1.

We first fit the UV continuum while changing only $\dot{M}$. In Fig. 1 we have plotted the observed (filled circles) and model (open circles) light curves and the $\dot{M}$ values obtained by fits to the IUE data. The solid lines in Figs. 1a,d represent the highest observed X-ray fluxes and mass transfer rates, and in Figs. 1c,e the average of previous observations. The dashed lines in Fig. 1a depict X-ray fluxes at 1/2, 1/4, and 1/40th of the expected values and in Fig. 1d mass transfer rates at 1/2 and 1/4 of the highest observed value. Overall while we get rather good fits to the IUE continuum with changes in $\dot{M}$ only, the predicted reduction in X-ray flux is a fraction of that observed.

To further reduce the X-ray flux we can change the inclination angle of the disk with respect to our line-of-sight as is done to explain the 35-day effect. Going to a point in the 35-day cycle that totally obscures the X-ray flux we find that disk inclination angles above 81° can achieve this. We assume that the X-ray flux when unobscured is proportional to $\dot{M}$ and determine the disk inclination angle required to reduce the X-ray flux to the level observed. We then fit the IUE data for that inclination angle. We find that reductions in $\dot{M}$ by about a factor of 2 and changes in disk inclination angle of up to 20° give good fits to the IUE continuum *and* reduces the X-ray flux appropriately (Fig. 2).

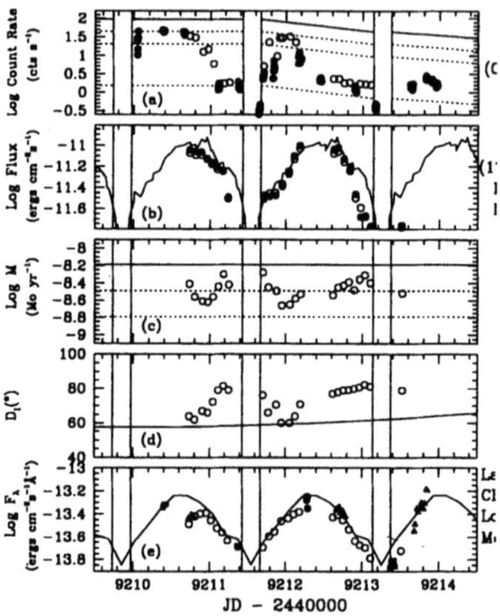

Figure 2. Model with changes in mass accretion rate and disk inclination angle

4. Summary

We interpret the initial drop in count rate accompanied by a change from spin-up to spin-down during our campaign in terms of a reduction in $\dot{M}$. Fits to UV/optical data show that the light curves can be reproduced by assuming a reduction in $\dot{M}$. The decrease in $\dot{M}$ is insufficient to account for the drop in X-ray flux, so disk obscuration must be taking place. The UV/optical/X-ray light curves can be reproduced by a reduction in $\dot{M}$ *and* accompanied changes in disk inclination angle. The change in torque on the neutron star implied by the switch from spin-up to spin-down can act to shift the disk. The lack of pulsations at energies greater than 1 keV is consistent with the interpretation that the hard X-rays are largely occulted. The changes in disk inclination angle may be related to shifts in 35-day period. Simultaneous ultraviolet, optical, and X-ray observations over several 35-day cycles should confirm this effect.

References

Cheng, F.H., Vrtilek, S.D., Raymond, J.C., 1995, Ap. J., **452**, 825
Cheng, F.H., Vrtilek, S.D., Raymond, J.C., 1996, these proceedings, p381
Ghosh, P., Lamb, F.K., 1979, Ap. J., **234**, 296
Howarth, I.D., Wilson, B., 1983, MNRAS, **202**, 347
Vrtilek, S.D., Mihara, T., Primini, F.A., et al., 1994, Ap. J., **436**, L9
Vrtilek, S.D., Raymond, J.C., Garcia, M.R., et al., 1990, A&A, **235**, 162

AN ARCHIVAL STUDY OF HST OBSERVATIONS OF HER X–1/HZ HER

F. H. CHENG[1], S. D. VRTILEK[1,2], J. C. RAYMOND[2]
1. University of Maryland, USA
2. Harvard-Smithsonian Center for Astrophysics, USA

1. Introduction

Her X–1 is an X-ray pulsar with a rotation period of 1.24 s and a binary period of 1.7 d (Tananbaum et al. 1972). The 1.7 d variations in optical and ultraviolet flux are attributed to X-ray heating of the companion star and disk (e.g. Howarth & Wilson 1983, hereafter HW83). The system displays a 35 d period, attributed to the effects of a tilted, precessing, accretion disk. Optical and ultraviolet flux variations continue *unchanged* throughout. This work is motivated by the following reasons:

• The observed IUE spectra have significantly flatter slopes than those predicted by previous models (e.g. HW83).
• The observed strength of the Balmer jump is anomalously low compared to that expected for a normal B star (Anderson et al. 1994).
• HST observations obtained by Anderson et al. (1994) in order to study emission lines have yielded high quality spectra of the *continuum* emission from HZ Her, enabling for the first time detailed model fitting efforts.

2. Model

• Adopt the system geometry of HW83 and X-ray heating code described in Vrtilek et al. (1990, 1991).
• Assume both disk and star have an albedo of 50%.
• Estimate $E(B-V) < 0.05$ from the lack of a 2200 Å absorption feature.
• Use $L_x = 0.5GM\dot{M}/r_1$, where r_1 is the radius of the neutron star.
• Calculate the temperature at star and disk surfaces due to X-ray heating. Use IUE and Kitt Peak stellar fluxes (referred to as *star-type spectra*) for that temperature to determine UV/optical continuum flux (HW83 used model stellar atmospheres).

A. Evans and J. H. Wood (eds.), Cataclysmic Variables and Related Objects, 381–382.

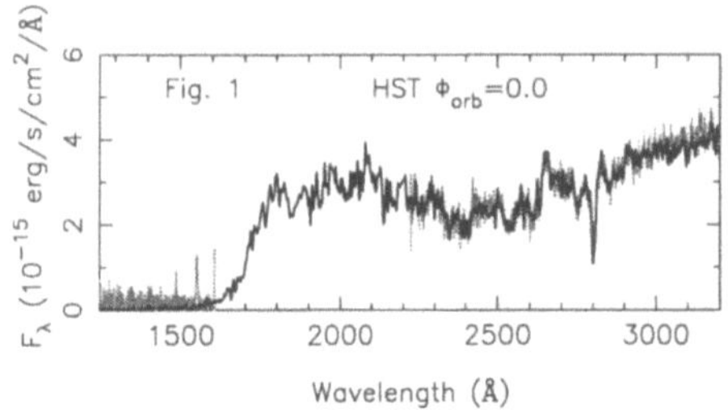

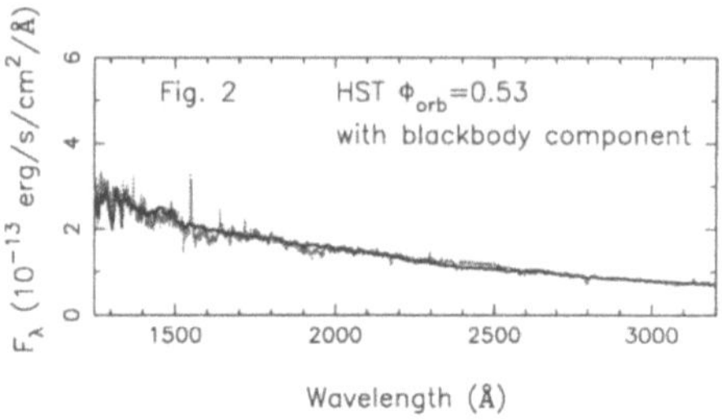

- Include blackbody component. Introduce two critical temperatures: T_{sc} and T_{dc}. For a given area element in the disk and star, if the temperatures T_d and T_s are higher than T_{dc} or T_{sc}, compute the flux using a blackbody, otherwise use *star-type* spectrum.
- Free parameters are $\dot{M}$, T_{dc} and T_{sc}, ϕ_{orb}, ϕ_{35}. Fixed parameters are q, β, θ_d, α_d, $\Delta\psi$, i, a, d, r_1 and r_2 (a detailed description of the model can be seen in Cheng, Vrtilek & Raymond 1995).

3. Results

- At $\phi_{orb} \sim 0.0$, we obtain the average temperature of the unheated star surface $T_{s0} = 8\,100 \pm 240$ K (Fig. 1).
- At $\phi_{orb} \sim 0.5$, we obtain $T_{dc} \sim 10\,000$ K, $T_{sc} \sim 18\,900$ K, and $\dot{M} = (6.5 \pm 0.9)\,10^{-9}\,M_\odot\,yr^{-1}$ (Fig. 2).
- Balmer Jump: see Table 1.

TABLE 1. **Comparison of the Balmer Jump**

ϕ_{orb}	Observed BJ	Model BJ
0.0	3.0	2.8
0.53	1.3	1.8 (no blackbody)
0.53	1.3	1.4 (with blackbody)

References

Anderson, S.F., Wachter, S., Margon, B., et al., 1994, Ap. J., **436**, 319
Cheng, F.H., Vrtilek, S.D., Raymond, J.C., 1995, Ap. J., **452**, 825
Howarth, I.D., Wilson, B., 1983, MNRAS, **204**, 347 (HW83)
Tananbaum, H., Gursky, H., Kellogg, E.M., et al., 1972, Ap. J., **174**, L143
Vrtilek, S.D., Penninx, W., Raymond, J.C., et al., 1991, Ap. J., **376**, 278
Vrtilek, S.D., Raymond, J. C., Garcia, M.R., et al., 1990, A&A, **235**, 162

THE 35 DAY CYCLE OF HER X-1

S. SCHANDL, F. MEYER
Max-Planck Institut für Astrophysik, Karl Schwarzschildstr. 1, 85748 Garching, Germany

Abstract. The 35 d period in the observed X-ray and optical light curves of Her X-1 can be understood in terms of a warped accretion disk precessing in the tidal field of the companion. We propose that repulsive forces of the X-ray driven coronal wind acting on the disk causes and maintains the warp of the disk (Fig. 1). The resulting shape of the precessing disk reproduces the main features of the observed X-ray light curve [1]. Based on the calculated disk shape we model the dip structure in the X-ray observations as caused by the interaction between the accretion stream and the warped disk.

The interaction yields a thickening of the involved disk region which is carried through the disk at the corresponding radii because of the Keplerian motion. Therefore the whole disk 'rings' are involved in this thickening (Fig. 2) which gets smoothed out on the thermal time-scale $(\alpha\Omega)^{-1}$. Because of the warped shape of the disk each disk 'ring' gets thickened twice per orbit, once the accretion stream hits the disk at the upper surface, 0.5 in orbital phase later at the lower surface. At turn-on we look marginally over the disk surface and thus the additional small thickening can cover the X-ray source at those orbital phases where the disk ring which does the turn-on is thickened (**anomalous dip**). This explains naturally that anomalous dips occur only at turn-on and only at two special orbital phases around 0.2 or 0.7. The dip spectrum shows cold matter absorption as expected from the disk matter.

The interaction yields also a cold and clumpy spray which covers the source at orbital phases just before each eclipse (Fig. 2). Because of the twist these **pre-eclipse dips** march backwards in orbital phase, as is observed. The observations show also strong intensity variations on short time-scales ($\sim$ min) and cold matter absorption features in the spectrum which the assumed cold and clumpy spray can reproduce (submitted to A&A).

The observed optical data depend on the phase of the 35 d precession.

A. Evans and J. H. Wood (eds.), Cataclysmic Variables and Related Objects, 383–384.

The most striking feature is the relatively high luminosity around phases $\phi_{\text{orbit}} = 0.3$, $\phi_{\text{precession}} = 0.0$. The corresponding feature in the simulation is caused by the interaction of the accretion stream with the disk **rim.**

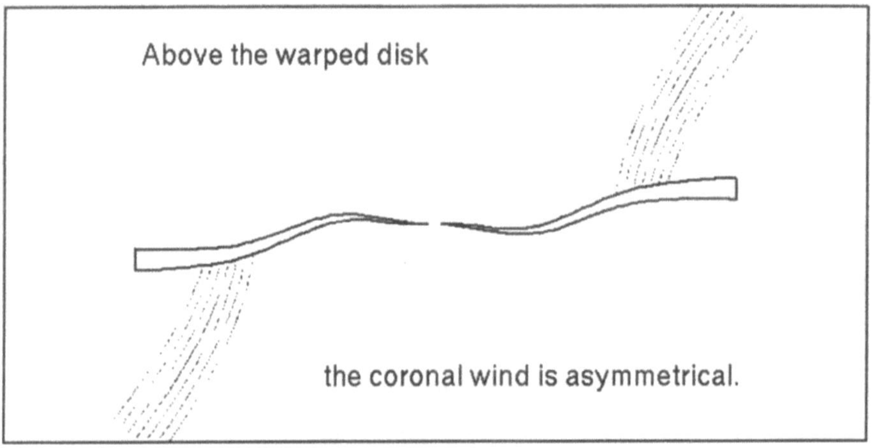

Figure 1. The vertical cut through the warped disk, with its X-ray driven coronal wind above the asymmetrical illuminated surface, shows schematically the mechanism which produces and maintains the tilt and twist of the accretion disk.

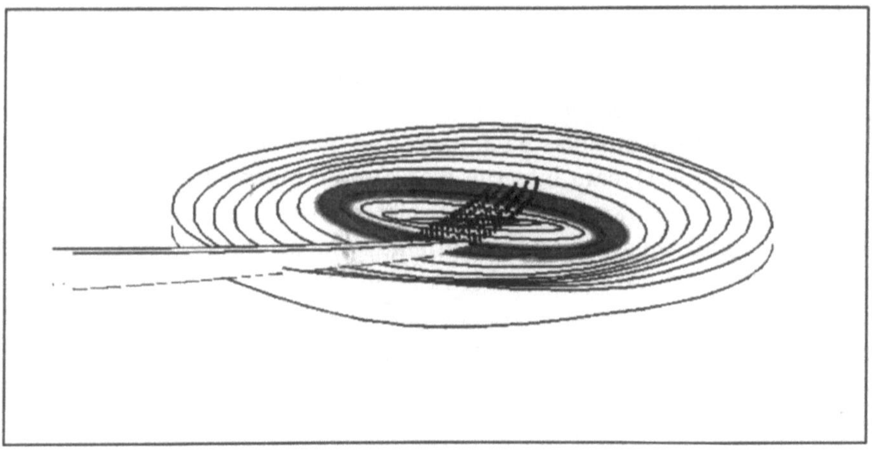

Figure 2. The impact of the accretion stream on the disk surface causes a spray of matter and a thickening of the disk which is carried through the disk along Keplerian orbits (shown grey). Both effects reproduce the observed pre-eclipse and anomalous dips.

References

1. Schandl, S., Meyer, F., 1994, A&A, **289**, 149

MODELS OF SOFT X-RAY TRANSIENTS AND DWARF NOVAE

J. P. LASOTA
UPR 176 du CNRS; DARC,
Observatoire de Paris, Section de Meudon,
92190 Meudon Cedex, France

Abstract. Models of Soft X-ray Transients are presented and compared with observations. The importance of inner advection–dominated flows in quiescent transient sources is discussed, as well as the problem of global stability of the standard outer accretion disc. A comparison is made with similar problems in dwarf nova models.

1. Introduction

Soft X-ray Transients (SXTs) are low–mass X-ray binaries (LMXBs) which spend most of the time in a very dim and quiescent state, but on rare occasions undergo outbursts during which they can become the brightest X-ray sources in the sky. Like other LMXBs, SXTs are close binaries in which a compact object – a black hole or a neutron star – accretes matter from a low mass companion: a lower main sequence star or a (sub)giant. The rise-times to outburst in SXTs are typically 2 to 10 d and the main outburst lasts from 20 to 90 d; but in some cases weaker outbursts can be observed for the year after the main event. The luminosity at outburst rises to $10^{37} \ldots 10^{38}$ erg s^{-1} and the total energy released is $\sim 10^{43} \ldots 10^{44}$ erg. It seems that the recurrence time for SXTs containing black holes is longer than in systems containing neutron stars.

The name 'Soft X-ray Transient' derives from the fact that most of the outburst energy is emitted in rather soft (a few keV) X-rays. It has been claimed that spectra of black-hole SXTs are characterised by an 'ultra–soft' component but, in a few cases – in which the accreting object is most probably a black hole – such a component has not been observed (Harmon et al. 1994). Black hole SXTs emit also very high energy photons; in the

A. Evans and J. H. Wood (eds.), Cataclysmic Variables and Related Objects, 385–394.

case of GRS 1124−684 an e^+–e^- annihilation line was observed (Goldwurm et al. 1993). On the other hand neutron star SXTs were observed to emit above 100 keV (Barret et al. 1993).

Black hole SXTs (BSXT) are sometimes called 'X-ray novae'. This is confusing, because the name 'Nova Mus', for example, is used to describe both the BSXT GRS 1124–684 and the classical nova GQ Mus, which was also an X-ray source.

The mechanism responsible for SXT outbursts is unknown. It is clear that the site of the eruption is the accretion disc, but it is still a subject of debate whether it is due to an instability in the disc or to an enhanced mass transfer from the companion, or to a combination of the two (see e.g. Lasota 1996). There is an obvious analogy between SXTs and dwarf novae (van Paradijs & Verbunt 1984), in which the black hole or the neutron star is replaced by a white dwarf. In particular dwarf nova superoutbursts, which have bigger amplitudes and last longer than the 'normal' (U Gem-type) outbursts are, in many respects, similar to SXT outbursts.

In the case of dwarf novae (DN) the disc instability model (DIM) seems to account for the most general properties of 'normal' outbursts (Cannizzo 1993), even if it requires some modifications and additions (Ludwig, Meyer-Hofmeister & Ritter 1994; Livio & Pringle 1992; Meyer & Meyer-Hofmeister 1994) which concern the inner disc structure.

In the case of superoutbursts, two kind of models have been considered. On the one hand Osaki (1989) has proposed a model in which superoutbursts result from a combination of thermal and tidal instabilities. On the other hand Osaki (1985), Whitehurst & King (1991), Whitehurst (1994) and Smak (1996) have suggested that enhanced mass transfer (EMT) from the companion is an important factor in the superoutburst mechanism.

As for DN, two kinds of models have been proposed to explain the SXT phenomenon: the DIM and the EMT model. The DIM (Huang & Wheeler 1989; Mineshige & Wheeler 1989; Cannizzo, Chen & Livio 1995) was an extension of the DN DIM to systems containing black holes (no model for neutron star SXTs has been proposed). Hameury, King & Lasota (1986, 1987, 1988, 1990) worked on a model in which EMT was triggered by X-ray illumination of the secondary during quiescence. Both models had difficulties with accounting for the observed SXT properties (Lasota 1996) but the X-ray EMT model can be ruled out for systems with orbital periods $P_{\rm orb} \lesssim 10\,{\rm h}$ (Gontikakis & Hameury 1993; Lasota 1996). The DIM for SXTs is contradicted by X-ray observations (Mineshige et al. 1992; Verbunt et al. 1994; Wagner et al. 1992; McClintock, Horne & Remillard 1995; Lasota 1996; Verbunt 1996) of transient systems in quiescence.

2. The standard disc model of SXTs

The DIM requires the whole disc in quiescence to be in a cold 'low state'; or in another words, it must be everywhere (at each radius) on the lower branch of the 'S–curve' representing the local equilibrium of the accretion disc. This requirement means that the surface density of the quiescent disc Σ has to be less than the critical surface density $\Sigma_{\max}(r)$ everywhere in the disc (see Fig. 1). This implies that an inequality has to be satisfied by the accretion rate in a quiescent disc (Lasota 1996):

$$\dot{M}_{\rm in} \lesssim 3\pi\nu\Sigma_{\rm max} \approx 3.11 \times 10^4 t_8^{-1} r_7^{3.11} m_{10}^{-0.37} \alpha^{-0.79}\, {\rm g\,s^{-1}} \tag{1}$$

where ν is the kinematic viscosity coefficient, m_{10} is the mass of the compact object in units of 10 $M_{\odot}$, t_8 is the recurrence time in units of 10^8 s, r_7 is the radius in 10^7 cm and α is the Shakura–Sunyaev viscosity parameter. Numerical models by Mineshige & Wheeler (1989; herefter MW) satisfy the inequality given by equation (1).

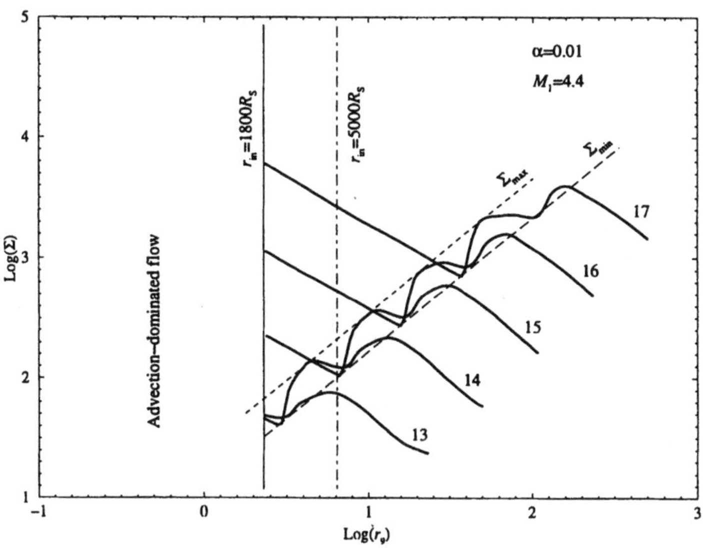

Figure 1. Surface density profiles of a truncated accretion disc around a black hole with mass $M_1 = 4.4$ $M_{\odot}$, for five values of accretion rates: $10^{13}, 10^{14}, 10^{15}, 10^{16}, 10^{17}$ g s^{-1}. r_9 is the radius in 10^9 cm. For a given $\dot{M}$ = const the disc can be in a globally stable *cold* equilibrium only for radii larger than the radius at which the density profile becomes tangent to the $\Sigma_{\max}$ line.

Several black-hole SXTs have been observed by Ginga (Mineshige et al. 1992) and ROSAT (see Verbunt 1996). Two systems, A0620–00 and V404 Cyg, have been detected at levels corresponding to accretion rates of at least $\sim 1.3\,10^{11}\,{\rm g\,s^{-1}}$ for A0620–00 and of $3\,10^{12}\,{\rm g\,s^{-1}}$ for V404 Cyg. Those accretion rates are several orders of magnitude larger than the values predicted by the disc instability model of MW. Detections of X-ray radiation from quiescent SXTs contradict the disc instability model, at least in the version proposed by MW (Mineshige et al. 1992).

In the case of A0620–00, the prototypical BSXT, the optical luminosity suggests a mass accretion rate of $\sim 6\,10^{15}$ g s^{-1} in the outer part of the accretion disc (McClintock et al. 1995), five orders of magnitude higher than the accretion rate at the inner disc as deduced from observations (and 10 orders of magnitude higher if one uses equation (1) as required by DIM!).

The accretion rate on to the compact object as given by X-ray observations of quiescent SXTs poses an insurmountable obstacle to models in which the inner accretion disc is of the Shakura–Sunyaev type. On the one hand the requirement of equation (1) can agree with observations only if one assumes a very low viscosity (e.g. $\alpha \sim 10^{-8}$), in fact no viscosity at all, in the inner disc. On the other hand X-rays have to be produced by a process connected to viscosity if one assumes a standard disc structure.

3. Advection–dominated quiescent SXT accretion discs

Since X-rays in quiescent SXTs are produced by accretion and not emitted by the companion (Verbunt 1996), one should consider other accretion flow structures that could describe quiescent SXTs.

Narayan, McClintock & Yi (1996) have proposed a model of quiescent BSXTs in which the inner ($r \lesssim 10^3 r_S$, where $r_S = 2GM/c^2$) accretion flow is advection–dominated. The standard, Shakura–Sunyaev disc model assumes that the energy released through viscous dissipation is radiated locally and efficiently. In such a disc the efficiency of converting gravitational potential energy into luminosity is close to r_{in}/r_S, which for neutron stars and black holes is ~ 0.1. In the case of inefficient cooling, for example when the accretion flow is optically thin, another mode of cooling may dominate: the viscously released energy is advected with the gas rather than being radiated. Such 'advection–dominated' flows have been studied first in the optically thick (radiation pressure dominated) case (Begelman & Meier 1982; Abramowicz et al. 1988) and more recently for the optically thin case (Abramowicz et al. 1995; Chen et al. 1995; Narayan & Yi 1994, 1995ab, Abramowicz & Lasota 1995).

The main difference between the description of 'standard' geometrically thin accretion discs and advection–dominated accretion flows is in the energy equation

$$Q_{\rm adv} \equiv \Sigma v_r T \frac{ds}{dr} = Q^+ - Q^- \qquad (2)$$

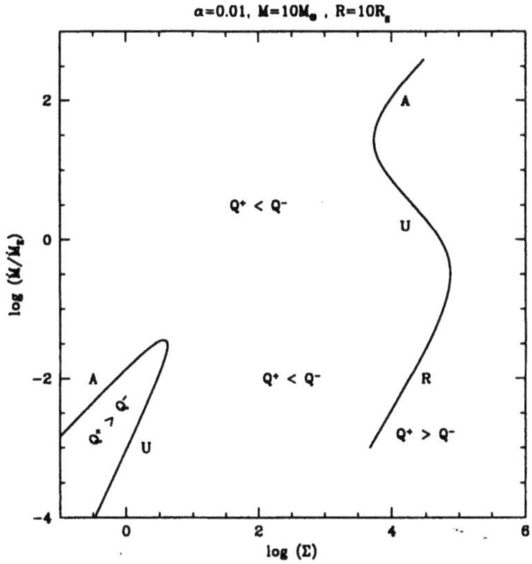

Figure 2. Equilibria of accretion flows on to a black hole in the surface density – accretion rate plane for a central mass $10\,\mathrm{M}_\odot$, and $\alpha = 0.01$ at a distance of $10r_S$. The advection–dominated solutions are indicated by 'A', the radiatively cooled, stable solutions by 'R' and the radiatively cooled, unstable solutions by 'U'. Viscously stable branches have positive slopes, thermally stable branches have $Q^+ > Q^-$ below and $Q^+ < Q^-$ above them. The total cooling rate Q^- includes all radiative processes *and* advection.

where v_r is the radial speed, T is the temperature and s the entropy. In the standard model of thin accretion disks (Frank, King & Raine 1992) advective cooling is neglected because the ratio of advective cooling, Q^{adv}, to the viscous heating, Q^+, is proportional to the square of the ratio of the disc thickness $H = H(r)$ to the radius r,

$$Q^{\mathrm{adv}} \sim \left(\frac{H}{r}\right)^2 Q^+ \tag{3}$$

and, of course, for thin disks $H \ll r$. Advection-dominated flows are much hotter than cooling-dominated thin disks and are therefore thicker in the vertical direction. For this reason they are also described as 'slim discs' or 'thick discs' depending on the relative vertical thickness.

Fig. 2 shows an example (Abramowicz et al. 1995; Abramowicz & Lasota 1995) of accretion flow equilibria with the energy equation in the form of equation (2). In this case the radiative cooling term includes only bremsstrahlung but the shape and form of the solutions represented here are very general (Chen et al. 1995). The S–shaped line corresponds to optically thick solutions, 'slim discs' (Abramowicz et al. 1988). The curve in the lower left corner represents optically thin solutions of which those marked by 'U' are radiatively cooled and thermally unstable. The advection–dominated solutions indicated by 'A' are both viscously and thermally stable so they represent the only physically possible hot, optically thin accretion flow configurations.

In an advection–dominated flow most of the energy that is realeased by viscous heating is retained in the gas and swallowed by the black hole. Only a very small fraction of this energy is radiated away so that the radiative efficiency is $\ll 0.1$. In this case one cannot deduce the accretion rate from the formula $\dot{M} \sim 10Lc^{-2}$.

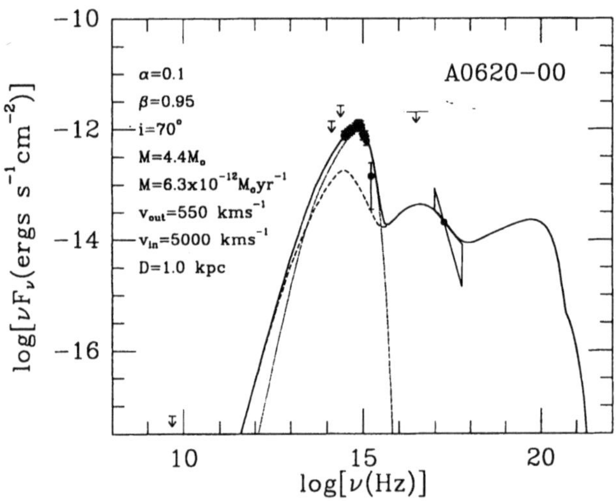

Figure 3. The spectrum of the quiescent BSXT A0620−00 according to Narayan et al. (1996). The filled circles show measured fluxes of A0620−00 plotted as νF_ν versus ν. The ROSAT point at $\nu \sim 10^{17}$ Hz is shown along with a 'bow-tie' to indicate the range of spectral slopes allowed by the data. The downward pointing arrows represent the various upper limits in radio, infrared and EUV. The lines show the spectrum predicted by the model. The parameters of the model are indicated on the left, where $\alpha = 0.1$ is the viscosity parameter in the advection–dominated flow, β is the ratio of the gas to the total (gas plus magnetic) pressure, i is the inclination, M the black hole mass and $\dot{M}$ the (constant through the flow) accretion rate. $v_{\rm in}$ and $v_{\rm out}$ correspond respectively to the inner and outer radii of the 'standard' thin disc. D is the distance to the system. The dotted line shows the contribution to the spectrum from the outer thin disc, the dashed line corresponds to the inner hot flow, and the solid line is the combined spectrum.

Narayan et al. (1996) have presented models of the BSXTs A0620−00, V404 Cyg and GRS 1124−684 in quiescence in which the accretion flow consists of two parts: an outer standard thin disc, extending outward from $\sim 3000 r_S$, and an inner, two-temperature, advection-dominated flow described by the Narayan & Yi (1995b) solution in which cooling by Comptonized synchrotron and bremsstrahlung radiation is included. The result for A0620−00 is shown in Fig. 2. The model is consistent with all the available data. In the hot, advection–dominated flow only a very small fraction ($10^{-4}\ldots10^{-3}$) of the thermal energy is radiated away in the inner flow regions. In this model the flow is stationary; the accretion rate ($\dot{M} = 4\,10^{14}$ g s^{-1} for A0620−00) is constant everywhere.

4. Equilibria of standard thin discs at very low accretion rates

In their model Narayan et al. (1996) assume that the outer standard disc is stationary. This disc extends outwards from a radius $\sim 10^3 r_{\rm S}$, which for the compact object masses of interest is $\sim 10^9$ cm. Since the accretion rates considered are $\lesssim 10^{15}$ g s^{-1}, the outer disc can be described as an accretion disc in a cataclysmic variable where the radius of the compact object (a white dwarf) is $\sim 10^9$ cm. Equilibria of accretion discs in CVs were studied in detail in the DIM for dwarf novae (see Cannizzo 1993 for a review).

At a given radius r, thermal equilibria of accretion discs in cataclysmic variables form, on the Σ–$\dot{M}$ (or Σ–$T_{\rm eff}$) plane, S–shaped curves, where the middle branch with the negative slope is thermally (and viscously) unstable. Hot stable equilibria therefore exist only for $\Sigma > \Sigma_{\rm min}$ and cold ones only for $\Sigma < \Sigma_{\rm max}$ (see e.g. Cannizzo 1993). For the disc to be globally unstable, i.e. unstable at some radius $r_{\rm in} \leq r \leq r_{\rm out}$, the mass transfer rate $\dot{M}_{\rm T}$ must satisfy the inequality

$$\dot{M}\left(\Sigma_{\rm max}\left(r_{\rm in}\right)\right) < \dot{M}_{\rm T} < \dot{M}\left[\Sigma_{\rm min}\left(r_{\rm out}\right)\right] \qquad (3)$$

The range of unstable mass transfer rates therefore depends on the size of the disc. Fig. 1 shows surface density profiles of (truncated) equilibrium accretion discs around a 4.4 $M_\odot$ black hole (the value in the Narayan et al. (1996) model of of A0620−00). The value of the viscosity parameter, α, is assumed to be 0.01. One should note that there is no reason for α to be the same in the advection–dominated and standard regions; on the contrary, one should expect the viscosity to be higher in the hot flow. The accretion rate $\dot{M}$ is constant for each curve. Models by Hameury, Huré & Lasota (1996) were used in this calculation.

The part of the curve which is above the $\Sigma_{\rm max}$ line cannot represent a cold thermal equilibrium so that the disc cannot be globally in equilibrium ($\dot{M}$ = const) for the corresponding mass transfer rate.

From Fig. 1 one can see that there exist two types of equilibria represented by the $\Sigma(r)$ curves: for $\dot{M} \lesssim \dot{M}'$ (where $\dot{M}'$ is some value of accretion rate depending on M_1 and α; in Fig. 1, $\dot{M}' \sim 10^{13}\ldots10^{14}\,{\rm g\,s^{-1}}$) the surface density is always less than $\Sigma_{\rm max}$. For such low $\dot{M}$ the disc is globally in cold stable equilibrium, from $r_{\rm out}$ down to $r_{\rm in}$. For $\dot{M} > \dot{M}'$ only the outer part of the disc can be in a cold stable equilibrium. Starting at some r_{out} from $\Sigma < \Sigma_{max}$, the $\Sigma(r)$ curve reaches, with decreasing r, the value Σ_{max} at some radius r_{crit}. The disc can be in cold equilibrium only for $r_{crit} < r < r_{out}$. The next segment of the curve between Σ_{max} and Σ_{min} is unstable while the curve to the left of Σ_{min} represents hot equilibria. For $\dot{M} > \dot{M}'$ the disc is therefore globally unstable down to r_{in}, and a cold steady state cannot exist.

There exists however a radius $r_{in} = r_{crit}$ for which the disc is globally stable for some $\dot{M}$ = const. From the example shown in Fig. 1 one can deduce that the global state of the disc will depend on the exact value of r_{in}. If one takes the 'best fit' values shown in Fig. 3 and the model of Narayan et al. (1996), i.e. $M_1 = 4.4\ \mathrm{M}_\odot$, and $r_{in} = 1800 r_S$ (corresponding to $v_{in} = 5000$ km s^{-1} and $\dot{M} = 4\,10^{14}\,\mathrm{g\,s^{-1}}$), the disc will be globally unstable in quiescence. For $r_{in} = 5000 r_S$, however, the disc would be globally stable. The uncertainty in the system's parameters (see e.g. Shahbaz, Naylor & Charles 1994) and in model parameters does not allow a definitive conclusion about the global stability of quiescent SXT outer discs.

A similar problem has been studied by Lasota et al. (1995) in the context of dwarf nova outbursts. Some SU UMa–type dwarf novae show only very rare and very long superoutbursts and no (or almost no) 'normal' outbursts. The best known system in this class is WZ Sge. On the other hand, DIM applied to dwarf novae requires some modifications to be applicable to the whole class of observed events. In particular the so–called UV–delay problem, and the X-ray and UV emissions from quiescent DN require a 'hole' in the inner disc regions (see e.g. Meyer & Meyer–Hofmeister 1994). This 'hole' can be produced by magnetic disruption if the white dwarf is (weakly) magnetized (Livio & Pringle 1992), or it can be the result of evaporation (Meyer & Meyer–Hofmeister 1994) of the flow which is unable to radiate the viscously released energy. The second mechanism was proposed by Narayan & Yi (1995) to explain the origin of advection–dominated flows around black holes (see below).

Lasota et al. (1995) show that if the inner regions of accretion discs in quiescent dwarf nova systems are removed, the remaining disc is globally stable for mass transfer rates of $\lesssim 10^{15}\,\mathrm{g\,s^{-1}}$. This implies that (super) outbursts in these systems have to be triggered by an enhanced mass transfer from the companion. They suggest that the lack of normal outbursts in WZ Sge results only from its low mass transfer rate: there are no outbursts because the disc is stable. A superoutburst would be triggered by an EMT which put the disc into a globally unstable state; in other words, superoutbursts would be due to a disc instability generated by an increased mass transfer. There are observational reasons to think that such a hybrid mechanism is at work in SU UMa stars (Smak 1996). The viscosity in WZ Sge's disc does not therefore have to be a few orders of magnitude lower than in other quiescent dwarf novae.

5. Soft X-ray transient outbursts

Fig. 2 shows that (at a given radius) for a whole range of accretion rates no stable equilibria can exist. In the example shown, above $\sim 0.03\dot{M}_{\rm E}$ no optically thin solution exists and above $\sim 0.3\dot{M}_{\rm E}$ no stable configuration is possible. An advection–dominated solution for a quiescent SXT will correspond to a point at the lower branch marked by an 'A'. If, for some reason (increased mass transfer rate or outer disc instability), the mass accretion rate in the inner flow is increased to a value for which no optically thin solution exists, the system will become unstable. The system point will have to move in the Σ–$\dot{M}$ plane but the whole process will be non–local and complicated by the fact that $r_{\rm in}$ will (probably) move inwards and the outer disc will invade the inner hot flow (see Narayan et al. 1996).

As mentioned above the outer disc may be stable or unstable, depending on the value of its inner radius. In the first case (Lasota, Narayan & Yi 1996) SXT outbursts would be generated by an EMT from the companion, as in the scheme proposed by Lasota et al. (1995) for WZ Sge: SXTs would be WZ Sge-type systems with a black hole instead of a white dwarf. The increased mass transfer is in this case only a 'detonator': it makes the disc unstable so that the outburst is finally due to a disc instability.

In the second case (Mineshige 1996, Lasota et al. 1996) the outer disc is unstable and undergoes dwarf nova type instabilities. In this case, however, contrary to the case of WZ Sge (Smak 1991), no abnormally low viscosity is required to get the right recurrence time etc. (Lasota et al. 1996).

Acknowledgements. I am grateful to Jean-Marc Huré for producing Fig. 1. I thank Didier Barret, John Cannizzo, Jeff McClintock, Jean-Marie Hameury, Emmi Meyer–Hofmeister, Andrew King, Friedrich Meyer, Ramesh Narayan, Joe Patterson, Rob Robinson, Joe Smak and Insu Yi for interesting and stimulating discussions.

References

Abramowicz, M.A., Lasota, J.P., 1995, Coments in Astrophys., in press
Abramowicz, M.A., Czerny, B., Lasota, J.P., Szuszkiewicz, E., 1988, Ap. J., **332**, 646
Abramowicz, M.A., Chen, X.M., Kato, S., et al., 1995, Ap. J., **438**, L37
Barret, D., Mandrou, P., Roques, J.P., et al., 1993, A&AS, **97**, 241
Begelman, M.C., Meier, D.L., 1982, Ap. J., **253**, 873
Cannizzo, J.K., 1993, in "Accretion Disks in Compact Stellar Systems", ed. J. Craig Wheeler, World Scientific Publishing, Singapore, p6
Cannizzo, J.K., Chen, W., Livio, M., 1995, Ap. J., **454**, 880
Chen, X.-M., Abramowicz, M.A., Lasota, J.P., et al., 1995, Ap. J., **443**, L61
Frank, J., King, A.R., Raine, D., 1992, Accretion Power in Astrophysics, Cambridge University Press, Cambridge

Goldwurm, A., Ballet, J., Laurent, P., et al. , 1993, A&AS, **97**, 293
Gontikakis, C., Hameury, J.M., 1993, A&A, **271**, 118
Hameury, J.M., King, A.R., Lasota, J.P., 1986, A&A, **162**, 71
Hameury, J.M., King, A.R., Lasota, J.P., 1987, A&A, **171**, 140
Hameury, J.M., King, A.R., Lasota, J.P., 1988, A&A, **192**, 187
Hameury, J.M., King, A.R., Lasota, J.P., 1990, Ap. J., **353**, 585
Hameury, J.M., Huré, J.M., Lasota, J.P., 1996, A&A, in preparation
Harmon, B.A., et al. , 1994, in "The Second Compton Symposium", eds C. Fichtel, N. Gehrels, J. P. Norris, AIP, New York, p210
Huang, M., Wheeler, J.C., 1989, Ap. J., **343**, 229
Lasota J.P., 1996, in "Compact Stars in Binaries", Proceedings of IAU Symposium 165, eds van den Heuvel E.P.J, van Paradijs J., Kluwer, Dordrecht, in press
Lasota, J.P., Hameury, J.M., Huré, J.M., 1995, A&A, **302**, L29
Lasota, J.P., Narayan, R., Yi, I., 1996, A&A, in press
Livio M., Pringle J.E., 1992, MNRAS, **259**, 23p
Ludwig K., Meyer-Hofmeister E., Ritter H., 1994, A&A, **290**, 473
McClintock, J.E., Horne, K., Remillard, R.A., 1995, Ap. J., **442**, 358
Meyer F., Meyer-Hofmeister E., 1994, A&A, **288**, 175
Mineshige, S., 1996, PASJ, submitted
Mineshige, S., Wheeler, J.C., 1989, Ap. J., **343**, 241
Mineshige, S., Ebisawa, K., Takizawa, M., et al., 1992, PASJ, **44**, 117
Narayan, R., Yi, I., 1994, Ap. J., **428**, L13
Narayan, R., Yi, I., 1995a, Ap. J., **444**, 231
Narayan, R., Yi, I., 1995b, Ap. J., **452**, 710
Narayan, R., McClintock, J. E., Yi, I., 1996, Ap. J., **457**, 821
Osaki, Y., 1985, A&A, **144**, 369
Osaki, Y., 1989, PASJ, **41**, 1005
van Paradijs J., Verbunt, F., 1984, in "High Energy Transients in Astrophysics", ed. S.E. Woosley, AIP Conference Proceedings No. 115, p49
Shahbaz, T., Naylor, T., Charles, P.A., 1994, MNRAS, **269**, 756
Smak, J., 1991, Acta Astron., **41**, 269
Smak, J., 1996, these proceedings, p45
Verbunt, F., 1996, in "Compact Stars in Binaries", Proceedings of IAU Symposium 165, eds van den Heuvel E. P. J, van Paradijs J., Kluwer, Dordrecht, in press
Verbunt, F., Belloni, T., Johnston, H.M., et al., 1994, A&A, **285**, 903
Wagner, R.M., Kreidt, T.J., Howell, S.B., Starrfield, S.G., 1992, Ap. J., **401**, L97
Whitehurst, R., 1994, MNRAS, **266**, 35
Whitehurst, R., King, A.R., 1991, MNRAS, **249**, 25

DOPPLER IMAGING OF V404 CYG IN QUIESCENCE

JORGE CASARES

University of Oxford, Department of Physics, Nuclear Physics Laboratory, Keble Road, Oxford OX1 3RH, UK

Abstract. We present Doppler images of the accretion disc in the black hole X-ray transient V404 Cyg during its quiescent phase. The Hα map shows evidence for a hot spot, indicating that mass transfer is continuing long after the 1989 outburst. Both the Hα and the IR Ca II emission maps exhibit intense bright spots at the position of the secondary star, which we take as a sign of chromospheric activity. The Ca II emission arises from the outer parts of the accretion disc, and its surface brightness distribution is consistent with dwarf novae discs ($I \propto R^{-3/2}$).

1. Introduction

V404 Cyg is the optical counterpart of the X-ray transient GS 2023+338, which was discovered by Ginga at the peak of its recent 1989 outburst [6]. Its striking X-ray behaviour received great attention, leading Tanaka [13] to propose V404 Cyg as a new black hole candidate. These include a hard power-law spectrum (without an ultra-soft component) with associated millisecond flickering, chaotic variability (varying by a factor of 200 in a few minutes) and a luminosity saturation effect at $L_x \simeq 1.5\,10^{38}\,(d/\mathrm{kpc})^2\,\mathrm{erg\,s^{-1}}$.

Since 1990 the system has settled down in quiescence at $V \sim 18.5$ mag. Follow-up spectroscopy led to the discovery of late-type absorption features moving at 211 km s^{-1} in a 6.5 d orbit. With a mass function of $f(M) = 6.3 \pm 0.3$ M$_\odot$, V404 Cyg provides the strongest dynamical evidence yet for a black hole [3]. The spectrum of the K0 secondary also showed the puzzling presence of the ^{7}Li resonance line [9], suggesting perhaps the action of spallation reactions during the outbursts (see [10]). V404 Cyg is also remarkable because of the exhibition of a strong 6 h variability in the continuum and emission lines [1, 4], the origin of which remains unknown.

A. Evans and J. H. Wood (eds.), Cataclysmic Variables and Related Objects, 395–398.

2. Database and updated system parameters

With a period of 6.5 d it is difficult to sample a full cycle of V404 Cyg adequately within a single campaign. Therefore, we have attempted to monitor the source continuously since the start of the quiescent phase. Our database now consists of 73 Hα spectra from 1990–91 (0.75 Å pixel^{-1}) and 84 spectra from 1992–94 (0.39 Å pixel^{-1}). In addition, during 1992 and 1994 we also acquired 58 spectra in the region of the IR Ca II triplet, at a dispersion of 0.75 Å pixel^{-1}. In Table 1 we summarize the updated system parameters and refined rotational broadening ($V \sin i$) as derived by using the K0IV template star HR8857 in both wavelength ranges independently (further details of the process can be found in [2]). Note the excellent agreement between all the quantities. As a matter of precaution, we initially avoided the Ca II triplet lines in the cross-correlation analysis of the IR spectra (there is evidence for broad Ca II emission). However, a further test using these lines alone has provided identical orbital parameters.

TABLE 1. System parameters

λ-range	γ (km s^{-1})	Period (d)	K (km s^{-1})	$V \sin i$ (km s^{-1})
Hα	0.3 ± 0.6	$6.47129 \pm 7\,10^{-5}$	208.4 ± 0.6	$40.8^{\dagger} \pm 0.9$
Ca II	0.2 ± 0.8	$6.47129 \pm 8\,10^{-5}$	207.6 ± 1.1	41.1 ± 1.0

†The slight difference with [2] is due to a small bug found in the broadening routine.

Our revised values for the mass function and mass ratio are $f(M) = 6.07 \pm 0.05\,\mathrm{M}_{\odot}$ and $q = M_c/M_x = 0.067 \pm 0.005$ respectively. Combining these with the limits to the inclination ($i = 52° \ldots 60°$ [12]) brings the black hole mass to the range 10.5... 14 $\mathrm{M}_{\odot}$.

In order to study the line emission regions in V404 Cyg we first need to correct for the contamination by the spectrum of the secondary star. This was performed by subtracting our HR8857 spectrum (broadened by 41 km s^{-1} and shifted to the secondary's frame) from each individual spectrum of V404 Cyg. The appropriate fraction to be subtracted was estimated following the prescription of Marsh et al. [8]. An outcome of this analysis is a possible dimming of the accretion disc contribution with time, particularly from 1991 to 1992. Therefore, we decided to exclude the lower resolution spectra (1990–1991) from our subsequent analysis. The spectral averages, with the secondary star removed, are presented in Fig. 1. The Ca II triplet lines appear to be saturated (all have EW$\simeq$20 Å) and hence they are optically thick, in contrast to the Balmer emission [4].

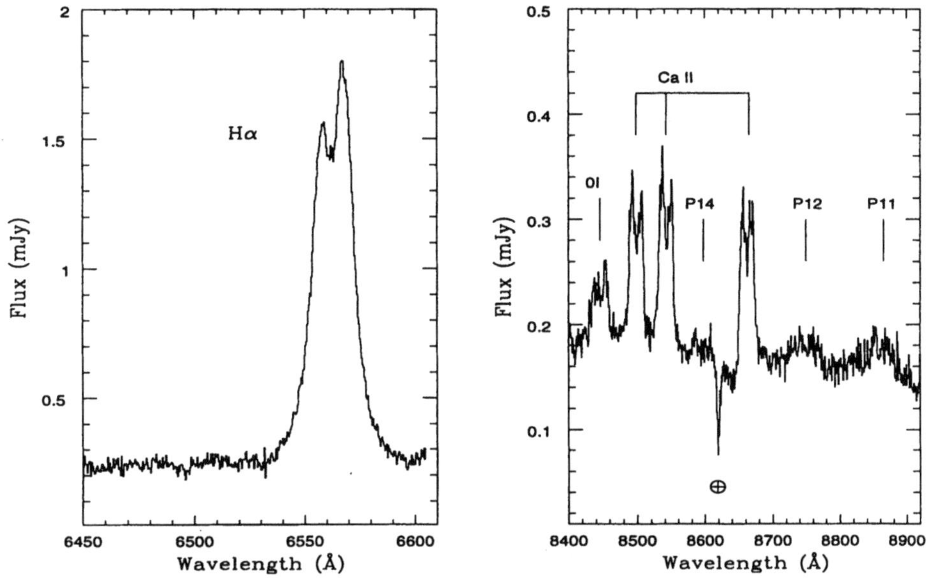

Figure 1. Averaged spectra after subtracting the contribution of the companion star.

3. Doppler imaging

Images of the Hα and Ca II emission regions in V404 Cyg have been constructed through combining the phase-resolved profiles using the Doppler tomography technique [7]. These are presented in Fig. 2. The Hα map shows extended emission close to where the gas stream impacts the outer edge of the accretion disc. This implies that mass transfer proceeds during quiescence and permits us to constrain the outer disc radius to $\sim 0.6 R_{\mathrm{L1}}$. Both the Hα and the Ca II maps exhibit bright spots at the position of the secondary. Although some of this might be a residual of the secondary subtraction (see [8]), we associate such emission with chromospheric activity, as is frequently observed in rapid rotators (e.g. RS CVn stars [11]). The possibility that some emission is powered by X-ray heating of the inner hemisphere is disproved by the negligible ellipticity of the radial velocity curve of the Ca II absorption lines.

The emission-line surface brightness (I) of the accretion disc can be extracted from the Doppler images after assuming that the velocity field is Keplerian. The disc surface brightness peaks further away for the Ca II lines than for Hα. In addition, $I_{\mathrm{Ca\,II}} \propto R^{-1.4}$, consistent with dwarf novae discs [5], whereas a less steep behaviour is found for Hα ($I_{\mathrm{H}\alpha} \propto R^{-0.7}$).

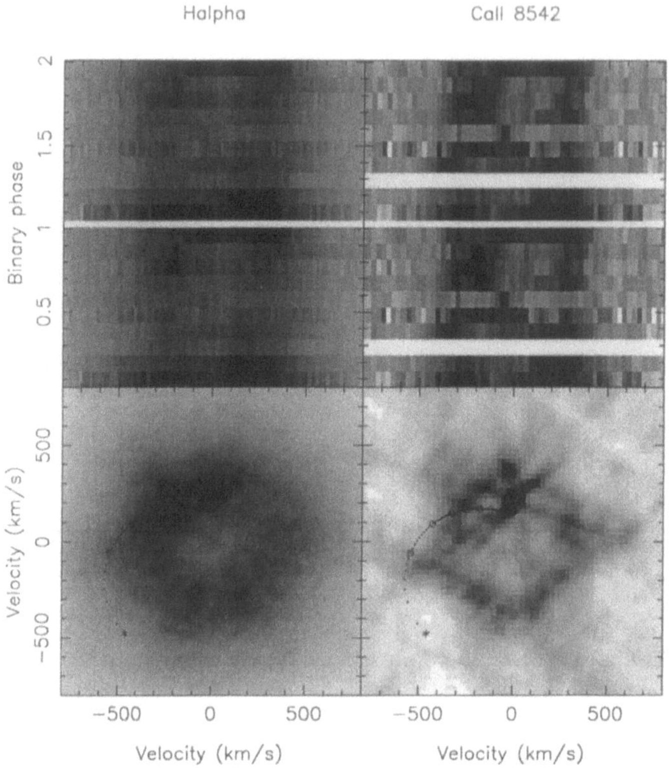

Figure 2. Trailed spectra and Doppler maps of Hα and Ca II λ8542.

References

1. Casares, J., Charles, P.A., 1992, MNRAS, **255**, 7
2. Casares, J., Charles, P.A., 1994, MNRAS, **271**, 5p
3. Casares, J., Charles, P.A., Naylor, T., 1992, Nature, **355**, 614
4. Casares, J., Charles, P.A., Naylor, T., Pavlenko, E.P., 1993, MNRAS, **265**, 834
5. Horne, K., Saar, S.H., 1991, Ap. J., **374**, L55
6. Marsden, B.G., 1989, IAUC 4783
7. Marsh, T.R., Horne, K., 1988, MNRAS, **235**, 269
8. Marsh, T.R., Robinson, E.L., Wood, J.H., 1994 MNRAS, **266**, 137
9. Martín, E. L., Rebolo, R., Casares, J., Charles, P.A., 1992, Nature, **358**, 129
10. Martín, E. L., Rebolo, R., Casares, J., Charles, P.A., 1994, Ap. J., **435**, 791
11. Rutten, R.G.M., 1987, A&A, **177**, 131
12. Shahbaz, T., Ringwald, F.A., Bunn, J.C., et al., 1994, MNRAS, **271**, 10p
13. Tanaka, Y., 1989, in "Proc. 23rd ESLAB Symp. on Two-Topics in X-Ray Astronomy", ed. J. Hunt & B. Battrick (ESA Publ. Division), p3

THE X-RAY NOVA GRO J0422+32 IN DECLINE AND QUIESCENCE

M. R. GARCIA, P. J. CALLANAN, J. E. MCCLINTOCK,
P. ZHAO
Smithsonian Astrophysical Observatory, Cambridge, MA, USA

We have followed the X-ray nova GRO J0422+32, spectroscopically and photometrically, throughout the decline to quiescence.

In the final stages of decay (days 430...880 after the outburst, see Callanan et al. (1995) for the epoch 1995), the equivalent width (EW) of the Hα emission increases monotonically and the R magnitude decreases monotonically. This suggests that the flux in the Hα line is constant, while the continuum fades. The Hα flux is the product of the R band flux ($F(R)$, arbitrarily scaled to 100 at $R = 19$ mag) and the EW, and is shown in the last column of the table below. The Hα flux varies by only $\sim 30\%$ while the continuum fades by a factor of eight (from $R = 19$ mag to $R = 21.3$ mag). So, to first order, the Hα luminosity is constant in the final stages of decay. While it is generally the case that the emission line EWs in individual dwarf novae also increase during the decay, the exact behavior seen in GRO J0422+32 is *not* what is seen for dwarf novae (on average). Using the relation between EW[Hβ] and M_v given in figure 6 of Patterson (1984) , we would expect a factor of ~ 5 variation in the Hα flux during days 430...880. The stability of the Hα flux implies that somehow the emission line region is 'disconnected' from the continuum (R–band) emission region.

Day	Date	EW(Å)	R	$F(R)$	Flux = EW×$F(R)$
430	1993 Oct 9/11	30	19.0	100	300
580	1994 March 10	100	20.0	40	400
790	1994 Sept 29/30	150	20.8	19	285
880	1994 Dec 30/31	210	21.3	12	250

Many authors have noted that in CVs, the width of the Balmer emission lines is an indicator of the inclination of the system. For any thin disk model this must be the case, as the observed line width will depend on the

A. Evans and J. H. Wood (eds.), Cataclysmic Variables and Related Objects, 399–400.

projected velocity (i.e. on sin i). If the disks are Keplerian and fill a constant fraction of their Roche lobe, one can easily derive the approximation that the projected velocity will also go like $(M/P)^{1/3}$ (see Frank, King & Raine 1992). In order to investigate this we correlated the Balmer line width measured by Williams (1983) versus the inclinations from Ritter (1994). A clear correlation is seen, although with a wide ($\sigma = 396\,\mathrm{km\,s^{-1}}$) scatter. If one includes the $\sim (M/P)^{1/3}$ factor, the correlation becomes more significant and the scatter drops to $284\,\mathrm{km\,s^{-1}}$. When we add GRO J0422+32 and points for other SXT in quiescence (taken from the literature) to the graph (in larger type) we find that they also seem to obey the same correlation (see figure below).

CV H-alpha FWHM vs. Orbital Inclination and Mass

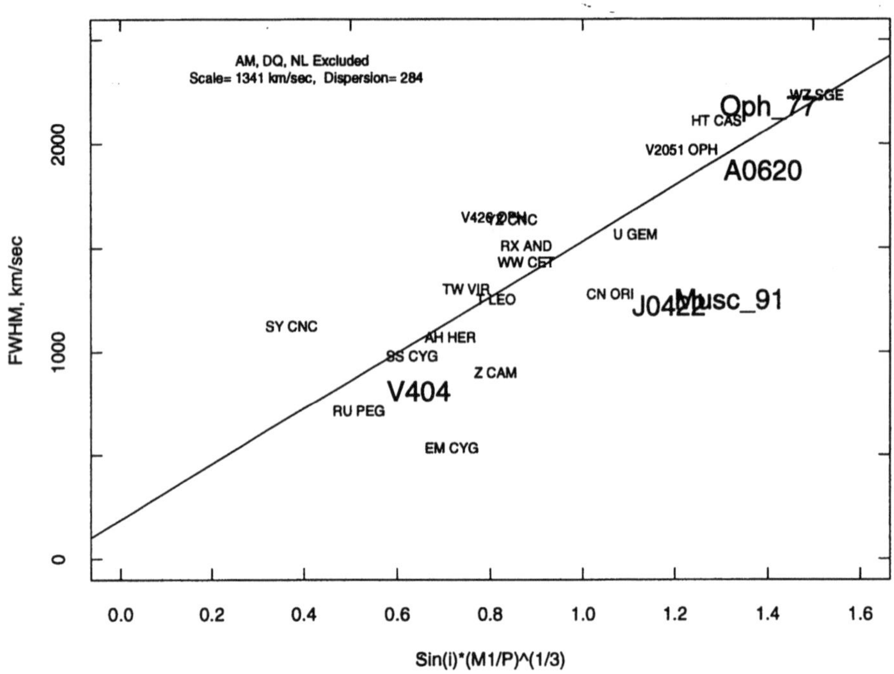

References

Callanan, P.J., Garcia, M.R., McClintock, J.E., et al., 1995, Ap. J., **441**, 786
Frank, J., King, A. R., Raine, D.J., 1992, "Accretion Power In Astrophysics", p54
Patterson, J., 1984, Ap. J. Supp., **54**, 443
Ritter, H., 1994, A&AS, **85**, 1179
Williams, G., 1983, Ap. J. Supp., **53**, 523

INVESTIGATION OF THE OPTICAL VARIABILITY OF GRO J0422+32.

M. POPOVA[1], G. BESKIN[1], S. MITRONOVA[1], S. NEIZVESTNY[1], I. PANFEROVA[1], V. PLOKHOTNICHENKO[1],C. BARTOLINI[2], A. GUARNIERI[2], R. MINARINI[2], A. PICCIONI[2]
1. Special Astrophysical Observatory (SAO), Nizhnij Arkhyz, Karachaevo-Circassia, 357147, Russia. pmar@sao.stavropol.su
2. Dipartimento di Astronomia, Universitá di Bologna, via Zamboni 33, 40126 Bologna, Italy. bartolini@alma02.bo.astro.it

1. Observations

Nova Per 1992 was observed in 1992–94 on the 6 m and 1 m telescopes at the Special Astrophysical Observatory (SAO) of the Russian Academy of Sciencies and on the 1.5 m telescope at the Bologna Astronomical Observatory (Italy). The apparatuses used were a TV spectrum scanner with the hard/software photometrical system MANIA (Multichannel Analysis of Nanoseconds Intensity Alterations) complex with a time resolution of 10^{-7} s [3] (on the 6 m telescope), a CCD photometer (on the 1 m telescope) and a double head fast photometer (on the 1.5 m telescope) [5].

Spectroscopy. Low dispersion spectra (resolution 2 Å pixel^{-1}) of Nova Per 1992 (GRO J0422+32) were obtained in the high ($V \leq 14$ mag) and low ($V \geq 18$ mag) optical states. In the high state the object had a blue continuum, showed blue variations on time-scales of hours to days and weak variable emissions of Hα, Hβ, He II 4686 and 5411, N III 4640. In the low optical state the spectra showed some emission features (Hα, Hβ and probably He I 5876) on a red continuum.

Photometry. In the high optical state the luminosity of Nova Per 1992 was irregularly variable in different colour bands on time-scales of 4 ms to 200 s with relative amplitudes of 0.5...4. The shortest flares had rise times of 4...40 ms (Fig. 1) which allows us to establish an upper limit of $10^8 \dots 10^9$ cm on the size of the region producing the flares. The brightness temperatures for these events exceed $10^8 \dots 10^9$ K. The power spectrum of the detected variability is flat from 0.01 to 250 Hz. In the low optical state Nova Per did not show any significant variability on times up to hours.

A. Evans and J. H. Wood (eds.), Cataclysmic Variables and Related Objects, 401–402.

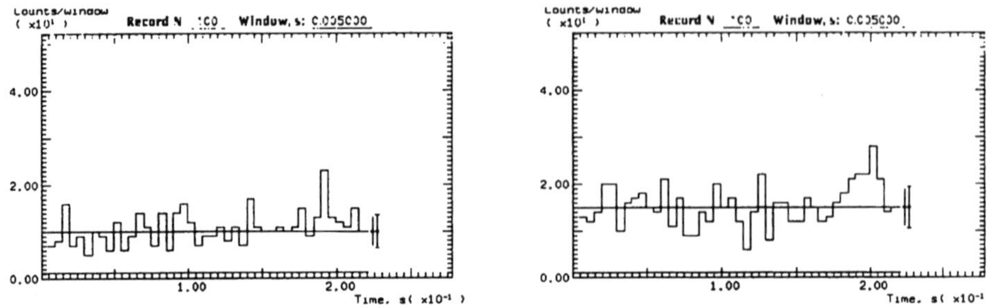

Figure 1. Flash of Nova Per 1992 in V (left) and B (right) bands, 1993 January 18.

2. Some conclusions

(i) We believe that the detected optical variability of GRO J0422+32 (in the high state) could be the result of a fragmentation in the accreting structure, where sizes of the plasma 'blobs', l, correspond to rise times of the flares, τ where $l \leq c\tau$.
(ii) All the detected shortest time-scale events have, with high probability, a nonthermal origin since, for brightness temperatures in the optical range exceeding $10^8 \ldots 10^9$ K, it is very difficult to propose a thermal mechanism for the production of the photons.
(iii) The differences between power spectra of the X-ray and optical variabilities suggest that the flares could be generated at equal distances from a compact object but by different mechanisms [4, 6].
Acknowledgements. This work was partially supported by a grant from the Italian Ministero per l'Universits e la Ricerca Scientifica e Tecnologica, by ESO's Support Programme for Central and East Europe (grant A-02-023), by the Scientific and Educational Center 'Cosmion', by the Soros ISF, by the Russian Ministry of Science and by the Russian Foundation of Fundamental Research (grant 95-02-0368).

References

1. Bartolini, C., Guarnieri, A., Piccioni, A., et al., 1994, Ap. J. Supp., **92**, 455
2. Beskin, G., Kajsin, S., Neizvestny, S., et al., 1993, IAUC, 5863
3. Beskin, G.M., Neizvestny, S.I., Pimonov, A.A., 1982, in Proc. IAU Colloq. 67, "Instrumentation for Astronomy with Large Optical Telescopes", ed. C.M. Humphries (Dordrecht: Reidel), p181
4. Ikhsanov, N.R., Pustil'nik, L.A., 1994, Ap. J. Supp., **90**, 959
5. Piccioni, A., Bartolini, C., Guarnieri, A., Giovanelli, F., 1979, Acta Astron., **29**, 463
6. Pustil'nik, L.A., Shvartsman, V.F. 1974, in Proc. IAU Symp. 64, "Gravitational Radiation and Gravitational Collapse", ed. C. Dewitt-Morette (Dordrecht), p213

INVESTIGATION OF THE OPTICAL VARIABILITY OF RELATIVISTIC OBJECTS WITH HIGH TIME RESOLUTION

S. N. MITRONOVA[1], G. M. BESKIN[1], S. I. NEIZVESTNY[1]
V. L. PLOKHOTNICHENKO[1], M. YU. POPOVA[1],
S. ZHARIKOV[1], O. G. BENVENUTO[2], C. FEINSTEIN[2],
M. MENDEZ[2], V. BORIAKOFF [3], D. DOSSA[3]

1. Special Astrophysical Observatory (SAO), Nizhnij Arkhyz, Karachaevo-Cherkesia, 357147, Russia

2. Observatorio Astronomico de la Plata, Paseo del Bosque s/n, 1900 La Plata, Argentina

3. Worcester Polytechnic Institute, USA

1. Detection of nonthermal optical flares in low mass X-ray binaries. Eleven low-mass X-ray binaries have been observed with the MANIA (Multichannel Analysis of Nanoseconds Intensity Alteration) complex; X0420+32, A0620−00, X1728−169, X1813−14, X1957+11 in 1986 at the 6 m telescope of SAO [7], and 2S0921−630, 4U1543−475, 4U1636−536, 4U1559−487, MXB1735−44, 4U1822−371 in 1991 at the 2.15 m telescope of CASLEO (Argentina) [1]. Analysis of the data using the special y_2 and d_2 function formalism [4], [6] has shown no brightness variations on time-scales of $10^{-7} \ldots 10$ s for all objects except A0620−00 and MXB1735−44.

A0620−00. The five millisecond flashes were detected on 1986 February 13 (Fig. 1). The first two flares have durations of 3 ms and 5 ms and their rise times were 1...2 ms. The other three events lasted 0.4...0.5 ms with rise times of about 0.1 ms. The lower limit of the flare amplitudes is 40 counts ms^{-1} and the corresponding brightness temperatures are $5\,10^9$ K

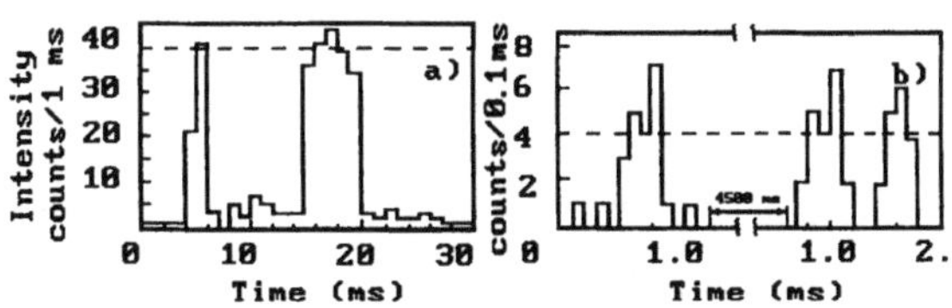

Figure 1. The ultrashort flares of A0620−00 on 1986 February 13, taken in white light with aperture 4.3 arcsec and seeing 1.5 arcsec; the dashed line marks the level of maximum intensity corresponding to the limiting transfer rate into the computer.

A. Evans and J. H. Wood (eds.), Cataclysmic Variables and Related Objects, 403–404.

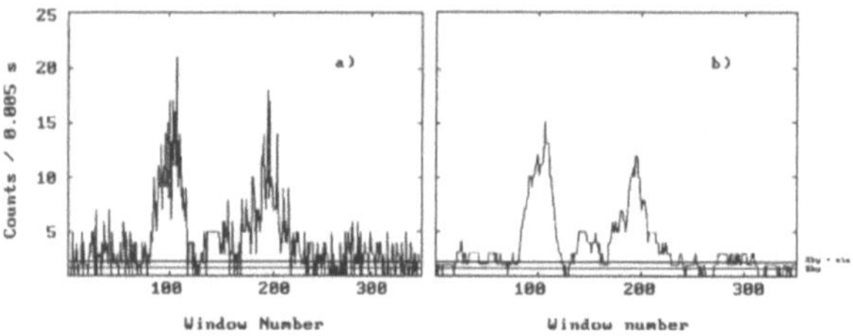

Figure 2. The white light curves (taken with aperture - 12 arcsec) of two flares of MXB1735−44. (a) Initial light curve, (b) smoothed light curve.

for the first two and $5\,10^{11}$ K for the others.

MXB1735−44. Two flares with duration of about 0.25 s were detected on 1991 May 9 (Fig. 2). The rise times of these flares are 110 ms and 120 ms (about 60 ms at the most steep parts) and their relative amplitudes are 32 and 17. The corresponding brightness temperatures are from $4\,10^7$ to $2\,10^8$ K. In order to study the fine structure of these events the detailed light curve, $I(t)$, was analyzed using the 'splash method' [1]. This method showed that both flares have fine structure on time-scales of 5...10 ms (Fig. 2a). The lower limit on the brightness temperatures of the fine structure is $2\,10^{10}$ K. We found a distance $D \approx 1$ kpc, $B \approx 19.3$ mag, $A_v \approx 1.2$ mag for A0620−00 [7] and $D \approx 7$ kpc, $B \approx 17.2$ mag, $A_v \approx 0.8$ mag for MXB1735−44 [1]. Thus, the brightness temperatures of the A0620−00 and MXB1735−44 flares ($10^8 \ldots 10^{11}$ K) contradict a thermal mechanism for their generation and support a nonthermal model.

2. A search for optical pulses of the 6.1 ms binary radio pulsar PSR 1953+290. The observations of the 6.1–ms radio pulsar PSR 1953+290 [2] were carried out in 1992 at the 6 m telescope with a CCD photometer to identify the optical counterpart of the source using its radio position and the MANIA complex (white light with 7 arcsec diaphragm) to search for optical pulses. No optical pulses or any stochastic variability [5] have been detected. The upper limit for brightness in the B−band is 26.5 mag with a significance level < 3%. By comparing the upper limit of the pulsar optical luminosity (taking into account the distance [2]) with the Crab pulsar luminosity gives us an upper limit for the magnetic field of $B \approx 10^8$ G.

References

1. Beskin, G.M., Neizvestny, S.I., Plokhotnichenko, V.L., et al., 1994, A&A, **289**, 141
2. Boriakoff, V., Buccheri, R., Fauci, F., 1983, Nature, **304**, 417
3. Elvis, M., Page, C.G., Pounds, K.A., et al., 1975, Nature, **257**, 656
4. Plokhotnichenko, V.L., 1983, Soobshchenija SAO, **38**, 29
5. Plokhotnichenko, V.L., 1992, Ph.D. thesis, Special Astrophysical Observatory
6. Shvartsman, V.F., 1977, Soobshchenija SAO, **19**, 5
7. Shvartsman, V.F., Beskin, G.M., Mitronova, S.N., et al., 1989, Sov. Ast. Let., **15**, 590

SUPERNOVAE & SUPERSOFT X-RAY SOURCES

ACCRETING WHITE DWARFS AND TYPE Ia SUPERNOVAE

MARIO LIVIO[1], DAVID BRANCH[2], L. R. YUNGELSON[3]
FRANCESCA BOFFI[1], E. BARON[2]
1. Space Telescope Science Institute, 3700 San Martin Drive, Baltimore, MD 21218, USA
2. Dept. of Physics and Astronomy, University of Oklahoma, Norman, OK 73019, USA
3. Institute of Astronomy of the Russian Acadamy of Sciences, 48 Pyatnitskaya Str., 109017 Moscow, Russia

Abstract. The question of the possible progenitors of supernovae Type Ia (SNe Ia) is examined. It is argued that SNe Ia are thermonuclear explosions of accreting C-O white dwarfs. The existing observational evidence favors somewhat models in which the exploding star ignites carbon upon reaching the Chandrasekhar mass. A careful examination of all the potential progenitor classes reveals that when realization frequencies are combined with a variety of observational charcteristics, no *single* class emerges as containing the obvious progenitors. It is argued that coalescing white dwarfs or supersoft X-ray sources are the most likely progenitor systems. A few critical observations which could help identify the progenitors unambiguously are discussed.

1. Characteristics and the basic model

The *defining* characteristics in the spectra of supernovae Type Ia (SNe Ia) are the lack of lines of hydrogen and the presence of a strong red Si II absorption feature (λ6355 shifted to ~ 6100 Å). The following are some of the observational characteristics of the class:

(i) nearly 90% of all SNe Ia form a homogeneous class in terms of their spectra, light curves and peak absolute magnitudes. The latter are given by $M_{\rm B} \simeq M_{\rm V} \simeq -19.75 + 5\log(H_0/50 \text{ km s}^{-1} \text{ Mpc}^{-1})$, with a dispersion of $\sigma(M_{\rm B}) \sim \sigma(M_{\rm V}) \sim 0.2$ (e.g. Tammannn & Sandage 1995);

A. Evans and J. H. Wood (eds.), Cataclysmic Variables and Related Objects, 407–415.

(ii) near maximum light, the spectra are characterized by high velocity ($8\,000\ldots30\,000\,\mathrm{km\,s^{-1}}$) intermediate mass elements (O–Ca). In the late, nebular phase, the spectra are dominated by forbidden lines of iron (e.g. Kirshner et al. 1993; Wheeler et al. 1995; Ruiz-Lapuente et al. 1995a);

(iii) in terms of explosion strength, SNe Ia can be roughly ordered as: SN 1991bg and SN 1992k represent the weakest events, followed by weak events like SN 1986g, followed by about 90% of all SNe Ia which are called 'normals', to the stronger-than-normal events like SN 1991t;

The luminosity function of SNe Ia declines very steeply on the bright side (e.g. Vaughan et al. 1995). Since selection effects cannot prevent the discovery of SNe which are brighter than the 'normals', this means that the normal SNe Ia are essentially the brightest;

(iv) fairly young populations appear to be most efficient at producing SNe Ia (they tend to be associated with spiral arms in spirals; Della Valle & Livio 1994; Bartunov, Tsvetkov & Filimonova 1994), but an old population ($\tau \gtrsim 4\,10^9\,\mathrm{yr}$) *can also* produce them (SNe Ia occur in ellipticals; Turatto, Cappellaro & Benetti 1994). The explosion strength appears to be inversely correlated with the age of the stellar population (Ruiz-Lapuente, Burkert & Canal 1995b).

As a consequence of the above points, and in particular point (iv), which shows that SNe Ia cannot be produced by core collapse of massive stars, the currently accepted model for SNe Ia is that they represent *thermonuclear disruptions of mass accreting white dwarfs* (WDs). Thus, the basic ingredient of the model for SNe Ia is the same type of object that is present in cataclysmic variables (CVs). The question that needs to be answered is: which binary system or systems are the immediate progenitors of SNe Ia? Before we address this question, we would like to explain the importance of identifying the progenitors.

2. Why is the identification of the progenitors of SNe Ia important?

Identifying the progenitor systems of SNe Ia is important for the following main reasons:

(i) in the absence of agreed upon detailed models for the explosion itself, a knowledge of the initial conditions and of the distribution of material in the vicinity of the exploding star are essential for the understanding of the explosion;

(ii) an identification of the progenitors will help (through the knowledge of the SNe Ia rate) to put constraints on models for binary evolution. In particular, it may help constrain the value of the common envelope efficiency parameter α_{CE} (see e.g. Iben & Livio 1993, Livio 1995 for reviews), which is presently very poorly known;

(iii) galaxy evolution depends on the input from supernovae in terms of energy (radiative and kinetic) and nucleosynthetic products and on the evolution of the SN rate with time;

(iv) the use of SNe Ia to determine H_0 and q_0 requires a knowledge of the evolution of the rate and the luminosity function of SNe Ia with cosmic time. Both of these quantities depend on the nature of the progenitors.

3. Constraints on the basic model

The basic model (thermonuclear disruption of an accreting WD) for SNe Ia can be further constrained. In particular, a combination of observations and theoretical calculations can be used in attempts to determine the required composition of the accreting white dwarf and the location in the WD (e.g. center versus shell) and the fuel where ignition first occurs.

3.1. THE COMPOSITION OF THE ACCRETING WD

The accreting WD could be, in principle, of He, CO or ONeMg composition. However, in the case of a He WD, the composition of the ejected matter would be primarily He and ^{56}Ni, which is inconsistent with the observations (see Sect. 1). In the case of ONeMg WDs there is no direct observational argument which can exclude them as potential progenitors. Theoretical calculations show, however, that the frequency of events that can be expected from such accretors is not high enough to explain the observed (rather deduced) SNe Ia rate (e.g. Livio 1993). Furthermore, detailed calculations show that accreting ONeMg WDs tend more to collapse quietly to form neutron stars, rather than to explode as SNe Ia (e.g. Gutierrez et al. 1995).

Consequently, it appears most likely that *the accreting WDs that produce SNe Ia are of a CO composition.*

3.2. WHERE IN THE WD AND IN WHICH FUEL DOES IGNITION TAKE PLACE?

There are presently two classes of models being considered: one in which *carbon* ignites at the WD *center*, when the WD reaches the *Chandrasekhar mass*, and the second, in which the accreted layer of *helium* ($\lesssim 0.2\,M_\odot$) on top of a CO WD, ignites *off-center*, typically at *sub-Chandrasekhar* masses. In the case of the carbon ignitors, numerical simulations have shown that the energetics, light curve and composition as a function of ejection velocity are all quite consistent with observations (e.g. Branch et al. 1985; Harkness 1991; Höflich, Khokhlov & Wheeler 1995). The fact that these models always explode at the Chandrasekhar mass has always been regarded as a potential explanation for the homogeneity of SNe Ia. The main difficulty with this class of models is related to statistics. Namely, it is not clear whether WDs in sufficient numbers to explain the SNe Ia rate can reach the Chandrasekhar mass (hydrogen and helium flashes which lead to mass loss make this difficult). A second, related difficulty, lies in the fact that WDs that are initially more massive than 1.2 $M_{\odot}$ (and therefore more likely, perhaps, to reach the Chandrasekhar mass), tend quietly to collapse to form neutron stars rather than to explode (Nomoto & Kondo 1991), although it is not clear how firm this result is.

The helium (sub-Chandrasekhar) ignitors may produce (in principle) three different outcomes, two of which are certainly not normal SNe Ia. In the case of the 'direct double detonation' (DDD) (namely, one detonation wave propagating outward through the helium and the second inward, through the CO), the whole WD is burned into ^{56}Ni and He (Nomoto 1982; Woosley, Weaver & Taam 1980). This is inconsistent with the observation of high-velocity intermediate mass elements.

The 'single detonation' (SD) case (the helium detonates but the CO core does not), in which the core may survive as a WD or disintegrate at low velocities, certainly does not correspond to normal SNe Ia (Nomoto 1982; Woosley, Taam & Weaver 1986).

The only sub-Chandrasekhar model which has some chance to produce an explosion resembling a SN Ia is the 'indirect double detonation' (IDD) model. In this scenario, one detonation propagates outward through the helium, while an inward moving pressure wave results in ignition at the center (through the compression of the core), followed by an outward moving detonation (Livne & Glasner 1990; Woosley & Weaver 1994; Livne & Arnett 1995; Höflich & Khokhlov 1995).

Models of this type have been shown to produce light curves which are in agreement with those observed in SNe Ia. The main problems currently still existing with this model are: (i) the highest velocity ejecta have the wrong composition (^{56}Ni and He; e.g. Livne & Arnett 1995), (ii) the light curve rises somewhat faster than observed (due to the presence of ^{56}Ni) and the early colors are inconsistent with observations (Höflich & Khokhlov 1995), (iii) one might expect this model to produce a gradual decline on the bright side of the luminosity function, in contradiction to the observed sharp decline. The last point would have been a consequence of the range of WD masses that the IDD model allows (more massive WDs produce brighter SNe).

As a consequence of the above discussion we feel that *carbon ignitors (at the Chandrasekhar mass) are presently favored by the observations*, but that the uncertainties that still exist in the calculations of the helium ignitor

IDD models suggest that this model should still be considered.

4. Candidate progenitor systems

We now return to the question of which binary systems could be the immediate progenitors of SNe Ia. Branch et al. (1995) examined this question in detail using realization frequencies of SNe Ia obtained from a variety of binaries. The calculations were performed with a population synthesis code of the type used by Yungelson et al. (1994, 1995). Of the many uncertainties which are involved in these calculations two are by far the most important ones in terms of their effects on the final results: the efficiency parameter in common envelope evolution, α_{CE}, and the fraction η of the accreted mass, which a WD is able to retain. The efficiency parameter α_{CE} describes the fraction of the gravitational energy change due to spiralling-in inside the common envelope which is deposited into envelope ejection. Since other sources of energy (e.g. recombination energy) may be involved, α_{CE} could in principle be even larger than 1 (see Iben & Livio 1993; Livio 1995 for reviews). At present, possible values of α_{CE} are not reliably constrained neither observationally nor theoretically. A similar uncertainty exists in relation to the accumulation ratio η. Different numerical codes result in rather different fractions of the accreted mass which remain atop the WD following shell flashes (e.g. Starrfield et al. 1992; Prialnik & Shara 1995). Furthermore, the accumulation ratio in a combination of hydrogen and helium flashes is even less certain (e.g. Iben & Tutukov 1994; Kato, Saio & Hachisu 1989).

Acknowledging the fact that these (and indeed other) uncertainties exist, Branch et al. (1995) adopted the following conservative assumption. A class of objects is *not* considered a major contributor to the SNe Ia rate, if the realization frequencies for young ($\sim 10^8$ yr) and old ($\lesssim 10^{10}$ yr) populations respectively satisfy: $\nu < 10^{-4}$ yr^{-1}, $\nu < 10^{-5}$ yr^{-1} (the deduced rate for the Galaxy is $\sim 4\,10^{-3}$ yr^{-1}; van den Bergh & McClure 1994). In Table 1 we give the realization frequencies for all the relevant candidate progenitors for carbon and helium ignitors. As can be seen from the table, if one insists on only *one class* of progenitors, then only WD mergers have a sufficiently high frequency among the carbon ignitors. No single class can in fact produce the necessary frequencies among the helium ignitors, although symbiotic systems come close.

We will therefore examine now in some detail these two classes of objects.

WD mergers have been originally suggested as SNe Ia progenitors by Iben & Tutukov (1984) and Webbink (1984). Recent hydrodynamic calculations have shown that when the lighter of the two WDs fills its Roche

TABLE 1. Candidate progenitor systems and their realization frequencies

Type of System	Realization Frequency (yr^{-1}) for Young ($\sim 10^8$ yr) Population	Realization Frequency (yr^{-1}) for Old ($\lesssim 10^{10}$ yr) Population
	a. Carbon Ignitors	
Cataclysmic Variable	10^{-5}	—
WD accreting from Roche lobe filling subgiant or giant	—	10^{-6}
Symbiotic System	10^{-6}	10^{-4}
Helium CV	10^{-4}	—
WD accreting helium from Roche lobe filling giant	10^{-5}	—
Merger of CO + He WDs	10^{-4}	10^{-5}
Merger of CO + CO WDs	10^{-3}	10^{-4}
	b. Helium Ignitors	
WD accreting from Roche lobe filling subgiant or giant	—	10^{-4}
Symbiotic System	10^{-5}	10^{-3}
Helium CV	10^{-3}	—
Merger of CO + He WDs, or accretion from He WD	?	?

lobe (the two WDs having been brought together by gravitational radiation), it is totally dissipated within a few orbital periods, to form a thick disk-type configuration (rotationally supported) around the primary (Rasio & Shapiro 1995; Benz et al. 1990). The subsequent evolution of the system is not entirely clear, since it is not known if central or off-center ignition will occur (e.g. Mochkovitch & Livio 1989, 1990; Mochkovitch et al. 1995). Two points which may give some confidence in the realization frequencies obtained for mergers are as follows. (i) The population synthesis calculations indicate that for populations younger than $\sim 10^{8.5}$ yr, the total (combined) mass of the merging WDs is of order 2.1 $M_\odot$, while it converges to the Chandrasekhar mass for older populations (Tutukov & Yungelson 1995). This may be consistent with the inverse correlation found between explosion strength and population age (Ruiz-Lapuente et al. 1995b). (ii) All the recently discovered double WD systems (Marsh 1995; Marsh, Dhillon & Duck 1995) fall into the peak of the period distribution predicted by the population synthesis calculations of Yungelson et al. (1994). On the other hand, it should be realized that some of the paths which lead to the formation of double WD systems involve *two* common envelope phases and thus they are necessarily uncertain (see e.g. Livio 1994).

It is important to note that what has been considered until recently as a fatal difficulty for the merger scenario, the tentative detection of hydrogen in the supernova 1990m (Polcaro & Viotti 1991), no longer poses a problem, since this detection has been shown to be erroneous (Della Valle, private communication). A difficulty that still exists is statistical in nature; among the nine double WD systems discovered so far (Saffer, Liebert & Olszewski 1988; Bragaglia et al. 1990; Marsh et al. 1995; Marsh 1995) not a single one is a SN Ia progenitor candidate, either because the orbital period is too long for the two WDs to merge in a Hubble time, or because the total mass is significantly lower than the Chandrasekhar mass. Other searches for double WD systems have given rather negative results (e.g. Robinson & Shafter 1987; Foss, Wade & Green 1991). While Yungelson et al. (1994) have shown that the results of any single one of these searches are still not in conflict with the possibility of producing SNe Ia at a rate compatible with observations ($\sim 3\,10^{-3}$ yr^{-1}), the combined negative results of all the searches starts to place uncomfortable constraints on this scenario (see also Renzini 1995).

As noted above, if helium ignitors represent the correct model (which presently appears difficult to accept), then symbiotic systems come the closest to producing the required frequency of SNe Ia. A real observational test for this scenario can come from deep radio observations performed immediately after the explosion. Due to the high wind mass loss rates observed from the giants in symbiotic systems ($\dot{M}_{\rm w} \sim 10^{-6}\,{\rm M}_\odot\,{\rm yr}^{-1}$, the circumstellar density in the vicinity of these systems should be high enough to produce detectable radio emission shortly after the SN explosion. Boffi & Branch (1995) showed that for $\dot{M}_{\rm w} \sim 10^{-6}\,{\rm M}_\odot\,{\rm yr}^{-1}$, the peak of the emission should occur about a week after the explosion and the expected peak flux is about 30 mJy for a distance of 4 Mpc. Thus, the upper limit to the radio flux from (the rather weak) SN 1986g (Eck et al. 1995), appears to indicate that the progenitor of this supernova at least, was probably not a symbiotic system. Finally, persistent supersoft X-ray sources, which involve a WD accreting at a rate $\gtrsim 10^{-7}\,{\rm M}_\odot\,{\rm yr}^{-1}$ from a subgiant companion (van den Heuvel et al. 1992; Rappaport, Di Stefano & Smith 1994; Yungelson et al. 1995) and burning hydrogen stably, include (in principle) the necessary ingredients for a system to reach the Chandrasekhar mass. While the realization frequency for such events appears presently to be too low (Yungelson et al. 1995), it is possible that this merely represents the uncertainties involved in population synthesis calculations. Hence, the class of supersoft X-ray sources should be kept in mind as potentially viable progenitor candidates.

5. General problems with current models

In the previous sections we described difficulties that arise in relation to the identification of SNe Ia progenitors. Some of the problems involved are well exemplified by the fact that whichever the preferred model may be, some questions remain unanswered. For example, if we decide that the helium (sub-Chandrasekhar) ignitors *are not* the correct model for the majority of SNe Ia, this may mean that either the realization frequencies given in Table 1 are very wrong (which is quite possible, in view of the many uncertainties), or that off-center helium ignition leads to consequences that are very different from IDDs (e.g. single detonations, although current models do not favor them). In the latter case, however, it may mean that a large population of these subluminous (as far as SNe go) explosions awaits detection. Conversly, if we decide that the helium ignitors *do* represent the correct model for SNe Ia, then it is not clear why current models predict that, for WD masses larger than $\sim 0.9\,M_{\odot}$, a direct double detonation should be obtained. The absence of these more luminous events in the observational sample seems to suggest that direct double detonations never occur in nature, in contradiction to theoretical expectations.

Acknowledgement. This research has been supported in part by NASA Grant NAGW-2678 at the Space Telescope Science Institute.

References

Bartunov, O.S., Tsvetkov, D.Yu., Filimonova, I.V., 1994, PASP, **106**, 1276
Benz, W., Bowers, R.L., Cameron, A.G.W., Press, W.H., 1990, Ap. J., **348**, 647
Boffi, F., Branch, D., 1995, PASP, **107**, 347
Bragaglia, A., Greggio, L., Renzini, A., D'Odorico, S., 1990, Ap. J., **365**, L13
Branch, D., Doggett, J.B., Nomoto, K., Thielemann, F.-K., 1985, Ap. J., **294**, 619
Branch, D., Livio, M., Yungelson, L.R., et al., 1995, PASP, in press
Della Valle, M., Livio, M., 1994, Ap. J., **423**, L31
Eck, C., Cowan, J.J., Roberts, D.A., et al., 1995, Ap. J., submitted
Foss, D., Wade, R.A., Green, R.F., 1991, Ap. J., **374**, 281
Gutierrez, J., Garcia-Berro, E., Iben, I., et al., 1995, preprint
Harkness, R.P., 1991, in "Supernovae", ed. S.E. Woosley (New York: Springer-Verlag), p454
Höflich, P., Khokhlov, A., 1995, Ap. J., in press
Höflich, P., Khokhlov, A., Wheeler, J.C., 1995, Ap. J., **444**, 831
Iben, I. Jr., Livio, M., 1993, PASP, **105**, 1373
Iben, I. Jr., Tutukov, A.V., 1984, Ap. J. Supp., **54**, 335

Iben, I. Jr., Tutukov, A.V., 1994, Ap. J., **431**, 264
Kato, M., Saio, H., Hachisu, I., 1989, Ap. J., **340**, 509
Kirshner, R.P., et al., 1993, Ap. J., **415**, 589
Livio, M., 1993, in "Cataclysmic Variables and Related Physics", eds O. Regev, G. Shaviv, Ann. Israel Phys. Soc., **10**, p57
Livio, M., 1994, Mem. S. A. It., **65**, 49
Livio, M., 1995, in "Evolutionary Processes in Binary Stars", eds. R.A.M.J. Wijers et al., (Dordrecht: Kluwer), in press
Livne, E., Arnett, D., 1995, Ap. J., in press
Livne, E., Glasner, A., 1990, Ap. J., **361**, 244
Marsh, T.R., 1995, MNRAS, in press
Marsh, T.R., Dhillon, V.S., Duck, S.R., 1995, MNRAS, in press
Mochkovitch, R., et al., 1995, in preparation
Mochkovitch, R., Livio, M., 1989, A&A, **209**, 111
Mochkovitch, R., Livio, M., 1990, A&A, **236**, 378
Nomoto, K., 1982, Ap. J., **257**, 780
Nomoto, K., Kondo, Y., 1991, Ap. J., **367**, L19
Polcaro, V.F., Viotti R., 1991, A&A, **242**, L9
Prialnik, D., Shara, M.M., 1995, AJ, **109**, 1735
Rappaport, S., Di Stefano, R., Smith, J.D., 1994, Ap. J., **426**, 692
Rasio, F.A., Shapiro, S., 1995, Ap. J., **438**, 887
Renzini, A., 1995, in "Supernovae", ed. R. McCroy (Cambridge: Cambridge University Press), in press
Robinson, E.L., Shafter, A.W., 1987, Ap. J., **322**, 296
Ruiz-Lapuente, P., Kirshner, R.P., Phillips, M.M., et al., 1995a, Ap. J., **439**, 60
Ruiz-Lapuente, P., Burkert, A., Canal, R., 1995b, Ap. J., **447**, L69
Saffer, R.A., Liebert, J., Olszewski, E., 1988, Ap. J., **334**, 947
Starrfield, S., Shore, S.N., Sparks, W.M., et al., 1992, Ap. J., **391**, L71
Tammannn, G.A., Sandage, A., 1995, Ap. J., in press
Tutukov, A.V., Yungelson, L.R., 1995, in "Cataclysmic Variables", eds A. Bianchini et al., Kluwer, p495
Turatto, M., Cappellaro, E., Benetti, S., 1994, AJ, **108**, 202
van den Bergh, S., McClure, R.D., 1994, Ap. J., **425**, 205
van den Heuvel, E.P.J., Bhattacharya, D., Nomoto, K., Rappaport, S.A., 1992, A&A, **262**, 97
Vaughan, T.E., Branch, D., Miller, D.L., Perlmutter, S., 1995, Ap. J., **439**, 558
Webbink, R.F., 1984, Ap. J., **277**, 355
Wheeler, J.C., Harkness, R.P., Khokhlov, A.V., Höflich, P., 1995, Phys. Rep., in press
Woosley, S.E., Taam, R.E., Weaver, T.A., 1986, Ap. J., **301**, 601
Woosley, S.E., Weaver, T.A., 1994, Ap. J., **423**, 371
Woosley, S.E., Weaver, T.A., Taam, R.E., 1980, in "Type Ia Supernovae", ed. J.C. Wheeler (Austin: University of Texas), p96
Yungelson, L.R., Livio, M., Tutukov, A.V., Saffer, R.A., 1994, Ap. J., **420**, 336
Yungelson, L.R., Livio, M., Truran, J.W., et al., 1995, Ap. J., submitted

A MODEL FOR THE GALACTIC POPULATION OF SUPERSOFT X-RAY SOURCES

L. YUNGELSON[1], A. TUTUKOV[1], A. FEDOROVA[1],
M. LIVIO[2], J. W. TRURAN[3]
1. Institute of Astronomy of the Russian Academy of Sciences, 48 Pyatnitskaya Str., 109017 Moscow, Russia
2. Space Telescope Science Institute, 3700 San Martin Drive, Baltimore, MD 21218, USA
3. Department of Astronomy and Astrophysics, Enrico Fermi Institute of Physics, University of Chicago, 5640 S. Ellis Avenue, Chicago, IL 60637, USA

Abstract. Three major sub-populations of Galactic supersoft X-ray sources may exist: semi-detached binaries with main sequence or subgiant donors, and symbiotic binaries ($\sim$ 550, $\sim$ 460, and $\sim$ 600 objects, respectively). Each group contains both permanent and transient sources. The intrinsic and interstellar absorptions reduce the number of observable sources to $\sim$ 25. We derive the distributions of the sources over orbital periods, masses of the components, and 'on'-times. The rate at which white dwarfs in Galactic binaries reach $M_{\rm Ch}$ is $\sim 3\,10^{-5}$ yr^{-1}. The rate of He-shell detonations which may lead to supernovae may be up to $3\,10^{-4}$ yr^{-1}.

1. Discussion

The common model of supersoft X-ray sources (hereafter SSS) relates the origin of their X-ray emission to steady hydrogen (van den Heuvel et al. 1992) or helium (Iben & Tutukov 1994) burning at the surface of accreting white dwarfs (WDs). By means of the population synthesis code for binary stars (e.g. Yungelson et al. 1995) we generate a population of systems in which accretion of H or He onto CO or ONe white dwarfs is possible via Roche lobe overflow or a stellar wind. Following the mass exchange history for each potential SSS (including the secular evolution of accretor mass), we estimate the number of SSS and their properties. Different to earlier

A. Evans and J. H. Wood (eds.), Cataclysmic Variables and Related Objects, 417–420.

TABLE 1. Numbers of supersoft X-ray sources and their parent systems

	CVs	Algols	Symbiotic stars	Double degenerates	Helium Algols	PNNi
			Parent Population			
Birthrate, yr^{-1}	0.0039	0.0006	0.047	0.013	0.0009	0.11
Total number	$4.1\,10^6$	$2.4\,10^5$	$1.6\,10^3$	$1.4\,10^8$	200	36
Novae, yr^{-1}	20	10	1.4	-	-	-
Permanent SSS	134	402	445	70	≤ 200	35
Transient SSS	412	59	162	-	-	-
Number of sources with correction for intrinsic absorption by spherical shell						
Permanent SSS	134	402	0			
Transient SSS	380	45	7			
			Number of 'detectable' sources			
Novae, yr^{-1}	0.7	0.3	0.1	-	-	-
Permanent SSS ($r \leq 2$ kpc)	2	7	0	1	≤ 4	≤ 1
Transient SSS ($V \leq 8$ mag)	8	1	7	-	-	-

models of SSS, we allow for both stable and unstable nuclear burning of the accreted matter. Subsequently, we consider selection effects and thus obtain the properties of the 'observed' population of SSS. We shall refer to the three main groups of SSS, for brevity, as CVs (cataclysmic variables), Algols (systems with subgiant donors), and SyS (symbiotic systems).

Hydrogen burns stably if $\dot{M}_{\rm acc}$ exceeds a certain limit. WDs accreting below the stability limit experience hydrogen burning flashes. Once the WD undergoes a flash it loses material, in nova events, via an optically thick wind and/or a common envelope. After the mass of the H-rich layer of the WD decreases below a certain limit, the WD settles in the region of the HR diagram which corresponds to stable burning, and there most of the remaining H is consumed. Stably burning WDs have $T_{\rm eff} \gtrsim 250\,000$ K if $M_{\rm wd} \gtrsim 0.54\,{\rm M}_\odot$ (Iben & Tutukov 1989). Hence, all the CO or ONe WD accretors may be candidate SSS. During this stage of evolution, the bolometric luminosity is $L \gtrsim 5\,10^{37}$ erg s^{-1}. For the estimate of the time during which the post-nova WD still emits in the supersoft X-ray range, we use the time it takes the WD to decline by 3 mag in its bolometric luminosity from the Prialnik & Kovetz (1995) grid of nova models. This estimate gives probably the upper bound of lifetimes.

There are 'permanent' sources, in which H burns steadily, and

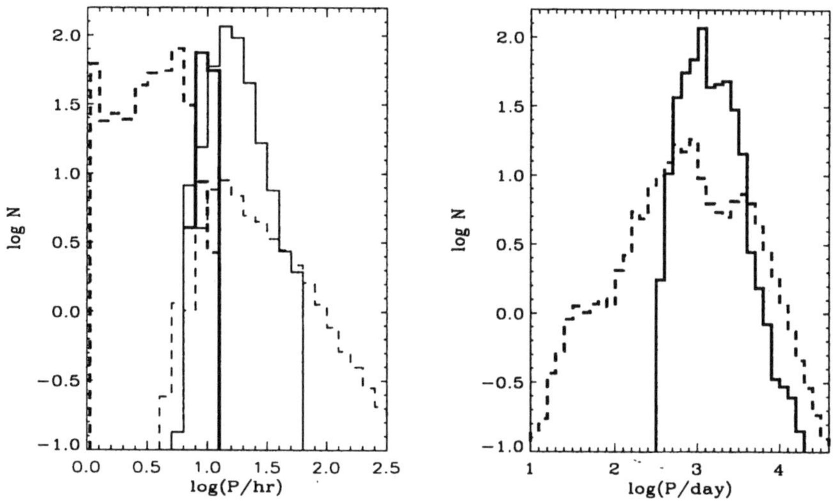

Figure 1. Distribution of SSS over orbital periods. Left panel—CVs (thick lines) and Algols (thin lines), right panel—SyS. Solid line—permanent sources, broken line—transient sources.

'transient' sources, in which supersoft X-rays are expected to be radiated for short ($\lesssim$ 150 yr) intervals of time following shell burning flashes. The different parent groups differ in terms of the number of objects which they contain of the two classes. High accretion rates result in a high proportion of permanent sources, while the low ones produce transient SSS.

The detectability of SSS may be influenced by intrinsic absorption. For the transient SSS the most efficient source of intrinsic absorption may be the shielding by the envelopes ejected during nova eruptions. For SyS, *all* the permanent sources have an optical depth in the *spherically symmetric* donor wind of $\tau > 1$ at the distance of the accretor from the donor, and only about 4% of the transient sources remain observable.

For high mass WDs, shielding times are longer than the lifetimes in the 'on' state. For example, a source with a mass exceeding $\sim 1.15\,M_\odot$ may never be observed. While a typical WD mass in observed nova systems is $\sim 1\,M_\odot$, in the model of the detectable sample of SSS it is between $0.6\,M_\odot$ and $0.8\,M_\odot$. There may be a higher chance to find a SSS in a survey of old ($\lesssim$ 150 yr) novae. This may explain why only one SSS was found in a survey of 26 novae, which exploded in the last 10 yr (Orio 1993).

For the detection of supersoft emission, the source has to be closer than ~ 2 kpc to the Sun. For the identification of a transient source with a nova, we apply the requirement for the nova to have $V \leq 8$ mag in the plateau luminosity stage. Combined, intrinsic and interstellar absorption leave only

$\sim$ 25 'detectable' Galactic SSS.

Other possible contributors to the Galactic population of SSS are semi-detached, double degenerate systems with He WD donors (Tutukov & Yungelson 1996) and semi-detached systems in which the donors have CO cores and thick helium mantles ('helium Algols'; Iben & Tutukov 1994). In these systems mass exchange may be stable, steady He burning at the surface of WDs occur and hence, emission of supersoft X-rays is possible. Finally, planetary nebula nuclei (PNNi) more massive than $0.7\,M_\odot$ spend a certain time at temperatures in excess of 250 000 K and have at this time $L \gtrsim 5\,10^3\,L_\odot$.

Fig. 1 shows the distribution of model SSS over $P_{\rm orb}$. Permanent sources appear in two distinct ranges of periods with peaks close to 10...20 h and 500...3000 d. Transient sources in CVs are evenly distributed between the 'minimum' period of CVs and 8 h. Algols cover the period range between 10 h and 100 h (with a 'tail' to 300 h), and in SyS the transient sources have $100 \lesssim P_{\rm orb}({\rm d}) \lesssim 10\,000$. Every class of model objects has its counterpart among the discovered Galactic SSS. For example, GQ Mus is a post-nova system, sources RX J0925.7−4756 and RX J0019+21 may be identified with systems with subgiant donors, and the symbiotic stars RR Tel and AG Dra confirm the contribution of this class.

Depending on the secular evolution of the accretion rate and initial mass of the WD, the latter may decrease or grow. We estimate that the rate at which accreting WDs in Galactic binaries reach the Chandrasekhar mass is only $3\,10^{-5}$ yr^{-1}. If the WD is not eroded by novae, a He layer accumulates, and it may detonate when the critical mass for ignition is accumulated. Helium detonation may result in the detonation of the core and a supernova (e.g. Livne & Glasner 1990). The only model systems in which the accumulation of a critical layer for detonation of He is possible are Algols. The estimated Galactic rate of possible detonations is $3\,10^{-4}$ yr^{-1}.

References

Iben, I., Tutukov, A.V., 1989, Ap. J., **342**, 430
Iben, I., Tutukov, A.V., 1994, Ap. J., **431**, 264
Livne, E., Glasner, A.S., 1990, Ap. J., **361**, 244
Orio, M., 1993, in "Cataclysmic Variables and Related Physics", eds O. Regev, G. Shaviv, Ann. Israel Phys. Soc., **10**, p219
Prialnik, D., Kovetz, A., 1995, Ap. J., **445**, 789
Tutukov, A., Yungelson, L., 1996, MNRAS, in press
van den Heuvel, E.P.J., Bhattacharya, D., Nomoto, K., Rappaport, S.A., 1992, A&A, **262**, 97
Yungelson, L., Livio, M., Tutukov, A., Kenyon, S.J., 1995, Ap. J., **447**, 665

THE PERIOD DISTRIBUTION OF SUPERSOFT SOURCES

P. KAHABKA
Astronomical Institute "Anton Pannekoek" and Center for High Energy Astrophysics, University of Amsterdam, Kruislaan 403, 1098 SJ Amsterdam, The Netherlands

Abstract. The observed and predicted orbital period distributions of supersoft sources are in good agreement.

1. Results and conclusions

Presently for 8 of 10 optically identified supersoft systems binary orbital periods have been determined (Kahabka & Trümper 1995, Kahabka 1995a). These systems can be classified into 3 groups: classical novae (with two members), unstable mass transfer (canonical) systems (with six members) and symbiotic novae (with one member). We consider only the period distribution of the canonical systems. The period distribution of this class extends from 4.1 h to 3.5 d. Rappaport, di Stephano & Smith (1994) simulated the orbital period distribution of the galactic supersoft systems. In model 7, in which the time evolution of the binary has explicitly been considered, a

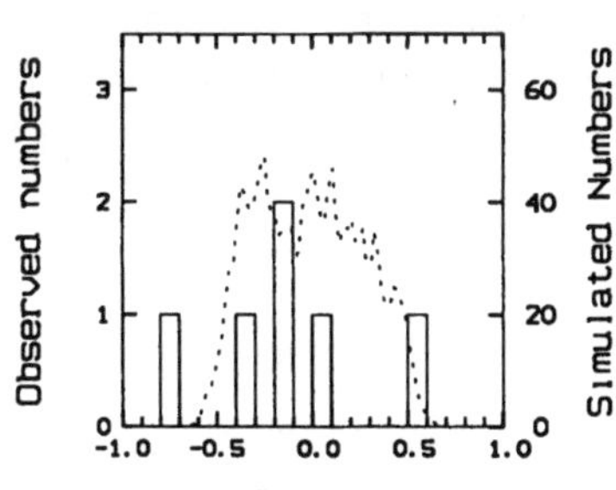

Figure 1. Orbital period distribution of supersoft sources. Only the distribution of the canonical systems (evolved main-sequence or post main-sequence star binaries) is shown.

A. Evans and J. H. Wood (eds.), *Cataclysmic Variables and Related Objects*, 421–422.

TABLE 1. Supersoft systems for which orbital periods have been determined

Name	P_{orb} [d]	log (P_{orb})	remarks
Nova Mus 1983 (GQ Mus)	0.0593	−1.22	classical nova
Nova V1974 Cyg (Nova Cyg 1992)	0.0812	−1.09	classical nova
1E0035.4−7230	0.172	−0.764	variable
CAL 87	0.442	−0.353	persistent
RX J0019.8+2156	0.660	−0.180	recurrent
RX J0513.9−6951	0.76	−0.119	recurrent
CAL 83	1.04	+0.0170	persistent
RX J0925.7−4758	3.4	+0.544	persistent
AG Dra	554	+2.74	symbiotic nova

distribution containing ~1800 members and extending from ~4.3 h to ~4 d is predicted. This can be considered to be a good overall agreement with the observed distribution and is in strong support for the steady-state nuclear burning model being applicable. Only the 4.1 h period, found in the (only) SMC binary system 1E0035.4−7230 may be of concern. This could be in favor for a recurrently nuclear burning and lower mass white dwarf (in order to keep it sufficiently long burning in agreement with the observations). Recurrent supersoft sources (cf. Kahabka 1995b) have been found to have slightly lower mass accretion rates ($\sim 8\,10^{-8} \ldots 4\,10^{-7}\,M_{\odot}\,yr^{-1}$) and higher white dwarf masses ($1.0 \ldots 1.4\,M_{\odot}$) compared to steady-state nuclear burning systems (cf. van den Heuvel et al. 1992). The observed systems most probably have already accumulated a significant amount of accreted hydrogen within evolutionary time-scales and gained in mass considerably. This may affect their further evolution. For a discussion of the period distribution of the other classes of supersoft X-ray binaries (symbiotic, CV-like systems, etc.) I refer to the work of Yungelson et al. (1996).

References

Kahabka, P., Trümper J., 1995, in "IAU Symposium 165, Compact Stars in Binaries", p425

Kahabka, P., 1995a, in "Cataclysmic Variables", eds A. Bianchini et al., Kluwer, p435

Kahabka, P., 1995b, A&A, **304**, 227

Rappaport, S., di Stephano, R., Smith, J.D., 1994, Ap. J., **270**, L9

van den Heuvel, E.P.J., Bhattacharya, D., Nomoto, K., et al., 1992, A&A, **262**, 97

Yungelson, L., Livio M., Tutukov A., Kenyon S.J., 1996, Ap. J., submitted

ROSAT AND OPTICAL OBSERVATIONS OF SUPERSOFT X-RAY SOURCES

K. REINSCH[1], A. VAN TEESELING[1], K. BEUERMANN[1,2], H.-C. THOMAS[3], T. M. C. ABBOTT[4]

1. Universitäts-Sternwarte, Geismarlandstr. 11, D-37083 Göttingen, Germany (reinsch@usw050.dnet.gwdg.de)

2. MPI für Extraterrestrische Physik, Garching, Germany

3. MPI für Astrophysik, Garching, Germany

4. ESO Chile, current address: CFHT, Hawaii, USA

Abstract. We present X-ray, UV and optical observations of the Galactic supersoft X-ray source (SSS) RX J0019.8+2156 and first results of our study of the X-ray and optical long-term variability of bright SSS.

1. Introduction

Luminous supersoft X-ray sources (SSS) have been established as a new and distinct class of objects by the discovery of some 20 objects during the ROSAT mission, which show properties similar to the prototype system CAL 83 (for a recent review see Hasinger 1994). Several models, including accretion onto black holes and neutron stars accreting above the Eddington rate, have been proposed to explain the nature of SSS. The most prominent interpretation considers a white dwarf which accretes matter from its more massive main sequence secondary at a rate just sufficient to permit (quasi-) stable nuclear burning near its surface (van den Heuvel et al. 1992).

2. The bright Galactic source RX J0019.8+2156

Our complete sample of bright, soft, high-galactic-latitude ROSAT sources contains just one Galactic object which bears strong similarity to the LMC SSS: RX J0019.8+2156 (Beuermann et al. 1995). The system displays quasi-sinusoidal modulations of the optical, UV and X-ray light curves, as well as regular radial-velocity variations with a 15.85 h period which we interpret as the binary orbital period. Time-resolved spectroscopy of RX J0019+21

A. Evans and J. H. Wood (eds.), Cataclysmic Variables and Related Objects, 423–424.

shows a complex structure and orbital variability of the Balmer and He II line profiles, and a P Cygni absorption component in the Balmer lines.

X-ray luminosity and temperature seem to be lower in RX J0019+21 than in CAL 83, RX J0513–69 and RX J0925–47. The simplest explanation is that the hot component in RX J0019+21 is a white dwarf with an accretion rate of $\sim 10^{-7}\,M_{\odot}\,yr^{-1}$, which just fails to allow stationary hydrogen burning. The system therefore may experience weak shell flashes, being currently in a luminosity minimum. This is supported by the optical long-term behaviour of RX J0019+21, which shows high and low states lasting about 20 yr each (Greiner & Wenzel 1995).

3. Long-term variability of supersoft sources

Besides orbital variability, most of the SSS are highly variable on different time-scales down to weeks (or even less). We have studied the X-ray and optical variability of several bright SSS to obtain constraints for different models and to assess the evolutionary status of individual systems.

While the soft X-ray flux of RX J0019+21 has been fairly constant on time-scales of years since the ROSAT All-Sky Survey (RASS) in 1990, the 0.1...2.4 keV flux of the LMC source RX J0549.9−7151 has decreased by more than a factor of 200 since the RASS observation and a pointed PSPC observation of CAL 87 containing RX J0549−71 as a serendipitous source (Cowley et al. 1993). The X-ray flux of RX J0439.8−6809 (Greiner, Hasinger & Thomas 1994) is variable by a factor of 2 on time-scales as short as a few seconds. On longer time-scales, however, the source has been remarkably constant since the RASS. The variability of the suspected faint optical counterpart is constrained to be less than ~ 0.5 mag.

References

Beuermann, K., Reinsch K., Barwig, H., et al., 1995, A&A, **294**, L1
Cowley, A.P., Schmidtke, P.C., Hutchings, J.B., et al., 1993, Ap. J., **418**, L63
Greiner, J., Hasinger, G., Thomas, H.-C., 1994, A&A, **281**, L61
Greiner, J., Wenzel, W., 1995, A&A, **294**, L5
Hasinger, G., 1994, Reviews in Modern Astron., **7**, 129
van den Heuvel, E.P.J., Bhattacharya, D., Nomoto, K., Rappaport, S.A., 1992, A&A, **262**, 97

VISUAL LIGHT FROM THE ECLIPSING SUPERSOFT X-RAY SOURCE CAL 87

S. SCHANDL, E. MEYER-HOFMEISTER, F. MEYER
Max-Planck Institut für Astrophysik, Karl Schwarzschildstr. 1, 85748 Garching, Germany

We calculate the visual light curve of CAL 87 based on the assumption that an accreting, steadily burning white dwarf irradiates the accretion disk and the secondary star, as suggested by van den Heuvel et al. (1992). The shape of the eclipse light curve and the absolute luminosities put strong constraints on the contributions of the various elements in the system. We find a reasonable fit is obtained with the following sources of visual light: a Roche lobe filling secondary star where the energy from irradiation is spread out over the surface, and an accretion disk with an optically thick, cold, clumpy spray which is caused by the massive accretion stream impinging on the disk (hot spot). This spray moving around the disk and illuminated by the white dwarf is found to be the brightest source of visual light in the binary and can account for the asymmetry in the light curve.

We calculate the emission of all contributing surfaces by assuming a blackbody radiation spectrum. For comparison with the observations we folded it with a visual filter function. At parts facing the white dwarf the irradiated flux is added to the viscous dissipation of the disk, the constant flux of the spray and the original stellar flux of the secondary. Energy transport on the stellar surface is approximated by spreading the irradiated flux weighted by a Gaussian kernel. For more details see Schandl, Meyer-Hofmeister & Meyer (1996).

The simulation yields an inclination of about 77° and supports the assumption that the white dwarf is permanently hidden by the disk.

References

Schandl S., Meyer-Hofmeister E., Meyer F., 1996, A&A, submitted (and Green Report No. 912 of the Max-Planck Institut für Astrophysik, Garching, Germany)
Schmidtke, P.C., McGrath, T.K., Cowley, A.P., Frattare, L.M., 1993, PASP, **105**, 863
van den Heuvel, E.P.J., Bhattacharya, D., Nomoto, K., Rappaport S.A., 1992, A&A, **262**, 97

A. Evans and J. H. Wood (eds.), Cataclysmic Variables and Related Objects, 425–426.

TABLE 1. The System Parameters

Facts:			
Orbital period	P	=	10.6 h
Distance	d	=	57.5 kpc
Assumed parameters (model of van den Heuvel et al. 1992):			
Primary star (white dwarf)	M_1	=	$0.75\,M_\odot$
Secondary star (Roche lobe filling main sequence star)	M_2	=	$1.5\,M_\odot$
Mass accretion rate	$\dot{M}$	=	$0.8\,10^{-7}\,M_\odot\,yr^{-1}$
Temperature of the non-irradiated Secondary star	$T_\star$	=	8000 K
⇒ disk size (0.8 Roche lobe)	s	=	$5.6\,10^{10}$ cm
⇒ luminosity of white dwarf	L	=	$2.2\,10^{37}\,erg\,s^{-1}$
Further parameters taken for the presented theoretical light curve:			
Albedo (for star and disk)	a	=	0.5
Inclination	i	=	77°

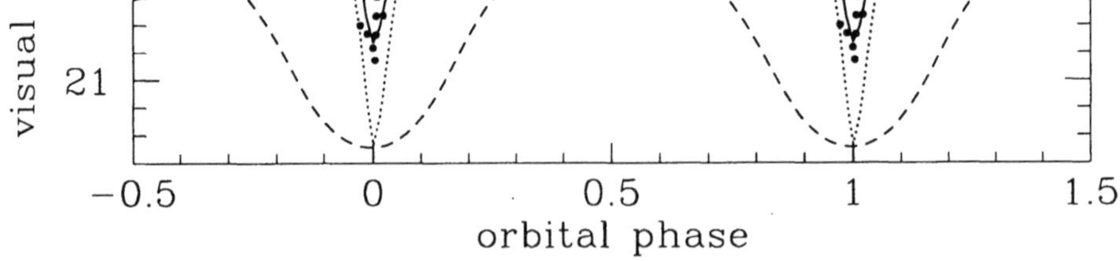

Figure 1. The visual light curve of CAL 87. At primary minimum the observer sees the dark side of the secondary and a non-eclipsed part of the disk. At secondary minimum one sees the partial occultation of the illuminated secondary by the disk. The dots show the composite V light curve where the photometry was done between 1985 November and 1992 December (from Schmidtke et al. 1993). The fit (solid line) consists of a secondary with sideways energy transport (dashed line) and the illuminated accretion disk with an additional hot spot region (dotted line).

THE SUPERSOFT SOURCE RXJ0019.8+2156: NEW PHOTOELECTRIC OBSERVATIONS

C. BARTOLINI, A. GUARNIERI, A. PICCIONI, L. SOLMI
Department of Astronomy, University of Bologna,
Via Zamboni 33, 40126 Bologna, Italy
bartolini@astbo3.bo.astro.it

1. The object

RXJ0019.8+2156 is a Supersoft X-ray Source (SSS) discovered by ROSAT and identified (Beuermann et al. 1995) as an object of magnitude V=12.2, brighter than any other known SSS. The mass function and the optical spectrum are similar to CAL 83, but the X-ray luminosity seems to be lower. The main light variations are: a cyclic modulation with a 15.8 h period and a 0.3 mag amplitude, a quasi-periodic pulsation less than 0.1 mag in 2 h (Beuermann et al. 1995), variations of about 1 mag over a time-scale of 20 yr and small irregular fluctuations on a time-scale of weeks to months (Greiner & Wenzel 1995).

The most promising model, at present, involves a binary system in which a hot 1 $M_{\odot}$ white dwarf with an accretion disk is accreting at a mean rate of $\dot{M} \approx 10^{-7}$ $M_{\odot}\,yr^{-1}$; sporadic hydrogen burning caused by an unstable mass transfer (Greiner & Wenzel 1995) seems to be present.

2. New observations

We observed RXJ0019 in the V band at the Bologna Observatory with a two-head photometer (Piccioni et al. 1979) during five nights: 1994 December 26, 27 and 1995 January 7, 8, 9. Although fainter, the comparison star was more stable than RXJ0019, which sometimes showed some flickering; the observed differences of magnitude on 1994 December 26, 27 are shown in Fig. 1a, folded with the period P = 0.6604565 d (Greiner & Wenzel 1995).

Starting from the same brightness as in December, on January 8 and 9 the star showed a progressive increase of luminosity of 1.7 mag (Fig. 1b).

A. Evans and J. H. Wood (eds.), Cataclysmic Variables and Related Objects, 427–428.

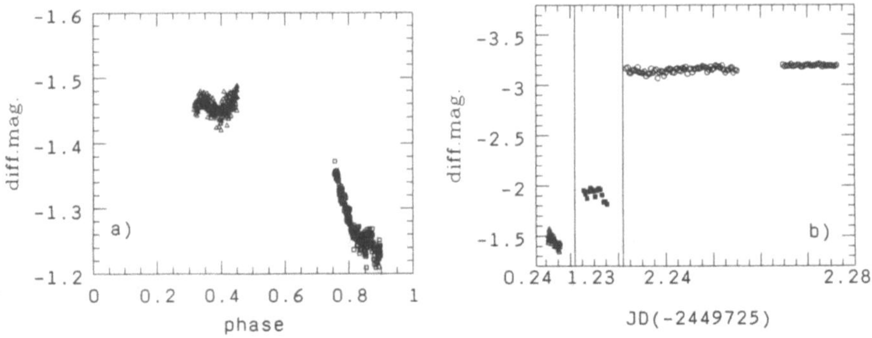

Figure 1. Bologna light curves: (a): on the nights 1994.12.26, 1994.12.27; (b): on the nights: 1995.1.7, 1995.1.8 and 1995.1.9

On January 7 and 8 the sky conditions were poor, with a quarter of moon, on January 9 the sky was photometric and the moon was further from the star, which showed fairly stable counts; therefore we are confident that the observed variation was real.

3. Discussion

We stress that the increase of brightness observed in 1995 January was stronger than any other previously observed variation and that it occurred in only 2 d, which is a typical time-scale for an outburst in a cataclysmic variable. However no outbursts seem to be present in the historical light curve (Greiner & Wenzel 1995). Such an increase of luminosity could be an indication that in RXJ0019 a CV phase is incipient.

The presence of flickering already noticed by Beuermann et al. (1995) and Greiner & Wenzel (1995) and confirmed by our observations, supports this idea.

References

Beuermann, K., Reinsch, K., Barwig, H., et al., 1995, A&A, **294**, L1
Greiner, J., Wenzel, W., 1995, A&A, **294**, L5
Piccioni, A., Bartolini, C., Guarnieri, A., Giovannelli, F., 1979, Acta Astronomica, **29**, 463

THE SUPERSOFT X-RAY SOURCE RXJ0537.6–7033

M. ORIO[1,2], M. DELLA VALLE[3], G. MASSONE[2],
H. ÖGELMAN[1]
1. Department of Physics, University of Wisconsin, Madison, USA
2. Osservatorio Astronomico di Torino, Italy
3. Dipartimento di Astronomia, Università di Padova, Italy

We present new X-ray data with an improved position for the supersoft X-ray source RXJ0537.6–7033, and propose a candidate for the optical identification.

The source was initially designated as RXJ0537.7–7034 because there was a high uncertainty in the position, at the edge of the field of the ROSAT PSPC (Orio & Ögelman 1993). It was later observed three more times by the ROSAT PSPC at intervals of approximately 6 months, from the beginning of 1992 to the end of 1993. The change of the name is due to the new, more precise position obtained in these observations. The count rate was 0.021 ± 0.005 count s^{-1} in the first 1992 observation and remained at the same level in the second 1992 observation. The source was not detected in the first half of 1993, with an upper limit of 0.002 count s^{-1}; finally in 1993 December it was detected with only 0.0035 ± 0.0007 count s^{-1}. It seems therefore to be very variable.

We propose the identification with an optical counterpart that is only 9 arcsec distant from the PSPC position, belongs to the LMC and shows a peculiar spectrum (as obtained with the ESO 3.6 m telescope at La Silla), with the He II λ4686 line in emission and most likely the hydrogen Balmer lines in absorption. A very preliminary analysis of optical photometric data indicates some variability. If the identification with this LMC object is correct, then RXJ0537.6–7033 is a 'supersoft' X-ray source at lower luminosity ($L_{bol} \simeq 10^{37}$ erg s^{-1}) and possibly more absorbed than other LMC supersoft X–ray sources.

References

Orio, M., Ögelman, H., 1993, A&A, **273**, L56

A. Evans and J. H. Wood (eds.), Cataclysmic Variables and Related Objects, 429.

POST COMMON ENVELOPE BINARIES & EVOLUTION

UNDERSTANDING THE LONG–TERM EVOLUTION OF CVS AND LMXBS: PRINCIPLES AND PROBLEMS

U. KOLB
Lick Observatory, UCSC, Santa Cruz CA 95064, USA, and Max-Planck-Institut für Astrophysik, Karl-Schwarzschild-Str. 1, 85740 Garching, Germany

Abstract. We introduce the standard model for formation and evolution of CVs, and show how observed collective properties compare with theoretically predicted population models. A short overview relates LMXBs to CVs.

1. Introduction

The standard model for formation and evolution of cataclysmic variables (CVs) rests on principles formulated decades ago: Kraft, Mathews & Greenstein (1962) realized the importance of gravitational wave emission for driving mass transfer in close binaries, an idea later refined by Faulkner (1971). Stimulated by Ritter's arguments (1976a,b) in favour of a wide binary origin of CVs, with a red giant as progenitor of the white dwarf, Paczyński (1976) suggested a common envelope phase as a mechanism to form a system with an orbital distance much smaller than the radius of the giant once was. Finally Spruit & Ritter (1983) and Rappaport, Verbunt & Joss (1983) independently explained the CV period gap by what is now known as the disrupted magnetic braking model. Low-mass X-ray binaries (LMXBs), the neutron star analogues of CVs, were long thought to evolve in a similar way. But although the models reproduce the main features of the observed collective properties of CVs fairly well (Sect. 2) and seem to be incomplete only in detail (Sect. 3), some more fundamental problems remain for LMXBs (Sect. 4).

We restrict this review to Galactic CVs and LMXBs with hydrogen donors. Compact binaries in globular clusters are likely to form via tidal capture and are discussed elsewhere (e.g. Livio 1996; Hut et al. 1992).

A. Evans and J. H. Wood (eds.), Cataclysmic Variables and Related Objects, 433–442.

2. Principles: formation and evolution of CVs

CVs are thought to form from rather wide main sequence binaries, with an initial separation such that the faster evolving, more massive primary fills its Roche lobe as a giant with predominantly convective envelope. The resulting mass transfer is dynamically unstable and leads to a common envelope (CE) phase, during which orbital energy is used to expel the giant's envelope on a very short time-scale. Left behind in a very close orbit are the exposed core, the later white dwarf (WD), and the almost unaffected low–mass, main sequence secondary. Loss of orbital angular momentum by magnetic stellar wind braking and gravitational wave emission, or slow nuclear expansion of the secondary, establish the semi–detached CV configuration. These angular momentum losses maintain continuous mass transfer from the secondary at a rate $\dot{M} \simeq 10^{-9} \ldots 10^{-8}\,\mathrm{M_\odot\,yr^{-1}}$ for long orbital period, P, and $\dot{M} \simeq 10^{-11} \ldots 10^{-10}\,\mathrm{M_\odot\,yr^{-1}}$ for short P, where only gravitational radiation operates. As a reaction to mass loss the secondary builds up a gravo–thermal luminosity, causing the stellar radius to become smaller (for a secondary mass $M_2 \gtrsim 0.6\,\mathrm{M_\odot}$) or larger ($M_2 \lesssim 0.6\,\mathrm{M_\odot}$) than the main sequence radius. With progressing mass transfer the secondary's outer convection zone penetrates deeper into the star, until at $M_2 \simeq 0.2\,\mathrm{M_\odot}$ the donor becomes fully convective and magnetic braking is assumed to vanish discontinuously. The system detaches since the secondary's contraction on a thermal time-scale is faster than the orbital shrinking on the much longer gravitational radiation time-scale. When mass transfer resumes the secondary has regained thermal equilibrium and the orbital period is correspondingly smaller. Systems in the detached phase are not observable as CVs, they are missing in the orbital period distribution of CVs and cause the well–known 2 . . . 3 h period gap. The donor eventually becomes a degenerate brown dwarf, causing P to increase, even though the orbital angular momentum further decreases. Accordingly, the system passes through a minimum orbital period P_{min} at $\simeq 80$ m. Further details of this evolutionary scheme are given by e.g. King (1988) and Kolb (1995b).

Along these lines detailed population synthesis calculations try to link observed properties of newly born main sequence binaries (the star formation rate, the distribution of primary mass, mass ratio and orbital separation) to observed collective properties of CVs like the CV period distribution by applying results from stellar evolution theory. Explicit computations of the CV formation rate were carried out by Politano (1988, 1996) and de Kool (1992). Kolb (1993) combined these models with detailed evolutionary sequences of CVs to obtain a multi-dimensional distribution function of system parameters representing the present Galactic CV population. Other, more extended, binary population synthesis calculations do not consider CVs in detail (e.g. Han, Podsiadlowski & Eggleton 1995; Yungelson et al. 1995).

Here we summarize the main results of CV population studies; for basic concepts and interpretations see Kolb (1995b):

Main sequence binaries qualify as CV progenitors if (i) the primary can form a WD core within the age of the Galaxy but avoids a supernova (primary mass $0.8 \lesssim M_\mathrm{p}/\mathrm{M}_\odot \lesssim 9$) and (ii) the mass ratio $q_\mathrm{i} = M_\mathrm{p}/M_2$ is large ($\gtrsim 3.5$), so that mass transfer is stable when the post–CE binary becomes semi–detached. Roughly, about 30 out of 100 main sequence binaries within the above mass range encounter dynamically unstable mass transfer leading to a CE. About 10 binaries survive the CE, 5 close enough to turn on mass transfer within a Hubble time, but only 2 in a stable configuration as a CV. Depending on the initial masses the possible initial separation range is $100 \dots 1200\,\mathrm{R}_\odot$ (period range $\simeq 60 \dots 2500\,\mathrm{d}$). The predicted CV formation rate depends most sensitively on the initial mass ratio distribution $g(q_\mathrm{i})$. Calibrated to the observed formation rate of isolated WDs the binary population models find for the number of CVs residing in the Galaxy today about $2\,10^6$ [for $g(q_\mathrm{i}) \propto q_\mathrm{i}^{-3}$] to $20\,10^6$ (for uncorrelated masses), i.e. an intrinsic CV space density $10^{-5} \dots 10^{-4}\ \mathrm{pc}^{-3}$.

To estimate the post–CE configuration usually a constant value ≤ 1 is assumed for the fraction, α_CE, of released orbital binding energy available to unbind the envelope. Given the complexity of the CE phase (see e.g. Iben & Livio 1993) it is not surprising that more recent numerical simulations cast some doubt on this concept (e.g. Yorke, Bodenheimer & Taam 1995; Terman & Taam 1996). Unless α_CE is a strong function of pre–CV binary parameters this should have no significant effect on synthesis results for CVs which are not very sensitive to the actual global value of α_CE.

Fig. 1 (left) depicts the period distribution of a visual magnitude (m_vis) limited sample obtained from one of the intrinsic population models by Kolb (1993; $\alpha_\mathrm{CE} = 1$, uncorrelated masses, magnetic braking according to Verbunt & Zwaan 1981). The system's visual magnitude L_vis is assumed to be that of a stationary accretion disk and can be estimated by $L_\mathrm{vis} \propto M_1^\alpha \dot{M}^\beta$, where M_1 is the WD mass, and $\alpha \simeq 1$, $\beta \simeq 3/4$ (Dünhuber 1994). Although intrinsically only 0.5% of all CVs have periods above the gap (and $\lesssim 70\%$ have evolved beyond the minimum period), the m_vis–limited sample is characterized by roughly an equal number of systems below and above the gap, just as the observed sample (Fig. 1, right). The minimum period in the computed distribution is shorter than 80 m due to the neglect of rotational and tidal corrections in the stellar models. The spike at $\log P/\mathrm{h} \simeq 0.02$ (Fig. 1, left) results from systems accumulating at P_min where $\dot{P} = 0$. The fact that it is absent in the observed sample points to additional selection effects not accounted for in the model, e.g. the very long dwarf nova recur-

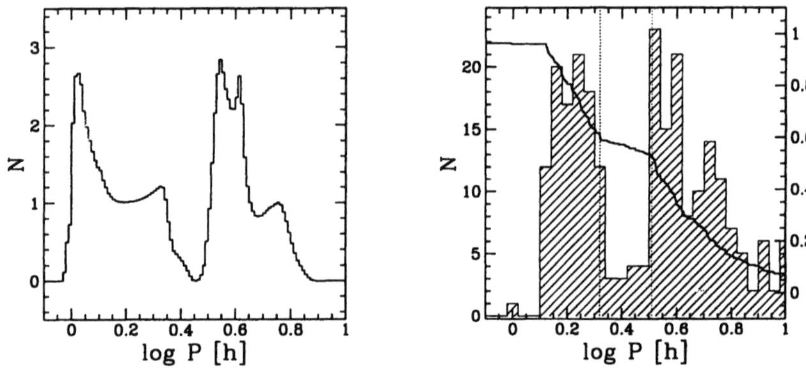

Figure 1. Left: m_{vis}–limited sample obtained from a CV population model by Kolb (1993). Right: Observed CV period histogram (hatched; scale on the left) and cumulative period distribution (scale on the right); data from Ritter (1995). The vertical dotted lines mark the period gap.

rence time at small P. The local minimum at $\log P/\mathrm{h} \simeq 0.67$ (Fig. 1, left) is due to an allowance for the formation of H_2 molecules in the outer layers of the secondary and should eventually show up in an observed sample with improved statistics, whereas the spike at $\log P/\mathrm{h} \simeq 0.62$ is just a numerical artefact (see Kolb 1993).

3. Problems: the period distribution of subtypes

The standard model introduced in the previous section seems to agree fairly well with the main features of the observed CV population, but a closer look at various CV subtypes reveals properties the models, as discussed so far, fail to reproduce.

Dwarf novae and nova-likes. Observations indicate that above the period gap the relative number of dwarf novae per period bin is an increasing function of orbital period (Fig. 2, left). In contrast, standard evolutionary models combined with predictions from the disk instability model for dwarf nova outbursts (see e.g. Cannizzo 1993 for a review) tend to predict the opposite. As an example, Fig. 2 (right) depicts different evolutionary sequences in the $P - \dot{M}$ plane, together with a limiting line separating stable (nova-likes, above the line) from unstable (dwarf novae, below) systems. The line specifies the location of CVs with a disk just cool enough for hydrogen recombination at its outer edge. Even though the precise run of $\dot{M}$ with P in CVs depends on the detailed mechanism extracting angular

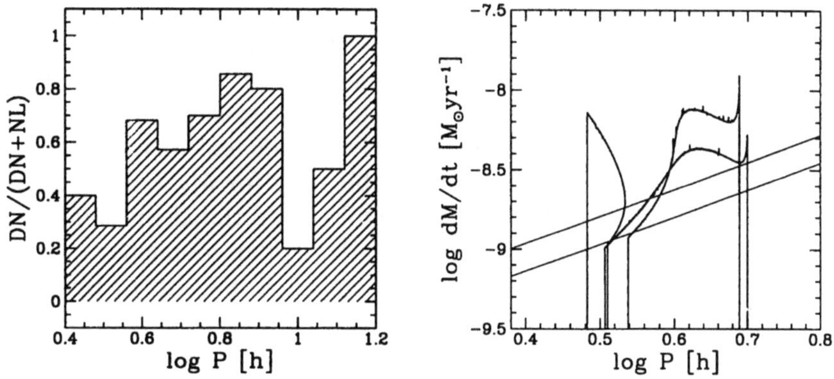

Figure 2. Left: The dwarf nova fraction in non-magnetic CVs above the period gap as a function of orbital period. Data from Ritter (1995). Right: Mass transfer rate versus orbital period for various evolutionary tracks of CVs (taken from Kolb & Ritter 1992). The initial WD/secondary masses (in $M_\odot$) are $(1.0, 0.6)$, $(0.7, 0.6)$ and $(1.0, 0.4)$, in order of decreasing turn-on period. The straight lines denote the critical mass transfer rate for DN outbursts for a disk radius of 0.7 (upper line) and 0.6 (lower line) times the WD's Roche lobe radius (following Shafter 1992).

momentum from the system, the mass transfer rate at the upper edge of the period gap can be deduced in a *model-independent way* from the observed width of the gap: The gap width measures the ratio $r = R_2/R_e$ of the radius of the secondary when it enters (R_2) and leaves (R_e) the detached phase. Since R_e is the thermal equilibrium radius, the value of r can be related to the ratio $\tau = t_{KH}/t_M$ of Kelvin–Helmholtz time-scale, t_{KH}, and mass loss time-scale, t_M (Stehle, Ritter & Kolb 1996), at the upper edge of the gap. Hence the mass transfer rate results from τ and the mass of the secondary in the detached phase, known from the location of the lower edge of the gap. The observed gap between 2.09 h and 3.24 h (limits taken from the cumulative period distribution shown in Fig. 1, right) requires $r \simeq 4/3$ and $\tau \simeq 5$, so that $\dot{M} \simeq 10^{-9}\,M_\odot yr^{-1}$. This value is below the critical limit for disk instability, even for an unusually small disk (see Fig. 2). Therefore close to the upper edge of the gap virtually all systems should have unstable disks, whereas only a few dwarf novae exist in this period regime, a fact already noted by Shafter (1992).

A related problem is the apparent scatter of the observed mass transfer rate at a given orbital period (e.g. Patterson 1984, Warner 1987) which seems to be at least an order of magnitude larger than the standard model can account for. Since the reaction of low-mass secondaries to mass loss demands a fairly uniform evolution (Stehle et al. 1996) this scatter was interpreted as a cyclic change of the instantaneous mass transfer rate around the secular mean, on a time-scale too long to be observable directly, but short enough in order not to change the secular mean (e.g. Hameury, King & Lasota 1989). Although weak irradiation of the secondary by the WD was identified as a promising mechanism causing such cycles (King et al. 1995a; Ritter et al. 1996a,b), a more recent model-independent analysis shows that purely irradiation driven cycles are possible only for $M_2 \gtrsim 0.7\,\mathrm{M}_\odot$ (Ritter, Zhang & Hameury 1996; King et al. 1995b). Additional mechanisms, e.g. a mass transfer dependent (consequential) angular momentum loss component (King & Kolb 1995), are needed to allow mass transfer cycles at periods close to the upper edge of the gap. These in turn could help to resolve the problem of the dwarf nova period distribution.

Systems in the period gap. As listed in Table 1, 19 systems out of a total number of 269 CVs have orbital periods in the period gap range 2.09...3.24 h (data from Ritter 1995). Apparently there is an overabundance of polars and — although this is clearly small number statistics — classical novae in the gap. Possible implications for the influence of the WD magnetic field on the long-term evolution were reviewed recently by several authors (King 1994; Chanmugam 1995; Kolb 1995b) and will not be repeated here. A gap population of $\simeq$ 7% is significantly above the prediction $\lesssim$ 3% from computed $m_{\rm vis}$-limited samples such as the one shown in Fig. 1, but the observed gap population of non-magnetic CVs is still at least marginally consistent with the models. For classical novae see below.

TABLE 1. CVs in the period gap

Subtype	systems in gap	% of gap population	% of CVs with known P belonging to subtype
non-magnetic dwarf novae	5	26	41
non-magnetic nova-likes	2	10	15
VY Scl systems	1	5	5
polars	6	32	16
intermediate polars	2	10	9
classical novae	3	16	8

Classical novae. A nova occurs whenever the hydrogen-rich envelope accreted on the WD ignites H-burning in a thermonuclear runaway (TNR). CV population models can be used to predict the distribution of the nova outburst frequency $\nu = \dot{M}/\Delta M_{\rm ign}$ over system parameters, provided the envelope mass at ignition $\Delta M_{\rm ign}$ is known. Roughly, the TNR begins when the pressure at the base of the envelope exceeds a critical value (e.g. Livio

1994). Correspondingly, the nova rate is large for high-mass WD CVs: $\Delta M_{\rm ign} \propto R_1^4 M_1^{-1} \rightarrow 0$ when M_1 approaches the Chandrasekhar limit. The main character of this dependency remains the same when results from an extended grid of more detailed TNR model computations, applying a diffusion description for envelope/WD material mixture prior to ignition (Prialnik & Kovetz 1995), replace this simple ignition criterion (Kolb 1995a). As an immediate consequence the mean WD mass obtained from the WD mass distribution of ν is very large, typically $\langle M_1 \rangle \simeq 1.3\,{\rm M}_\odot$, whereas observed values, though subject to considerable uncertainty, point to a much smaller value ($\simeq 0.8\,{\rm M}_\odot$). The period distribution of ν is very similar to the $m_{\rm vis}$-limited sample computed for CVs (Fig. 1) since in both cases the intrinsic distribution is weighted with a selection factor $\propto \dot{M}$. In particular, a period gap is predicted and the expected fraction of novae below the gap is typically of order 50%.

This is in marked contrast to the distribution of the small sample of 21 novae with determined orbital period (Fig. 3, left): 3 novae fall below the classical CV period gap, 3 within the gap (see Table 1), so that the observed distribution shows no signature of a gap at all (whereas polars do show a gap, see e.g. Kolb 1995b). It is not clear if this is due to small number statistics, selection effects operating against the detection of novae in systems with small mass transfer rate $\lesssim 10^{-10}\,{\rm M}_\odot{\rm yr}^{-1}$, or differences in the outburst behaviour of high-$\dot{M}$ and low-$\dot{M}$ novae. Alternatively, a closer inspection of the relation between mass transfer cycles and nova outbursts and the role of dynamical friction when the secondary orbits in the expanding nova envelope (see Schenker, Kolb & Ritter 1996 for further references) might resolve some of these problems.

On the other hand, the total Galactic nova rate predicted from the models, $1 \ldots 10\,{\rm yr}^{-1}$, is surprisingly close to the presently favoured observed value ($20\,{\rm yr}^{-1}$, Della Valle & Livio 1994; but see also Della Valle & Duerbeck 1993).

4. Relatives: low-mass X-ray binaries

The purpose of this final section is to emphasize the difference between evolutionary histories of LMXBs and CVs. No attempt is made to provide a review on formation and evolution of LMXBs (for which the reader is referred to e.g. Bhattacharya & van den Heuvel 1991; Verbunt 1993; D'Antona 1996).

The main additional obstacle for the formation of LMXBs is that the progenitor binary has to survive the supernova explosion required to produce a neutron star. The corresponding constraints on the initial parameter space limit the LMXB formation rate to a value well below the CV birth

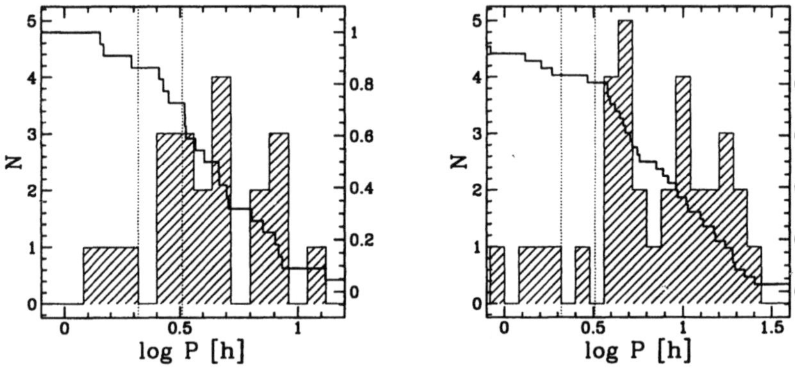

Figure 3. Observed period histogram (hatched; scale on the left) and cumulative period distribution (scale on the right) for classical novae (left) and low-mass X-ray binaries (right). Data from Ritter (1995).

rate.

A possible LMXB producing evolutionary channel very similar to the CV case is sometimes referred to as the standard model: here the initially very massive primary ($10 \lesssim M_p/\mathrm{M}_\odot \lesssim 50$) emerges from the orbit-shrinking CE phase as a helium star. Subsequently it passes through a Wolf–Rayet phase and explodes finally as a supernova (SN). For spherically symmetric mass ejection the binary remains bound only if less than half of the pre–SN mass leaves the system. Since this is typically not the case (the He star generally being more massive than its companion) many authors invoke kick velocities to circumvent the disruption of the binary. The still-detached post–SN binary could be observable as a binary radio pulsar if the neutron star were magnetic.

More exotic ways to generate LMXBs have been suggested, e.g. a triple star origin (Eggleton & Verbunt 1986), SN kicks causing small orbits without involving a CE (Kalogera 1996), or accretion induced collapse of a massive WD to produce the neutron star in an already existing close binary, thus avoiding major mass loss (Whelan & Iben 1973).

Once a LMXB is formed its long–term evolution should be determined by the same principles as for CVs. However, the LMXB orbital period distribution (see Fig. 3, right) extends to longer periods, i.e. the fraction of binaries with (sub)giant donors is significantly larger than for CVs, and there are only few systems below $P \simeq 4$ h. Accordingly, no well-defined period gap or minimum period exists. More indirect evidence for differences from a standard 'CV–type' evolution comes from estimates for the formation rate of low–mass binary pulsars, which are believed to descend from

LMXBs. Kulkarni & Narayan (1988) found a value too large by a factor 10...100 to be compatible with a characteristic LMXB lifetime of 10^9 yr. However, more recent estimates (Johnston & Bailes 1991) suggest a much more moderate discrepancy.

To explain these findings it was pointed out that irradiation of the secondary by either the X-ray radiation from the vicinity of the neutron star during the LMXB phase or pulsar radiation during a detached state might alter the long–term evolution significantly. Two model categories can be distinguished, both potentially leading to an accelerated evolution: the first group considers the effect X–ray heating has on the secondary's internal structure. Deposited below the photosphere the incoming flux blocks internally generated luminosity, thereby expanding the donor star and driving an elevated mass transfer rate (Podsiadlowski 1991). The main unknown here is the time-scale and amplitude of the expansion, determined essentially by how efficiently the star can cool through the unirradiated side (see e.g. Hameury et al. 1993). The second group considers evaporation of the star's atmospheric layers, i.e. substantial mass loss in a strong stellar wind (e.g. Tavani 1991). Again, the quantitative implications remain speculative.

In conclusion, both for the understanding of details in the observed inter-relation of CV subtypes and of the fundamental evolution of LMXBs more investigations considering effects of weak and strong irradiation on (nearly) Roche lobe filling low–mass stars are urgently needed.

Acknowledgements. I would like to thank R. Taam and J. Frank for stimulating conversations, and J. Faulkner for comments and improving the language of the manuscript. Support from an IAU travel grant and NASA grant NAGW 3599 is greatly acknowledged.

References

Bhattacharya, D., van den Heuvel, E. P. J., 1991, Phys. Rep., **203**, 1
Cannizzo, J. K., 1993, in "Accretion disks in compact stellar systems", ed. J. C. Wheeler, World Scientific, p6
Chanmugam, G. 1995, in "Cataclysmic Variables", eds A. Bianchini et al., Kluwer, p415
D'Antona, F., 1996, in "Evolutionary Processes in Binary Stars", eds R. A. M. J. Wijers, M. B. Davies, C. A. Tout, NATO ASI Series, Kluwer, Dordrecht, in press
Della Valle, M., Duerbeck, H., 1993, A&A, **271**, 175
Della Valle, M., Livio, M., 1994, Ap. J., **423**, 31
Dünhuber, H., 1994, Ph.D. thesis, Universität München
Eggleton, P. P., Verbunt, F., 1986, MNRAS, **220**, 13P
Faulkner, J., 1971, Ap. J., **170**, L99
Hameury, J.-M., King, A. R., Lasota, J.-P., 1989, MNRAS, **237**, 39
Hameury, J.-M., King, A. R., Lasota, J.-P., Raison, F., 1993, A&A, **277**, 81
Han, Z., Podsiadlowski, P., Eggleton, P. P., 1995, MNRAS, **272**, 800
Hut, P., McMillan, S., Goodman, J., et al., 1992, PASP, **104**, 981
Iben, I., Livio, M., 1993, PASP, **105**, 1373

Johnston, S., Bailes, M., 1991, MNRAS, **252**, 277
Kalogera, V., 1996, in "Proceedings of the 2nd Hellenistic Astronomical Conference", Thessaloniki, Greece, in press
King, A. R., 1988, QJRAS, **29**, 1
King, A. R., Kolb, U., 1995, Ap. J., **439**, 330
King, A. R., Frank, J., Kolb, U., Ritter, H., 1995a, Ap. J., **444**, L37
King, A. R., Frank, J., Kolb, U., Ritter, H., 1995b, Ap. J., submitted
Kolb, U. 1995a, in "Cataclysmic Variables", eds A. Bianchini et al., Kluwer, p511
Kolb, U. 1995b, in "Magnetic Cataclysmic Variables", eds D. Buckley, B. Warner, ASP Conf. Ser. **85**, p440
Kolb, U., 1993, A&A, **271**, 149
Kolb, U., Ritter, H., 1992, A&A, **254**, 213
de Kool, M., 1992, A&A, **261**, 188
Kraft, R. P., Mathews, J., Greenstein, J. L., 1962, Ap. J., **136**, 312
Kulkarni, S. R., Narayan, R., 1988, Ap. J., **335**, 755
Livio, M., 1994, in "Interacting Binaries", eds S. N. Shore, M. Livio, E. P. J. van den Heuvel, Springer, Berlin, p135
Livio, M., 1996, in "Binary Stars in Clusters", ed. E. Milone, ASP Conf. Series, in press
Paczyński, B., 1976, in "Structure and evolution of close binary systems", eds P. Eggleton, S. Mitton, J. Whelan, Reidel, Dordrecht, p75
Patterson, J., 1984, Ap. J. Supp., **54**, 443
Podsiadlowski, P., 1991, Nature, **350**, 136
Politano, M., 1988, Ph.D. thesis, University of Illinois, Urbana
Politano, M., 1996, Ap. J., in press
Prialnik, D., Kovetz, A., 1995, Ap. J., **445**, 789
Rappaport, S., Verbunt, F., Joss, P. C., 1983, Ap. J., **275**, 713
Ritter, H., 1976a, in "Structure and evolution of close binary systems", eds P. Eggleton, S. Mitton, J. Whelan, Reidel, Dordrecht, p205
Ritter, H., 1976b, MNRAS, **175**, 279
Ritter, H., 1990, A&AS, **85**, 1179
Ritter, H., 1995, A compilation of cataclysmic binaries and related objects with known or suspected orbital periods, version of June 1st, 1995, private communication; last published version: Ritter & Kolb 1995; Ritter 1990
Ritter, H., Kolb, U., 1995, in "X-ray Binaries", eds W. H. G. Lewin, J. van Paradijs, E. P. J. van den Heuvel, Cambridge University Press, p578
Ritter, H., Zhang, Z., Hameury, J.-M., 1996, these proceedings, p449
Ritter, H., Zhang, Z., Kolb, U., 1996a, in "Compact Stars in Binaries", IAU Symposium 165, ed E. P. J. van den Heuvel, in press
Ritter, H., Zhang, Z., Kolb, U., 1996b, in preparation
Schenker, K., Kolb, U., Ritter, H., 1996, these proceedings, p447
Shafter, A. W., 1992, Ap. J., **394**, 268
Spruit, H. C., Ritter, H., 1983, A&A, **124**, 267
Stehle, R., Ritter, H., Kolb, U., 1996, MNRAS, in press
Tavani, M., 1991, Ap. J. Lett., **366**, L27
Terman, J. L., Taam, R. E., 1996, Ap. J., in press
Verbunt, F., 1993, ARA&A, **31**, 93
Verbunt, F., Zwaan, C., 1981, A&A, **100**, L7
Warner, B., 1987, MNRAS, **227**, 23
Whelan, J., Iben, I., 1973, Ap. J., **186**, 1007
Yorke, H. W., Bodenheimer, P., Taam, R. E., 1995, Ap. J., **451**, 308
Yungelson, L. R., Livio, M., Tutukov, A. Kenyon, S. J., 1995, Ap. J., **447**, 656

SIMULATIONS OF ANGULAR MOMENTUM EVOLUTION IN WIND-FED CV PRECURSORS

K. J. PEARSON, G. A. WYNN, A. R. KING
Astronomy Group, Leicester University, Leicester, LE1 7RH, UK

1. Introduction

It is thought that the secondary stars in cataclysmic variables (CVs) may undergo a period of mass loss in the form of a wind during the evolution of the system (Mullan et al. 1992). This wind is thought to magnetically brake the secondary star with a time-scale $\sim 10^8$ yr (e.g. van Paradijs 1986). When the secondary's spin has been brought close to synchronism with the orbit it is possible for tidal torques to lock the secondary in synchronous rotation.

We have modelled this process by adapting a method that has proved successful in explaining the properties of the intermediate polars (Wynn & King 1995). We model the secondary magnetic field as a pure dipole which drags on the wind material as it crosses field lines. This has the effect of trying to force material to move along field lines. This does not model the magnetic interaction completely since it does not allow for the back reaction of the material on the field structure. However, this will only become significant in regions where the field is weak and hence where material is not exerting a large torque on the star.

We can compare our results to previous calculations to enable us to calibrate the strength of the drag term in the equations of motion. Future work will involve extending this treatment to the white dwarf (WD) in such systems to examine the effect of the wind material on the WD spin.

2. The simulation

The simulations were carried out with an adapted version of HYDISC (Whitehurst 1988a,b). We simulated the effect of the magnetic field by a drag force per unit mass:

$$\mathbf{f}_{\mathrm{drag}} = -k[\mathbf{v}' - (\mathbf{v}'.\mathbf{b})\mathbf{b}] \tag{1}$$

A. Evans and J. H. Wood (eds.), Cataclysmic Variables and Related Objects, 443–444.

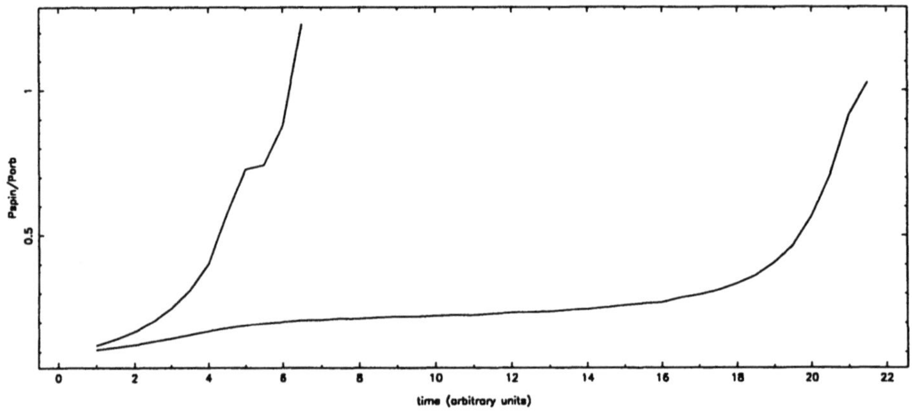

Figure 1. Graph of $P_{\rm spin}/P_{\rm orb}$ for two evolutions with $k = 1.40\ {\rm s}^{-1}$ (faster) and $k = 0.140\ {\rm s}^{-1}$.

where $\mathbf{v}'$ is the velocity of the particle relative to the (rotating) field and $\mathbf{b}$ is a unit vector in the local field direction. The parameter k is the strength of the drag effect. In general it will be related to the magnetic field strength at any point.

3. Results

We followed a scheme where the gravitational attraction of the secondary was assumed everywhere to be balanced by the mechanism causing the wind mass loss. The wind was ejected initially from the surface of the secondary which was assumed to be spherical with a radius inside the Roche lobe. The orbital period was 8 h. Fig. 1 shows the effect on the secondary spin of two different values of k. The pole of the dipole was taken to be in the plane of the orbit for both of these simulations. The initial speed of the material was $10\,c_{\rm s}$. The time to evolve $P_{\rm spin}/P_{\rm orb}$ from 0.1 to 1 was $\sim 2\,10^8$ yr for the faster (high field) evolution, assuming a mass loss rate of $\sim 10^{-10}\ {\rm M}_\odot\,{\rm yr}^{-1}$.

References

Mullan, D.J, Doyle, J.G., Redman, R.O., Mathioudakis, M., 1992, Ap. J., **397**, 225
Theuns, T., Jorissen, A., 1993, MNRAS, **265**, 946
van Paradijs, J., 1986, MNRAS, **218**, 31p
Whitehurst, R., 1988a, MNRAS, **232**, 35
Whitehurst, R., 1988b, MNRAS, **233**, 529
Wynn, G.A., King, A.R., 1995, MNRAS, **275**, 9

A CORE-ENVELOPE DECOUPLING DYNAMO MODEL FOR CATACLYSMIC VARIABLES

L. ZANGRILLI[1], C. A. TOUT[2], A. BIANCHINI[1]
1. Department of Astronomy, University of Padova
Vicolo dell'Ossevatorio 5, 35122 Padova, Italy
2. Institute of Astronomy, University of Cambridge
Madingley Road, Cambridge CB3 0HA, UK

The progenitors of cataclysmic variables (CVs) undergo a common envelope phase, during which their initial orbital period is reduced to a few hours. After this phase the radiative core of the secondary might rotate at a lower rate than the tidally coupled envelope. A shear is then generated at the core-envelope interface which can give rise to a boundary layer dynamo and, consequently, to the magnetic braking of the binary system. The temporal variation of the shear energy content can be obtained by comparing the rate of magnetic energy production with the input rate of orbital energy in the convective envelope. We assume for the flux tubes a filling factor, f_V, owing to the expected size of the tubes and the fact that the dynamo action may be concentrated inside an equatorial belt of the boundary layer shell. We also assume that the production of toroidal field is efficient only inside the boundary layer, while the poloidal component of the field can be enhanced also across the outer convective envelope (see also Zangrilli & Bianchini 1995). We find that the decrease in the shear energy, caused by the continuous production of magnetic field and by the progressive reduction of the core moment of inertia, leads the core coupling with the envelope, thus switching off the dynamo at the core-envelope boundary. Further details of this *core-envelope decoupling dynamo model* will be given in a forthcoming paper (Zangrilli, Tout & Bianchini 1996).

Table 1

Model	$\mathrm{Period_{core}}$(initial)	$\mathrm{Period_{env}}$(initial)	f_V
MOD1	100 h	6 h	0.1
MOD2	100 h	6 h	0.01
MOD3	10 h	6 h	0.01

The evolution of a CV has been computed using a double polytrope code for the secondary star (Rappaport, Verbunt & Joss 1983) and the magnetic

A. Evans and J. H. Wood (eds.), Cataclysmic Variables and Related Objects, 445–446.

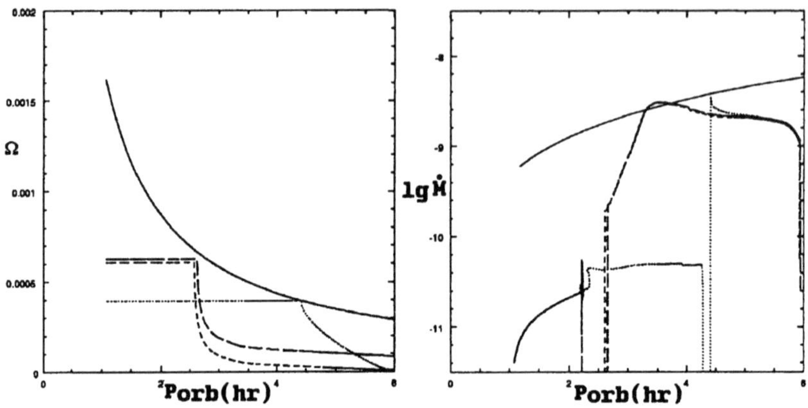

Figure 1. (a: left) Core angular velocity evolution versus orbital period. Dotted line corresponds to MOD1; long dashed and short dashed lines correspond to MOD2 and MOD3, respectively. Continuous line is the envelope angular velocity, which is the same for the three models. (b: right) Mass transfer rate versus orbital period. Line styles are as in (a). Continuous line is the critical mass transfer rate curve.

braking model of Mestel & Spruit (1987). The evolutionary tracks shown in Fig. 1 refer to the three models of Table 1. In Fig. 1a we see that, in MOD1, the core couples with the envelope at about 4.5 h, thus switching off the dynamo. This is caused by the large magnetic energy production, favoured by a large filling factor. In MOD2 and MOD3 this happens at just about 3 h, owing to the small moment of inertia of the core at this stage. In Fig. 1b we see that in MOD1 mass transfer ceases much earlier than the 3 h upper limit of the gap. Instead, MOD2 and MOD3 reproduce it well. Moreover, we can see that both MOD2 and MOD3, in the interval 3...3.5 h, are above the critical mass transfer curve (Shafter 1992), which separates stable accretion discs (above) from the unstable ones (below). This appears to be in agreement with the observed lack of dwarf nova systems in this interval of orbital periods. It is worth noticing that $f_V = 0.01$ gives the better agreement with the observed statistics of CVs, while little dependence of the binary evolution on the initial rotational regime of the core is found.

References

Mestel, L., Spruit, H.C., 1987, MNRAS, **226**, 57
Rappaport, S., Verbunt, F., Joss, P.C., 1983, Ap. J., **275**, 713
Shafter, A.W., 1992, Ap. J., **394**, 268
Zangrilli, L., Bianchini, A., 1995, in "Cataclysmic Variables", eds A. Bianchini et al., Kluwer, p405
Zangrilli, L., Tout, C.A., Bianchini, A., 1996, in preparation

NOVA OUTBURSTS AND THE SECULAR EVOLUTION OF CATACLYSMIC VARIABLES

K. SCHENKER[1], U. KOLB[1,2], H. RITTER[1]
1. MPI für Astrophysik, Garching, Germany
2. Lick Observatory, UCSC, Santa Cruz, CA 95064, USA

Abstract. We present calculations of the long-term evolution of CVs which include the influence of nova outbursts. In particular we investigate the consequences of the discontinuous mass loss due to recurring outburst events and the effects of *frictional angular momentum loss* (FAML), i.e. the interaction of the expanding nova envelope with the secondary. We show that a description assuming continuous mass loss – averaged over a complete nova cycle – is applicable for determining the mean mass transfer rate and the secular evolution both with and without FAML. Between two subsequent outbursts, deviations from the mean evolution depend on the strength of FAML and on the mass ejected during the outburst. Formally FAML is a consequential angular momentum loss [1] and therefore increases the mean mass transfer rate by pushing the systems closer to mass transfer instability. Depending on the actual strenghth of FAML the long-term evolution of CVs could be significantly different from the standard model predictions.

1. Physical situation and model calculations

Typical model computations of the long-term evolution of CVs take into account classical nova outbursts only as a source of a continuous isotropic stellar wind from the white dwarf (WD). In contrast to that, the real physical situation is as follows. During an outburst a mass $\Delta M_{\rm ejc}$ is removed from the system with the WD's specific angular momentum j_1. Although this event is practically instantaneous compared to the recurrence time of nova outbursts ($\tau_{\rm rec} \sim 10^3 \ldots 10^6$ yr) it results in a short common envelope phase, during which angular momentum is transferred from the orbit to the envelope [5]. In the simplified model used here [4] this additional angular momentum loss depends mainly on the ratio $K_1 = v_{\rm exp}/v_{\rm sec}$ of the velocity of the expanding envelope near the secondary, $v_{\rm exp}$, and the secondary's orbital velocity, $v_{\rm sec}$. The final specific angular momentum of the ejected matter is then $j_{\rm ejc} = j_1 + j\,\mathcal{A}(q)\sqrt{1+K_1^2}/K_1$, where j is the specific orbital

A. Evans and J. H. Wood (eds.), Cataclysmic Variables and Related Objects, 447–448.

angular momentum of the system and the function $\mathcal{A} \sim 0.1 \ldots 0.2$ depends only weakly on the mass ratio, $q = M_1/M_2$.

Full resolution of nova outbursts requires a time step significantly smaller than the recurrence time, whereas a complete evolution can last for more than 10^{10} yr. Thus complete secular evolution calculations have been performed with a fast bipolytrope code [3], short sections and averaged models also with a full stellar evolution code [6].

2. Results and discussion

In the fully resolved calculations all outburst parameters were taken from nova models [7] and the strength of FAML was calculated with constant values of $v_{\rm exp}$. These calculations yield two types of cyclic behavior, either (i) a sudden decrease in the mass transfer rate, $\dot{M}$, due to the outburst followed by a slow increase on a time-scale of $\sim 10^4$ yr until the next outburst (no or weak FAML), or (ii) a sudden increase followed by a slow decrease (strong FAML). Both types can maintain stable mass transfer for the whole secular evolution of the system. Comparing the explicitly resolved calculations to sequences obtained with averaged continuous mass loss confirms the validity of the latter to determine the mean evolution and mass transfer rate. The amplitude of the modulation of $\dot{M}$ can be estimated from outburst parameters [8].

A realistic description should also consider how FAML varies during the secular evolution. Accordingly we have performed averaged calculations using a variable $v_{\rm exp}$. If FAML grows sufficiently in strength with decreasing orbital separation (mainly because $v_{\rm exp}$ is expected to be smaller closer to the WD) mass transfer becomes adiabatically unstable, leading to a runaway where $\dot{M}$ grows orders of magnitude within a very short time until FAML disappears or a common envelope forms [2]. Furthermore a dependence of $v_{\rm exp}$ on outburst characteristics introduces a feedback mechanism, e.g. in the sense that a higher $\dot{M}$ causes less violent novae with slower expansion and thus stronger FAML, unless $\dot{M}$ is so high that the FAML effect disappears altogether because the nova envelope does not reach the secondary. This can lead to a very complex evolution with various detached phases.

References

1. King, A.R., Kolb, U., 1995, Ap. J., **439**, 330
2. Kolb, U., Livio, M., Schenker, K., Ritter, H., 1996, in preparation
3. Kolb, U., Ritter, H., 1992, A&A, **254**, 213
4. Livio, M., Govarie, A., Ritter, H., 1991, A&A, **246**, 84
5. MacDonald, J., 1986, Ap. J., **305**, 251
6. Mazzitelli, I., 1989, Ap. J., **340**, 249
7. Prialnik, D., Kovetz, A., 1995, Ap. J., **445**, 789
8. Schenker, K., Kolb, U., Ritter, H., 1996., in preparation

SECULAR EVOLUTION OF CATACLYSMIC VARIABLES WITH IRRADIATION–INDUCED MASS TRANSFER

H. RITTER[1], Z. ZHANG[1], J. M. HAMEURY[2]
1. MPI für Astrophysik, D–85748 Garching, Germany
2. Observatoire de Strasbourg, F–6700 Strasbourg, France

1. Introduction

The possible importance of the reaction of a low–mass star to external irradiation for the long–term evolution of compact binaries has been noted only rather recently; first in the context of the evolution of low–mass X–ray binaries (e.g. Podsiadlowski 1991; Harpaz & Rappaport 1991; Frank, King & Lasota 1992; Hameury et al. 1993) and subsequently by Ritter, Zhang & Kolb (1995a,b, hereafter RZK) also for the evolution of cataclysmic variables (CVs). Based on a simple model for describing the reaction of a low–mass star to irradiation RZK showed that CVs can be dynamically unstable against irradiation–induced mass transfer and that, as a consequence of this, mass transfer could occur via cycles in which phases of high, irradiation–enhanced mass transfer alternate with phases of little or no mass transfer. The occurrence of such mass transfer cycles in CVs was subsequently discussed from a more general point of view by King (1995) and King et al. (1995). Whereas the possibility of mass transfer cycles in CVs is now fully recognised, the question as to which systems can undergo such cycles and which cannot has not yet been addressed in detail. It is the purpose of this contribution to provide at least a partial answer to this question.

2. Stability against irradiation–induced mass transfer

The criterion for dynamical stability against irradiation–induced mass transfer as derived by RZK is

$$\zeta_{\rm s} - \zeta_R > \zeta_{\rm irr} = -M_{\rm s}\frac{\partial}{\partial \dot{M}_{\rm s}}\left(\frac{\partial \ln R_{\rm s}}{\partial t}\right)_{\rm irr} . \qquad (1)$$

Here $M_{\rm s}$, $R_{\rm s}$, $\dot{M}_{\rm s}$ and $\zeta_{\rm s}$ are respectively the mass, the radius, mass loss rate and adiabatic mass radius exponent of the secondary (the mass donor). ζ_R

A. Evans and J. H. Wood (eds.), Cataclysmic Variables and Related Objects, 449–452.

is the mass radius exponent of the critical Roche radius and $(\partial \ln R_s/\partial t)_{\rm irr}$ the thermal relaxation term due to irradiation.

In the following we shall first discuss the consequences of (1) in the framework of the simple model adopted by RZK in which it is assumed that a fraction s of the secondary's surface is exposed to a uniform external irradiation flux $F_{\rm irr}$. Combining equations (5) and (11) of RZK, (1) can be rewritten as follows:

$$C = 2(\zeta_s - \zeta_R)\frac{\tau_{\rm KH,0}}{\tau_{\rm M}}\mathcal{F}^{-1} > 2s\; xg(x), \tag{2}$$

where $\tau_{\rm KH,0} = \frac{GM_s^2}{R_s L_{s,0}}$ is the Kelvin–Helmholtz time of the unirradiated star, $\tau_{\rm M} = -M_s/\dot{M}_s$ the mass transfer time-scale and $\mathcal{F}$ a function which scales roughly as the inverse of the relative mass of the secondary's convective envelope. In the bipolytrope approximation (Kolb & Ritter 1992) $\mathcal{F} = \{\}\{\}^{-1}$, where the expressions in the curly brackets are defined in equation (10) of RZK. Furthermore, in (2), $x = F_{\rm irr}/\{\sigma T_0^4\}$ is the dimensionless irradiating flux, T_0 being the effective temperature of the unirradiated star, and

$$g = -\frac{\partial}{\partial F_{\rm irr}}[\sigma T_{\rm irr}^4 - F_{\rm irr}] = -\frac{\partial}{\partial x}\left[\left(\frac{T_{\rm irr}}{T_0}\right)^4 - x\right] = g(x), \tag{3}$$

where $T_{\rm irr}$ is the effective temperature on the irradiated parts of the star. The virtue of rewriting (1) in the form of (2) is that in (2) terms which depend on irradiation (right hand side) are separated from terms which do not (left hand side). In fact, the function C defined in (2) depends only on binary parameters and via $\tau_{\rm M}$ on the adopted model for the long–term evolution. Thus, along an evolutionary sequence the value of C is always known. The question of whether (2) is fulfilled or not thus depends on the value of $2s\; xg(x)$.

3. Reaction of the stellar surface to external irradiation

Using a simple one–zone model for the superadiabatic surface layer of low–mass stars, RZK derived an explicit expression for g (their equation 7). Here we shall not make use of this result because it does not apply to the whole mass range of interest. Rather we shall base our discussion on very general properties of the function g and on results of numerical calculations of g performed using the code described by Hameury et al. (1993).

First of all it is important to notice the following general properties of the function $g(x)$ which are entirely model independent:

(i) $\int_0^\infty g(x)dx = 1,$ (4)

(ii) $g(x=0) \leq 1$ for positive albedos, and (5)

(iii) $\lim g(x \to \infty) = 0.$ (6)

Because of these properties, $g(x)$ must be a rather steeply decreasing function of x with $0 < g(x) \ll 1$ if $x >$ a few. This is true, in particular, for stars with a mass $M \lesssim 0.6\,\mathrm{M}_\odot$, for which $g(x=0) \approx 1$. As a consequence, the function $xg(x) = 0$ for $x = 0$ and $x \to \infty$, and $xg(x)$ attains a maximum $\mathrm{Max}[xg(x)] \approx 0.5$ for values $x \approx 1$ to a few. This holds in the whole mass range of interest, i.e. $0.1 \lesssim M/\,\mathrm{M}_\odot \lesssim 1$. Based on these results we can now resume the discussion of (2).

4. Secular evolution of CVs with irradiation

Violation of (2) is only a necessary but not a sufficient condition for the occurrence of mass transfer cycles. On the other hand, fulfillment of (2) is a sufficient condition for the absence of mass transfer cycles. Now, because the function $xg(x)$ has a maximum value, we can state that *no mass transfer cycles will occur if*

$$C > 2s\,\mathrm{Max}[xg(x)] \approx s. \tag{7}$$

So far, we have carried out our discussion in the framework of the simple model of RZK. However, even if we adopt a much more realistic and elaborate model, i.e. the point source model, in which the secondary is assumed to be irradiated by a point source at the location of the white dwarf primary, an analogous argument can be made to yield [in place of (7)]

$$C > \mathrm{Max}\left[\int_0^{\theta_{\max}} x(\theta) g\{x(\theta)\} \sin\theta d\theta\right], \tag{8}$$

where now the irradiating flux F_{irr} depends on θ, the colatitude of a given point on the stellar surface as measured from the substellar point. Using numerical results for $g(x)$ we find the maximum of the integral in (8) to be smaller than $2s\,\mathrm{Max}[xg(x)]$ by about a factor of 2 if $s \approx 1/2$. This is clearly seen in Fig. 1 where we plot as a full line the run of C as a function of M_s for a standard CV evolution ($M_1 = 1\,\mathrm{M}_\odot$, $M_{2,\mathrm{i}} = 1\,\mathrm{M}_\odot$, disrupted magnetic braking model), the line corresponding to $2s\,\mathrm{Max}[xg(x)]$ for $s = 1/2$ (dashed line) and the right hand side of (8) as a dash–dotted line. Eq. (8) and Fig. 1 lead to the conclusion that in the framework of standard CV evolution no mass transfer cycles can occur if $M_\mathrm{s} \lesssim 0.7\,\mathrm{M}_\odot$. The reason for the stability of systems with $M_\mathrm{s} \lesssim 0.7\,\mathrm{M}_\odot$ is that the function C increases strongly with decreasing M_s whereas the right hand sides of (7) and (8) remain almost constant. The strong increase of C with decreasing M_s has two main reasons: first, with decreasing M_s the stellar luminosity decreases steeply (mass-luminosity relation!), i.e. $\tau_{\mathrm{KH,0}}$ increases correspondingly; second, with decreasing M_s the factor $\mathcal{F}$ grows strongly because the (relative) mass of the convective envelope grows until the star becomes fully convective. In other words, C reflects the thermal inertia of the convective envelope

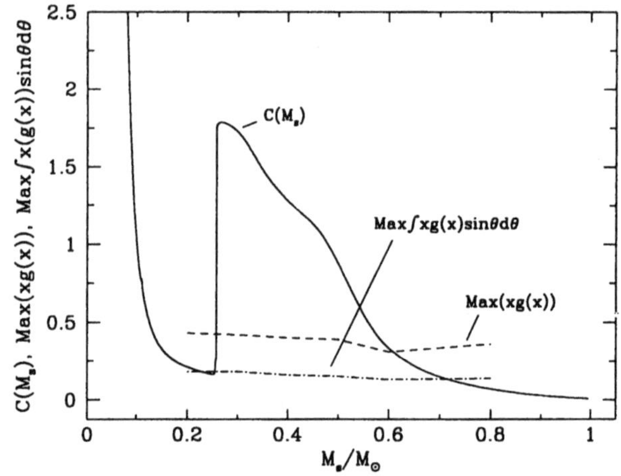

Figure 1. The run of C, Max$[xg(x)]$ and Max$[\int xg(x)\sin\theta d\theta]$ as a function of the secondary's mass M_s. For details see text.

which increases with decreasing M_s because its heat capacity increases and the luminosity available for changing the entropy of the envelope decreases.

In principle there are two possibilities for reducing the value of C and thus widening the mass range in which mass transfer cycles are possible. However, both these possibilities are not compatible with standard CV evolution. The first possibility consists of increasing τ_M, i.e. of reducing the secular mean mass transfer rate. However, doing so by a significant factor would destroy the period gap, the width of which sets an upper limit of $\tau_M \sim 10^8$ yr. The second possibility consists of reducing the value of $\zeta_s - \zeta_R$ which is possible only if a CV suffers significant consequential angular momentum loss. The difficulties arising from such a proposition have been examined in detail by King & Kolb (1995).

References

Frank, J., King, A.R., Lasota, J.P., 1992, Ap. J., **385**, L45

Hameury, J.M., King, A.R., Lasota, J.P., Raison, F., 1993, A&A, **277**, 81

Harpaz, A., Rappaport, S., 1991, Ap. J., **383**, 739

King, A.R., 1995, in "Cataclysmic Variables", eds A. Bianchini et al., Kluwer, p523

King, A.R., Kolb, U., 1995, Ap. J., **439**, 330

King, A.R., Frank, J., Kolb, U., Ritter, H., 1995, Ap. J., **444**, L37

Kolb, U., Ritter, H., 1992, A&A, **254**, 213

Podsiadlowski, P., 1991, Nature, **350**, 136

Ritter, H., Zhang, Z., Kolb, U., 1995a, IAU Symp. No. 165, ed. E.P.J. van den Heuvel, in press

Ritter, H., Zhang, Z., Kolb, U., 1995b, in "Cataclysmic Variables", eds A. Bianchini et al., Kluwer, p479

FROM COMMON ENVELOPE TO PRE–CATACLYSMIC VARIABLES: AN OBSERVATIONAL TEST OF COMMON ENVELOPE EVOLUTION

M. J. SARNA
N. Copernicus Astronomical Center, Polish Academy of Sciences, ul. Bartycka 18, 00–716 Warsaw, Poland

1. Evolutionary scheme for cataclysmic variables

The generally accepted evolutionary scenario for cataclysmic variables (CVs) is common envelope (CE) evolution (Iben & Livio 1993) proposed by Paczyński (1976). The secondary spirals towards the giant's compact core converting orbital energy into kinetic energy of the giant's envelope and the envelope is ejected. The dynamics of the red dwarf and red giant envelope interaction have been studied by several groups (Livio & Soker 1988; Taam & Bodenheimer 1991). After the ejection of the red giant envelope the post common envelope detached binaries (PCEBs) are formed. These can be divided into three groups:

- Hot subdwarf with a red dwarf star inside a planetary nebula.
- Hot subdwarf with a red dwarf star without a planetary nebula.
- Hot white dwarf with a red dwarf star without a planetary nebula.

Next, due to angular momentum loss by magnetic braking and/or gravitational radiation, the red dwarf component fills its Roche lobe and a cataclysmic variable is formed.

2. Spherically symmetric accretion

CE evolution is a very important factor in the structure of the secondary component in PCEBs. Let us first consider spherically symmetric accretion onto the main sequence star. It is a well known fact that a star with a radiative envelope or a thin convective zone ($M_* > 0.55\, M_\odot$) expands during accretion (Kippenhahn & Meyer–Hofmeister 1977). For a star which has a thick convective envelope, or is fully convective, the effectiveness of heating is crucial:

A. Evans and J. H. Wood (eds.), Cataclysmic Variables and Related Objects, 453–456.

• in the case where the entropy of the infalling matter is greater than that in the outer layer of the accreting star the reaction of the star is determined by the formation of a radiative temperature inversion layer below the stellar surface (Prialnik & Livio 1985; Sarna & Ziółkowski 1987). The radius of the star grows rapidly in response to accretion;

• in the case where the entropy of the infalling matter is less than or comparable to that of the matter in the outer layers of the accreting star, the star should shrink upon accretion (Webbink 1976; Whyte & Eggleton 1985).

The response of an accreting main sequence star within a CE will be characterized by the first case, where entropy is higher.

3. Accretion during common envelope evolution

Hjellming & Taam (1991) considered the response of a $1.25\,M_\odot$ main sequence star to the conditions within the envelope of a 2 or $5\,M_\odot$ red giant. There are two phases characterizing the evolution of a main sequence star inside the red giant's envelope: the mass accretion and the mass loss phases. During accretion, the original surface layers become nearly isothermal while the outer layers of accreted material have normal temperature gradients. Very high luminosities are produced in the outer layers which drive expansion, increasing the radius by more than a factor of 10. The star fills its Roche lobe and mass loss begins. The calculation ends when the radius decreases to its initial value and the giant's envelope has been expelled. The net addition of secondary mass during the CE phase is minimal. Note that during the mass loss stage of the secondary star during the CE phase the separation and period of the system decrease. This happens because the mass of the carbon–oxygen core is less than that of the secondary star.

A quite different situation occurs for a lower mass secondary. For example, if we consider a $0.3\,M_\odot$ secondary, its response to accretion will be the same as for a $1.25\,M_\odot$ star, but its internal structure will be different (Prialnik & Livio 1985; Sarna & Ziółkowski 1987). After accretion of $0.1\ldots0.2\,M_\odot$ the radius will expand by a factor of a few and the secondary fills its Roche lobe and mass loss begins. But now mass loss (transfer) is unstable, as pointed out by Hameury et al. (1993). The separation and period at this stage start to increase, because now the mass of the carbon–oxygen core is greater than the mass of the secondary star. The extended secondary shrinks below its Roche lobe. We suggest that the net addition to the secondary's mass will not be negligible.

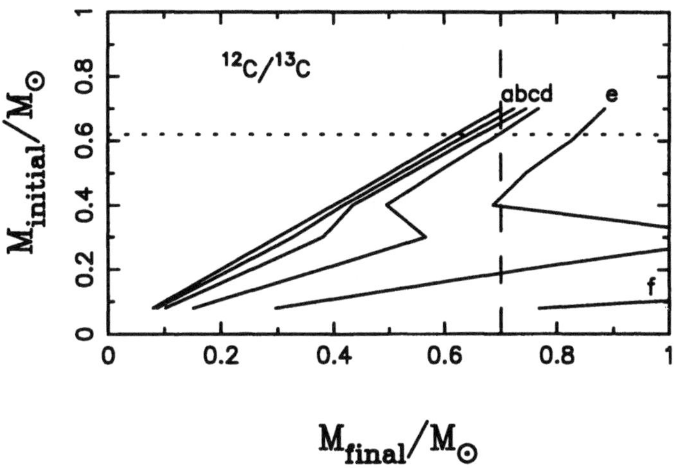

Figure 1. The $^{12}C/^{13}C$ ratio as a function of the pre-CE and PCEB red dwarf masses. The curves of constant isotopic ratio indicated by the letters a, b, c, d, e and f correspond to values of 91, 70, 50, 30, 20 and 16, respectively. The vertical, dashed, line is for a $0.7\,M_\odot$ red dwarf; if its isotopic ratio is 50, then its initial mass was $0.62\,M_\odot$.

4. Isotopic ratios for dwarf and giant stars

It is a well known fact that the ratios of the various isotopes change during different phases of stellar evolution (Iben 1967; El Eid 1994). If the star evolves towards the red giant region a thick convective envelope develops changing the distribution of isotopes. Typical values (Harris, Lambert & Smith 1988) of the $^{12}C/^{13}C$ ratio for main sequence stars and red giants are respectively 90 and 14 ± 1.5. The error for the red giant values is typical of determinations of the isotopic ratios for particular giants.

5. Basic idea

The basic assumptions are that during CE evolution the main sequence star accretes some amount of the giant's envelope, and after the CE phase, when the secondary relaxes to its thermal equilibrium configuration, the accreted material is mixed with the convective envelope of the red dwarf. The $^{12}C/^{13}C$ ratios after relaxation are shown in Fig. 1, where initial mass means the mass of the red dwarf before the common envelope phase, and final after (Sarna et al. 1995). If we have a mass determination for the red dwarf component in the PCEB, and also have an isotopic ratio determination for carbon, using this graph we can determine the initial mass of the red dwarf and hence the amount of mass which is accreted during the CE phase.

6. Infrared observation of cataclysmic variables

We propose (Sarna et al. 1995) a simple observational test based on the infrared measurement of the $^{12}C/^{13}C$ ratio using the CO bands. Dhillon & Marsh (1995) observed the K–band echelle spectra of the detached PCEB V471 Tau. The spectrum of V471 Tau shows ^{12}CO but this is rotationally broadened due to the 12.51 h orbital period, which smears the absorption profile. The ^{12}CO band in V471 Tau is also red-shifted due to the orbital motion. There is a marginal detection of ^{13}CO around 2.346 microns.

7. Prediction

The dwarf stars in PCEBs with planetary nebulae, in which the secondary is not in thermal equilibrium due to the short time since the CE phase, will exhibit $^{12}C/^{13}C$ ratios typical of giants. The best candidates are: V477 Lyr, UU Sge.

The dwarf stars in PCEBs without planetary nebulae, in which the secondary is in thermal equilibrium given the long time since the CE phase, should exhibit $^{12}C/^{13}C$ ratios intermediate between giants and solar values. The best candidates are Feige 24, HZ 9, V471 Tau.

References

Dhillon, V.S., Marsh, T.R., 1995, MNRAS, **275**, 89
El Eid, M.F., 1994, A&A, **285**, 915
Hameury, J.-M., King, A.R., Lasota J.-P., Raison F., 1993, A&A, **277**, 81
Harris, M.L., Lambert, D.L., Smith, V.V., 1988, Ap. J., **325**, 768
Hjellming, M.S., Taam, R.E., 1991, Ap. J., **370**, 709
Iben, I., 1967, ARAA, **5**, 571
Iben, I., Livio, M., 1993, PASP, **105**, 1373
Kippenhahn, R., Meyer-Hofmeister, E., 1977, A&A, **54**, 539
Livio, M., Soker, N., 1988, Ap. J., **329**, 764
Paczyński, B., 1976, in "The Structure and Evolution of Close Binary Systems", eds. P.P. Eggleton, S. Mitton, J. Whelan (Dordrecht, Reidel), p75
Prialnik, D., Livio, M., 1985, MNRAS, **216**, 37
Sarna, M.J., Ziółkowski, J., 1987, Acta Astron., **38**, 89
Sarna, M.J., Dhillon, V.S., Marsh, T.R., Marks, P.B., 1995, MNRAS, **272**, l41
Taam, R.E., Bodenheimer, P., 1991, Ap. J., **373**, 246
Webbink, R.F., 1976, Ap. J. Supp., **32**, 583
Whyte, C., Eggleton, P.P., 1985, MNRAS, **214**, 357

THE $^{12}C/^{13}C$ RATIO AS A TRACER OF THE EVOLUTION OF POST COMMON ENVELOPE SYSTEMS AND CATACLYSMIC VARIABLES

M. J. SARNA[1,2], P. B. MARKS[2], R. C. SMITH[2]
1. N. Copernicus Astronomical Center, Warsaw, Poland
2. Astronomy Centre, University of Sussex, Brighton, UK

To provide a direct test of common envelope (CE) evolution which can be easily confirmed by observations, we (Sarna et al. 1995) recently modelled the change in the abundance ratio of $^{12}C/^{13}C$ on the surface of the lower mass star of a binary during the CE phase. The model is based on the fact that it is probable that the dwarf star accretes material during the CE phase. Since, during the CE phase, the dwarf secondary effectively exists within the atmosphere/envelope of the giant or supergiant primary, the accreted material has the abundances/composition of a giant/supergiant star. The $^{12}C/^{13}C$ ratio is known to decrease from approximately 90 in dwarf stars (in which the ^{13}CO band at 2.3448 microns is barely visible) to approximately 10 in giants (in which the ^{13}CO band at 2.3448 microns is fairly prominent). Hence, by measuring the $^{12}C/^{13}C$ ratio in post common envelope binaries (PCEBs) and comparing it to our models we would be able not only to confirm the CE theory but also to determine the amount of mass accreted during the CE phase and hence the initial mass of the dwarf component prior to the CE phase. We also propose an evolutionary scenario in which PCEBs with secondary component mass near $1.0\,M_\odot$ start semi-detached evolution almost immediately after the CE phase. The progenitor system is a wide binary consisting of a $3\,M_\odot$ primary with a $1.0\,M_\odot$ secondary star. The primary evolves to fill its Roche lobe when it has a $0.6\,M_\odot$ C–O core, with two shell burning regions. Such a star has a thick convective envelope, mass transfer is dynamically unstable and a common envelope forms. After the CE phase we are left with a close detached binary consisting of the primary's core ($0.6\,M_\odot$) and the secondary ($1.0\,M_\odot$) main sequence star. Shortly afterwards the secondary fills its Roche lobe and mass transfer occurs (Sarna, Marks & Smith 1995). The system now evolves as a semi–detached binary (CV), transferring material to the white dwarf which undergoes nova outbursts. Figs. 1 and 2 show the isotopic ratios of $^{12}C/^{13}C$ and $^{16}O/^{17}O$ during the semi–detached evolution. In Fig. 1

A. Evans and J. H. Wood (eds.), Cataclysmic Variables and Related Objects, 457–458.

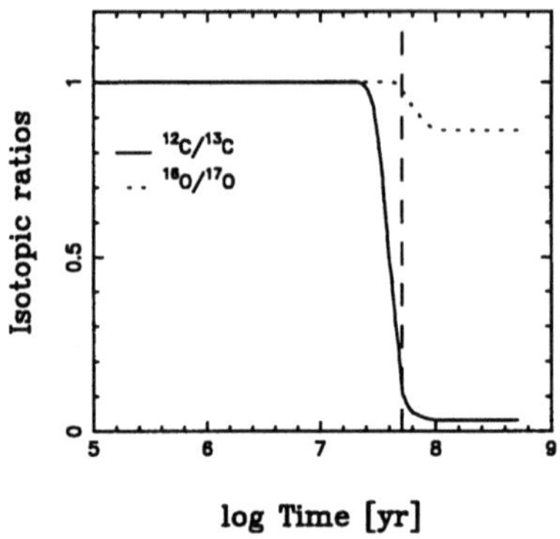

Figure 1. The isotopic ratios in the mass–losing secondary star are shown, as a function of time, for a secondary which accreted no material during the CE phase. The solid line represents the evolution of the carbon isotopic ratio. The dotted line represents the evolution of the oxygen isotopic ratio. In both cases the ratios are normalised to the cosmic value which we assume to be 91 and 2724 for carbon and oxygen ratios respectively. The vertical (dashed) line shows where the system detaches.

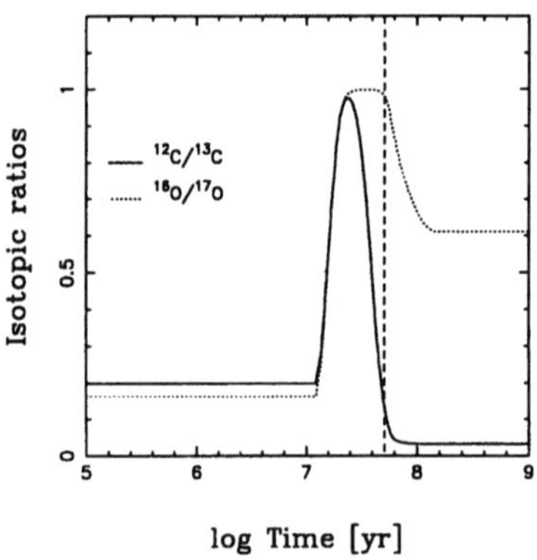

Figure 2. The isotopic ratios versus time are shown for a secondary which accreted $0.2\,M_{\odot}$ during the CE stage. All annotation is as in Fig. 1.

the secondary did not accrete any material during CE evolution whilst in Fig. 2 the secondary accreted $0.2\,M_{\odot}$ during the CE stage.

References

Sarna, M.J., Dhillon, V.S., Marsh, T.R., Marks, P.B., 1995, MNRAS, **272**, 41P
Sarna, M.J., Marks, P.B., Smith, R.C., 1995, MNRAS, **276**, 1336

EVOLUTIONARY SCENARIOS FOR DOUBLE DEGENERATE SYSTEMS

P. B. MARKS[1], M. J. SARNA[1,2], R. C. SMITH[1]

1. Astronomy Centre, University of Sussex, Brighton, UK
2. N. Copernicus Astronomical Center, Warsaw, Poland

1. Observations

There are presently eight double degenerate systems with well determined orbital parameters, their periods being either a few hours or a few days (Marsh, Dhillon & Duck 1995; Marsh 1995). The masses of the primaries and secondaries lie in the range $0.15\ldots0.45\,M_{\odot}$.

2. Calculations

We calculate two evolutionary scenarios (Sarna, Marks & Smith 1996); the first is Algol-type evolution with two phases of stable mass transfer, and the second involves first a stage of common envelope (CE) evolution followed by a stage of stable mass transfer. In both calculations we assume non–conservative mass transfer by which we mean that the total mass and angular momentum of the system are not conserved. For both scenarios we start our calculations after the first stage of mass transfer has finished. In all calculations the primary is the initially more massive star that filled its Roche lobe and transferred material to the secondary during the first phase of mass transfer, hence the secondary is the star that fills its Roche lobe in our calculations. The system's orbital period decreases and then increases until the system detaches; we are left with a detached white dwarf/white dwarf binary with an orbital period of the order of hours or of days (see Table 1). There must exist some bifurcation period below which the systems evolve towards orbital periods of the order of hours and above which the systems evolve to periods of the order of several days.

A. Evans and J. H. Wood (eds.), Cataclysmic Variables and Related Objects, 459–460.

TABLE 1. Results of model computations

Model No.	1	2	3	4	Algol
P_i [d]	0.950	1.000	1.263	1.612	2.664
$M_{1i}/M_\odot$	0.250	0.250	0.250	0.250	0.207
$M_{2i}/M_\odot$	1.750	1.750	1.750	1.750	1.920
$M_{He2i}/M_\odot$	0.000	0.000	0.032	0.072	0.098
P_f [d]	0.176	0.395	1.620	2.556	2.636
$M_{1f}/M_\odot$	0.411	0.409	0.404	0.405	0.409
$M_{2f}/M_\odot$	0.145	0.158	0.206	0.198	0.286
$M_{He2f}/M_\odot$	0.068	0.086	0.097	0.147	0.125
log Time [yr]	8.692	8.607	7.900	7.906	6.901

Subscript i denotes the initial parameters, subscript f denotes detachment and subscript He2 denotes the secondary's helium core. The duration of the semi–detached evolution is given by log Time.

3. Conclusions

We can explain the separation/segregation of the orbital periods of double degenerate systems using the bifurcation 'mechanism'. This 'mechanism' also works in a similar way for low–mass binaries with millisecond pulsars and white dwarf components (Ergma & Sarna 1996). We can see that our theoretical models reproduce the general features of the observational data very well, but more accurate orbital parameters will be necessary before we can model the evolutionary history of particular systems.

References

Ergma, E., Sarna, M.J., 1996, MNRAS, in press
Marsh, T.R., Dhillon, V.S., Duck, S.R, 1995, MNRAS, **275**, 828
Marsh, T.R., 1995, MNRAS, **275**, L1
Sarna, M.J., Marks, P.B., Smith, R.C., 1996, MNRAS, in press

ON THE ORIGIN OF THE ABUNDANCE ANOMALIES IN IK PEG

B. SMALLEY[1], K. C. SMITH[2], D. WONNACOTT[3]

1. Dept of Physics, Keele University, Staffordshire, ST5 5BG

2. Dept of Physics & Astronomy, UCL, London, WC1E 6BT

3. MSSL, Holmbury St. Mary, Dorking, Surrey, RH5 6NT

Abstract. IK Peg is a binary system comprising a pulsating A-type star in orbit with a massive white dwarf. A detailed abundance analysis of IK Peg A has been performed. It is found that the Ca and Sc abundances are approximately solar, and the Fe-group elements slightly enhanced. IK Peg is not a classical Am star but the results are not inconsistent with its spectroscopic classification as a marginal Am star. An excess of Ba and Sr are found. These anomalies could be explained by radiative diffusion processes operating in the atmosphere of IK Peg A, even though it is undergoing small-amplitude pulsations. Alternatively, since the companion is a massive white dwarf, these anomalies could be the result of mass transfer during the common envelope phase of the binary system's evolution.

1. Introduction

The sixth-magnitude A-type star IK Peg (HR 8210) has been the subject of much interest over the last few years. It is a well-known single-lined spectroscopic binary with a period of 21.9 d, but only recently has the companion (IK Peg B) been positively identified as a massive ($1.15\,M_{\odot}$) hot (35 000 K) white dwarf (Wonnacott, Kellett & Stickland 1993). The primary itself (IK Peg A) is by no means an uninteresting main-sequence A-type star. It is known to be undergoing small amplitude pulsations (Kurtz 1978; Wonnacott et al. 1994) and exhibits mild spectroscopic peculiarities similar to those associated with the metallic-lined A-type stars (Cowley et al. 1969).

Because the system is a post common envelope binary (Sarna et al. 1995), we have performed a detailed elemental abundance analysis, in order to see if we can find any anomalies that may be the result of mass transfer

A. Evans and J. H. Wood (eds.), Cataclysmic Variables and Related Objects, 461–462.

from the white dwarf progenitor. Here we present a summary of the results given by Smalley et al. (1996).

2. Results

The results of the detailed elemental abundance analysis by Smalley et al. (1996) are summarized below:

EL	[EL/H]	EL	[EL/H]	EL	[EL/H]	EL	[EL/H]
C	-0.34 ± 0.18	Si	$+0.13 \pm 0.16$	V	$+0.07 \pm 0.16$	Cu	$+0.05 \pm 0.21$
N	$+0.03 \pm 0.19$	S	-0.01 ± 0.15	Cr	$+0.13 \pm 0.19$	Zn	$+0.08 \pm 0.17$
O	-0.19 ± 0.19	K	-0.24 ± 0.14	Mn	$+0.05 \pm 0.18$	Sr	$+0.49 \pm 0.17$
Na	$+0.07 \pm 0.13$	Ca	-0.11 ± 0.25	Fe	$+0.17 \pm 0.17$	Y	$+0.08 \pm 0.15$
Mg	-0.13 ± 0.23	Sc	-0.19 ± 0.20	Co	$+0.22 \pm 0.15$	Zr	-0.07 ± 0.18
Al	-0.17 ± 0.17	Ti	$+0.26 \pm 0.24$	Ni	$+0.29 \pm 0.19$	Ba	$+1.13 \pm 0.17$

3. Conclusion

A detailed abundance analysis of the pulsating A-type star IK Peg A has revealed that the Ca and Sc abundances are approximately solar and the Fe-group elements slightly enhanced. This star is *not* a classical Am star, but the results are not inconsistent with its *spectroscopic classification* as a marginal Am star. Whether this marginal Am character is due to the effects of radiative diffusion is unclear, because there is the possibility that some, or all, of the abundance anomalies may be the result of mass transfer.

The obvious excess of Ba and Sr may be evidence for contamination by the white dwarf progenitor during the common envelope phase of the binary system evolution. Smalley et al. (1996) suggested that IK Peg A is a hot F str λ 4077 star. The F str λ 4077 stars are, however, a heterogeneous class, with the cooler members being related to Barium stars, while the hotter members resembling the Am or δ Del stars (North, Berthet & Lanz 1994). Thus, it is unclear as to the origin of the anomalies in the *hot* F str λ 4077 stars, since their similarities to Am stars raises the possibility that the anomalies are due to radiative diffusion and *not* mass transfer. In fact, even if there were any anomalies due to mass transfer they may well have been masked by the effects of subsequent radiative diffusion (North et al. 1994).

References

Cowley A., Cowley C., Jaschek H., Jaschek C., 1969, AJ, **74**, 375
Kurtz, D.W., 1978, Ap. J., **221**, 869
North P., Berthet S., Lanz T., 1995, A&A, **281**, 775
Sarna M.J., Dhillon V.S., Marsh T.R., Marks P.B., 1995, MNRAS, **272**, L41
Smalley B., Smith K.C., Wonnacott D., Allen C.S., 1996, MNRAS, **278**, 688
Wonnacott D., Kellett B.J., Stickland D.J., 1993, MNRAS, **262**, 277
Wonnacott D., Kellett B.J., Smalley B., Lloyd C., 1994, MNRAS, **267**, 1045

THE EXTENDED DUST EMISSION AROUND GK PER

S. M. DOUGHERTY[1], L. B. F. M. WATERS[2,3], M. F. BODE[1],
H. M. LLOYD[1], D. J. M. KESTER[3], TJ. R. BONTEKOE[3,4]
1. Chemical and Physical Sciences, John Moores University of Liverpool, Byrom Street, Liverpool, L3 3AF, UK
2. Astronomical Institute "Anton Pannekoek", University of Amsterdam, Kruislaan 403, 1098 SJ Amsterdam, The Netherlands
3. SRON Laboratory for Space Research, P.O. Box 800, 9700 AV Groningen, The Netherlands
4. Bontekoe Data Consultancy, Herengracht 47, NL-2312 LC Leiden, The Netherlands

GK Per, a classical nova system that erupted in 1901, is one of the more unusual examples of its type. It has the longest known orbital period for a classical nova (1.997 d; Crampton, Cowley & Fisher 1986); and it contains a white dwarf primary with an evolved K2 sub-giant secondary. Most remarkably, the IRAS Sky Survey (1991) reveals that GK Per exhibits far-IR dust emission extending $\sim 17'$ to the NW and SE of the nova (Bode et al. 1987; Seaquist et al. 1989). We have re-analysed the IRAS data using maximum entropy reconstruction (Bontekoe et al. 1991; Bontekoe, Koper & Kester 1994) to resolve structures at a spatial resolution approaching the diffraction limit of IRAS, which is $1'$ at 60 μm and $1.7'$ at 100 μm.

The analysis of the resulting HIRAS images of GK Per has led us to reconsider the origin of the extended envelope around the binary. For a full discussion see Dougherty et al. (1996). Our conclusions can be summarised as follows:

1. The extended dust emitting region observed by IRAS has a strongly symmetric distribution about the binary. This rules out an interstellar origin. The region is at an approximately uniform temperature of 22 K, which implies the observed discrete emission features are due to density enhancements. The dust mass in the envelope is $0.02 \ldots 0.04\,M_\odot$, compared to a gas mass of $1.0 \ldots 1.2\,M_\odot$, implying a dust-to-gas ratio of $0.02 \ldots 0.04$. The estimated age of the envelope is $1.1\,10^5$ or $1.1\,10^6$ yr, dependent on the adopted expansion velocity.

A. Evans and J. H. Wood (eds.), Cataclysmic Variables and Related Objects, 463–464.

2. The age of the binary compared to the age of the envelope requires that the material in the extended emission region originated from the evolved secondary via Roche lobe overflow to the white dwarf primary, and subsequent ejection by the white dwarf. The initial mass transfer rate was sufficiently high to convert the white dwarf into a 'born-again' AGB star. Thus, the ejection will have taken place in a common envelope phase. The present luminosity and spectral type of the secondary indicate a main sequence mass of $1.3 \ldots 1.5\,M_\odot$.
3. We argue that the discrete emission regions result from multi-epoch mass ejection on a time-scale of 4 or $40\,10^4$ yr. We suggest that they *may* be the result of thermal pulses of the born-again AGB primary star.
4. The present mass transfer rate and the time that has elapsed since the high mass transfer rate stopped imply that the classical nova outburst in 1901 was the first such event.
5. The shape of the extended emission region strongly suggests a bipolar outflow. CO observations of Scott, Rawlings & Evans (1994) and the warped appearance of the IR emitting region in the HIRAS images support such a model. However, such a geometry is difficult to understand in the context of common envelope evolution.

The HIRAS images have provided new insight into the evolution of the GK Per binary system. To verify the model outlined above, accurate determination of the expansion velocity of the extended region and improved binary parameters are required. In addition, the detailed three-dimensional geometry of the extended emission region needs to be clarified, which we aim to achieve via mapping of the CO and H I line emission. The small-scale structure of the emission region will be a target for the Infrared Space Observatory (ISO).

References

Bode M.F., Seaquist, E.R., Frail, D.A., et al., 1987, Nature, **329**, 519
Bontekoe, Tj.R., Kester, D.J.M., Price, S.D., et al., 1991, A&A, **248**, 328
Bontekoe, Tj.R., Koper, E., Kester, D.J.M., 1994, A&A, **284**, 1037
Crampton, D., Cowley, A.P., Fisher, W.A. 1986, Ap. J., **300**, 788
Dougherty, S.M., Waters, L.B.F.M., Bode, M.F., et al., 1996, A&A, in press
IRAS Sky Survey Atlas Explanatory Supplement, 1991, Wheelock S., Gautier T.N., Chillemi J., et al., (eds.)
Scott, A.D., Rawlings, J.M.C., Evans, A., 1994, MNRAS, **269**, 707
Seaquist, E.R., Bode, M.F., Frail, D.A., et al., 1989, Ap. J., **344**, 805

LOW MASS WHITE DWARFS IN BINARIES

T. R. MARSH[1], V. S. DHILLON[2], S. R. DUCK[3]
1. University of Southampton, Dept. of Physics, Highfield, Southampton, SO17 1BJ, UK
2. Royal Greenwich Observatory, Apartado de Correos 321, Santa Cruz de La Palma, 38780 Tenerife, Canary Islands, Spain
3. University of Oxford, Dept. of Physics, Nuclear and Astrophysics Laboratory, Keble Road, Oxford, OX1 3RH, UK

Abstract. The lowest mass white dwarfs that can have been produced by single star evolution in the lifetime of the Galaxy have masses of about $0.53\,M_{\odot}$. There are however several white dwarfs known with significantly lower masses. Evolution in a binary provides a straightforward explanation as the star can lose its envelope before ever burning helium. The products are expected to be short-period binary stars. To test this we have looked at 15 such stars and have found that 8 of them are close binaries. Thus binary evolution does have a major role in the formation of low-mass white dwarfs. We discuss whether the non-detections could also be binary stars.

1. Introduction

A consistent prediction of binary star evolution is the existence of white dwarf/white dwarf binary stars with orbital periods of order hours to days. Such systems which are descended from much more widely separated main-sequence star binaries are expected to exist in significant numbers, and of order 10 to 20% of all white dwarfs may reside in them (Iben 1991). Until recently though, out of over 1000 confirmed white dwarfs only 2 short period, double-degenerate systems were known [L870−2: $P = 1.56$ d, (Saffer et al. 1988); and WD 0957−666: $P = 1.18$ d (Bragaglia et al. 1990)].

The paucity of known short-period, double-degenerate binaries exists in spite of several surveys devoted to finding them (Robinson & Shafter (1987); Foss et al. (1991); Bragaglia et al. (1990)). However, all of these surveys were designed to detect short-period degenerate pairs with a com-

A. Evans and J. H. Wood (eds.), Cataclysmic Variables and Related Objects, 465–468.

bined mass exceeding the Chandrasekar limit as such systems are possible progenitors of the Type Ia supernovae. The combination of short-period and high mass produces high velocities which do not require great sensitivity to detect. However, the same features are rare amongst the expected population of double-degenerates leaving the majority of lower mass, longer period systems undetected.

In an effort to test the theory more stringently, we decided to concentrate upon the low mass tail of the white dwarf distribution (Bergeron, Saffer & Liebert 1992) since this should be dominated by the products of binary star evolution. Detailed write-ups of some of this work are already appearing in the literature (Marsh 1995) so in this paper we take more of an overview of the results so far.

2. Observations and their analysis

We have now had 2 runs of 3 nights on the 4.2 m WHT and 2 runs of 1 week each on the 2.5 m INT (the second of which has yet to be analysed). On the WHT we are able to cover simultaneously Hα for radial velocity measurement and the I−band for late-type features. On the INT we can only cover one or other of these at a time. We use a dispersion of 0.4 Å pixel^{-1} at Hα with a FWHM of 2 pixels in every case.

This moderately high dispersion is needed to make the most of the narrow core in Hα displayed by many DA (hydrogen lines only) white dwarfs. All of our discoveries have come from the Hα region, even the white dwarf/M dwarf binary (Table 1). An important part of the analysis has been the development of a multi-Gaussian fitting routine which permits varied constraints to be applied.

The combination of the Gaussian fit velocities, the high dispersion and the narrow core of Hα have produced typical radial velocity uncertainties in the range 2 to 5 km s^{-1}. These are low values given the huge breadth of the pressure broadened lines in white dwarfs and we believe that much of our success can be put down to them.

3. Results

We have now analysed data on 15 white dwarfs. Table 1 summarises our results on these systems. The table shows that out of the 15 systems, 8 are binaries. Of the binaries, 7 are double-degenerates and one is a detached white dwarf/M dwarf binary. The latter system is certainly interesting as it is the shortest period such system known and falls exactly in the centre of the period gap of cataclysmic variable stars, however we will not discuss it further here. The new systems have extended the double-degenerate period range to shorter and longer periods with the range now extending from

TABLE 1. Low mass white dwarfs studied to date.

Object WD	M_W $M_\odot$	Prob	Binary	Period (d)	Comments
0136+768	0.44	0.82	Yes	?	Double-lined WD/WD
0147+674	0.45	0.72	No	–	
0316+345	0.40	0.94	No	–	
0339+523	0.34	0.89	No	–	
0453+418	0.43	0.94	No	–	
0710+741	0.35	0.90	Yes	0.103	WD + 0.1 $M_\odot$ M dwarf
0839+231	0.48	0.89	No	–	
1101+364	0.31	0.84	Yes	0.145	Double-lined WD/WD
1241−010	0.31	0.93	Yes	3.347	Single-lined WD/WD
1317+453	0.33	0.96	Yes	4.872	Single-lined WD/WD
1353+409	0.40	0.87	No	–	
1614+136	0.33	0.90	No	–	
1713+332	0.35	0.95	Yes	1.127	Double-lined WD/WD
2032+188	0.36	0.92	Yes	?	Single-lined WD/WD
2331+290	0.39	0.62	Yes	0.167	Single-lined WD/WD

3.47 h to 4.87 d. This is significant when we compare with the theoretically expected distribution which is limited in extent.

In Fig. 1 we show the 7 systems of known orbital period (including L870−2 and WD 0957−666) along with the expected distribution from Yungelson et al. (1994). The observations and theory are in surprising agreement given the many uncertainties contained in the theoretical prediction. Double white dwarf binaries pass through a stage in which they orbit within a single envelope. The efficiency with which the envelope is ejected has a significant effect upon the final orbital period. If the envelope is ejected inefficiently, the binary shrinks greatly and the orbital period becomes very short. The observed periods suggest a fairly efficient ejection; Yungelson et al. (1994) used $\alpha_{CE} = 1$ for the distribution plotted in Fig. 1.

We have no evidence for companions in 7/15 systems. Table 1 lists our detection probabilities based upon two 0.4 $M_\odot$ white dwarfs in randomly inclined orbits following Yungelson et al.'s period distribution. We should have detected more than 8 systems. These stars may yet be binaries however

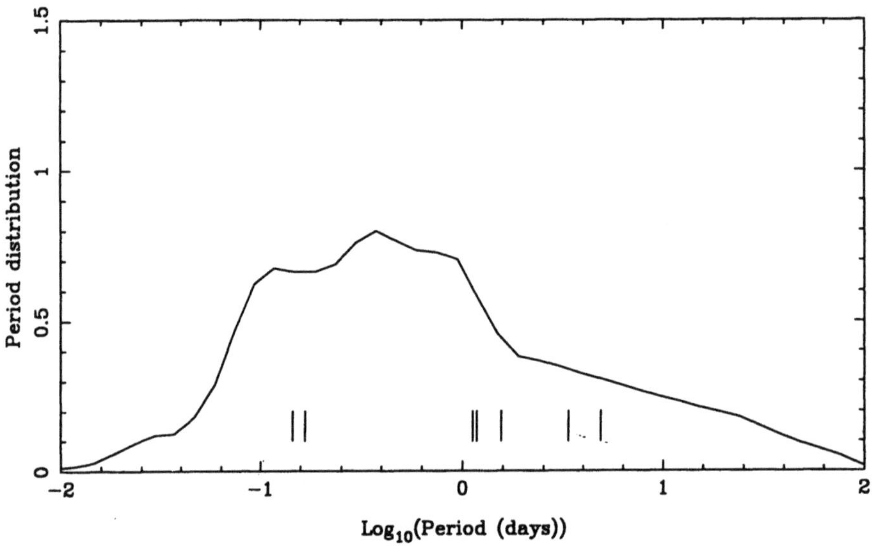

Figure 1. Known orbital periods compared with theoretical prediction.

because symmetric double-lined systems are less likely to be detected and there may be a longer period tail to the distribution than the theoretical distribution suggests. However, there is considerable room for improving the constraints upon these systems and more work is needed on them.

The only other indications of discrepancy between theory and observations are the two measured mass ratios for L870−2 and WD 1101+364 of 0.90 ± 0.04 and 0.87 ± 0.03 which are both in the wings of the theoretical prediction (Han, Podsiadlowski & Eggleton 1995).

4. Conclusion

Low mass white dwarfs are predominantly formed in binary stars.

References

Bragaglia, A., Greggio, L., Renzini, A., D'Odorico, S., 1990, Ap. J., **365**, L13
Bergeron, P., Saffer, R.A., Liebert, J., 1992, Ap. J., **394**, 228
Foss, Ð., Wade, R.A., Green, R.F., 1991, Ap. J., **374**, 281
Han, Z., Podsiadlowski, P., Eggleton, P.P., 1995, MNRAS, **272**, 800
Iben, I., 1991, Ap. J. Supp., **76**, 55
Marsh, T.R., 1995, MNRAS, **275**, L1
Robinson, E.L., Shafter, A.W., 1987, Ap. J., **322**, 296
Saffer, R.A., Liebert, J., Olszewski, E., 1988, Ap. J., **334**, 947
Yungelson, L.R., Livio, M., Tutukov, A.V., Saffer, R., 1994, Ap. J., **420**, 336

HOT SUBDWARFS IN BINARIES: MULTIWAVELENGTH OBSERVATIONS AND EVOLUTIONARY IMPLICATIONS

A. ULLA[1,2], P. THEJLL[1], S. VENNES[3]
C. S. HANSEN-RUIZ[4], J. L. RASILLA[5]
1. Niels Bohr Institute, Copenhagen, Denmark
2. LAEFF-INTA, Madrid, Spain
3. Center for Extreme Ultraviolet Astrophysics, Berkeley, USA
4. Institute of Astronomy, Cambridge, UK
5. Instituto de Astrofísica de Canarias, Spain

Abstract. We are looking for stellar companions to very hot helium-rich subluminous (sdO) stars by using infrared photometry, optical CCD imaging and spectroscopy. The work is complemented with a comprehensive bibliographic and space-mission databanks search for the objects in our sample (Thejll, Ulla & MacDonald 1995). Companions can be used to indirectly measure the distance to the hot subdwarfs, and give clues about possible binary evolution scenarios. We have therefore gathered and analysed data for 57 hot subdwarfs in a search for IR excesses and companions. 43 of the stars are observed for the first time in the *JHK* bands and 28 objects are found to have larger IR fluxes than expected from a single sdO.

1. Introduction

Hot helium-rich subdwarf stars (sdOs) are situated above the very blue end of the horizontal branch in the field and cluster populations. Beyond knowing that they are hot (40 000 K to 80 000 K) and helium-rich (photospheres are 10% to 100% He, by number), little is known about the evolutionary status, let alone origins, of these stars. The surface gravity, found in NLTE spectral analysis, is uncertainly known but tends to lie in the range $\log g = 5.0 \ldots 6.5$. In the spectra, metals can be seen and these typically include C, N and Fe, but the abundances may not be original in that radiative levitation, gravitational settling and wind-driven mass loss are likely to alter the composition of the photospheres.

The spatial distribution of the sdOs is only tentatively known from the

A. Evans and J. H. Wood (eds.), Cataclysmic Variables and Related Objects, 469–470.

NLTE analysis of about 40 hot sdO stars coupled to the assumption of a mass near $0.5\,M_{\odot}$. Depending on the limiting magnitude of the observational material scale heights of at least 500 pc for the population of sdOs can be surmised. The kinematics of hot sdOs are being investigated (Thejll & Flynn, in preparation) for a large sample of stars. Previously, smaller samples of bright sdO kinematics have been analysed and divergent results have been found – but the results found are consistent with the sample limiting magnitudes, i.e. bright samples yield 'young' population memberships while the age increases with the limiting magnitude of the sample, thus indicating consistency with a large age for the population as a whole.

The binarity of the sdOs remains unknown, while that of their neighbors, the hot hydrogen-rich subdwarfs (sdBs), is observable at 50% and this implies a 100% binarity rate, as may be calculated from the assumed distribution over spectral classes of the companions. For the sdBs a scale height of 500 pc has been reported (Saffer & Liebert 1994). The sdB stars lie on the extended blue horizontal branch (EHB) and are currently identified with the helium-core burning stage of stellar evolution. Evolutionary tracks of such models lead from the EHB into the sdO region and thus suggest an evolutionary link that in turn requires similarity of scale heights, kinematics and rates of binarity between the sdBs and sdOs.

In a collaboration between photometrists, spectroscopists and modelers, and using available sources of information, we reported in this poster on a project designed to measure the rate of binarity of sdO stars. Beyond comparing the rate of binarity to that of the sdBs we can compare it to that of other potential progenitors and descendents such as the hydrogen-deficient R CrB stars and the helium-rich DB white dwarfs. It is also possible to use detected binary companions for derivation of atmospheric parameters of the hot subdwarf by measuring relative radii from multiwavelength photometry. A third interest lies in compiling a list of binary sdOs for separate investigations of the separation between the two stars; this has relevance for questions related to the possible co-evolution scenarios.

We have found that about 50% of the mainly spectroscopically selected sample of sdO stars are binary and we continue the analysis of these data.

References

Saffer, R.A., Liebert, J., 1994, in "White Dwarfs", Koester D. and Werner K. (eds.), Springer Verlag, p221

Thejll, P.A., Ulla A., MacDonald J., 1995, A&A, **303**, 773

THE ECLIPSES OF THE CLOSE BINARY STAR BE UMA

JANET H. WOOD[1], E. L. ROBINSON[2], E.-H. ZHANG[2]
1. Physics Department, Keele University, Keele, ST5 5BG, UK
2. Department of Astronomy, University of Texas, Austin, TX 78712, USA

1. Introduction

BE UMa is a close binary star, not transferring mass, with an extremely hot primary star irradiating the inner face of the cool secondary star. The light curve shows a large-amplitude, sinusoidal variation with a period of 2.29 d, and an eclipse that is centered on the minimum of the variation [1], [3]. According to [1], the eclipse is partial, not total. However, it has been argued [2] that the eclipse was really flat bottomed and thus total. This has important repercussions for the deduced model of the system. To resolve this issue we obtained simultaneous *UBVR* photometry of BE UMa using the Stiening 4-channel, high-speed photometer on the 82-inch telescope at McDonald Observatory. The mean light curves are shown in Fig. 1. The eclipse in all colours is round bottomed and partial. The different depths are caused by the different contribution from the red secondary star in each bandpass.

2. The system parameters

Along with the inclination, i, the primary star radius, R_1/a, and the secondary star radius, R_2/a, for each bandpass the flux from the primary star, the limb darkening of the primary star, and the secondary star flux, determine the light curve. For a given inclination we fit the four light curves simultaneously. We find our best fit for $i = 82.4° \pm 0.1°$, $R_1/a = 0.0781 \pm 0.0001$ and $R_2/a = 0.138 \pm 0.001$. The number of fitted data points is 6768 and for this fit $\chi^2 = 7202$. The deduced limb darkening coefficients are very unusual: to obtain the best fit, limb brightening of the primary was required in the U band with almost full limb darkening in the B band.

Additional information is necessary to determine the stellar masses, M_1 and M_2, the mass ratio, $q = M_2/M_1$, and the orbital separation, a. Using

A. Evans and J. H. Wood (eds.), Cataclysmic Variables and Related Objects, 471–472.

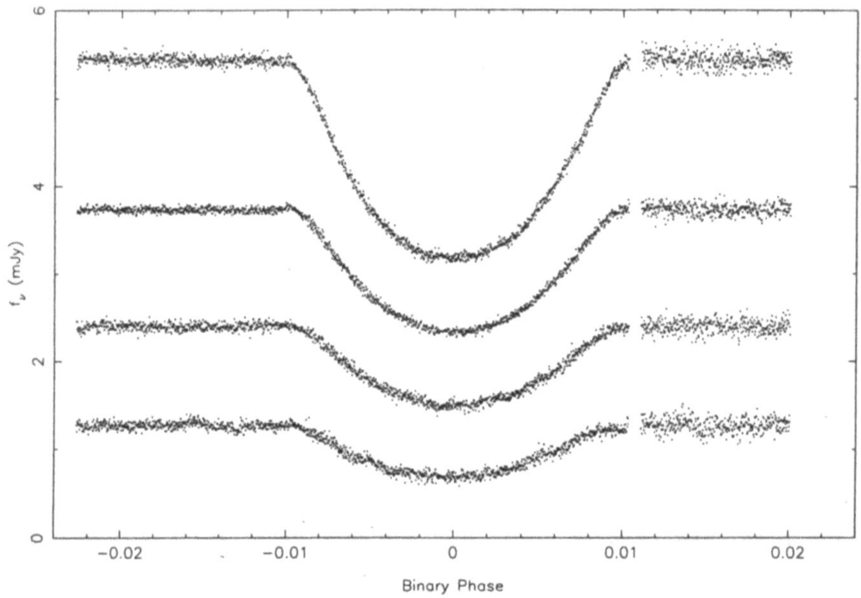

Figure 1. From bottom to top, the mean R, V, B and U light curves at 5 s time resolution. Each light curve has been offset upwards by 1 mJy with the R light curve at its true flux density.

the radial velocity of the secondary star [2], corrected for the effects of irradiation, and assuming that the secondary star lies on the main sequence, we can determine the system parameters for each inclination. For $i = 82.4°$ we find $M_1 = 1.43 \pm 0.15\,M_\odot$, $M_2 = 1.39 \pm 0.15\,M_\odot$ and $a = (7.2 \pm 0.3)\,10^{11}$ cm. We note that the primary star is a subdwarf: it is not small enough to be a white dwarf. The secondary star lies well inside its Roche lobe.

At phase zero we are seeing the unirradiated face of the secondary star and can use the measured fluxes to determine its spectral type. We find that G6–K4 is acceptable. We fit pure hydrogen, $\log g = 6$, model atmospheres [4] to our primary star fluxes, The fit for $T_e = 125\,000$ K corresponds to a distance of $986 < D(\mathrm{pc}) < 1205$. If the secondary star is a K3–K4 star it has a distance consistent with the 1σ limits deduced from the subdwarf. For a K dwarf secondary star, we therefore have a self-consistent model.

References

1. Ando, H., Okazaki, A., Nishimura, S., 1982, PASJ, **34**, 141
2. Crampton, D., Cowley, A.P., Hutchings, J.B., 1983, Ap. J., **272**, 202
3. Kurochkin, N.E., 1971, Perem. Zvezdy., **18**, 85
4. Wesemael, F., Auer, L.H., van Horn, H.M., et al., 1980, Ap. J. Supp., **43**, 159

AN EVOLUTIONARY SCENARIO FOR SHORT PERIOD ($\leq$10 DAYS) MILLISECOND BINARY PULSARS

E. ERGMA[1,2], M. J. SARNA[1,3], J. ANTIPOVA[1]
1. Tartu University, Physics Department, Ülikooli 18, EE2400, Tartu, Estonia
2. NORDITA, Blegdamsvej 17, DK2100, Copenhagen 0, Denmark
3. N. Copernicus Astronomical Center, Polish Academy of Science, ul. Bartycka 18, 00–716 Warsaw, Poland

1. Discussion

We present numerical calculations that simulate the evolution of a low mass ($1\,M_\odot$) star transfering mass to a compact object (Muslimov & Sarna 1993; Ergma & Sarna 1996). Mass transfer starts when the secondary turns off the main sequence (having a small helium core). We have calculated 14 evolutionary sequences with the assumption of non-conservative or conservative evolution. We can conclude that near the bifurcation point the evolution is very sensitive to: (i) the assumption of conservative or non-conservative evolution, (ii) the structure of the mass losing star. Small changes in the initial period when the secondary fills its Roche lobe will lead to large changes in the final period and final mass of the remnant. Presently there are 40 known low-mass binary pulsars (LMBP). The evolutionary scenario for the wider systems ($10 < P_{\rm orb}({\rm d}) < 1000$) is connected with that of wide low-mass X-ray binaries (LMXB) in which the donor star will fill its Roche lobe after helium core formation. For such stars there is a simple relation between the mass of the degenerate helium core and the stellar radius (Refsdal & Weigert 1971). The final orbital period is uniquely determined by the core mass of the final helium white dwarf (Joss, Rappaport & Webbink 1987; Rappaport et al. 1995). As was pointed out by van den Heuvel & Bitzaraki (1994) the picture described above is too simplified for systems with initial orbital periods ranging from approximately 0.5 to 2 d. The recent discovery of the very short orbital period systems PSR J1012+5307 ($P_{\rm orb} = 0.604\,{\rm d}$, $P_{\rm pul} = 5.26\,{\rm ms}$, $M = 0.11\,M_\odot$; Nicastro et al. 1995) and PSR J0751+18 ($P_{\rm orb} = 0.263\,{\rm d}$, $P_{\rm pul} = 3.48\,{\rm ms}$, $M = 0.12\,M_\odot$; Lundgren, Zepka & Cordes

A. Evans and J. H. Wood (eds.), Cataclysmic Variables and Related Objects, 473–474.

1995) has raised the question of how to explain their past evolution. Our calculations support the model of Kluźniak, Czerny & Ray (1992) in which the secondary star of PSR 1744−24A fills its Roche lobe when it turns off the main sequence. Due to the low accretion rate the system is in the 'propeller' phase which may explain its very erratic eclipsing behaviour. The orbital parameters of PSR J0751+18 may be explained if we assume non-conservative evolution starting from an initial period, P_i, of 1.05 d. After the accretion stops near 0.343 d the system will evolve due to gravitation wave radiation (GWR) and after $\sim 7.5\,10^9$ yr its orbital parameters will be as observed. We may explain the orbital parameters of PSR J1012+5307 if we assume again non-conservative evolution but this time starting with an initial period between 0.53...0.63 d. Since the orbital period is very large GWR is ineffective, and the system's orbital period will be as it was when the system detached.

Acknowledgements. EE would like to thank Drs. D. Lorimer, F. Camillo and S. Lundgren for providing the data for PSR J0751+18 and PSR J1012 +5307.

References

Ergma, E., Sarna, M.J., 1996, MNRAS, in press
Joss, P.C., Rappaport, S.A., Webbink, R.F., 1987, Ap. J., **319**, 180
Kluźniak, W., Czerny, M., Ray, A., 1992, in "X-ray Binaries and Recycled Pulsars" eds, E.P.J. van den Heuvel, S.A. Rappaport, p375
Lundgren, S.C., Zepka, A.F., Cordes, J.M., 1995, Ap. J., **453**, 419
Muslimov, A.G., Sarna M.J., 1993, MNRAS, **262**, 164
Nicastro, L., Lyne, A.G., Lorimer, D.R., et al., 1995, MNRAS, **273**, L68
Rappaport, S.A., Podsiadlowski, P., Joss, P.C., et al., 1995, MNRAS, **273**, 731
Refsdal, S., Weigert, A., 1971, A&A, **6**, 426
van den Heuvel, E.P.J., Bitzaraki, O., 1994, in "Evolutionary Relations in the Zoo of Interacting Binaries" eds, F. D"Antona, Mem. Soc. Astron. Ital., p237

ON THE ORBITAL PERIOD DISTRIBUTION OF RECYCLED PULSARS

L. A. NELSON[1], S. RAPPAPORT[2]
1. Physics Dept., Bishop's Univ., Lennoxville, QC, Canada
2. CSR, Room 37-551, MIT, Cambridge, MA 02139, USA

Abstract. Using population synthesis techniques, we investigate the orbital period distribution of 'recycled' pulsars. We find that the relative distribution of systems with periods between approximately one and fifty days are reasonably well reproduced. However, the existence of binary pulsars with periods between 70 and ~ 1000 d cannot be explained within the evolutionary framework considered here.

1. Introduction

More than 50 recycled radio pulsars have been discovered. Of the pulsars known to be members of binary systems, most are believed to contain low-mass degenerate companions in nearly circular orbits having orbital periods in the range of several hours to 1000 d. According to one evolutionary scenario, an evolved star initially undergoes stable mass transfer onto a neutron star companion and leaves behind a degenerate core [3]. During the process, the neutron star is 'spun up'. In our calculations, we explicitly follow the evolution of these systems from the moment when Roche lobe overflow occurs until the cessation of mass transfer. The initial conditions were determined by using population synthesis techniques. In all cases, the neutron star was assumed to have a mass of $1.4\,M_\odot$. Details of the evolutionary calculations will be presented elsewhere [1].

2. Conclusions

If the secondary (donor) is sufficiently evolved at the onset of mass transfer, nuclear evolution of the secondary dominates the evolution of the binary. As a result, the orbital period generally increases until the mass in the envelope of the (sub)giant is exhausted. The end product is a detached binary consisting of a 'recycled' pulsar and a degenerate dwarf companion. Most of these systems seem to obey a well defined $P_{\rm orb} - M_{\rm wd}$ relationship

A. Evans and J. H. Wood (eds.), Cataclysmic Variables and Related Objects, 475–476.

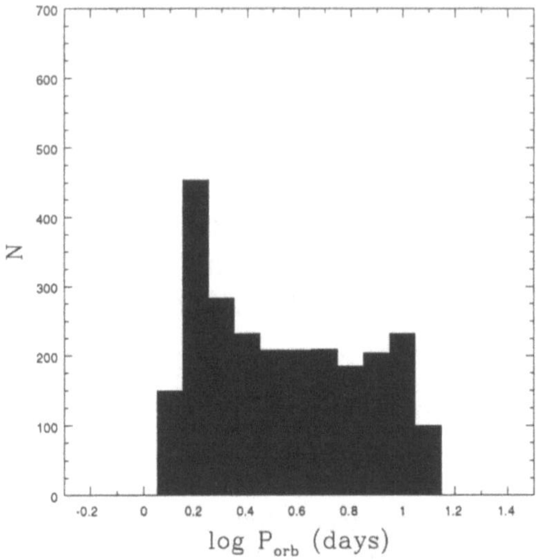

Figure 1. Relative orbital period distribution of simulated binary pulsars containing low-mass degenerate companions in the Galactic disk at the current epoch. The original neutron star progenitor was chosen to have a mass between 10 and 30 $M_\odot$ and the mass of the secondary was taken to be uncorrelated with respect to the primary's mass (a fit to the Miller & Scalo IMF was used). The efficiency (α) of the removal of the envelope during the common envelope phase was taken to be unity. Largely unevolved secondaries (i.e. those that did not develop a significant helium core) and systems suffering dynamical instability were removed from the population.

(the exceptions being systems with small $M_{\rm wd}$). We find that most of the binary pulsar systems that are formed have orbital periods between 1 and 15 d (see Fig. 1). This accounts for the paucity of systems in the 'gap' of the orbital period distribution between 15 and 70 d. The apparent decrease in the number of observed systems with periods of less than 1 d can be understood in terms of the bifurcation [2] of the theoretical tracks. The existence of binary pulsars with periods between 70 and $\sim$ 1000 d cannot be explained within the evolutionary framework considered here. This problem requires further investigation [1].

References

1. Nelson, L. A., Rappaport, S., 1996, in preparation
2. Pylyser, E. H. P., Savonije, G. J., 1988, A&A, **191**, 57
3. Rappaport, S., Podsiadlowski, P., Joss, P. C., et al., 1995, MNRAS, **273**, 731

CONFERENCE SUMMARY

KEELE CATACLYSMIC COMMENTARY

BRIAN WARNER
Department of Astronomy, University of Cape Town,
7700 Rondebosch, South Africa

What kind of person is invited to make Conference summaries? For conference *introductions* it's easy – you just invite some past, long-retired Directorsaurus Rex to make a few deathbed corrections to his memories of how he founded the field and anticipated all future developments. For conference *after-dinner speeches* it's easy to find a sharp-elbowed antagonist for whom a few more unexpected remarks and double entendres can't hurt an already notorious reputation. But in a conference summariser one is looking for even-handedness, wisdom, unsurpassed knowledge, ability to pick winners and, above all, brevity.

This is why conference summaries are an impossible dream; and why conference summarizers often resort to the antics of introducers or after-dinner speakers.

What I am about to say is more of a coda than summary. In a conference like the one we have experienced this week a summary could not do justice. The fact is, the field of CVs has become so large that much of what we have heard have been summaries of specialised areas, with glimpses of work in progress. Many fields have not been touched on at all. The posters fill in some but not all of the gaps. I was tempted to illustrate the point by providing a summary of everything *not* said – but realised that I was hardly qualified to speak for the silent majority.

What has become quite evident is that most progress continues to be the result of improvements in equipment power. Historians of science analyse scientists of the past centuries into 'instrumentalists' and others. Hipparchus and Galileo are examples of the former, Copernicus and Newton are examples of the latter.

Looking at the history of CVs, observational techniques changed from visual and photographic provision of long-term light curves, through the spectroscopy and photoelectric photometry of the early 1960s which established that CVs were close binaries. Progress then came in a series of jumps, with the introduction of high time resolution photometry and spectroscopy in the late 1960s, the impact of satellite UV and X-ray observations in the

A. Evans and J. H. Wood (eds.), Cataclysmic Variables and Related Objects, 479–482.

1970s, the increased sensitivity of infrared detectors and the detection of CVs at radio wavelengths and eventually even at γ-ray energies.

The most important results at this conference have conspicuously come from the latest instrumental developments – from ROSAT, EUVE, ASCA and HST; from Merlin and VLA; and the promise of new types of behaviour to be found by ALEXIS. Equally important as the new hardware are the analysis techniques:– again some of the finest new results are from applications of eclipse deconvolution[1] and Doppler tomogram methods to observational material of high signal-to-noise ratio.

It is now possible to image directly or indirectly the early stages of eruptions of novae, the distribution of emission measure in non-magnetic and magnetic CVs, the asymmetric distribution of absorption or emission line strengths over the surfaces of CV secondaries, the stream of gas propellered out of AE Aqr, the relativistic ejection of collimated blobs of gas from low mass X-ray binaries such as GRS 1915+105, etc. Details are being learnt of the heating and subsequent cooling of the white dwarf primaries of dwarf novae in outburst; of the structure of CV winds; of the inhomogeneities of nova ejecta; of the field geometry and positions of accretion regions in polars. Possibilities have been raised of rapidly learning more about the X-ray emitting regions in polars and the boundary layers in non-magnetic CVs by examining the ratios of reflected to direct hard X-ray spectra from white dwarf and accretion disc surfaces. This may help to clear the confusion still existing in boundary layer structures – from most observational evidence, boundary layers are much cooler ($\sim 10^5$ K) and of larger extent than the classical predictions.

The last CV conference, in Cape Town, was very specialised. This one has allowed discussions in all fields, including related stars. I think that it is clear that both kinds are necessary in order to avoid the danger of losing the broad picture. CV observers will be envious of the richness and understanding of QPOs in low-mass X-ray binaries; LMXB observers must envy the ease with which the primary and central regions of the disc are observed in some CVs. Both are concerned about disc rim structures and the existence of emission in unexpected regions of discs.

The number of astronomers and the number of publications has been growing exponentially for some decades. Fortunately the increase in advanced instrumentation has largely kept pace. And there is more already on the horizon: the impact of 8-metre class telescopes and of AXAF and other satellites on the field of CV studies is going to be tremendous. The success of the High Speed Photometer on HST was such that we should agitate for

[1]If this technique has allowed some astronomers to see further into CVs it is because they have looked over the shoulders of dwarfs.

its reinstatement. VLBI from space can hope to image the structures of the nearest CVs.

Let me now touch on a few subjects where I think I may make useful comments or ask helpful questions. These serve to establish that I was conscious for many of the sessions.

1. It seems to me clear that there are two sources of flickering in non-magnetic CVs: the bright spot and the inner disc. It was an historical accident that the first system to be studied in detail was U Gem, in which most of the optical continuum comes from the bright spot, which was shown to be the source of flickering. It would be interesting to see if the amplitude of flickering of the bright spot in CVs in general is roughly a constant fraction of the bright spot luminosity. It would also be useful to try to understand why disc flickering is different from object to object: is this a signature of interaction between inner disc and weakly magnetised primaries?
2. If the hibernation scheme is correct then it is not possible to say that GR provides a lower boundary to the rate of mass transfer – this may be the secular average for short orbital period CVs but the current rate may be as low as zero. The spread of $\dot{M}$ at a given orbital period is probably bounded from the top by saturation of irradiation-driven mass transfer and from the bottom by the selection effect of the necessity of recognising CVs from their disc emissions or outbursts. The full range is easily detected below the $P_{\rm orb}$ gap – at long periods the secondary dominates and low $\dot{M}$ systems would be difficult currently to find, but they should show up as white dwarf companions to rapidly rotating K stars.
3. There is increasing evidence, only some of which has been mentioned at this conference, that many supposedly non-magnetic CV discs are truncated. When discussing the reasons for this it is important to keep all of the evidence in mind, including the behaviour of Dwarf Nova Oscillations during outburst, which I think favours the effects of low magnetic fields in the primaries.
4. The magnetic field geometry in the half dozen polars that have been studied in detail and in the isolated magnetic white dwarfs all depart greatly from simple centred dipoles. They are off-centre dipoles or mixtures of dipole and quadrupole.

   The current interpretation, given at this conference, of DQ Her requires a sustained equality of accretion rate onto two magnetic poles. Does this demand a centred dipole geometry? Rosen et al. (1991) point out that in an intermediate polar the effects of asymmetric fields are reduced in the region where gas is threaded by the field lines. Yet at this conference we have seen that YY Dra can at times show equally luminous accretion spots and at other times not. Clearly there is more complexity here than had been realised.

5. Deep imaging of old novae has uncovered several shells not before known. Noting the spectacular shell still easily visible in GK Per nearly a century after eruption, it is important to carry out imaging of *older* novae – i.e. novae that went unnoticed in the nineteenth century which now will be counted among nova-like variables and perhaps even some dwarf novae (noting that some nova remnants have been found to start dwarf nova outbursts within 50 years of nova eruption).

To return to the exponential growth of astronomy, and the popularity of the CV industry. For students hoping to find permanent employment in this research area I point out a book called 'The Perfect CV: How to get the job you really want', by Tom Jackson. The bad news is that it states that 500 000 copies have already been sold, so the competition will be fierce.

The Keele Conference was well organised, thanks particularly to the efforts of Janet Wood and Nye Evans, ably supported by C. Hellier, J. Lockley, T. Naylor, F. Ringwald, B. Smalley, M. Somers, W. Welsh and numerous members of the Physics Department.

References

Rosen, S., Mason, K.O., Mukai, K., Williams, O.R., 1991, MNRAS, **249**, 417

SUBJECT INDEX

Page numbers in **bold type** indicate a review, page numbers in *italic type* indicate that the reference is primarily to a figure or a table.

OBJECT INDEX

Variable star names; objects are listed alphabetically by constellation, and alphabetically within constellation. Page numbers in *italic type* indicate that reference is primarily to a figure or a table.

Other designations. Objects having co-ordinate-based designation are listed first; thereafter the listing is alphabetical. Page numbers in *italic type* indicate that the reference is primarily to a figure or a table.

MIX
Papier aus verantwortungsvollen Quellen
Paper from responsible sources
FSC® C105338

If you have any concerns about our products,
you can contact us on
ProductSafety@springernature.com

In case Publisher is established outside the EU,
the EU authorized representative is:
Springer Nature Customer Service Center GmbH
Europaplatz 3, 69115 Heidelberg, Germany

Printed by Libri Plureos GmbH
in Hamburg, Germany